Fourth Edition

Seeing Through
Statistics

Jessica M. Utts
University of California, Irvine

CENGAGE
Learning·

Australia • Brazil • Mexico • Singapore • United Kingdom • United States

Seeing Through Statistics, Fourth Edition
Jessica M. Utts

Product Director: Liz Covello

Senior Product Team Manager: Richard Stratton

Senior Product Manager: Molly Taylor

Senior Content Developer: Jay Campbell

Content Coordinator: Danielle Hallock

Media Developer: Andrew Coppola

Senior Marketing Manager: Gordon Lee

Content Project Manager: Alison Eigel Zade

Senior Art Director: Linda May

Manufacturing Planner: Sandee Milewski

Rights Acquisition Specialist:
    Shalice Shah-Caldwell

Production Service and Compositor:
    Graphic World, Inc.

Text and Cover Designer: Rokusek Design

Cover Image: ©Aleksey Stemmer/
    shutterstock.com

For product information and technology assistance, contact us at
Cengage Learning Customer & Sales Support, 1-800-354-9706
For permission to use material from this text or product,
submit all requests online at **www.cengage.com/permissions**
Further permissions questions can be emailed to
permissionrequest@cengage.com

Library of Congress Control Number: 2013952119

Student Edition:

ISBN-13: 978-1-285-05088-1

ISBN-10: 1-285-05088-6

Cengage Learning
200 First Stamford Place, 4th Floor
Stamford, CT 06902
USA

Cengage Learning is a leading provider of customized learning solutions with office locations around the globe, including Singapore, the United Kingdom, Australia, Mexico, Brazil, and Japan. Locate your local office at **www.cengage.com/global**

Cengage Learning products are represented in Canada by Nelson Education, Ltd.

To learn more about Cengage Learning, visit **www.cengage.com**

Purchase any of our products at your local college store or at our preferred online store **www.cengagebrain.com**

Instructors: Please visit login.cengage.com and log in to access instructor-specific resources.

Printed in the United States of America
3 4 5 6 7 19 18 17 16 15

To my ancestors, without whom this book would not exist:

Allan
Benner
Blackburn
Davis
Dorney
Engstrand
Gessner/Ghesner
Glockner
Grimshaw
Haire
Henry
Highberger/Heuberger
Hons
Hutchinson
Johnson
Kiefer
Miller
Noland
Peoples
Rood
Schoener
Shrader
Shrum
Simpson
Sprenckel
Stark
Utts/Utz
Wells
Whaley/Whalley
Woods

And many more, some of whom I have yet to discover!

Contents

CHAPTER **3**

Measurements, Mistakes, and Misunderstandings   **40**

CHAPTER **4**

How to Get a Good Sample   **64**

PART 2

Finding Life in Data   135

CHAPTER **10**

Relationships Between Measurement Variables   202

CHAPTER **11**

Relationships Can Be Deceiving   227

CHAPTER **15**

Understanding Uncertainty through Simulation   328

CHAPTER **16**

Psychological Influences on Personal Probability   344

PART **4**

Making Judgments from Surveys and Experiments   **405**

CHAPTER **19**

CHAPTER **20**

CHAPTER 24

Significance, Importance, and Undetected Differences 519

CHAPTER 25

Meta-Analysis: Resolving Inconsistencies across Studies 534

Preface

If you have never studied statistics, you are probably unaware of the impact the science of statistics has on your everyday life. From knowing which medical treatments work best to choosing which television programs remain on the air, decision makers in almost every line of work rely on data and statistical studies to help them make wise choices. Statistics deals with complex situations involving uncertainty. We are exposed daily to information from surveys and scientific studies concerning our health, behavior, attitudes, and beliefs, or revealing scientific and technological breakthroughs. This book's first objective is to help you understand this information and to sift the useful and the accurate from the useless and the misleading. My aims are to allow you to rely on your own interpretation of results emerging from surveys and studies and to help you read them with a critical eye so that you can make your own judgments.

A second purpose of this book is to demystify statistical methods. Traditional statistics courses often place emphasis on how to compute rather than on how to understand. This book focuses on statistical ideas and their use in real life.

Finally, the book contains information that can help you make better decisions when faced with uncertainty. You will learn how psychological influences can keep you from making the best decisions, as well as new ways to think about coincidences, gambling, and other circumstances that involve chance events.

Philosophical Approach

If you are like most readers of this book, you will never have to *produce* statistical results in your professional life, and, if you do, a single statistics book or course would be inadequate preparation anyway. But certainly in your personal life and possibly in your professional life, you will have to *consume* statistical results produced by others. Therefore, the focus of this book is on understanding the use of statistical methods in the real world rather than on producing statistical results. There are dozens of real-life, in-depth case studies drawn from various media sources as well as scores of additional real-life examples. The emphasis is on understanding rather than computing, but the book also contains examples of how to compute important numbers when necessary, especially when the computation is useful for understanding.

Although this book is written as a textbook, it is also intended to be readable without the guidance of an instructor. Each concept or method is explained in plain language and is supported with numerous examples.

Organization

There are 27 chapters divided into four parts. Each chapter covers material more or less equivalent to a one-hour college lecture. The final chapters of Part 1 and Part 4 consist solely of case studies and are designed to illustrate the thought process you should follow when you read studies on your own.

By the end of Part 1, "Finding Data in Life," you will have the tools to determine whether or not the results of a study should be taken seriously, and you will be able to detect false conclusions and biased results. In Part 2, "Finding Life in Data," you will learn how to turn numbers into useful information and to quantify relationships between such factors as aspirin consumption and heart attack rates or meditation and test scores. You will also learn how to detect misleading graphs and figures and to interpret trends over time.

Part 3 is called "Understanding Uncertainty in Life" and is designed to help you do exactly that. Every day we have to make decisions in the face of uncertainty. This part of the book will help you understand what probability and chance are all about and presents techniques that can help you make better decisions. You will also learn how to interpret common economic statistics and how to use the power of computers to simulate probabilities. The material on probability will also be useful when you read Part 4, "Making Judgments from Surveys and Experiments." Some of the chapters in Part 4 are slightly more technical than the rest of the book, but once you have mastered them you will truly understand the beauty of statistical methods. Henceforth, when you read the results of a statistical study, you will be able to tell whether the results represent valuable advice or flawed reasoning. Unless things have changed drastically by the time you read this, you will be amazed at the number of news reports that exhibit flawed reasoning.

Thought Questions: Using Your Common Sense

All of the chapters, except the one on ethics and those that consist solely of case studies, begin with a series of Thought Questions that are designed to be answered before you read the chapter.

Most of the answers are based on common sense, perhaps combined with knowledge from previous chapters. Answering them *before* reading the chapter will reinforce the idea that most information in this book is based on common sense. You will find answers to the thought questions—or to similar questions—embedded in the chapter.

In the classroom, the thought questions can be used for discussion at the beginning of each class. For relatively small classes, groups of students can be assigned to discuss one question each, then to report back to the class. If you are taking a class in which one of these formats is used, try to answer the questions on your own before class. By doing so, you will build confidence as you learn that the material is not difficult to understand if you give it some thought.

Case Studies and Examples: Collect Your Own

The book is filled with real-life Case Studies and Examples covering a wide range of disciplines. These studies and examples are intended to appeal to a broad audience. In the rare instance in which technical subject-matter knowledge is required, it is given with the example. Sometimes, the conclusion presented in the book will be different from the one given in the original news report. This happens because many news reports misinterpret statistical results.

I hope you find the case studies and examples interesting and informative; however, you will learn the most by examining current examples on topics of interest to you. Follow any newspaper, magazine, or Internet news site for a while and you are sure to find plenty of illustrations of the use of surveys and studies. If you start collecting them now, you can watch your understanding increase as you work your way through this book.

Formulas: It's Your Choice

If you dread mathematical formulas, you should find this book comfortably readable. In most cases in which computations are required, they are presented step by step rather than in a formula. The steps are accompanied by worked examples so that you can see exactly how to carry them out.

On the other hand, if you prefer to work with formulas, each relevant chapter ends with a section called Focus On Formulas. The section includes all the mathematical notation and formulas pertaining to the material in that chapter.

Exercises and Mini-Projects

Numerous exercises appear at the end of each chapter. Many of them are similar to the Thought Questions and require an explanation for which there is no one correct answer. Answers to every third exercise (3, 6, 9, etc.) are provided at the back of the book. These are indicated with an asterisk next to the exercise number. *Teaching Seeing Through Statistics: An Instructor's Resource Manual,* which is available for download from the companion website (http://www.cengage.com/UttsSTS4e), explains what is expected for each exercise.

In most chapters, the exercises contain many real-life examples. However, with the idea that you learn best by doing, most chapters also contain mini-projects. Some of these ask you to find examples of studies of interest to you; others ask you to conduct your own small-scale study. If you are reading this book without the benefit of a class or instructor, I encourage you to try some of the projects on your own.

Covering the Book in a Quarter, in a Semester, or on Your Own

I wrote this book for a one-quarter course taught three times a week at the University of California at Davis as part of the general education curriculum. My aim was to allow one lecture for each chapter, thus allowing for completion of the book (and a midterm or two) in the usual 29- or 30-lecture quarter. When I teach the course, I do not cover every detail from each chapter; I expect students to read some material on their own.

If the book is used for a semester course, it can be covered at a more leisurely pace and in more depth. For instance, two classes a week can be used for covering new material and a third class for discussion, additional examples, or laboratory work. Alternatively, with three regular lectures a week, some chapters can be covered in two sessions instead of one.

Instructors can download a variety of instructor resources from the companion website www.cengage.com/UttsSTS4e. The website includes additional information on how to cover the material in one quarter or semester. The website also includes tips on teaching this material, ideas on how to cover each chapter, sample lectures, additional examples, and exercise solutions. See below for a full description of the companion website.

Instructors who want to focus on more in-depth coverage of specific topics may wish to exclude others. Certain chapters can be omitted without interrupting the flow of the material or causing serious consequences in later chapters. These include Chapter 9 on plots and graphs, Chapters 16 and 17 on psychological and intuitive misunderstandings of probability, Chapter 18 on understanding economic data, Chapter 25 on meta-analysis, and Chapter 26 on ethics.

If you are reading this book on your own, you may want to concentrate on selected topics only. Parts 1 and 3 can be read alone, as can Chapters 9 and 18. Part 4 relies most heavily on Chapters 8, 12, 13, and 14. Although Part 4 is the most technically challenging part of the book, I strongly recommend reading it because it is there that you will truly learn the beauty as well as the pitfalls of statistical reasoning.

If you get stuck, try to step back and reclaim the big picture. Remember that although statistical methods are very powerful and are subject to abuse, they were developed using the collective common sense of researchers whose goal was to figure out how to find and interpret information to understand the world. They have done the hard work; this book is intended to help you make sense of it all.

A Summary of Changes from the First to Third Editions

In case you are comparing this fourth edition with an earlier edition, here is a summary of changes that were made from the first to second, and second to third editions. For the second edition, over 100 new exercises were added, many based on news stories. In the short time between the first and second editions, Internet use skyrocketed, and so for the second edition many examples from, and references to, websites with interesting data were added. The most substantial structural change from the first to the second edition was in Part 3. Using feedback from instructors,

Chapters 15 and 16 from the first edition were combined and altered to make the material more relevant to daily life. Some of that material was moved to the subsequent two chapters (Chapters 16 and 17 in the second edition). Box plots were added to Chapter 7, and Chapter 13 was rewritten to reflect changes in the Consumer Price Index. Wording and data were updated throughout the book as needed.

There were major changes made from the second to third edition. First, an appendix was added containing 20 news stories, which are used in examples and exercises throughout the book. These are tied to full journal articles, most of which were on a CD accompanying the third edition. The CD (which has now been replaced by a website) contained interactive applets as well. Material was reorganized and expanded, including a new chapter on ethics. New exercises and mini-projects were added, most of which take advantage of the news stories and journal articles in the appendix and on the CD/website.

Again in response to feedback from users, Chapter 12 from the second edition was expanded and divided into Chapters 12 and 13 in the third edition. As a consequence, all of the remaining chapters were renumbered. As mentioned, there was also a new chapter, Chapter 26, called "Ethics in Statistical Studies." As you have probably heard, some people think that you can use statistics to prove (or disprove) anything. That's not quite true, but it is true that there are multiple ways that researchers can naively or intentionally bias the results of their studies. Ethical researchers have a responsibility to make sure that doesn't happen. As an educated consumer, you have a responsibility to ask the right questions to determine if something unethical has occurred. Chapter 26 illustrates some subtle (and not so subtle) ways in which ethics play a role in research. New sections were also added to Chapters 2, 5, 7, 12, and 22 (formerly Chapter 21).

New for the Fourth Edition

There are three major content changes for this edition. First, a new chapter has been added in Part 3, called "Understanding Uncertainty through Simulation" (Chapter 15). The material from that chapter is then used in Part 4 to show how some of the inference procedures can be simulated, expanding the range of problems that can be solved. The former Chapter 14, "Reading the Economic News," has been rewritten and moved to Part 3 with the new title "Understanding the Economic News." And finally, the former Chapter 15, "Understanding and Reporting Trends over Time" has been assimilated into other chapters—the material on time series has been moved to Chapters 9 and 10, and the section on seasonal adjustments to economic data has been moved to the new Chapter 18.

In addition to the major content changes, additional topics have been included in some chapters. The chapters on intuition and probability have been expanded. A section on multiple testing and multiple comparisons has been added to Chapter 24. More information on how to do t-tests and confidence intervals has been added in the relevant chapters. Instructions on using the computer to solve problems have been added in many chapters. A summary called "Thinking about Key Concepts" has been added to the end of most chapters. Throughout the book examples, case studies and exercises have been updated, and new studies have been added. Some of the older examples and case studies have been replaced, but can be found on the companion website.

In previous editions, answers to some homework problems were available in the back of the book, but in this edition that feature has been made more systematic. The answers to every exercise with a number divisible by 3 (3, 6, 9, etc.) are found in the back of the book. In order to implement this feature, the exercises in most chapters have been reordered. In doing so, they also have been rearranged so that the order now more closely follows the order of the content in most chapters.

News Stories and Journal Articles in the Appendix and on the Companion Website

One of the goals of this book is to help you understand news stories based on statistical studies. To enhance that goal, an appendix with 20 news stories was added beginning with the third edition. In this edition, some of the stories have been replaced with instructions on how to find them on the Internet rather than printing the full story in the book. But the news stories present only part of the information you need to understand the studies. When journalists write such stories, they rely on original sources, which in most cases include an article in a technical journal or a technical report prepared by researchers. To give you that same exposure, the companion website for the book contains the full text of the original sources for most of the news stories. Because these articles include hundreds of pages, it would not have been possible to append the printed versions of them. Having immediate access to these reports allows you to learn much more about how the research was conducted, what statistical methods were used, and what conclusions the original researchers drew. You can then compare these to the news stories derived from them and determine whether you think those stories are accurate and complete. In some cases, an additional news story or press release is included on the companion website as well.

Aplia[TM] is an online interactive learning solution that helps students improve comprehension—and their grade—by integrating a variety of mediums and tools such as video, tutorials, practice tests, and an interactive eBook. Created by a professor to enhance his own courses, Aplia provides automatically graded assignments with detailed, immediate feedback on every question, and innovative teaching materials. More than 1 million students have used Aplia at over 1800 institutions.

New for the fourth edition, available via Aplia, is **MindTap**[TM] **Reader**, Cengage Learning's next generation eBook. MindTap Reader provides robust opportunities for students to annotate, take notes, navigate, and interact with the text (e.g. ReadSpeaker). Annotations captured in MindTap Reader are automatically tied to the Notepad app, where they can be viewed chronologically and in a cogent, linear fashion. Instructors also can edit the text and assets in the Reader, as well as add videos or URLs. Go to http://www.cengage.com/mindtap for more information.

Companion Website for *Seeing Through Statistics*

The companion website (www.cengage.com/UttsSTS4e) has been established for users of this book.

Two of the major features of the website are the **journal articles** mentioned previously and a collection of **computer applets**. The applets will allow you to explore some of the concepts in this book in an interactive way, and the website includes suggestions for how to do so.

This site also includes a variety of resources and information of interest to users of this book:

- Microsoft® PowerPoint® lecture slides and figures from the book
- information on how to cover the material in one quarter or semester
- tips on teaching this material
- ideas on how to cover each chapter
- sample lectures
- additional examples
- exercise solutions

Cengage Learning Testing Powered by Cognero

Cengage Learning Testing is a flexible, online system that allows you to:

- author, edit, and manage test bank content from multiple Cengage Learning solutions
- create multiple test versions in an instant
- deliver tests from your LMS, your classroom, or wherever you want

Acknowledgments

I would like to thank Robert Heckard and William Harkness from the Statistics Department at Penn State University, as well as their students, for providing data collected in their classes. I would also like to thank the following individuals for providing valuable insights for the first to third editions of this book: Mary L. Baggett, Northern Kentucky University; Dale Bowman, University of Mississippi; Paul Cantor, formerly of Lehman College, City University of New York; James Casebolt, Ohio University–Eastern Campus; Deborah Delanoy, University College Northampton (England); Monica Halka, Portland State University; Hariharan K. Iyer, Colorado State University; Richard G. Krutchkoff, Virginia Polytechnic Institute and State University; Lawrence M. Lesser, Armstrong Atlantic State University; Vivian Lew, University of California–Los Angeles; Maggie McBride, Montana State University–Billings; Monnie McGee, Hunter College; Nancy Pfenning, University of Pittsburgh; Scott Plous, Wesleyan University; Lawrence Ries, University of Missouri–Columbia; Larry Ringer, Texas A&M University; Barb Rouse, University of Wyoming; Ralph R. Russo, University of Iowa; Laura J. Simon, Pennsylvania State University; Daniel Stuhlsatz, Mary Baldwin College; Eric Suess, California State University–Hayward; Larry Wasserman, Carnegie

Mellon University; Sheila Weaver, University of Vermont; Farroll T. Wright, University of Missouri–Columbia; and Arthur B. Yeh, Bowling Green State University.

In addition, I want to express my gratitude to the following people for their many helpful comments and suggestions for this fourth edition: Peter Dunn, University of the Sunshine Coast (Australia); Golde Holtzman, Virginia Tech University; Abraham Ayebo, Purdue University–Calumet; Laura Chihara, Carleton College; Samuel Cook, Wheelock College; Lawrence Lesser, The University of Texas–El Paso; Jamis Perrett, Texas A&M University; Mamunur Rashid, Indiana University Purdue University–Indianapolis; and Dennis Wacker, Saint Louis University.

Finally, I want to thank my mother Patricia Utts and my sister Claudia Utts-Smith, for helping me realize the need for this book because of their unpleasant experiences with statistics courses; Alex Kugushev, former Publisher of Duxbury Press, for persisting until I agreed to write it (and beyond); Carolyn Crockett, Cengage Learning, for her encouragement and support during the writing of the second and third editions; Jay Campbell and Molly Taylor, Cengage Learning, for their encouragement and support in writing this edition, and Robert Heckard, Penn State University, for instilling in me many years ago, by example, the enthusiasm for teaching and educating that led to the development of this material.

Jessica M. Utts

Finding Data in Life

By the time you finish reading Part 1 of this book, you will be reading studies reported in the news with a whole new perspective. In these chapters, you will learn how researchers should go about collecting information for surveys and experiments. You will learn to ask questions, such as who funded the research, that could be important in deciding whether the results are accurate and unbiased.

Chapter 1 is designed to give you some appreciation for how statistics helps to answer interesting questions. Chapters 2 to 5 provide an in-depth, behind-the-scenes look at how surveys and experiments are supposed to be done. In Chapter 6, you will learn how to tie together the information from the previous chapters, including seven steps to follow when reading about studies.

These steps all lead to the final step, which is the one you should care about the most. You will have learned how to *determine whether the results of a study are meaningful enough to encourage you to change your lifestyle, attitudes, or beliefs.*

The Benefits and Risks of Using Statistics

Thought Questions

1. A news story about drug use and grades concluded that smoking marijuana at least three times a week resulted in lower grades in college. How do you think the researchers came to this conclusion? Do you believe it? Is there a more reasonable conclusion?

2. It is obvious to most people that, on average, men are taller than women, and yet there are some women who are taller than some men. Therefore, if you wanted to "prove" that men were taller, you would need to measure many people of each sex. Here is a theory: On average, men have lower resting pulse rates than women do. How could you go about trying to prove or disprove that? Would it be sufficient to measure the pulse rates of one member of each sex? Two members of each sex? What information about men's and women's pulse rates would help you decide how many people to measure?

3. Suppose you were to learn that the large state university in a particular state graduated more students who eventually went on to become millionaires than any of the small liberal arts colleges in the state. Would that be a fair comparison? How should the numbers be presented in order to make it a fair comparison?

4. In a survey done in September, 2012, employers were asked a series of questions about whether colleges were preparing students adequately for careers in their companies. Of the 50,000 employers contacted, 704 responded. One of the questions asked was "How difficult is it to find recent college graduates who are qualified for jobs at your organization?" Over half (53%) of the respondents said that it was difficult or very difficult. Based on these results, can you conclude that about 53% of all employers feel this way? Why or why not?
(Source: http://chronicle.com/items/biz/pdf/Employers%20Survey.pdf.)

1.1   Why Bother Reading This Book?

If you have never studied statistics, you are probably unaware of the impact the science of statistics has on your everyday life. From knowing which medical treatments work best to choosing which television programs remain on the air, decision makers in almost every line of work rely on data and statistical studies to help them make wise choices.

We are exposed daily to information from surveys and scientific studies concerning our health, behavior, attitudes, and beliefs or revealing scientific and technological breakthroughs. This book's first objective is to help you understand this information and to sift the useful and the accurate from the useless and the misleading. (And there are plenty of both out there!) By the time you finish reading the book, you should be a statistical detective—able to read with a critical eye and to rely on your own interpretation of results emerging from surveys and statistical studies.

Another purpose of this book is to demystify statistical methods. Traditional statistics courses often place emphasis on how to compute rather than on how to understand. This book focuses on statistical ideas and their use in real life. Lastly, the book also contains information that can help you make better decisions. You will learn how psychological influences can keep you from making the best decisions, as well as new ways to think about coincidences, gambling, and other circumstances that involve chance events.

1.2   What is Statistics All About?

If we were all exactly the same—had the same physical makeup, the same behaviors and opinions, liked the same music and movies, and so on—most statistical methods would be of little use. But fortunately we aren't all the same! Statistical methods are used to analyze situations involving uncertainty and natural variation. They can help us understand our differences as well as find patterns and relationships that apply to all of us.

When you hear the word *statistics,* you probably either get an attack of math anxiety or think about lifeless numbers, such as the population of the city or town where you live, as measured by the latest census, or the per capita income in Japan. The goal of this book is to open a whole new world of understanding of the term *statistics,* and to help you realize that the invention of statistical methods is one of the most important developments of modern times. These methods influence everything from life-saving medical advances to the percent salary increase given to millions of people every year.

The word **statistics** is actually used to mean two different things. The better-known definition is that statistics are numbers measured for some purpose. A more appropriate, complete definition is the following:

*Statistics is a collection of procedures and principles for gaining and
analyzing information to educate people and help them make better decisions
when faced with uncertainty.*

Using this definition, you have undoubtedly used statistics in your own life. For example, if you were faced with a choice of routes to get to school or work, or to get between one classroom building and the next, how would you decide which one to take? You would probably try each of them a number of times (thus gaining information) and then choose the best one according to criteria important to you, such as speed, fewer red lights, more interesting scenery, and so on. You might even use different criteria on different days—such as when the weather is pleasant versus when it is not. In any case, by sampling the various routes and comparing them, you would have gained and analyzed useful information to help you make a decision.

In addition to helping us make decisions, statistical studies also help us satisfy our curiosity about other people and the world around us. Do other people have the same opinions we do? Are they behaving the way we do—good or bad? Here is an example that may answer one of those questions for you.

EXAMPLE 1.1   **Look Who's Talking!**

Texting or talking on a cell phone (other than hands-free) while driving is illegal in most of the United States and in many other countries. How many law-breakers are there? Does it differ by country? Does it differ by age group? The answers are lots, yes, and yes! Statistical surveys conducted in the United States and Europe in 2011 asked people, "In the past 30 days, how often have you talked on your cell phone while you were driving?" Response choices were "never," "just once," "rarely," "fairly often," and "regularly." The results showed that over two-thirds (68.7%) of adults surveyed in the United States admitted talking on their cell phone at least once in the past 30 days, but just over one-fifth (20.5%) of those in the United Kingdom admitting doing so. In the United States, almost 70% of respondents aged 18 to 24 admitted that they had talked on their phone, but only about 60% of those aged 55 to 64 did so. The age differences were even more striking for those who admitted texting, with slightly over 50% of the youngest age group doing so but under 10% of the oldest age group doing so.

Although it is interesting to know whether others are behaving like we are, these statistics also have implications for public health officials and lawmakers who are studying them to determine why drivers in the United Kingdom are so much more law-abiding than are drivers in the United States in this instance. Or are they? In this book, you will learn to ask other probing questions, such as whether the two groups are equally likely to be telling the truth in their responses and whether the difference could be partially explained by the fact that people in the United States drive more than people in the United Kingdom.

(Source: *Morbidity and Mortality Weekly Report,* March 15, 2013 Vol. 62, No. 10, Centers for Disease Control and Prevention.) ∎

1.3   Detecting Patterns and Relationships

How do scientists decide what questions to investigate? Often, they start with observing something and become curious about whether it's a unique circumstance or something that is part of a larger pattern. In Case Study 1.1, we will see how one researcher followed a casual observation to a fascinating conclusion.

| CASE STUDY 1.1 | **Heart or Hypothalamus?** |

SOURCE: Salk (1973), pp. 26–29.

You can learn a lot about nature by observation. You can learn even more by conducting a carefully controlled experiment. This case study has both. It all began when psychologist Lee Salk noticed that despite his knowledge that the hypothalamus plays an important role in emotion, it was the heart that seemed to occupy the thoughts of poets and songwriters. There were no everyday expressions or song titles such as "I love you from the bottom of my hypothalamus" or "My hypothalamus longs for you." Yet, there was no physiological reason for suspecting that the heart should be the center of such attention. Why had it always been the designated choice?

Salk began wondering about the role of the heart in human relationships. He also noticed that when on 42 separate occasions he watched a rhesus monkey at the zoo holding her baby, she held the baby on the left side, close to her heart, on 40 of those occasions. He then observed 287 human mothers within 4 days after giving birth and noticed that 237, or 83%, held their babies on the left. Handedness did not explain it; 83% of the right-handed mothers and 78% of the left-handed mothers exhibited the left-side preference. When asked why they chose the left side, the right-handed mothers said it was so their right hand would be free. The left-handed mothers said it was because they could hold the baby better with their dominant hand. In other words, both groups were able to rationalize holding the baby on the left based on their own preferred hand.

Salk wondered if the left side would be favored when carrying something other than a newborn baby. He found a study in which shoppers were observed leaving a supermarket carrying a single bag; exactly half of the 438 adults carried the bag on the left. But when stress was involved, the results were different. Patients at a dentist's office were asked to hold a 5-inch rubber ball while the dentist worked on their teeth. Substantially more than half held the ball on the left.

Salk speculated, "It is not in the nature of nature to provide living organisms with biological tendencies unless such tendencies have survival value." He surmised that there must indeed be survival value to having a newborn infant placed close to the sound of its mother's heartbeat.

To test this conjecture, Salk designed a study in a baby nursery at a New York City hospital. He arranged for the nursery to have the continuous sound of a human heartbeat played over a loudspeaker. At the end of 4 days, he measured how much weight the babies had gained or lost. Later, with a new group of babies in the nursery, no sound was played. Weight gains were again measured after 4 days.

The results confirmed what Salk suspected. Although they did not eat more than the control group, the infants treated to the sound of the heartbeat gained more weight (or lost less). Further, they spent much less time crying. Salk's conclusion was that "newborn infants are soothed by the sound of the normal adult heartbeat." Somehow, mothers intuitively know that it is important to hold their babies on the left side. What had started as a simple observation of nature led to a further understanding of an important biological response of a mother to her newborn infant. ∎

How to Move From Noticing to Knowing

Some differences are obvious to the naked eye, such as the fact that the average man is taller than the average woman. If we were content to know about only such obvious relationships, we would not need the power of statistical methods. But had you noticed that babies who listen to the sound of a heartbeat gain more weight? Have you ever noticed that taking aspirin helps prevent heart attacks? How about the fact that people are more likely to buy jeans in certain months of the year than in others? The fact that men have lower resting pulse rates than women do? The fact that listening to Mozart improves performance on the spatial reasoning questions of an IQ test? All of these are relationships that have been demonstrated in studies using proper statistical methods, yet none of them are obvious to the naked eye.

Let's take the simplest of these examples—one you can test yourself—and see what's needed to properly demonstrate the relationship. Suppose you wanted to verify the claim that, on average, men have lower resting pulse rates than women do. Would it be sufficient to measure only your own pulse rate and that of a friend of the opposite sex? Obviously not. Even if the pair came out in the predicted direction, the singular measurements would certainly not speak for all members of each sex.

It is not easy to conduct a statistical study properly, but it is easy to understand much of how it should be done. We will examine each of the following concepts in great detail in the remainder of this book; here we just introduce them, using the simple example of comparing male and female pulse rates.

To conduct a statistical study properly, one must:

1. Get a representative sample.

2. Get a large enough sample.

3. Decide whether the study should be an observational study or a randomized experiment.

---

1. Get a representative sample. Most researchers hope to extend their results beyond just the participants in their research. Therefore, it is important that the people or objects in a study be representative of the larger group for which conclusions are to be drawn. We call those who are actually studied a **sample** and the larger group from which they were chosen a **population.** (In Chapter 4, we will learn some ways to select a proper sample.) For comparing pulse rates, it may be convenient to use the members of your class. But this sample would not be valid if there were something about your class that would relate pulse rates and sex, such as if the entire men's track team happened to be in the class. It would also be unacceptable if you wanted to extend your results to an age group much different from the distribution of ages in your class. Often researchers are constrained to using such "convenience" samples, and we will discuss the implications of this later in the book.

2. Get a large enough sample. Even experienced researchers often fail to recognize the importance of this concept. In Part 4 of this book, you will learn how to detect the problem of a sample that is too small; you will also learn that such a sample can sometimes lead to erroneous conclusions. In comparing pulse rates, collecting one pulse rate from each sex obviously does not tell us much. Is two enough? Four? One hundred? The answer to that question depends on how much *natural variability* there is among pulse rates. If all men had pulse rates of 65 and all women had pulse rates of 75, it wouldn't take long before you recognized a difference. However, if men's pulse rates ranged from 50 to 80 and women's pulse rates ranged from 52 to 82, it would take many more measurements to convince you of a difference. The question of how large is "large enough" is closely tied to how diverse the measurements are likely to be within each group. The more diverse, or variable, the individuals within each group, the larger the sample needs to be to detect a real difference between the groups.

3. Decide whether the study should be an observational study or a randomized experiment. For comparing pulse rates, it would be sufficient to measure or "observe" both the pulse rate and the sex of the people in our sample. When we merely observe things about our sample, we are conducting an **observational study.** However, if we were interested in whether frequent use of aspirin would help prevent heart attacks, it would not be sufficient to simply observe whether people frequently took aspirin and then whether they had a heart attack. It could be that people who were more concerned with their health were both more likely to take aspirin and less likely to have a heart attack, or vice versa. Or, it could be that drinking the extra glass of water required to take the aspirin contributes to better health.

To be able to make a causal connection, we would have to conduct a **randomized experiment** in which we *randomly* assigned people to one of two groups. **Random assignments** are made by doing something akin to flipping a coin to determine the group membership for each person. In one group, people would be given aspirin, and in the other, they would be given a dummy pill that looked like aspirin. So as not to influence people with our expectations, we would not tell people which one they were taking until the experiment was concluded. In Case Study 1.2, we briefly examine the experiment that initially established the causal link between aspirin use and reduction of heart attacks. In Chapter 5, we discuss these ideas in much more detail.

CASE STUDY 1.2	**Does Aspirin Prevent Heart Attacks?**

In 1988, the Steering Committee of the Physicians' Health Study Research Group released the results of a 5-year randomized experiment conducted using 22,071 male physicians between the ages of 40 and 84. The physicians had been randomly assigned to two groups. One group took an ordinary aspirin tablet every other day, whereas the other group took a "placebo," a pill designed to look just like an aspirin but with no active ingredients. Neither group knew whether they were taking the active ingredient.

The results, shown in Table 1.1, support the conclusion that taking aspirin does indeed help reduce the risk of having a heart attack. The rate of heart attacks in the

TABLE 1.1	The Effect of Aspirin on Heart Attacks		
Condition	**Heart Attack**	**No Heart Attack**	**Attacks per 1000**
Aspirin	104	10,933	9.42
Placebo	189	10,845	17.13

group taking aspirin was only 55% of the rate of heart attacks in the placebo group, or just slightly more than half as big. Because the men were randomly assigned to the two conditions, other factors, such as amount of exercise, should have been similar for both groups. The only substantial difference in the two groups should have been whether they took the aspirin or the placebo. Therefore, we can conclude that taking aspirin caused the lower rate of heart attacks for that group.

Notice that because the participants were all male physicians, these conclusions may not apply to the general population of men. They may not apply to women at all because no women were included in the study. More recent evidence has provided even more support for this effect, however—something we will examine in more detail in an example in Chapter 27. ∎

1.4   Don't Be Deceived by Improper Use of Statistics

Let's look at some examples representative of the kinds of abuses of statistics you may see in the media. The first example illustrates the danger of not getting a representative sample; in the second example, the statistics have been taken out of their proper context; and in the third and fourth examples, you will see how to stop short of making too strong a conclusion on the basis of an observational study.

EXAMPLE 1.2   Robotic Polls and Representative Samples

Methods for polling voters to predict election results have become more complicated as more and more people rely exclusively on cell phones instead of landlines. But some polling firms were slow to change their methods and, consequently, made significant blunders in predicting the outcome of the 2012 Presidential Election. In particular, polling companies that exclusively used "robopolls," which choose their participants using randomized computer dialing and then use an automated script, missed the mark by an average of 4.3 percentage points in favor of the Republican candidate Mitt Romney. One firm even had a 15.7% Republican bias! What went wrong? By 2012, about one-third of Americans relied solely on cell phones and had no landline. But it was illegal to call cell phones using robopolls, so organizations that used them only reached homes with landlines. Younger voters, those less well-off financially and non-Caucasians, were less likely to have landlines and also more likely to vote for the Democratic candidate, Barack Obama. Therefore, the robopolls underestimated the support for Obama.

(*Source:* http://fivethirtyeight.blogs.nytimes.com/2012/11/10/which-polls-fared-best-and-worst-in-the-2012-presidential-race/#more-37396.) ∎

EXAMPLE 1.3 Toxic Chemical Statistics

When a federal air report ranked the state of New Jersey as 22nd in the nation in its release of toxic chemicals, the New Jersey Department of Environmental Protection happily took credit (Wang, 1993, p. 170). The statistic was based on a reliable source, a study by the U.S. Environmental Protection Agency. However, the ranking had been made based on total pounds released, which was 38.6 million for New Jersey. When this total was turned into pounds per square mile in the state, it became apparent New Jersey was one of the worst—fourth on the list. Because New Jersey is one of the smallest states by area, the figures were quite misleading until adjusted for size.   ■

EXAMPLE 1.4 Mom's Smoking and Kid's IQ

Read the article in Figure 1.1, and then read the headline again. Notice that the headline stops short of making a causal connection between smoking during pregnancy and lower IQs in children. Reading the article, you can see that the results are based on an observational study and not an experiment—with good reason: It would clearly be unethical to randomly assign pregnant women to either smoke or not. With studies like this, the best that can be done is to try to measure and statistically adjust for other factors that might be related to both smoking behavior and children's IQ scores. Notice that when the researchers did so, the gap in IQ between the children of smokers and nonsmokers narrowed from nine points down to four points. There may be even more factors that the researchers did not measure that

Figure 1.1

Don't make causal connections from observational studies

Source: "Study: Smoking May Lower Kids' IQs." Associated Press, February 11, 1994. Reprinted with permission.

Study: Smoking May Lower Kids' IQs

ROCHESTER, N.Y. (AP)—Secondhand smoke has little impact on the intelligence scores of young children, researchers found.

But women who light up while pregnant could be dooming their babies to lower IQs, according to a study released Thursday.

Children ages 3 and 4 whose mothers smoked 10 or more cigarettes a day during pregnancy scored about 9 points lower on the intelligence tests than the offspring of nonsmokers, researchers at Cornell University and the University of Rochester reported in this month's *Pediatrics* journal.

That gap narrowed to 4 points against children of nonsmokers when a wide range of interrelated factors were controlled. The study took into account secondhand smoke as well as diet, education, age, drug use, parents' IQ, quality of parental care and duration of breast feeding.

"It is comparable to the effects that moderate levels of lead exposure have on children's IQ scores," said Charles Henderson, senior research associate at Cornell's College of Human Ecology in Ithaca.

would account for the remaining four-point difference. Unfortunately, with an observational study, we simply cannot make causal conclusions. We will explore this particular example in more detail in Chapter 6. ■

EXAMPLE 1.5   Does Marijuana Impair the Brain?

An article headlined "New study confirms too much pot impairs brain" read as follows:

> More evidence that chronic marijuana smoking impairs mental ability: Researchers at the University of Iowa College of Medicine say a test shows those who smoke seven or more marijuana joints per week had lower math, verbal and memory scores than non-marijuana users. Scores were particularly reduced when marijuana users held a joint's smoke in their lungs for longer periods. (*San Francisco Examiner*, 13 March 1993, p. D-1)

This research was clearly based on an observational study because people cannot be randomly assigned to either smoke marijuana or not. The headline is misleading because it implies that there is a causal connection between smoking marijuana and brain functioning. All we can conclude from an observational study is that there is a relationship. It could be the case that people who choose to smoke marijuana are those who would score lower on the tests anyway. ■

In addition to learning how to evaluate statistical studies, in this book you will learn some simple methods for computing probabilities. Case Study 1.3 not only describes one practical application of probability but also illustrates that it is important to question what assumptions are made before computing the probability of a particular event.

CASE STUDY 1.3   **Using Probability to Detect Cheating**

Professors and other students hate it when students cheat on exams, and some professors have devised methods that make it relatively easy to detect cheating on essays and similar questions. But detecting cheating on multiple choice exams is not as easy. That's where probability and statistics can help. Professor Robert Mogull thought he detected cheating by two students on multiple choice exams in his statistics class at Sacramento State University because they had identical questions wrong on all four 25-question multiple choice exams. He calculated the probability of that happening to be extremely small, failed the two students, and published a paper explaining his method (Mogull, 2003). But others have criticized his probability calculations because he assumed that all students were equally likely to miss any particular question (Actuarial Outpost, 2013). The critics pointed out that there are all sorts of reasons why two particular students might miss the same questions as each other, especially if they were friends. Perhaps they studied together, they had the same major and thus had similar knowledge, had the same statistics course in high school, and so on.

In a similar case, Klein (1992) described a situation in which two students were accused of cheating on a multiple-choice medical licensing exam. They had been observed whispering during one part of the 3-day exam and their answers to the questions they got wrong very often matched each other. The licensing board determined that the statistical evidence for cheating was overwhelming. They estimated that the odds of two people having answers as close as these two did were less than 1 in 10,000. Further, the students were husband and wife. Their tests were invalidated.

The case went to trial, and upon further investigation, the couple was exonerated. They hired a statistician who was able to show that the agreement in their answers during the session in which they were whispering was no higher than it was in the other sessions. What happened? The board assumed students who picked the wrong answer were simply guessing among the other choices. This couple had grown up together and had been educated together in India. Answers that would have been correct for their culture and training were incorrect for the American culture (for example, whether a set of symptoms was more indicative of tuberculosis or a common cold). Their common mistakes often would have been the right answers for India. So, the licensing board erred in calculating the odds of getting such a close match by using the assumption that they were just guessing. And, according to Klein, "with regard to their whispering, it was very brief and had to do with the status of their sick child" (p. 26). ∎

1.5   Summary and Conclusions

In this chapter, we have just begun to examine both the advantages and the dangers of using statistical methods. We have seen that it is not enough to know the results of a study, survey, or experiment. We also need to know how those numbers were collected and who was asked. In the upcoming chapters, you will learn much more about how to collect and process this kind of information properly and how to detect problems in what others have done. You will learn that a relationship between two characteristics (such as smoking marijuana and lower grades) does not necessarily mean that one causes the other, and you will learn how to determine other plausible explanations. In short, you will become an educated consumer of statistical information.

Thinking About Key Concepts

- Cause and effect conclusions cannot generally be made based on results of observational studies. The problem is that the groups being studied are likely to differ in lots of other ways, any of which could be causing the observed difference in outcome.
- In randomized experiments, cause and effect conclusions *can* generally be made. By randomly assigning individuals to receive different treatments, the groups should be similar on everything except the treatment they are given.

- If a study is conducted using a sample that is representative of a larger group for the question of interest, then the results from the sample can be considered to apply to the larger group (population) as well.
- If there is little natural variability in the responses being measured, then a sample with only a small number of individuals should be adequate for detecting differences among groups. But if there is lots of variability, differences are only likely to be detected if a large number of individuals are included in the measured sample.
- When comparing statistics for groups or locations of different sizes, it is more informative to report the results per person or per unit, rather than for the group as a whole.

Exercises

Exercises with numbers divisible by 3 (3, 6, 9, etc.) are included in the Solutions at the back of the book. They are marked with an asterisk (*).

1. Explain why the relationship shown in Table 1.1, concerning the use of aspirin and heart attack rates, can be used as evidence that aspirin actually prevents heart attacks.

2. "People who often attend cultural activities, such as movies, sports events and concerts, are more likely than their less cultured cousins to survive the next eight to nine years, even when education and income are taken into account, according to a survey by the University of Umea in Sweden" (*American Health,* April 1997, p. 20).

   a. Can this claim be tested by conducting a randomized experiment? Explain.

   b. On the basis of the study that was conducted, can we conclude that attending cultural events causes people to be likely to live longer? Explain.

   c. The article continued, "No one's sure how Mel Gibson and Mozart help health, but the activities may enhance immunity or coping skills." Comment on the validity of this statement.

   d. The article notes that education and income were taken into account. Give two examples of other factors about the people surveyed that you think should also have been taken into account.

*3. Explain why the number of people in a sample is an important factor to consider when designing a study.

4. Explain what problems may arise in trying to make conclusions based on a survey mailed to the subscribers of a specialty magazine. Find or construct an example.

5. "If you have borderline high blood pressure, taking magnesium supplements may help, Japanese researchers report. Blood pressure fell significantly in subjects who got 400–500 milligrams of magnesium a day for four weeks, but not in those getting a placebo" (*USA Weekend,* 22–24 May 1998, p. 11).

   a. Do you think this was a randomized experiment or an observational study? Explain.

   b. Do you think the relationship found in this study is a causal one, in which taking magnesium actually causes blood pressure to be lowered? Explain.

*6. A study reported in the May 17, 2012, issue of *The New England Journal of Medicine* followed people for an average of 13 years and found that people who consumed two or more cups of coffee a day were less likely to die during the course of the study than those who drank no coffee.

   *a. Was this study a randomized experiment or an observational study? Explain how you know.

   *b. Based on this study, can it be concluded that drinking coffee *causes* people to live longer?

   *c. The following headlines appeared on news websites reporting these results. In each case, explain whether or not the conclusion in the headline is justified.

   i. *"Coffee positively associated with life expectancy"* (Source: http://www.coffee-andhealth.org/2012/05/21/coffee-positively-associated-with-life-expectancy/)

   ii. *"NIH Study: Coffee Really Does Make You Live Longer, After All"* (Source: http://www.theatlantic.com/health/archive/2012/05/nih-study-coffee-really-does-make-you-live-longer-after-all/257302/)

7. Refer to Case Study 1.1. When Salk measured the results, he divided the babies into three groups based on whether they had low (2510 to 3000 g), medium (3010 to 3500 g), or high (3510 g and over) birthweights. He then compared the infants from the heartbeat and silent nurseries separately within each birth-weight group. Why do you think he did that? (*Hint:* Remember that it would be easier to detect a difference in male and female pulse rates if all males measured 65 beats per minute and all females measured 75 than it would be if both groups were quite diverse.)

8. A psychology department is interested in comparing two methods for teaching introductory psychology. Four hundred students plan to enroll for the course at 10:00 A.M., and another 200 plan to enroll for the course at 4:00 P.M. The registrar will allow the department to teach multiple sections at each time slot and to assign students to any one of the sections taught at the student's desired time. Design a study to compare the two teaching methods. For example, would it be a good idea to use one method on all of the 10:00 sections and the other method on all of the 4:00 sections? Explain your reasoning.

*9. Suppose you have a choice of two grocery stores in your neighborhood. Because you hate waiting, you want to choose the one for which there is generally a shorter wait in the checkout line. How would you gather information to determine which one is faster? Would it be sufficient to visit each store once and time how long you had to wait in line? Explain.

10. Universities are sometimes ranked for prestige according to the amount of research funding their faculty members are able to obtain from outside sources. Explain why it would not be fair to simply use total dollar amounts for each university, and describe what should be used instead.

11. Refer to Case Study 1.3, in which two students were accused of cheating because the licensing board determined the odds of such similar answers were less than 1 in 10,000. Further investigation revealed that over 20% of all pairs of students had matches giving low odds like these (Klein, 1992, p. 26). Clearly, something was wrong with the method used by the board. Read the case study and explain what erroneous assumption they made in their determination of the odds. (*Hint:* Use your own experience with answering multiple-choice questions.)

*12. Suppose researchers want to know whether smoking cigars increases the risk of esophageal cancer.

   *a. Could they conduct a randomized experiment to test this? Explain.

   *b. If they conducted an observational study and found that cigar smokers had a higher rate of esophageal cancer than those who did not smoke cigars, could they conclude that smoking cigars increases the risk of esophageal cancer? Explain why or why not.

13. Suppose the officials in the city or town where you live would like to ask questions of a "representative sample" of the adult population. Explain some of the characteristics this sample should have. For example, would it be sufficient to include only homeowners?

14. Give an example of a decision in your own life, such as which route to take to school, for which you think statistics would be useful in making the decision. Explain how you could collect and process information to help make the decision.

*15. Suppose you have 20 tomato plants and want to know if fertilizing them will help them produce more fruit. You randomly assign 10 of them to receive fertilizer and the remaining 10 to receive none. You otherwise treat the plants in an identical manner.

   *a. Explain whether this would be an observational study or a randomized experiment.

   *b. If the fertilized plants produce 30% more fruit than the unfertilized plants, can you conclude that the fertilizer caused the plants to produce more? Explain.

16. Sometimes television news programs ask viewers to call and register their opinions about an issue. One number is to be called for a "yes" opinion and another number for a "no" vote. Do you think viewers who call are a representative sample of all viewers? Explain.

17. Suppose a researcher would like to determine whether one grade of gasoline produces better gas mileage than another grade. Twenty cars are randomly divided into two groups, with 10 cars receiving one grade and 10 receiving the other. After many trips, average mileage is computed for each car.

   a. Would it be easier to detect a difference in gas mileage for the two grades if the 20 cars were all the same size, or would it be easier if they covered a wide range of sizes and weights? Explain.

   b. What would be one disadvantage to using cars that were all the same size?

*18. National polls are often conducted by asking the opinions of a few thousand adults nationwide and using them to infer the opinions of all adults in the nation. Explain who is in the sample and who is in the population for such polls.

19. Suppose the administration at your school wants to know how students feel about a policy banning smoking on campus. Because they can't ask all students, they must rely on a sample.

   a. Give an example of a sample they could choose that would *not* be representative of all students.

   b. Explain how you think they could get a representative sample.

20. A news headline read, "Study finds walking a key to good health: Six brisk outings a month cut death risk." Comment on what type of study you think was done and whether this is a good headline.

*21. Suppose a study first asked people whether they meditate regularly and then measured their blood pressures. The idea would be to see if those who meditate have lower blood pressure than those who do not do so.

   *a. Explain whether this would be an observational study or a randomized experiment.

   *b. If it were found that meditators had lower-than-average blood pressures, can we conclude that meditation *causes* lower blood pressure? Explain.

22. Refer to the definitions of sample and population on page 7. For each of the following, explain who is in the sample, and what population you think is represented.

   a. Case Study 1.1, testing the effect of a heartbeat sound in a baby nursery.

   b. Case Study 1.2, testing the effect of taking aspirin on heart attack rates.

23. In its March 3–5, 1995 issue, *USA Weekend* magazine asked readers to return a survey with a variety of questions about sex and violence on

television. Of the 65,142 readers who responded, 97% were "very or somewhat concerned about violence on TV" (*USA Weekend*, 2–4 June 1995, p. 5). Based on this survey, can you conclude that about 97% of all U.S. citizens are concerned about violence on TV? Why or why not?

*24. Refer to Thought Question 4 on page 3. A survey was sent to 50,000 employers but only 704 responded. The purpose of the survey was to find out how employers feel about the preparation of college graduates to work in their companies.

*a. Who do you think would be more likely to take the time to respond to such a survey—employers who are satisfied with the preparation of college graduates or employers who are not satisfied?

*b. One of the findings of the study was that 31% of respondents thought that college graduates were unprepared for a job search, while only 20% thought they were well prepared or very well prepared. (The remaining 49% thought they were "prepared.") Using your answer to part (a), explain whether or not the 704 survey respondents are likely to represent how all employers feel on this question.

Mini-Projects

1. Design and carry out a study to test the proposition that men have lower resting pulse rates than women.

2. Find a news story that discusses a recent study involving statistical methods. Identify the study as either an observational study or a randomized experiment. Comment on how well the simple concepts discussed in this chapter have been applied in the study. Comment on whether the news article, including the headline, accurately reports the conclusions that can legitimately be made from the study. Finally, discuss whether any information is missing from the news article that would have helped you answer the previous questions.

References

Actuarial Outpost, http://www.actuarialoutpost.com/actuarial_discussion_forum/archive/index.php/t-42328.html, accessed May 10, 2013.

Klein, Stephen P. (1992). Statistical evidence of cheating on multiple-choice tests. *Chance* 5, no. 3–4, pp. 23–27.

Mogull, Robert G. (2003). A device to detect student cheating. *Journal of College Teaching and Learning*, 1, no. 9, pp. 17-21.

Salk, Lee. (May 1973). The role of the heartbeat in the relations between mother and infant. *Scientific American,* pp. 26–29.

Steering Committee. Physicians' Health Study Research Group. (28 January 1988). Preliminary report: Findings from the aspirin component of the ongoing Physicians' Health Study. *New England Journal of Medicine* 318, no. 4, pp. 262–264.

Wang, Chamont. (1993). *Sense and nonsense of statistical inference.* New York: Marcel Dekker.

CHAPTER **2**

Reading the News

Thought Questions

1. Advice columnists sometimes ask readers to write and express their feelings about certain topics. For instance, Ann Landers once asked readers whether they thought engineers made good husbands. Do you think the responses are representative of public opinion? Explain why or why not.

2. Taste tests of new products are often done by having people taste both the new product and an old familiar standard. Do you think the results would be biased if the person handing the products to the respondents knew which was which? Explain why or why not.

3. Nicotine patches attached to the arm of someone who is trying to quit smoking dispense nicotine into the blood. Suppose you read about a study showing that nicotine patches were twice as effective in getting people to quit smoking as "placebo" patches (made to look like the real thing). Further, suppose you are a smoker trying to quit. What questions would you want answered about how the study was done and its results before you decided whether to try the patches yourself?

4. For a door-to-door survey on opinions about various political issues, do you think it matters who conducts the interviews? Give an example of how it might make a difference.

2.1 The Educated Consumer of Data

Visit any news-based website, pick up a popular magazine or tune into a news broadcast, and you are almost certain to find a story containing conclusions based on data. Should you believe what you read? Not always. It depends on how the data were collected, measured, and summarized. In this chapter, we discuss seven critical components of statistical studies. We examine the kinds of questions you should ask before you believe what you read. We go into further detail about these issues in subsequent chapters. The goal in this chapter is to give you an overview of how to be a more educated consumer of the data you encounter in your everyday life.

What Are Data?

In statistical parlance, **data** is a plural word referring to a collection of numbers or other pieces of information to which meaning has been attached. For example, the numbers 1, 3, and 10 are not necessarily data, but they become so when we are told that these were the weight gains in grams of three of the infants in Salk's heartbeat study, discussed in Chapter 1. In Case Study 1.2, the data consisted of two pieces of information measured for each participant: (1) whether they took aspirin or a placebo, and (2) whether they had a heart attack.

Don't Always Believe What You Read

When you encounter the results of a study in the news, you are rarely presented with the actual data. Someone has usually summarized the information for you, and he or she has probably already drawn conclusions and presented them to you. Don't always believe them. The meaning we can attach to data, and to the resulting conclusions, depends on how well the information was acquired and summarized.

In the remaining chapters of Part 1, we look at proper ways to obtain data. In Part 2, we turn our attention to how it should be summarized. In Part 4, we learn the power as well as the limitations of using the data collected from a sample to make conclusions about the larger population. In this chapter, we address seven features of statistical studies that you should think about when you read a news article. You will begin to be able to think critically and make your own conclusions about what you read.

2.2 Origins of News Stories

Where do news stories originate? How do reporters hear about events and determine that they are newsworthy? For stories based on statistical studies, there are several possible sources. The two most common of these sources are also the most common outlets for researchers to present the results of their work: academic conferences and scholarly journals.

News from Academic Conferences

Every academic discipline holds conferences, usually annually, in which researchers can share their results with others. Reporters routinely attend these academic conferences and look for interesting news stories. For larger conferences, there is usually a "press room" where researchers can leave press releases for the media. If you pay attention, you will notice that in certain weeks of the year, there will be several news stories about studies with related themes. For instance, the American Psychological Association meets in August, and there are generally some news stories emerging from results presented there. The American Association for the Advancement of Science meets in February, and news stories related to various areas of science will appear in the news that week.

One problem with news stories based on conference presentations is that there is unlikely to be a corresponding written report by the researchers, so it is difficult for readers of the news story to obtain further information. News stories based on conference reports generally mention the name and date of the conference as well as the name and institution of the lead researcher, so sometimes it is possible to contact the researcher for further information. Some researchers make conference presentations available on their websites.

News from Published Reports

Many news stories about statistical studies are based on published articles in scholarly journals. Reporters routinely read these journals when they are published, or they get advance press releases from the journal offices. News stories based on journal articles usually mention the journal and date of publication, so if you are interested in learning more about the study, you can obtain the original journal article. Journal articles are sometimes available on the journal's website or on the website of the author(s). You can also write to the lead author and request that the article be sent to you.

As a third source of news stories about statistical studies, some government and private agencies release in-depth research reports. Unlike journal articles, these reports are not necessarily "peer-reviewed" or checked by neutral experts on the topic. An advantage of these reports is that they are not restricted by space limitations imposed by journals and often provide much more in-depth information than do journal articles.

A supplementary source from which news stories may originate is a university media office. Most research universities have an office that provides press releases when faculty members have completed research that may be of interest to the public. The timing of these news releases usually corresponds to a presentation at an academic conference or publication of results in an academic journal, but the news release summarizes the information so that journalists don't have to be as versed in the technical aspects of the research to write a good story. When you read about a study in the news and would like more information, the news office of the lead researcher's institution is a good place to start looking. They may have issued a press release on which the story was based.

News Stories and Original Sources in the Appendix and on the Companion Website, www.cengage.com/stats/UttsSTS4e

To illustrate how the concepts in this book are used in research and eventually converted into news stories, there is a collection of examples included with this book. In each case, the example includes a story from a newspaper, magazine, or website, and these are printed or referenced in the Appendix and on the companion website accompanying the book. Sometimes there is also a press release. These are provided as an additional "News Story" and included on the companion website. Most of the news stories are based on articles from scholarly journals or detailed reports. Many of these articles are printed in full on the companion website, labeled as the "Original Source." Throughout this book, you will find a website icon when you need to refer to the material on the companion website. By comparing the news story and the original source, you will learn how to evaluate what is reported in the news.

2.3   How to Be a Statistics Sleuth: Seven Critical Components

Reading and interpreting the results of surveys or experiments is not much different from reading and interpreting the results of other events of interest, such as sports competitions or criminal investigations. If you are a sports fan, then you know what information should be included in reports of competitions and you know when crucial information is missing. If you have ever been involved in an event that was later reported in the news, you know that missing information can lead readers to erroneous conclusions.

In this section, you are going to learn what information should be included in news reports of statistical studies. Unfortunately, crucial information is often missing. With some practice, you can learn to figure out what's missing, as well as how to interpret what's reported. You will no longer be at the mercy of someone else's conclusions. You will be able to reach your own conclusions. To provide structure to our examination of news reports, let's list Seven Critical Components that determine the soundness of statistical studies. A good news report should provide you with information about all of the components that are relevant to that study.

Component 1: The *source* of the research and of the *funding*.

Component 2: The *researchers* who had *contact* with the participants.

Component 3: The *individuals* or objects studied and how they were *selected*.

Component 4: The exact nature of the *measurements* made or *questions* asked.

> **Component 5:** The *setting* in which the measurements were taken.
>
> **Component 6:** *Differences* in the groups being compared, *in addition* to the factor of interest.
>
> **Component 7:** The *extent* or *size* of any claimed effects or differences.

Before delving into some examples, let's examine each component more closely. You will find that most of the problems with studies are easy to identify. Listing these components simply provides a framework for using your common sense.

Component 1: The *source* of the research and of the *funding*   Studies are conducted for three major reasons. First, governments and private companies need to have data in order to make wise policy decisions. Information such as unemployment rates and consumer spending patterns are measured for this reason. Second, researchers at universities and other institutions are paid to ask and answer interesting questions about the world around us. The curious questioning and experimentation of such researchers have resulted in many social, medical, and scientific advances. Much of this research is funded by government agencies, such as the National Institutes of Health. Third, companies want to convince consumers that their programs and products work better than the competition's, or special-interest groups want to prove that their point of view is held by the majority.

Unfortunately, it is not always easy to discover who funded research. Many university researchers are now funded by private companies. In her book *Tainted Truth* (1994), Cynthia Crossen warns us:

> *Private companies, meanwhile, have found it both cheaper and more prestigious to retain academic, government, or commercial researchers than to set up in-house operations that some might suspect of fraud. Corporations, litigants, political candidates, trade associations, lobbyists, special interest groups—all can buy research to use as they like. (p. 19)*

If you discover that a study was funded by an organization that would be likely to have a strong preference for a particular outcome, it is especially important to be sure that correct scientific procedures were followed. In other words, be sure the remaining components have sound explanations.

Component 2: The *researchers* who had *contact* with the participants   It is important to know who actually had contact with the participants and what message those people conveyed. Participants often give answers or behave in ways to comply with the desires of the researchers. Consider, for example, a study done at a shopping mall to compare a new brand of a certain product to an old familiar brand. Shoppers are asked to taste each brand and state their preference. It is crucial that both the person presenting the two brands and the respondents be kept entirely blind as to which is which until after the preferences have been selected. Any clues might bias the respondent to choose the old familiar brand. Or, if the interviewer is clearly eager to have them choose one brand over the other, the respondents will most likely oblige in order to please. As another example, if you discovered that a study on the prevalence of illegal drug use was conducted by sending uniformed police officers door-to-door, you would probably not have much faith in the results. We will discuss other ways in which researchers influence participants in Chapters 4 and 5.

Component 3: The *individuals* or objects studied and how they were *selected* It is important to know to whom the results can be extended. In general, the results of a study apply only to individuals similar to those in the study. For example, until recently, many medical studies included men only, so the results were of little value to women. When determining who is similar to those in the study, it is also important to know how participants were enlisted for the study. Many studies rely on volunteers recruited through newspapers or websites, who are usually paid a small amount for their participation. People who would respond to such recruitment efforts may differ in relevant ways from those who would not. Surveys relying on voluntary responses are likely to be biased because only those who feel strongly about the issues are likely to respond. For instance, some websites have a "question of the day" to which people are asked to voluntarily respond by clicking on their preferred answer. Only those who have strong opinions are likely to participate, so the results cannot be extended to any larger group.

Component 4: The exact nature of the *measurements* made or *questions* asked As you will see in Chapter 3, precisely defining and measuring most of the things researchers study isn't easy. For example, if you wanted to measure whether people "eat breakfast," how would you do so? What if they just have juice? What if they work until midmorning and then eat a meal that satisfies them until dinner? You need to understand exactly what the various definitions mean when you read about someone else's measurements.

In polls and surveys, the "measurements" are usually answers to specific questions. Both the wording and the ordering of the questions can influence answers. For example, a question about "street people" would probably elicit different responses than a question about "families who have no home." Ideally, you should be given the exact wording that was used in a survey or poll.

Component 5: The *setting* in which the measurements were taken The setting in which measurements were taken includes factors such as when and where they were taken and whether respondents were contacted by phone, mail, e-mail, or in person. A study can be easily biased by timing. For example, opinions on whether criminals should be locked away for life may change drastically following a highly publicized murder or kidnapping case. If a study is conducted by landline telephone and calls are made only in the evening, even certain groups of people who have landlines would be excluded, such as those who work the evening shift or who routinely eat dinner in restaurants, and people without landlines would not be represented at all.

Where the measurements were taken can also influence the results. Questions about sensitive topics, such as sexual behavior or income, might be more readily answered on a website, where respondents feel more anonymous. Sometimes research is done in a laboratory or university office, and the results may not readily extend to a natural setting. For example, studies of communication between two people are sometimes done by asking them to conduct a conversation in a university office with a voice recorder present. Such conditions almost certainly produce more limited conversation than would occur in a more natural setting.

Component 6: *Differences* in the groups being compared, *in addition* to the factor of interest   If two or more groups are being compared on a factor of interest, it is important to consider other ways in which the groups may differ that might influence the comparison. For example, suppose researchers want to know if smoking marijuana is related to academic performance. If the group of people who smoke marijuana has lower test scores than the group of people who don't, researchers may want to conclude that the lower test scores are due to smoking marijuana. Often, however, other disparities in the groups can explain the observed difference just as well. For example, people who smoke marijuana may simply be the type of people who are less motivated to study and thus would score lower on tests whether they smoked or not. Reports of research should include an explanation of any such possible differences that might account for the results. We will explore the issue of these kinds of extraneous factors, and how to control for them, in much more detail in Chapter 5.

Component 7: The *extent* or *size* of any claimed effects or differences   Media reports about statistical studies often fail to tell you how large the observed effects were. Without that knowledge, it is hard for you to assess whether you think the results are of any practical importance. For example, if, based on Case Study 1.2, you were told simply that taking aspirin every other day reduced the risk of heart attacks, you would not be able to determine whether it would be worthwhile to take aspirin. You should instead be told that for the men in the study, the rate was reduced from about 17 heart attacks per 1000 participants without aspirin to about 9.4 heart attacks per 1000 with aspirin. Often news reports simply report that a treatment had an effect or that a difference was observed, but don't tell you the size of the difference or effect. We will investigate this issue in great detail in Part 4 of this book.

2.4   Four Hypothetical Examples of Bad Reports

Throughout this book, you will see numerous examples of real studies and news reports. So that you can get some practice finding problems without having to read unnecessarily long news articles, let's examine some hypothetical reports. These are admittedly more problematic than many real reports because they serve to illustrate several difficulties at once.

Read each article (at the top of pages 24, 26, 27 and 29) and see if your common sense gives you some reasons why the headline is misleading. Then read the commentary accompanying the article, discussing the Seven Critical Components.

Study Shows Psychology Majors Are Smarter than Chemistry Majors

A fourth-year psychology student, for her senior thesis, conducted a study to see if students in her major were smarter than those majoring in chemistry. She handed out questionnaires in five advanced psychology classes and five advanced chemistry labs. She asked the students who were in class to record their grade-point averages (GPAs) and their majors. Using the data only from those who were actually majors in these fields in each set of classes, she found that the psychology majors had an average GPA of 3.05, whereas the chemistry majors had an average GPA of only 2.91. The study was conducted last Wednesday, the day before students were home enjoying Thanksgiving dinner.

Hypothetical News Article 1: "Study Shows Psychology Majors Are Smarter than Chemistry Majors"

Component 1: The *source* of the research and of the *funding* The study was a senior thesis project conducted by a psychology major. Presumably, it was cheap to run and was paid for by the student. One could argue that she would have a reason to want the results to come out as they did, although with a properly conducted study, the motives of the experimenter should be minimized. As we shall see, there were additional problems with this study.

Component 2: The *researchers* who had *contact* with the participants Presumably, only the student conducting the study had contact with the respondents. Crucial missing information is whether she told them the purpose of the study. Even if she did not tell them, many of the psychology majors may have known her and known what she was doing. Any clues as to desired outcomes on the part of experimenters can bias the results.

Component 3: The *individuals* or objects studied and how they were *selected* The individuals selected are the crux of the problem here. The measurements were taken on advanced psychology and chemistry students, which would have been fine if they had been sampled correctly. However, only those who were in the psychology classes or in the chemistry labs that day were actually measured. Less conscientious students are more likely to leave early before a holiday, but a missed class is probably easier to make up than a missed lab. Therefore, perhaps a larger proportion of the students with low GPAs were absent from the psychology classes than from the chemistry labs. Due to the missing students, the investigator's results would overestimate the average GPA for psychology students more so than for chemistry students.

Component 4: The exact nature of the *measurements* made or *questions* asked Students were asked to give a "self-report" of their GPAs. A more accurate method would have been to obtain this information from the registrar at the university. Students may not know their exact GPA. Also, one group may be more likely to know the exact value than the other. For example, if many of the chemistry majors were planning to apply to medical school in the near future, they may be only too aware of their grades. Further, the headline implies that GPA is a measure of intelligence. Finally, the research assumes that GPA is a standard measure. Perhaps grading is more competitive in the chemistry department.

Component 5: The *setting* in which the measurements were taken Notice that the article specifies that the measurements were taken on the day before a major holiday. Unless the university consisted mainly of commuters, many students may have left early for the holiday, further aggravating the problem that the students with lower grades were more likely to be missing from the psychology classes than from the chemistry labs. Further, because students turned in their questionnaires anonymously, there was presumably no accountability for incorrect answers.

Component 6: *Differences* in the groups being compared, *in addition* to the factor of interest The factor of interest is the student's major, and the two groups being compared are psychology majors and chemistry majors. This component considers whether the students who were interviewed for the study may differ in ways other than their choice of major. It is difficult to know what differences might exist without knowing more about the particular university. For example, because psychology is such a popular major, at some universities students are required to have a certain GPA before they are admitted to the major. A university with a separate premedical major might have the best of the science students enrolled in that major instead of chemistry. Those kinds of extraneous factors would be relevant to interpreting the results of the study.

Component 7: The *extent* or *size* of any claimed effects or differences The news report does present this information, by noting that the average GPAs for the two groups were 3.05 and 2.91. Additional useful information would be to know how many students were included in each of the averages given, what percentage of all students in each major were represented in the sample, and how much variation there was among GPAs within each of the two groups.

Hypothetical News Article 2: "Per Capita Income of U.S. Shrinks Relative to Other Countries"

Component 1: The *source* of the research and of the *funding* We are told nothing except the name of the group that conducted the study, which should be fair warning. Being called "an independent research group" in the story does not mean that it is an unbiased research group. In fact, the last line of the story illustrates the probable motive for their research.

Hypothetical News
Article 2

Per Capita Income of U.S. Shrinks Relative to Other Countries

An independent research group, the Institute for Foreign Investment, has noted that the per capita income of Americans has been shrinking relative to some other countries. Using per capita income figures from The World Bank (most recent available, from two years ago) and exchange rates from last Friday's financial pages, the organization warned that per capita income for the United States has risen only 10% during the past 5 years, whereas per capita income for certain other countries has risen 50%. The researchers concluded that more foreign investment should be allowed in the United States to bolster the sagging economy.

Component 2: The *researchers* who had *contact* with the participants   This component is not relevant because there were no participants in the study.

Component 3: The *individuals* or objects studied and how they were *selected*   The objects in this study were the countries used for comparison with the United States. We should have been told which countries were used, and why.

Component 4: The exact nature of the *measurements* made or *questions* asked   This is the major problem with this study. First, as mentioned, we are not even told which countries were used for comparison. Second, current exchange rates but older per capita income figures were used. If the rate of inflation in a country had recently been very high, so that a large rise in per capita income did not reflect a concomitant rise in spending power, then we should not be surprised to see a large increase in per capita income in terms of actual dollars. In order to make a valid comparison, all figures would have to be adjusted to comparable measures of spending power, taking inflation into account. We will learn how to do that in Chapter 18.

Components 5, 6, and 7: The *setting* in which the measurements were taken. *Differences* in the groups being compared, *in addition* to the factor of interest. The *extent* or *size* of any claimed effects or differences   These issues are not relevant here, except as they have already been discussed. For example, although the size of the difference between the United States and the other countries is reported, it is meaningless without an inflation adjustment.

Hypothetical News
Article 3

Researchers Find Drug to Cure Excessive Barking in Dogs

Barking dogs can be a real problem, as anyone who has been kept awake at night by the barking of a neighbor's canine companion will know. Researchers at a local university have tested a new drug that they hope will put all concerned to rest. Twenty dog owners responded to an email from their veterinarian asking for volunteers with problem barking dogs to participate in a study. The dogs were randomly assigned to two groups. One group of dogs was given the drug, administered as a shot, and the other dogs were not. Both groups were kept overnight at the research facility and frequency of barking was observed. The researchers deliberately tried to provoke the dogs into barking by doing things like ringing the doorbell of the facility and having a mail carrier walk up to the door. The two groups were treated on separate weekends because the facility was only large enough to hold ten dogs. The researchers left an audio recorder running and measured the amount of time during which any barking was heard. The dogs who had been given the drug spent only half as much time barking as did the dogs in the control group.

Hypothetical News Article 3: "Researchers Find Drug to Cure Excessive Barking in Dogs"

Component 1: The *source* of the research and of the *funding*   We are not told why this study was conducted. Presumably, it was because the researchers were interested in helping to solve a societal problem, but perhaps not. It is not uncommon for drug companies to fund research to test a new product or a new use for a current product. If that were the case, the researchers would have added incentive for the results to come out favorable to the drug. If everything were done correctly, such an incentive wouldn't be a major factor; however, when research is funded by a private source, that information should be provided when the results are announced.

Component 2: The *researchers* who had *contact* with the participants   We are not given any information about who actually had contact with the dogs. One important question is whether the same handlers were used with both groups of dogs. If not, the difference in handlers could explain the results. Further, we are not told whether the dogs were primarily left alone or were attended most of the time. If researchers were present most of the time, their behavior toward the dogs could have had a major impact on the amount of barking.

Component 3: The *individuals* or objects studied and how they were *selected*
We are told that the study used dogs whose owners volunteered them as problem dogs for the study. Although the report does not mention payment, it is quite common for volunteers to receive monetary compensation for their participation. The volunteers presumably lived in the area of the university. The dog owners had to be willing to be separated from their pets for the weekend. These and other factors mean that the owners and dogs who participated may differ from the general population. Further, the initial reasons for the problem behavior may vary from one participant to the next, yet the dogs were measured together. Therefore, there is no way to ascertain if, for example, dogs who bark only because they are lonely would be helped. For these reasons, we cannot extend the results of this study to conclude that the drug would work similarly on the entire population of dogs or even on all problem dogs. However, the dogs were randomly assigned to the two groups, so we should be able to extend the results to all dogs similar to those who participated.

Component 4: The exact nature of the *measurements* made or *questions* asked
The researchers measured each group of dogs as a group, by listening to an audio recording and measuring the amount of time during which there was any barking. Because dogs are quite responsive to group behavior, one barking dog could set the whole group barking for a long time. Therefore, just one particularly obnoxious dog in the control group alone could explain the results. It would have been better to separate the dogs and measure each one individually.

Component 5: The *setting* in which the measurements were taken   The two groups of dogs (drug and no drug) were measured on separate weekends. This creates another problem. First, the researchers knew which group was which and may have unconsciously provoked the control group slightly more than the group receiving the drug. Further, conditions differed over the two weeks. Perhaps it was sunny one weekend and raining the next, or there were other subtle differences, such as more traffic one weekend than the next, small planes overhead, and so on. All of these could change the behavior of the dogs but might go unnoticed or unreported by the experimenters.

The measurements were also taken outside of the dogs' natural environments. The dogs in the experimental group in particular would have reason to be upset because they were first given a shot and then put together with nine other dogs in the research facility. It would have been better to put them back into their natural environment because that's where the problem barking was known to occur.

Component 6: *Differences* in the groups being compared, *in addition* to the factor of interest   The dogs were randomly assigned to the two groups (drug or no drug), which should have minimized overall differences in size, temperament, and so on for the dogs in the two groups. However, differences were induced between the two groups by the way the experiment was conducted. Recall that the groups were measured on different weekends—this could have created the difference in behavior. Also, the drug-treated dogs were given a shot to administer the drug, whereas the

Hypothetical News
Article 4

Survey Finds Most Women Unhappy in Their Choice of Husbands

A popular women's magazine, in a survey of its subscribers, found that over 90% of them are unhappy in their choice of whom they married. Copies of the survey were mailed to the magazine's 100,000 subscribers. Surveys were returned by 5000 readers. Of those responding, 4520, or slightly over 90%, answered no to the question: "If you had it to do over again, would you marry the same man?" To keep the survey simple so that people would return it, only two other questions were asked. The second question was, "Do you think being married is better than being single?" Despite their unhappiness with their choice of spouse, 70% answered yes to this. The final question, "Do you think you will outlive your husband?" received a yes answer from 80% of the respondents. Because women generally live longer than men, and tend to marry men somewhat older than themselves, this response was not surprising. The magazine editors were at a loss to explain the huge proportion of women who would choose differently. The editor could only speculate: "I guess finding Mr. Right is much harder than anyone realized."

control group was given no shot. It could be that the very act of getting a shot made the drug group lethargic. A better design would have been to administer a placebo shot—that is, a shot with an inert substance—to the control group.

Component 7: The *extent* or *size* of any claimed effects or differences   We are told only that the treated group barked half as much as the control group. We are not told how much time either group spent barking. If one group barked 8 hours a day but the other group only 4 hours a day, that would not be a satisfactory solution to the problem of barking dogs.

Hypothetical News Article 4: "Survey Finds Most Women Unhappy in Their Choice of Husbands"

Components 1 through 7   We don't even need to consider the details of this study because it contains a fatal flaw from the outset. The survey is an example of what is called a "volunteer sample" or a "self-selected sample." Of the 100,000 who received the survey, only 5% responded. The people who are most likely to respond to such a survey are those who have a strong emotional response to the question. In this case, it would be women who are unhappy with their current situation who would probably respond. Notice that the other two questions are more general and, therefore, not likely to arouse much emotion either way. Thus, it is the strong reaction to the first question that would drive people to respond. The results would certainly not be representative of "most women" or even of most subscribers to the magazine.

| CASE STUDY 2.1 | **Who Suffers from Hangovers?** |

SOURCE: News Story 2 in the Appendix and Original Source 2 on the companion website, www.cengage.com/stats/UttsSTS4e.

Read News Story 2 in the Appendix, "Research shows women harder hit by hangovers" and access the original source of the story on the companion website, the journal article "Development and Initial Validation of the Hangover Symptoms Scale: Prevalence and Correlates of Hangover Symptoms in College Students." Let's examine the seven critical components based on the news story and, where necessary, additional information provided in the journal article.

Component 1: The *source* of the research and of the *funding*   The news story covers this aspect well. The researchers were "a team at the University of Missouri-Columbia" and the study was "supported by the National Institutes of Health."

Component 2: The *researchers* who had *contact* with the participants   This aspect of the study is not clear from the news article, which simply mentions, "The researchers asked 1,230 drinking college students . . . ." However, the journal article says that the participants were enrolled in Introduction to Psychology courses and were asked to fill out a questionnaire. So it can be assumed that professors or research assistants in psychology had contact with the participants.

Component 3: The *individuals* or objects studied and how they were *selected*   The news story describes the participants as "1,230 drinking college students, only 5 percent of whom were of legal drinking age." The journal article provides much more information, including the important fact that the participants were all enrolled in introductory psychology classes and were participating in the research to fulfill a requirement for the course. The reader must decide whether this group of participants is likely to be representative of all drinking college students, or some larger population, for severity of hangover symptoms. The journal article also provides information on the sex, ethnicity, and age of participants.

Component 4: The exact nature of the *measurements* made or *questions* asked   The news story provides some detail about what was asked, noting that the participants were asked *"to describe how often they experienced any of 13 symptoms after drinking. The symptoms ranged from headaches and vomiting to feeling weak and unable to concentrate."* The journal article again provides much more detail, listing the 13 symptoms and explaining that participants were asked to indicate how often they were experienced on a 5-point scale (p. 1444 of the journal article). Further, participants were asked to provide a "hangover count" in which they noted how many times they had experienced at least one of the 13 symptoms in the past year, using a 5-point scale. This scale ranged from "never" to "52 times or more." Additional questions were asked about alcoholism in the participant's family and early experience with alcohol. Detailed information about all of these questions is included in the journal article.

Component 5: The *setting* in which the measurements were taken   This information is not provided explicitly, but it can be assumed that measurements were taken in the Psychology Department at the University of Columbia-Missouri. One missing fact that may be helpful in interpreting the results is if the questions were administered to a large group of students at once, or individually, and whether students could be identified when the researchers read their responses.

Component 6: *Differences* in the groups being compared, *in addition* to the factor of interest   The purpose of the research was to develop and test a "Hangover Symptoms Scale" but two interesting differences in groups emerged when the researchers made comparisons. The groups being compared in the first instance were males and females; thus, Male/Female was the factor of interest. The researchers found that females suffered more from hangovers. This component is asking if there may be other differences between males and females, other than "Male" and "Female" that could help account for the difference. One possibility mentioned in the news article is body weight. Males tend to weigh more than females on average. An interesting question, not answered by the research, is if a group of males and females of the same weight, say 130 pounds, were to consume the same amount of alcohol, would the females suffer more hangover symptoms? The difference in weight between the two groups is in addition to the factor of interest, which is Male/Female. It may be the weight difference, and not the sex difference, that accounts for the difference in hangover severity.

The other comparison mentioned in the news article is between students who had alcohol-related problems, or whose biological parents had such problems, and students who did not have that history. In this case, the alcohol-related problems (of the student or parents) is the factor of interest. However, you can probably think of other differences in the two groups (those with problems and those without) that may help account for the difference in hangover severity between the two groups. For instance, students with a history of problems may not have as healthful diets in the past or present as students without such problems, and that may contribute to hangover severity. So the comparison of interest, between those with an alcohol problem in their background and those without, may be complicated by other differences in these two groups.

Component 7: The *extent* or *size* of any claimed effects or differences   The news story does not report how much difference in hangover severity was found between men and women, or between those with and without a history of alcohol problems. Reading the journal article may explain why this is so—the article itself does not report a simple difference. In fact, simple comparisons don't yield much difference; for instance, 11% of men and 14% of women never experienced any hangover symptoms in the previous year. Differences only emerged when complicating factors such as amount of alcohol consumed were factored in. The researchers report, "After controlling for the frequency of drinking and getting drunk and for the typical quantity of alcohol consumed when drinking, women were significantly more likely than men to experience at least one of the hangover symptoms" (p. 1446). The article does not elaborate, such as explaining what would be the difference for a male and female who drank the same amount and equally often. ■

2.5 Planning Your Own Study: Defining the Components in Advance

Although you may never have to design your own survey or experiment, it will help you understand how difficult it can be if we illustrate the Seven Critical Components for a very simple hypothetical study you might want to conduct. Suppose you are interested in determining which of three local supermarkets has the best prices so you can decide where to shop. Because you obviously can't record and summarize the prices for all available items, you would have to use some sort of sample.

To obtain meaningful data, you would need to make many decisions. Some of the components need to be reworded because they are being answered in advance of the study, and obviously not all of the components are relevant for this simple example. However, by going through them for such a simple case, you can see how many ambiguities and decisions can arise when designing a study.

Component 1: The *source* of the research and of the *funding* Presumably you would be funding the study yourself, but before you start you need to decide why you are doing the study. Are you only interested in items you routinely buy, or are you interested in comparing the stores on the multitude of possible items?

Component 2: The *researchers* who had *contact* with the participants In this example, the question would be who is going to visit the stores and record the prices. Will you personally visit each store and record the prices? Will you send friends to two of the stores and visit the third yourself? If you use other people, you would need to train them so there would be no ambiguities. For example, if there are multiple brands and/or sizes of the same item, which price gets recorded? If there are organic and non-organic versions of an item, which price gets recorded?

Component 3: The *individuals* or objects studied and how they were *selected* In this case, the "objects studied" are items in the grocery store. The correct question is, "On what items should prices be recorded?" Do you want to use exactly the same items at all stores? What if one store offers its own brand but another only offers name brands? Do you want to choose a representative sampling of items you are likely to buy or choose from all possible items? Do you want to include nonfood items? How many items should you include? How should you choose which ones to select? If you are simply trying to minimize your own shopping bill, it is probably best to list the 20 or 30 items you buy most often. However, if you are interested in sharing your results with others, you might prefer to choose a representative sample of items from a long list of possibilities.

Component 4: The exact nature of the *measurements* made or *questions* asked
You may think that the cost of an item in a supermarket is a well-defined measurement. But if a store is having a sale on a particular item on your list, should you use the sale price or the regular price? Should you use the price of the smallest possible size of the product? The largest? What if a store always has a sale on one brand or another of something, such as laundry soap, and you don't really care which brand you buy? Should you then record the price of the brand on sale that week? Should you record the prices listed on the shelves or actually purchase the items and see if the prices listed were accurate?

Component 5: The *setting* in which the measurements were taken   When will you conduct the study? Supermarkets in university towns may offer sale prices on items typically bought by students at certain times of the year—for example, just after students have returned from vacation. Many stores also offer sale items related to certain holidays, such as ham or turkey just before Christmas or eggs just before Easter. Should you take that kind of timing into account?

Component 6: *Differences* in the groups being compared, *in addition* to the factor of interest   The groups being compared are the groups of items from the three stores. There should be no additional differences related to the direct costs of the items. However, if you were conducting the study in order to minimize your shopping costs, you might ask if there are hidden costs for shopping at one store versus another. For example, do you always have to wait in line at one store and not at another, and should you therefore put a value on your time? Does one store make mistakes at check-out more often than another? Does one store charge for bags or give a discount for bringing your own bags? Does it cost more to drive to one store than another?

Component 7: The *extent* or *size* of any claimed effects or differences   This component should enter into your decision about where to shop after you have finished the study. Even if you find that items in one store cost less than in another, the amount of the difference may not convince you to shop there. You would probably want to figure out approximately how much shopping in a particular store would save you over the course of a year. You can see why knowing the amount of a difference found in a study is an important component for using that study to make future decisions.

CASE STUDY 2.2	**Flawed Surveys in the Courtroom**

SOURCES: http://openjurist.org/716/f2d/854, accessed May 11, 2013.
http://openjurist.org/615/f2d/252, accessed May 11, 2013.
Gastwirth (1988), pp. 517–520.

Companies sometimes sue other companies for alleged copying of names or designs that the first company claims are used to identify its products. In these kinds of cases, the court often wants to know whether it's likely that consumers are confusing the two companies based on the name or design. There is only one way to find that out—ask the consumers! But in the two instances presented here, the companies bringing the lawsuit made a crucial mistake—they did not ask a representative sample of people who were likely to buy the product.

In the first example, Brooks Shoe Manufacturing Company sued Suave Shoe Corporation for manufacturing shoes incorporating a "V" design used in Brooks's athletic shoes. Brooks claimed that the design was an unregistered trademark that people used to identify Brooks shoes. One of the roles of the court was to determine whether the design had "secondary meaning," which is a legal term indicating that "the primary significance of the term [or design] in the minds of the consuming public is not the product but the producer." (http://openjurist.org/596/f2d/111)

To show that the design had "secondary meaning" to buyers, Brooks conducted a survey of 121 spectators and participants at three track meets. Interviewers approached people and asked them a series of questions that included showing them a Brooks shoe with the name masked and asking them to identify it. Of those surveyed, 71% were able to identify it as a Brooks shoe, and 33% of those people said it was because they recognized the "V." When shown a Suave shoe, 39% of them thought it was a Brooks shoe, with 48% of those people saying it was because of the "V" design on the Suave shoe. Brooks Company argued that this was sufficient evidence that people might be confused and think Suave shoes were manufactured by Brooks.

Suave had a statistician as an expert witness who pointed out a number of flaws in the Brooks survey. Let's examine them using the Seven Critical Components as a guide. First, the survey was funded and conducted by Brooks, and the company's lawyer was instrumental in designing it. Second, the court determined that the interviewers who had contact with the respondents were inadequately trained in how to conduct an unbiased survey. Third, the individuals asked were not selected to be representative of the general public in the area (Baltimore/Washington, D.C.). For example, 78% had some college education, compared with 18.4% in Baltimore and 37.7% in Washington, D.C. Further, the settings for the interviews were track meets, where people were likely to be more familiar with athletic shoes. The questions asked were biased. For example, the exact wording used when a person was handed the shoes was: "I am going to hand you a shoe. Please tell me what brand you think it is." The way the question is framed would presumably lead respondents to think the shoe has a well-known brand name, which Brooks had at the time but Suave did not. Later in the questioning, respondents were asked, "How long have you known about Brooks Running Shoes?"

Suave introduced its own survey conducted on 404 respondents properly sampled from households in Baltimore, Maryland, and Greensboro, North Carolina, in which one or more individuals had purchased any type of athletic shoe during the previous year. Of those, only 2.7% recognized a Brooks shoe on the basis of the "V" design. The combination of the poor survey methods by Brooks and the proper survey by Suave convinced the court that the public did not make enough of an association between Brooks and the "V" design to allow Brooks to claim legal rights to the design. Brooks appealed the case to the Eleventh Circuit Court of Appeals, but the original decision was upheld, partly on the basis of the quality of the two surveys. (See http://openjurist.org/716/f2d/854.)

In the second example, Amstar Corporation, the makers of Domino sugar, sued Domino's Pizza, claiming that the use of the name was a trademark infringement. The original ruling was in favor of Amstar, and Domino's Pizza was told that they could no longer use the name, but the decision was overturned on appeal. Both companies had presented consumer surveys as part of their arguments. Although the first court thought that the Amstar survey was adequate and the Domino's Pizza survey was not, the appeals court bluntly stated, "Our own examination of the survey evidence convinces us that both surveys are substantially defective" (http://openjurist.org/615/f2d/252, note 38).

Who was surveyed? Amstar's survey consisted of female heads of households who were home during the day. They were shown a Domino's Pizza box and asked if they thought the company made any other products. If the answer was yes, they were asked "What products other than pizza do you think are made by the company that makes Domino's Pizza?" to which 71% responded "sugar." The court reasonably noted that surveying women who were home during the day was surely a biased sample, in favor of consumers who recognize Domino as the maker of sugar. But the appeals court noted an even more striking flaw: Eight of the 10 cities in which the surveys were conducted did not have a Domino's Pizza outlet, and in the other two cities, it had been open for less than 3 months. Therefore, participants in this survey were clearly more likely to know about Domino sugar than Domino's Pizza.

What about the survey conducted by Domino's Pizza? That one was conducted at Domino's Pizza outlets, and not surprisingly, found that almost no one was confusing the pizza company with the sugar maker.   ■

Thinking About Key Concepts

- Surveys that are open to anyone and rely completely on those who voluntarily respond are unlikely to represent any larger group. The only people likely to volunteer are those who feel strongly about the issues in the survey.
- People like to please others, and if they know the desired outcome in a survey or study, they are likely to comply and produce that response. Therefore, it's best if the person collecting the data does not know what the desired outcome is.

- It's important to know the details about how a study was conducted so that you can decide whether the results would apply to you. For instance, results may not apply to you if the study was done on a different age group or on people that differ from you in other crucial ways.
- It's important to know how large a difference was found in a study that might apply to you. For instance, if you hear that nicotine patches help people quit smoking or a certain diet helps people lose weight, you would probably be more interested in trying those things if the effects were large than if they were very small.

Exercises

Exercises with numbers divisible by 3 (3, 6, 9, etc.) are included in the Solutions at the back of the book. They are marked with an asterisk (*).

1. Suppose that a television network wants to know how daytime television viewers feel about a new soap opera the network is broadcasting. A staff member suggests that just after the show ends they give two phone numbers, one for viewers to call if they like the show and the other to call if they don't. Give two reasons why this method would not produce the desired information. (*Hint:* The network is interested in all daytime television viewers. Who is likely to be watching just after the show, and who is likely to call in?)

2. The April 24, 1997, issue of "UC-Davis Lifestyle Newstips" reported that a professor of veterinary medicine was conducting a study to see if a drug called clomipramine, an anti-anxiety medication used for humans, could reduce "canine aggression toward family members." The newsletter said, "Dogs demonstrating this type of aggression are needed to participate in the study. . . . Half of the participating dogs will receive clomipramine, while the others will be given a placebo." A phone number was given for dog owners to call to volunteer their dogs for the study. To what group could the results of this study be applied? Explain.

*3. A prison administration wants to know whether the prisoners think the guards treat them fairly. Explain how each of the following components could be used to produce biased results, versus how each could be used to produce unbiased results:

   *a. Component 2: The *researchers* who had *contact* with the participants.

   *b. Component 4: The exact nature of the *measurements* made or *questions* asked.

4. According to Cynthia Crossen (1994, p. 106): "It is a poller's business to press for an opinion whether people have one or not. 'Don't knows' are worthless to pollers, whose product is opinion, not ignorance. That's why so many polls do not even offer a 'don't know' alternative."

   a. Explain how this problem might lead to bias in a survey.

   b. Which of the Seven Critical Components would bring this problem to light?

5. Many research organizations give their interviewers an exact script to follow when conducting interviews to measure opinions on controversial issues. Why do you think they do so?

*6. The student who conducted the study in "Hypothetical News Article 1" in this chapter collected two pieces of data from each participant. What were the two pieces of data?

7. Refer to Case Study 1.1, "Heart or Hypothalamus?" Discuss each of the following components, including whether you think the way it was handled would detract from Salk's conclusion:

   a. Component 3

   b. Component 4

   c. Component 5

   d. Component 6

8. Suppose a tobacco company is planning to fund a telephone survey of attitudes about banning smoking in restaurants. In each of the following phases of the survey, should the company disclose who is funding the study? Explain your answer in each case.

   a. When respondents answer the phone, before they are interviewed.

   b. When the survey results are reported in the news.

   c. When the interviewers are trained and told how to conduct the interviews.

*9. Suppose a study were to find that twice as many users of nicotine patches quit smoking than nonusers. Suppose you are a smoker trying to quit. Which version of an answer to each of the following components would be more compelling evidence for you to try the nicotine patches? Explain.

   *a. Component 3. Version 1 is that the nicotine patch users were lung cancer patients, whereas the nonusers were healthy. Version 2 is that participants were randomly assigned to use the patch or not after answering an advertisement in the newspaper asking for volunteers who wanted to quit smoking.

   *b. Component 7. Version 1 is that 25% of nonusers quit, whereas 50% of users quit. Version 2 is that 1% of nonusers quit, whereas 2% of users quit.

10. In most studies involving human participants, researchers are required to fully disclose the purpose of the study to the participants. Do you think people should always be informed about the purpose *before* they participate? Explain.

11. Suppose a study were to find that drinking coffee raised cholesterol levels. Further, suppose you drink two cups of coffee a day and have a family history of heart problems related to high cholesterol. Pick three of the Seven Critical Components and discuss why knowledge of them would be useful in terms of deciding whether to change your coffee-drinking habits based on the results of the study.

*12. Explain why news reports should give the extent or size of the claimed effects or differences from a study instead of just reporting that an effect or difference was found.

13. Refer to the definition of *data* on page 18. In Hypothetical News Article 3 on page 27, what two pieces of data were collected on each dog? For each one, explain whether it was collected separately for each dog or for the dogs as a group.

14. Holden (1991, p. 934) discusses the methods used to rank high school math performance among various countries. She notes, "According to the International Association for the Evaluation of Educational Achievement, Hungary ranks near the top in 8th-grade math achievement. But by the 12th grade, the country falls to the bottom of the list because it enrolls more students than any other country—50%—in advanced math. Hong Kong, in contrast, comes in first, but only 3% of its 12th graders take math." Explain why answers to Components 3 and 6 would be most useful when interpreting the results of rankings of high school math performance in various countries, and describe how your interpretation of the results would be affected by knowing the answers.

*15. Moore and Notz (2014, p. 25) reported the following contradictory evidence: "The advice columnist Ann Landers once asked her readers, 'If you had it to do over again, would you have children?' She received nearly 10,000 responses, almost 70% saying 'No!'. . . A professional nationwide random sample commissioned by *Newsday* . . . polled 1373 parents and found that 91% would have children again." Using the most relevant one of the Seven Critical Components, explain the contradiction in the two sets of answers.

16. An advertisement for a cross-country ski machine, NordicTrack, claimed, "In just 12 weeks, research shows that *people who used a Nordic-Track lost an average of 18 pounds.*" Explain how each of the following components should have been addressed if the research results are fair and unbiased.

  a. Component 3: The *individuals* or objects studied and how they were *selected.*

  b. Component 4: The exact nature of the *measurements* made or *questions* asked.

  c. Component 5: The *setting* in which the measurements were taken.

  d. Component 6: *Differences* in the groups being compared, *in addition* to the factor of interest.

17. In a court case similar to those described in Case Study 2.2, the American Basketball Association (ABA) sued AMF Voit, Inc. for manufacturing basketballs with the same red, white, and blue coloring as the ABA basketballs. The ABA commissioned a survey that found 61% of those asked associated the coloring with ABA basketballs. The sample consisted of males aged 12 to 23 who had played basketball within the past year. Do you think the court accepted the survey results as good evidence that the population of people likely to buy basketballs knew the coloring of ABA basketballs? Explain. (If you are curious, do a web search of American Basketball Association v. AMF Voit. The case ultimately went to the Supreme Court.)

*18. Is it necessary that "data" consist of numbers? Explain.

Mini-Projects

1. Scientists publish their findings in technical magazines called journals. Most university libraries have hundreds of journals available for browsing, many of them accessible electronically. Find out where the medical journals are located. Browse the shelves or electronic journals until you find an article with a study that sounds interesting to you. (The *New England Journal of Medicine* and the *Journal of the American Medical Association* often have articles of broad interest, but there are also numerous specialized journals on medicine, psychology, environmental issues, and so on.) Read the article and write a report that discusses each of the Seven Critical Components for that particular study. Argue for or against the believability of the results on the basis of your discussion. Be sure you find an article discussing a single study and not a collection or "meta-analysis" of numerous studies.

2. Explain how you would design and carry out a study to find out how students at your school feel about an issue of interest to you. Be explicit enough that someone would actually be able to follow your instructions and implement the study. Be sure to consider each of the Seven Critical Components when you design and explain how to do the study.

3. Find an example of a statistical study reported in the news for which information about one of the Seven Critical Components is missing. Write two hypothetical reports addressing the missing component that would lead you to two different conclusions about the applicability of the results of the study.

References

Crossen, Cynthia. (1994). *Tainted truth: The manipulation of fact in America.* New York: Simon & Schuster.

Gastwirth, Joseph L. (1988). *Statistical reasoning in law and public policy.* Vol. 2. *Tort law, evidence and health.* Boston: Academic Press.

Holden, Constance. (1991). Questions raised on math rankings. *Science* 254, p. 934.

Moore, David S. and William I. Notz (2014). *Statistics: Concepts and controversies.* 8th ed. New York: W.H. Freeman.

CHAPTER 3

Measurements, Mistakes, and Misunderstandings

Thought Questions

1. Suppose you were interested in finding out what people felt to be the most important problem facing society today. Do you think it would be better to give them a fixed set of choices from which they must choose or an open-ended question that allowed them to specify whatever they wished? What would be the advantages and disadvantages of each approach?

2. You and a friend are each doing a survey to see if there is a relationship between height and happiness. Without discussing in advance how you will do so, you both attempt to measure the height and happiness of the same 100 people. Are you more likely to agree on your measurement of height or on your measurement of happiness? Explain, discussing how you would measure each characteristic.

3. A newsletter distributed by a politician to his constituents gave the results of a "nationwide survey on Americans' attitudes about a variety of educational issues." One of the questions asked was, "Should your legislature adopt a policy to assist children in failing schools to opt out of that school and attend an alternative school—public, private, or parochial—of the parents' choosing?" From the wording of this question, can you speculate on what answer was desired? Explain.

4. You are at a swimming pool with a friend and become curious about the width of the pool. Your friend has a 12-inch ruler, with which he sets about measuring the width. He reports that the width is 15.771 feet. Do you believe the pool is exactly that width? What is the problem? (Note that .771 feet is 9 1/4 inches.)

5. If you were to have your intelligence, or IQ, measured twice using a standard IQ test, do you think it would be exactly the same both times? What factors might account for any changes?

3.1   Simple Measures Don't Exist

In the last chapter, we listed Seven Critical Components that need to be considered when someone conducts a study. You saw that many decisions need to be made, and many potential problems can arise when you try to use data to answer a question. One of the hardest decisions is contained in Component 4—that is, in deciding exactly what to measure or what questions to ask. In this chapter, we focus on problems with defining measurements and on the subsequent misunderstandings and mistakes that can result. When you read the results of a study, it is important that you understand exactly how the information was collected and what was measured or asked. Consider something as apparently simple as trying to measure your own height. Try it a few times and see if you get the measurement to within a quarter of an inch from one time to the next. Now imagine trying to measure something much more complex, such as the amount of fat in someone's diet or the degree of happiness in someone's life. Researchers routinely attempt to measure these kinds of factors.

3.2   It's All in the Wording

You may be surprised at how much answers to questions can change based on simple changes in wording. Here are two examples.

EXAMPLE 3.1 You Get What You Ask for: Two Polls on Immigration

In the spring of 2013, the question of immigration into the United States was at the forefront of many political discussions. So, polling organizations naturally wanted to know how the public felt about increasing versus decreasing the number of legal immigrants allowed into the United States. Just ask, right? Well, it's not quite so simple. Here are two polls conducted within a few days of each other, with similar wording:

- A Fox News poll conducted April 20–22, 2013, asked, *"Do you think the United States should increase or decrease the number of LEGAL immigrants allowed to move to this country?"*
- A CBS News/New York Times poll conducted April 24–28, 2013, asked, *"Should LEGAL immigration into the United States be kept at its present level, increased, or decreased?"*

The results of the two polls are shown in the following table.

Poll	Increase	Decrease	Stay Same	Unsure
Fox News	28%	55%	10%	7%
CBS/NYTimes	25%	31%	35%	8%

While the percentage responding in favor of increasing legal immigration is fairly similar in the two polls, the percentage in favor of decreasing it or keeping it the same are much different. Why? A close look at the wording reveals the answer. When the option "kept at its present level" was given in the question (CBS/New York Times Poll), 35% of respondents chose it. But when that option was not given and respondents had to come up with it on their own (Fox News Poll), only 10% did so. The lesson is that small changes in wording can make a big difference in how people respond in surveys. You should always find out exactly what was asked when reading the results of a survey. (*Source*: http://pollingreport.com/immigration.htm, accessed May 11, 2013.)   ■

EXAMPLE 3.2   **Is Marijuana Easy to Buy but Hard to Get?**

Refer to the detailed report on the companion website labeled as Original Source 13: "2003 CASA National Survey of American Attitudes on Substance Abuse VIII: Teens and Parents," which describes a survey of teens and drug use. One of the questions (number 36, p. 44) asked teens about the relative ease of getting cigarettes, beer, and marijuana. About half of the teens were asked about "buying" these items and the other half about "obtaining" them. The questions and percent giving each response were:

"Which is easiest for someone of your age to buy: cigarettes, beer or marijuana?"
"Which is easiest for someone of your age to obtain: cigarettes, beer or marijuana?"

Response	Version with "buy"	Version with "obtain"
Cigarettes	35%	39%
Beer	18%	27%
Marijuana	34%	19%
The Same	4%	5%
Don't know/no response	9%	10%

Notice that the responses indicate that beer is easier to "obtain" than is marijuana, but marijuana is easier to "buy" than beer. The subtle difference in wording reflects a very important difference in real life. Regulations and oversight authorities have made it difficult for teenagers to buy alcohol, but not to obtain it in other ways.   ■

Many pitfalls can be encountered when asking questions in a survey or experiment. Here are some of them; each will be discussed in turn:

1. Deliberate bias
2. Unintentional bias
3. Desire to please
4. Asking the uninformed
5. Unnecessary complexity
6. Ordering of questions
7. Confidentiality versus anonymity

Deliberate Bias

Sometimes, if a survey is being conducted to support a certain cause, questions are deliberately worded in a biased manner. Be careful about survey questions that begin with phrases like "Do you agree that. . . ." Most people want to be agreeable and will be inclined to answer "yes" unless they have strong feelings the other way. For example, suppose an anti-abortion group and a pro-choice group each wanted to conduct a survey in which they would find the best possible agreement with their position. Here are two questions that would each produce an estimate of the proportion of people who think abortion should be completely illegal. Each question is almost certain to produce a different estimate:

1. Do you agree that abortion, the murder of innocent beings, should be outlawed?
2. Do you agree that there are circumstances under which abortion should be legal, to protect the rights of the mother?

Appropriate wording should not indicate a desired answer. For instance, a Gallup Poll conducted in May 2013 (Saad, 2013) contained the question "Do you think abortion should be legal under any circumstances, legal only under certain circumstances, or illegal in all circumstances?" Notice that the question does not indicate which answer is preferable. In case you're curious, 26% of respondents thought it should always be legal, 20% thought it should always be illegal, and 52% thought it depends on the circumstance. (The remaining 2% had no opinion.).

Unintentional Bias

Sometimes questions are worded in such a way that the meaning is misinterpreted by a large percentage of the respondents. For example, if you were to ask people what drugs they use, you would need to specify if you mean prescription drugs, illegal drugs, over-the-counter drugs, or common substances such as caffeine. If you were to ask people to recall the most important date in their life, you would need to clarify if you meant the most important calendar date or the most important social engagement with a potential partner. (It is unlikely that anyone would mistake the question as being about the shriveled fruit, but you can see that the same word can have multiple meanings.)

Desire to Please

Most survey respondents have a desire to please the person who is asking the question. They tend to understate their responses about undesirable social habits and opinions, and vice versa. For example, in recent years estimates of the prevalence of cigarette smoking based on surveys do not match those based on cigarette sales. Either people are not being completely truthful or lots of cigarettes are ending up in the garbage. Political pollsters, who are interested in surveying only those who will actually vote, learned long ago that it is useless to simply ask people if they plan to vote. Most of them will say yes, because that's the socially correct answer to give. Instead, the pollsters ask questions to establish a history of voting, such as "Where did you go to vote in the last election?"

Asking the Uninformed

People do not like to admit that they don't know what you are talking about when you ask them a question. In a paper on the topic, Graeff (2007, p. 682) summarizes much of the research by noting that "survey respondents have freely given opinions about fictitious governmental agencies, congressional bills, nonexistent nationalities, fictitious political figures, and have even given directions to places that do not exist." The following example illustrates some ways that survey researchers have tried to reduce this problem, but you will see that they have not been completely successful.

EXAMPLE 3.3　Giving Opinions on Fictional Brands

To test the extent to which people don't want to admit they don't know something, Graeff (2007) asked students at his university to provide numerical ratings of their opinion of six brands of running shoes. Included in the list were five real brands and a fictional brand named Pontrey. Different versions of the survey were given to different groups of students. One version provided the option of "Don't Know/No Opinion," and in that version, only 18% of the respondents expressed an opinion about Pontrey shoes. But another version provided no such option, and in that case, 88% of respondents gave an opinion about the fictional Pontrey shoes. They did not want to admit that they had never heard of them. (Not surprisingly, the average rating for Pontrey shoes was lower than the average for any of the real brands.) In other versions of the survey, students were first asked to rate their knowledge of each brand on a scale from 1 (not familiar at all) to 7 (very familiar) before giving their opinion of the brand. Admitting that they had little knowledge of Pontrey shoes reduced the uninformed responses somewhat, but not as much as offering the "don't know" option.　▪

Unnecessary Complexity

If questions are to be understood, they must be kept simple. A question such as "Shouldn't former drug dealers not be allowed to work in hospitals after they are released from prison?" is sure to lead to confusion. Does a yes answer mean they should or they shouldn't be allowed to work in hospitals? It would take a few readings to figure that out.

Another way in which a question can be unnecessarily complex is to actually ask more than one question at once. An example would be a question such as "Do you support the president's health care plan because it would ensure that all Americans receive health coverage?" If you agree with the idea that all Americans should receive health coverage, but disagree with the remainder of the plan, do you answer yes or no? Or what if you support the president's plan, but not for that reason?

Ordering of Questions

If one question requires respondents to think about something that they may not have otherwise considered, then the order in which questions are presented can change the results. For example, suppose a survey were to ask, "To what extent do you think teenagers today worry about peer pressure related to drinking alcohol?" and then ask, "Name the top five pressures you think face teenagers today." It is quite likely that respondents would use the idea they had just been given and name peer pressure related to drinking alcohol as one of the five choices.

In general, survey respondents assume that questions on the survey are related to each other, so they will interpret subsequent questions in the context of questions that preceded them. Here is an amusing example.

EXAMPLE 3.4 **Is Happiness Related to Dating?**

Clark and Schober (1992, p. 41) report on a survey that asked the following two questions:

1. How happy are you with life in general?
2. How often do you normally go out on a date? (about ___ times a month)

When the questions were asked in this order, there was almost no relationship between the two answers. But when question 2 was asked first, the answers were highly related. Clark and Schober speculate that in that case, respondents consequently interpreted question 1 to mean; "Now, considering what you just told me about dating, how happy are you with life in general?" ∎

Confidentiality versus Anonymity

People sometimes answer questions differently based on the degree to which they believe they are anonymous. Because researchers often need to perform follow-up surveys, it is easier to try to ensure confidentiality than true anonymity. In ensuring confidentiality, the researcher promises not to release identifying information about respondents. In a truly anonymous survey, the researcher does not know the identity of the respondents.

Questions on issues such as sexual behavior and income are particularly difficult because people consider those to be private matters. A variety of techniques have been developed to help ensure confidentiality, but surveys on such issues are hard to conduct accurately.

CASE STUDY 3.1 **No Opinion of Your Own? Let Politics Decide**

SOURCES: Morin (10–16 April 1995), p. 36.
http://www.huffingtonpost.com, 11 April 2013

This is an excellent example of how people will respond to survey questions, even when they do not know about the issues, and how the wording of questions can influence responses. In 1995, the *Washington Post* decided to expand on a 1978 poll taken in Cincinnati, Ohio, in which people were asked whether they "favored or opposed repealing the 1975 Public Affairs Act." There was no such act, but about one-third of the respondents expressed an opinion about it.

In February 1995, the *Washington Post* added this fictitious question to its weekly poll of 1000 randomly selected respondents: "Some people say the 1975 Public Affairs Act should be repealed. Do you agree or disagree that it should be repealed?" Almost half (43%) of the sample expressed an opinion, with 24% agreeing that it should be repealed and 19% disagreeing. The *Post* then tried another trick that produced even more disturbing results. This time, they polled two separate groups of 500 randomly selected adults. The first group was asked: "President Clinton [a Democrat] said that the 1975 Public Affairs Act should be repealed. Do you agree or disagree?" The second group was asked: "The Republicans in Congress said that the 1975 Public Affairs Act should be repealed. Do you agree or disagree?" Respondents were also asked about their party affiliation. Overall, 53% of the respondents expressed an opinion about repealing this fictional act. The results by party affiliation were striking: For the "Clinton" version, 36% of the Democrats but only 16% of the Republicans agreed that the act should be repealed. For the "Republicans in Congress" version, 36% of the Republicans but only 19% of the Democrats agreed that the act should be repealed.

In April 2013, the *Huffington Post* repeated this poll, replacing "Clinton" with "Obama." The results were similar. (Sources: http://www.huffingtonpost.com/2013/04/11/survey-questions-fiction_n_2994363.html and http://big.assets.huffingtonpost.com/toplines_full.pdf) ■

3.3   Open or Closed Questions: Should Choices Be Given?

An **open question** is one in which respondents are allowed to answer in their own words, whereas a **closed question** is one in which they are given a list of alternatives from which to choose their answer. Usually the latter form offers a choice of "other," in which the respondent is allowed to fill in the blank.

Advantages and Disadvantages of Open Questions

As we have seen in Examples 3.1 and 3.3, when people respond to surveys they rarely volunteer answers that aren't among the choices given, even when that option is offered. Therefore, one advantage of open questions is that respondents are free to say whatever they choose, rather than limiting themselves to the choices provided.

The main disadvantage of open questions is that the responses are difficult to summarize. If a survey includes thousands of respondents, it can be a major chore to categorize their responses. Another problem with open questions is that logical responses may not readily come to mind, and the wording of the question might unintentionally exclude answers that would have been appealing had they been included in a list of choices. Schuman and Scott (May 22, 1987) demonstrated this problem with the following example.

EXAMPLE 3.5   **Is the Invention of the Computer Important?**

In the 1980s, Schuman and Scott asked 347 people to "name one or two of the most important national or world event(s) or change(s) during the past 50 years." When asked as an open question, the most common response was World War II (14%), followed by the Vietnam War (10%), the exploration of space (7%), and the assassination of John F. Kennedy (5%). But 54% of respondents didn't give any of those most common answers, and 10% didn't know what to say. Only five people (1.4%) mentioned the invention of the computer, which in retrospect was probably the most important change from that time period. It seems that respondents thought of events and not changes, although the question mentioned both.

The survey was then repeated with a new sample of 354 people who were given a closed form question with five choices. The choices were the four most common ones from the open form, plus "Invention of the computer." When asked as a closed form question, 30% chose the invention of the computer, more than any other response. The common responses from the open form poll also received more support in the closed form question, when they were provided as options. For instance, 23% chose World War II, and 16% chose the exploration of space. ■

Advantages and Disadvantages of Closed Form Questions

One advantage of closed form questions is that they are easier to administer and to analyze than open form questions. Another advantage should be obvious to you if you have taken multiple choice tests—you don't have to come up with the answer on your own; you only need to select from the choices given.

A disadvantage of closed form questions is that they may limit the options because respondents will rarely volunteer a choice that isn't presented, even if they are given the option to choose "Other, please specify." Therefore, it is very important that survey authors think carefully about the choices offered, and it is important that you know what choices were given when you read the results of a survey. Another disadvantage, especially with phone interviews, is that respondents tend to choose the options given later in the list. To compensate for this "recency effect," pollsters often randomize the order of the response options, giving them in different orders to different participants.

Pilot Studies and Pilot Surveys

One compromise that takes into account the advantages and disadvantages of both methods is to conduct an open form **pilot study**, or **pilot survey**, before creating choices for a closed form survey. (No, this doesn't mean that surveys should be conducted on airline pilots!) In a pilot study, a small group of people are asked the questions in open form (the "pilot survey"), and their responses are used to create the choices for the closed form. Often the pilot study will include a focus group discussion to find out what thought process respondents used to arrive at their answers. Other features can be tested in a pilot study as well, such as whether the order of the questions influenced the responses and whether participants understood the questions in the way the survey designers intended.

EXAMPLE 3.6 Questions in Advertising

In her excellent book, *Tainted Truth*, Wall Street Journal reporter Cynthia Crossen explains how advertisers often present results of surveys without giving the full story. As an example, an advertisement for Triumph cigarettes boasted: "TRIUMPH BEATS MERIT—an amazing 60% said Triumph tastes as good as or better than Merit." In truth, three choices were offered to respondents, including "no preference." The results were: 36% preferred Triumph, 40% preferred Merit, and 24% said the brands were equal. So, although the wording of the advertisement is not false, it is also true that 64% said Merit tastes as good as or better than Triumph (Crossen, 1994, pp. 74–75). Which brand do you think wins? ∎

CASE STUDY 3.2 | **How is the President Supposed to Know What People Think?**

Sources:
http://www.people-press.org/2008/11/13/section-4-early-voting-campaign-outreach-and-the-issues/, released November 13, 2008; accessed May 12, 2013.
http://www.people-press.org/methodology/questionnaire-design/open-and-closed-ended-questions/, accessed May 12, 2013,

If you had just won your first term as president of the United States, you might be curious to know what prompted people to vote for you instead of your opponent. Just get a polling agency to ask them, right? It's not as simple as you might think! In November, 2008, shortly after Barack Obama beat John McCain in the presidential election, the Pew Research organization conducted two polls asking people why they voted as they did. One poll was closed form and the other was open form.

In both the open and closed form of the survey, people were asked "What one issue mattered most to you in deciding how you voted for President?" Results were also compared with an exit poll taken on the day of the election as voters were leaving their polling places. The exit poll was asked in closed form and in person. The other two surveys were conducted by telephone. The results of the open and closed form surveys were strikingly different, as shown in the table below.

"What one issue mattered most to you in deciding how you voted for President?"

Issue:	Open Form	Closed Form	Exit Poll
The economy	35%	58%	63%
The war in Iraq	5%	10%	10%
Health care	4%	8%	9%
Terrorism	6%	8%	9%
Energy policy	0%	6%	7%
Other	43%	8%	0%
Don't know	7%	2%	2%
Total	100%	100%	100%

When presented with the five options shown in the table, over half of respondents chose "The economy." Although "The economy" was still the most frequent response in the open form, only 35% of respondents chose it. Because 43% of respondents in the open choice form chose something not included in the closed form list, all five closed form choices received less support than they did when they were the only options explicitly provided. In case you are curious about what responses were offered by the remaining 43% in the open form, they included things like moral issues, taxes, hope for change, and specific mentions of the candidates by name. ■

Remember that, as the reader, you have an important role in interpreting the results of any survey. You should always be informed as to whether questions were asked in open or closed form, and if the latter, you should be told what the choices were. You should also be told whether "don't know" or "no opinion" was offered as a choice in either case.

3.4 Defining What Is Being Measured

EXAMPLE 3.7 Teenage Sex

To understand the results of a survey or an experiment, we need to know exactly what was measured. Consider this example. A letter to advice columnist Ann Landers stated: "According to a report from the University of California at San Francisco . . . sexual activity among adolescents is on the rise. There is no indication that this trend is slowing down or reversing itself." The letter went on to explain that these results were based on a national survey (*Davis* (CA) *Enterprise*, 19 February 1990, p. B-4). On the same day, in the same newspaper, an article entitled "Survey: Americans conservative with sex" reported that "teenage boys are not living up to their reputations. [A study by the Urban Institute in Washington] found that adolescents seem to be having sex less often, with fewer girls and at a later age than teenagers did a decade ago" (p. A-9).

Here we have two apparently conflicting reports on adolescent sexuality, both reported on the same day in the same newspaper. One indicated that teenage sex was on the rise; the other indicated that it was on the decline. Although neither report specified exactly what was measured, the letter to Ann Landers proceeded to note that "national statistics show the average age of first intercourse is 17.2 for females and 16.5 for males." The article stating that adolescent sex was on the decline measured it in terms of frequency. The result was based on interviews with 1880 boys between the ages of 15 and 19, in which "the boys said they had had six sex partners, compared with seven a decade earlier. They reported having had sex an average of three times during the previous month, compared with almost five times in the earlier survey." Thus, it is not enough to note that both surveys were measuring adolescent or teenage sexual behavior. In one case, the author was, at least partially, discussing the age of first intercourse, whereas in the other case the author was discussing the frequency. ■

EXAMPLE 3.8 Out of Work, Discouraged, but Not Unemployed!

Ask people whether they know anyone who is unemployed; they will invariably say yes. But most people don't realize that in order to be officially unemployed, and included in the unemployment statistics given by the U.S. government, you must meet very stringent criteria. The Bureau of Labor Statistics uses this definition when computing the official United States unemployment rate (http://www.bls.gov/cps/lfcharacteristics.htm#unemp, accessed May 12, 2013):

> *Persons are classified as unemployed if they do not have a job, have actively looked for work in the prior 4 weeks, and are currently available for work.*

To find the unemployment rate, the number of people who meet this definition is divided by the total number of people "in the labor force," which includes these individuals and people classified as employed. But "discouraged workers" are not included at all. "Discouraged workers" are defined as:

> *Persons not in the labor force who want and are available for a job and who have looked for work sometime in the past 12 months (or since the end of their last job if they held one within the past 12 months), but who are not currently looking because they believe there are no jobs available or there are none for which they would qualify.* (http://www.bls.gov/bls/glossary.htm; accessed May 12, 2013)

If you know someone who fits that definition, you would undoubtedly think of that person as unemployed even though they hadn't looked for work in the past 4 weeks. However, he or she would not be included in the official statistics. You can see that the true number of people who are not working is higher than government statistics indicate. ■

These two examples illustrate that when you read about measurements taken by someone else, you should not automatically assume you are speaking a common language. A precise definition of what is meant by "adolescent sexuality" or "unemployment" should be provided.

Some Concepts Are Hard to Define Precisely

Sometimes it is not the language but the concept itself that is ill-defined. For example, there is still no universal agreement on what should be measured with intelligence, or IQ, tests. The tests were originated at the beginning of the 20th century in order to determine the mental level of school children. The intelligence quotient (IQ) of a child was found by dividing the child's "mental level" by his or her chronological age. The "mental level" was determined by comparing the child's performance on the test with that of a large group of "normal" children, to find the age group the individual's performance matched. Thus, if an 8-year-old child performed as well on the test as a "normal" group of 10-year-old children, he or she would have an IQ of $100 \times (10/8) = 125$.

IQ tests have been expanded and refined since the early days, but they continue to be surrounded by controversy. One reason is that it is very difficult to define what is meant by intelligence. It is difficult to measure something if you can't even agree on what it is you are trying to measure. If you are interested in knowing more about these tests and the surrounding controversies, you can find numerous books on the subject. A classic book on this topic is by Anastasi and Urbina (1997). It provides a detailed discussion of a large variety of psychological tests, including IQ tests.

EXAMPLE 3.9

Stress in Kids

The studies reported in News Stories 13 and 15 both included "stress" as one of the important measurements used. But they differed in how they measured stress. In Original Source 13, "2003 CASA National Survey of American Attitudes on Substance Abuse VIII: Teens and Parents," teenage respondents were asked:

> *How much stress is there in your life? Think of a scale between 0 and 10, where 0 means you usually have no stress at all and 10 means you usually have a very great deal of stress, which number would you pick to indicate how much stress there is in your life?* (p. 40)

Categorizing responses as low stress (0 to 3), moderate stress (4 to 6), and high stress (7 to 10), the researchers found that low, medium, and high stress were reported by 29%, 45%, and 26% of teens, respectively.

For News Story 15, the children were asked more specific questions to measure stress. According to Additional News Story 15, "To gauge their stress, the children were given a standard questionnaire that included questions like: 'How often have you felt that you couldn't control the important things in your life?'"

There is no way to know which method is more likely to produce an accurate measure of "stress," partly because there is no fixed definition of stress. Stress in one scenario might mean that someone is working hard to finish an exciting project with a tight deadline. In another scenario, it might mean that someone feels helpless and out of control. Those two versions are likely to have very different consequences on someone's health and well-being. What is important is that as a reader, you are informed about how the researchers measured stress in any given study. ■

Measuring Attitudes and Emotions

Similar problems exist with trying to measure attitudes and emotions such as self-esteem and happiness. The most common method for trying to measure such things is to have respondents read statements and determine the extent to which they agree with the statement. For example, a test for measuring happiness might ask respondents to indicate their level of agreement, from "strongly disagree" to "strongly agree," with statements such as "I generally feel optimistic when I get up in the morning." To produce agreement on what is meant by characteristics such as "introversion," psychologists have developed standardized tests that claim to measure those attributes.

3.5 Defining a Common Language

So that we're all speaking a common language for the rest of this book, we need to define some terms. We can perform different manipulations on different types of data, so we need a common understanding of what those types are. Other terms defined in this section are those that are well known in everyday usage but that have a slightly different technical meaning.

Categorical versus Measurement Variables

Thus far in this book, we have seen examples of measuring opinions (such as whether you think abortion should be legal), numerical information (such as weight gain in infants), and attributes that can be transformed into numerical information (such as IQ). To understand what we can do with these measurements, we need definitions to distinguish numerical, quantitative measures from categorical, qualitative ones. Although statisticians make numerous fine distinctions among types of measurements, for our purposes it will be sufficient to distinguish between just two main types: categorical variables and measurement variables. Subcategories of these types will be defined for those who want more detail.

Categorical Variables

Categorical variables are those we can place into a category but that may not have any logical ordering. For example, you could be categorized as male or female. You could also be categorized based on which option you choose as your reason for voting for a particular candidate, as in Case Study 3.2. Notice that we are limited in how we can manipulate this kind of information numerically. For example, we cannot talk about the average reason for voting for a candidate in the same way as we can talk about the average weight gain of infants during the first few days of life.

If the possible categories have a natural ordering, the term **ordinal variable** is sometimes used. For instance, in a public opinion poll respondents may be asked to give an opinion chosen from "strongly agree, agree, neutral, disagree, strongly disagree." Level of education attained may be categorized as "less than high school, high school graduate, college graduate, postgraduate degree." To distinguish them from ordinal variables, categorical variables for which the categories do not have a natural ordering are sometimes called **nominal variables.**

Measurement Variables

Measurement variables, also called **quantitative variables,** are those for which we can record a numerical value and then order respondents according to those values. For example, IQ is a measurement variable because it can be expressed as a single number. An IQ of 130 is higher than an IQ of 100. Age, height, and number of cigarettes smoked per day are other examples of measurement variables. Notice that these can be worked with numerically. Of course, not all numerical summaries will make sense even with measurement variables. For example, if one person in your family smokes 20 cigarettes a day and the remaining three members smoke none, it is accurate but misleading to say that the average number of cigarettes smoked by your family per day is 5 per person. We will learn about reasonable numerical summaries in Chapter 7.

Interval and Ratio Variables

Occasionally a further distinction is made for measurement variables based on whether ratios make sense. An **interval variable** is a measurement variable in which it makes sense to talk about differences, but not about ratios. Temperature is a good example of an interval variable. If it was 20 degrees last night and it's 40 degrees today, we wouldn't say it is twice as warm today as it was last night. But it would be reasonable to say that it is 20 degrees warmer, and it would mean the same thing as saying that when it's 60 degrees it's 20 degrees warmer than when it's 40 degrees. A **ratio variable** has a meaningful value of zero, and it makes sense to talk about the ratio of one value to another. Pulse rate is a good example. For instance, if your pulse rate is 60 before you exercise and 120 after you exercise, it makes sense to say that your pulse rate doubled during exercise. (And of course having a pulse rate of 0 is *extremely* meaningful!)

Continuous versus Discrete Measurement Variables

Even when we can measure something with a number, we may need to distinguish further whether it can fall on a continuum. A **discrete variable** is one for which you could actually count the possible responses. For example, if we measure the number of automobile accidents on a certain stretch of highway, the answer could be 0, 1, 2, 3, and so on. It could not be 2 1/2 or 3.8. Conversely, a **continuous variable** can be anything within a given interval. Age, for example, falls on a continuum.

Something of a gray area exists between these definitions. For example, if we measure age to the nearest year, it may seem as though it should be called a discrete variable. But the real difference is conceptual. With a discrete variable you can count the possible responses without having to round off. With a continuous variable you can't. In case you are confused by this, note that long ago you probably figured out the difference between the phrases "the number of" and "the amount of." You wouldn't say, "the amount of cigarettes smoked," nor would you say, "the number of water consumed." Discrete variables are analogous to numbers of things, and continuous variables are analogous to amounts. You still need to be careful about wording, however, because we have a tendency to express continuous variables in discrete units. Although you wouldn't say, "the number of water consumed," you might say, "the number of glasses of water consumed." That's why it's the *concept* of number versus amount that you need to think about.

Validity, Reliability, and Bias

The words we define in this section are commonly used in the English language, but they also have specific definitions when applied to measurements. Although these definitions are close to the general usage of the words, to avoid confusion we will spell them out.

Validity

When you talk about something being *valid,* you generally mean that it makes sense to you; it is sound and defensible. The same can be said for a measurement. A **valid measurement** is one that actually measures what it claims to measure. Thus, if you tried to measure happiness with an IQ test, you would not get a valid measure of happiness.

A more realistic example would be trying to determine the selling price of a home. Getting a valid measurement of the actual sales price of a home is tricky because the purchase often involves bargaining on what items are to be left behind by the old owners, what repairs will be made before the house is sold, and so on. These items can change the recorded sales price by thousands of dollars. If we were to define the "selling price" as the price recorded in public records, it may not actually reflect the price the buyer and seller had agreed was the true worth of the home.

To determine whether a measurement is valid, you need to know exactly what was measured. For example, many readers, once they are informed of the definition, do not think the unemployment figures provided by the U.S. government are a valid measure of unemployment, as the term is generally understood. Remember (from Example 3.8) that the figures do not include "discouraged workers." However, the government statistics are a valid measure of the percentage of the "labor force" that is currently "unemployed," according to the precise definitions supplied by the Bureau of Labor Statistics. The problem is that most people do not understand exactly what the government has measured.

Reliability

When we say something or someone is *reliable,* we mean that that thing or person can be depended upon time after time. A reliable car is one that will start every time and get us where we are going without worry. A reliable friend is one who is always there for us, not one who is sometimes too busy to bother with us. Similarly, a **reliable measurement** is one that will give you or anyone else approximately the same result time after time when taken on the same object or individual. In other words, it is *consistent.* For example, a reliable way to define the selling price of a home would be the officially recorded amount. This may not be valid, but it would give us a consistent figure without any ambiguity.

Reliability is a useful concept in psychological and aptitude testing. An IQ test is obviously not much use if it measures the same person's IQ to be 80 one time and 130 the next. Whether we agree that the test is measuring what we really mean by "intelligence" (that is, whether it is really valid), it should at least be reliable enough to give us approximately the same number each time. Commonly used IQ tests are fairly reliable: About two-thirds of the time, taking the test a second time gives a reading within 2 or 3 points of the first test, and, most of the time, it gives a reading within about 5 points.

The most reliable measurements are physical ones taken with a precise measuring instrument. For example, it is much easier to get a reliable measurement of height than of happiness, assuming you have an accurate tape measure. However, you should be cautious of measurements given with greater precision than you think the measuring tool would be capable of providing. The degree of precision probably exceeds the reliability of the measurement. For example, if your friend measures the width of a swimming pool with a ruler and reports that it is 15.771 feet wide, which is 15' 9 1/4", you should be suspicious. It would be very difficult to measure a distance that large reliably with a 12-inch ruler. A second measuring attempt would undoubtedly give a different number.

Bias

A systematic prejudice in one direction is called a *bias*. Similarly, a measurement that is systematically off the mark in the same direction is called a **biased measurement.** If you were trying to weigh yourself with a scale that was not satisfactorily adjusted at the factory and was always a few pounds under, you would get a biased view of your own weight. When we used the term earlier in discussing the wording of questions, we noted that either intentional or unintentional bias could enter into the responses of a poorly worded survey question. Notice that a biased measurement differs from an unreliable measurement because it is consistently off the mark in the same direction.

Connections between Validity, Reliability, and Bias

The differences and connections among the meanings of validity, reliability, and bias can be confusing. For instance, it is not possible for a measurement to be unreliable but still valid in every individual instance. If it's truly measuring what it's supposed to measure every time (i.e., is always valid), it would have to be consistent (i.e. reliable). Let's look at some examples:

- Suppose a woman's weight varies between 140 and 150 pounds, but when asked her weight she always answers (optimistically!) that it's 140 pounds. Then her answer is *reliable*, but it is *not valid* (except on the days when she really does weight 140). Her response is *biased* in the low direction.
- Suppose you take a multiple choice test that truly does measure what you know. Then the test is a *valid* measurement of your knowledge. If you retook a similar test, you should do about equally well, so the test is *reliable*. (Any professor will tell you that designing such tests is a difficult task!)
- A highway patrol officer parks on the side of an open stretch of highway and uses radar to measure the speed of cars as they pass her car. She takes an average of the speeds of 100 cars. The average is a *reliable* measure of the speed of cars passing that point in the sense that she will probably get fairly consistent data, but it is *not a valid measurement* of the average speed of cars for that highway in general. Cars would surely slow down when they saw her parked there, so the measurement is *biased* in the low direction.
- A campus has an academic honesty policy in which students are asked to report all observed cheating incidences to a centralized office. The number of reported incidences per year would probably be relatively consistent and a *reliable* measure of how many incidences students would report, but would *not be a valid measure* of the amount of cheating that occurs. Students are reluctant to turn in other students if they observe them cheating. The measurement would be *biased* in the low direction.
- A thermometer always overestimates the ambient temperature by 5 degrees when it's in the direct sun and always underestimates it by 5 degrees when it's in the shade, but averages out to the correct average daily temperature. At any given

point in time, the thermometer is *not a valid* measure of the ambient temperature, but it does produce a *valid measure* of the daily temperature. As a measuring instrument for daily temperature, the thermometer is *unbiased*. But at any given point in time, it is *biased* either too high (when sunny) or too low (when shady). It is also *reliably* too high or too low at any given time, because it is always off by a consistent amount.

Variability across Measurements

If someone has *variable* moods, we mean that that person has unpredictable swings in mood. When we say the weather is quite variable, we mean it changes without any consistent pattern. Most measurements are prone to some degree of **variability.** By that, we mean that they are likely to differ from one time to the next or from one individual to the next because of unpredictable errors, discrepancies, or natural differences that are not readily explained. If you tried to measure your height as instructed at the beginning of this chapter, you probably found some unexplainable variability from one time to the next. If you tried to measure the length of a table by laying a ruler end to end, you would undoubtedly get a slightly different answer each time.

Unlike the other terms we have defined, which are used to characterize a single measurement, variability is a concept used when we talk about two or more measurements in relation to each other. Sometimes two measurements vary because the measuring device produces unreliable results—for example, when we try to measure a large distance with a small ruler. The amount by which each measurement differs from the true value is called **measurement error.**

Variability can also result from changes across time in the system being measured. For example, even with a very precise measuring device your recorded blood pressure will differ from one moment to the next. Unemployment rates vary from one month to the next because people move in and out of jobs and the workforce. These differences represent **natural variability** across time in the individual or system being measured.

Natural variability also explains why many measurements differ across individuals. Even if we could measure everyone's height precisely, we wouldn't get the same value for everyone because people naturally come in different heights. If we measured unemployment rates in different states of the United States at the same time, they would vary because of natural variability in conditions and individuals across states. If we measure the annual rainfall in one location for each of many years, it will vary because weather conditions naturally differ from one year to the next.

Understanding Natural Variability is the Key to Understanding Statistics

Understanding the concept of natural variability is crucial to understanding modern statistical methods. When we measure the same quantity across several individuals, such as the weight gain of newborn babies, we are bound to get some variability. Although some of this may be due to our measuring instrument, most of it is simply due to the fact that everyone is different. Variability is plainly inherent in nature. Babies all gain weight at their own pace. If we want to compare the weight gain of a group of babies who have consistently listened to a heartbeat to the weight gain of a group of babies who have not, we first need to know how much variability to expect due to natural causes.

We encountered the idea of natural variability when we discussed comparing resting pulse rates of men and women in Chapter 1. If there were no variability within each sex, it would be easy to detect a difference between males and females. The more variability there is within each group, the more difficult it is to detect a difference between groups. Natural variability can occur when taking repeated measurements on the same individual as well. Even if it could be measured precisely, your pulse rate is not likely to remain constant throughout the day. Some measurements are more likely to exhibit this variability than others. For example, height (if it could be measured precisely) and your opinion on issues like gun control and abortion are likely to remain constant over short time periods.

In summary, variability among measurements can occur for at least the following three reasons:

- Measurements are imprecise, and *measurement error* is a source of variability.
- There is *natural variability across individuals* at any given time.
- There may be *natural variability across time* in a characteristic on the same individual.

The Key to Statistical Discoveries: Comparing Natural Variability to Created Variability

In Part 4, we will learn how to sort out differences due to natural variability from differences due to features we can define, measure, and possibly manipulate, such as variability in blood pressure due to amount of salt consumed, or variability in weight loss due to time spent exercising. In this way, we can study the effects of diet or lifestyle choices on disease, of advertising campaigns on consumer choices, of exercise on weight loss, and so on.

This one basic idea, comparing natural variability to the variability created by different behaviors, interventions, or group memberships, forms the heart of modern statistics. It has allowed Salk to conclude that heartbeats are soothing to infants and the medical community to conclude that aspirin helps prevent heart attacks. We will see numerous other conclusions based on this idea throughout this book.

Thinking About Key Concepts

- Subtle changes in wording, ordering of questions and whether questions are asked as open or closed form can make a big difference in the outcome of a survey.
- People like to please others, so they will give socially acceptable answers in surveys and will even provide opinions on topics they know nothing about, pretending that they do know.

- Be suspicious of statistics in which the degree of precision of what is reported would be impossible to obtain with available measuring instruments.
- Valid measurements accurately reflect what they are trying to measure, reliable measurements are consistent when repeated (but not necessarily valid), and biased measurements systematically err in the same direction.
- Comparing natural person-to-person variation with variation created by different group memberships (natural or imposed) is a key to making new discoveries with statistical methods. For instance, in Case Study 1.1, Lee Salk noticed that the increased weight gain for infants who heard a heartbeat (compared to those who did not) was large enough that it could not be attributed to natural variation in infant weight gain.

Exercises

Exercises with numbers divisible by 3 (3, 6, 9, etc.) are included in the Solutions at the back of the book. They are marked with an asterisk (*).

1. Give an example of a survey question that is
   a. Deliberately biased
   b. Unintentionally biased
   c. Unnecessarily complex
   d. Likely to cause respondents to lie

2. Here is a potential survey question: "Do you agree that marijuana should be legal?"
   a. Explain which one of the seven pitfalls listed in Section 3.2 applies to this question.
   b. Reword the question so that it avoids the seven pitfalls.

*3. Here is a potential survey question: "Do you support banning prayers in schools so that teachers have more time to spend teaching?"
   *a. Explain which two of the seven pitfalls listed in Section 3.2 applies to this question.
   *b. Reword the question so that it avoids the seven pitfalls.

4. Here is a potential survey question: "Studies have shown that consuming one alcoholic drink daily helps reduce heart disease. How many alcoholic drinks do you consume daily?"
   a. Explain which two of the seven pitfalls listed in Section 3.2 are most problematic for this question.
   b. Reword the question so that it avoids the seven pitfalls.

5. Refer to Case Study 2.2, "Flawed Surveys in the Courtroom." Discuss the study conducted by Brooks Shoe Manufacturing Company in the context of the seven pitfalls, listed in Section 3.2, that can be encountered when asking questions in a survey.

*6. Schuman and Presser (1981, p. 277) report a study in which one set of respondents was asked question A, and the other set was asked question B:
   A. Do you think the United States should forbid public speeches against democracy?
   B. Do you think the United States should allow public speeches against democracy?

For one version of the question, only about one-fifth of the respondents were against such freedom of speech, whereas for the other version almost half were against such freedom of speech. Which question do you think elicited which response? Explain.

7. Give an example of two questions in which the order in which they are presented would determine whether the responses were likely to be biased.

8. In February 1998, U.S. President Bill Clinton was under investigation for allegedly having had an extramarital affair. A Gallup Poll asked the following two questions: "Do you think most presidents have or have not had extramarital affairs while they were president?" and then "Would you describe Bill Clinton's faults as worse than most other presidents, or as no worse than most other presidents?" For the first question, 59% said "have had," 33% said "have not," and the remaining 8% had no opinion. For the second question, 24% said "worse," 75% said "no worse," and only 1% had no opinion. Do you think the order of these two questions influenced the results? Explain.

*9. Sometimes medical tests, such as those for detecting HIV, are so sensitive that people do not want to give their names when they take the test. Instead, they are given a number or code, which they use to obtain their results later. Is this procedure anonymous testing or is it confidential testing? Explain.

10. Give three versions of a question to determine whether people think smoking should be completely banned in airports. Word the question to be as follows:

   a. Version 1: As unbiased as possible

   b. Version 2: Likely to get people to respond that smoking should be forbidden

   c. Version 3: Likely to get people to respond that smoking should not be forbidden

11. Refer to the detailed report labeled as Original Source 13: "2003 CASA National Survey of American Attitudes on Substance Abuse VIII: Teens and Parents" on the companion website.

   a. Locate the questions asked of the teens, in Appendix D. Two questions asked as "open questions" were Question 1 and Question 11. Explain which of the two questions was less likely to cause problems in categorizing the answers.

   b. The most common response to Question 11 was "Sports team." Read the question and explain why this might have been the case.

   c. Two versions of Question 28 were asked, one using the word *sold* and one using the word *used*. Did the wording of the question appear to affect the responses? Explain.

*12. In 2011 there was a dispute between the Governor of Wisconsin and the teacher's union about whether teachers should have the right to go on strike. In February of that year, a Rasmussen Poll asked a nationwide sample "Should teachers, fireman and policemen be allowed to go on strike?"

   *a. Do you think this question is a valid way to determine whether people think teachers should be allowed to go on strike? Explain.

   *b. Do you think the responses to this question would be biased in favor of supporting teachers being allowed to strike, not supporting that view, or not biased in either direction? Explain.

   *c. Write a valid question for determining opinions on this issue.

   (Source: http://www.rasmussenreports.com/public_content/politics/questions/pt_survey_questions/february_2011/questions_unions_february_18_19_2011, accessed May 12, 2013)

13. Give an example of a survey question that is

   a. Most appropriately asked as an open question

   b. Most appropriately asked as a closed question

14. An advertiser of a certain brand of aspirin (let's call it Brand B) claims that it is the preferred painkiller for headaches, based on the results of a survey of headache sufferers. The choices given to respondents were: Tylenol, Extra-Strength Tylenol, Brand B aspirin, Advil.

   a. Is this an open- or closed-form question? Explain.

b. Comment on the variety of choices given to respondents.

c. Comment on the advertiser's claim.

*15. Explain how "depression" was measured for the research discussed in News Story 19 in the Appendix, "Young romance may lead to depression, study says."

16. Read Original Source 4 on the companion website, *Duke Health Briefs:* Positive Outlook Linked to Longer Life in Heart Patients." Explain how the researchers measured "happiness."

17. Locate Original Source 11, "Driving impairment due to sleepiness is exacerbated by low alcohol intake." It is not on the companion website but may be available at http://www.ncbi.nlm.nih.gov/pmc/articles/PMC1740622/pdf/v060p00689.pdf. Find the description of how the researchers measured "subjective sleepiness."

a. Explain how "subjective sleepiness" was measured.

b. Was "subjective sleepiness" measured as a nominal, ordinal, or measurement variable? Explain.

*18. Give an example of a measure that is

*a. Valid and categorical

*b. Reliable but biased

*c. Unbiased but not reliable

19. If we were interested in knowing whether the average price of homes in a certain county had gone up or down this year in comparison with last year, would we be more interested in having a valid measure or a reliable measure of sales price? Explain.

20. Do you think the crime statistics reported by the police are a valid measure of the amount of crime in a given city? Are they a reliable measure? Discuss.

*21. Specify whether each of the following is a categorical or measurement variable. If you think the variable is ambiguous, discuss why.

*a. Years of formal education

*b. Highest level of education completed (grade school, high school, college, higher than college)

22. Specify whether each of the following is a categorical or measurement variable. If you think the variable is ambiguous, discuss why.

a. Brand of car owned.

b. Price paid for the last car purchased.

c. Type of car owned (subcompact, compact, mid-size, full-size, sports, pickup).

23. Refer to exercises 21 and 22, which asked whether variables were categorical or measurement variables. The variables are repeated below. In each case, if the variable is categorical, specify whether it is ordinal or nominal. If it is a measurement variable, specify whether it is an interval or a ratio variable. Explain your answers.

a. Years of formal education.

b. Highest level of education completed (grade school, high school, college, higher than college).

c. Brand of car owned.

d. Price paid for the last car purchased.

e. Type of car owned (subcompact, compact, mid-size, full-size, sports, pickup).

*24. Specify whether each of the following measurements is discrete or continuous. If you think the measurement is ambiguous, discuss why.

*a. The number of floors in a building

*b. The height of a building measured as precisely as possible

25. Specify whether each of the following measurements is discrete or continuous. If you think the measurement is ambiguous, discuss why.

a. The number of words in a book.

b. The weight of a book.

c. A person's IQ.

26. Refer to Exercises 24 and 25. In each case, the measurement is repeated below. Explain whether it is an interval or a ratio variable.

a. The number of floors in a building.

b. The height of a building measured as precisely as possible.

c. The number of words in a book.

d. The weight of a book.

e. A person's IQ.

*27. Can a variable be both nominal and categorical? Explain.

28. Explain whether a variable can be both

   a. Nominal and ordinal

   b. Interval and categorical

   c. Discrete and interval

29. Explain the difference between a discrete variable and a categorical variable. Give an example of each type.

*30. Suppose you were to compare two routes to school or work by timing yourself on each route for five days. Suppose the times on one route were (in minutes) 10, 12, 13, 15, 20, and on the other route they were 10, 15, 16, 18, 21.

   *a. The average times for the two routes are 14 minutes and 16 minutes. Would you be willing to conclude that the first route is faster, on average, based on these sample measurements?

   *b. Give an example of two sets of times, where the first has an average of 14 minutes and the second an average of 16 minutes, for which you *would* be willing to conclude that the first route is faster.

31. Refer to Exercise 30. Explain how the concept of natural variability would enter into your conclusion about whether or not it could be concluded that the first route is faster, on average, than the second route.

32. Airlines compute the percentage of flights that are on time to be the percentage that arrive no later than 15 minutes after their scheduled arrival time. Is this a valid measure of on-time performance? Is it a reliable measure? Explain.

*33. If each of the following measurements were to be taken on a group of 50 college students (once only for each student), it is unlikely that all 50 of them would yield the same value. In other words, there would be variability in the measurements. In each case, explain whether

their values are likely to differ because of natural variability across time, natural variability across individuals, measurement error, or some combination of these three causes.

   *a. Systolic blood pressure

   *b. Time on the student's watch when the actual time is 12 noon

34. Refer to Exercise 33. For each of the following, explain whether their values are likely to differ because of natural variability across time, natural variability across individuals, measurement error, or some combination of these three causes.

   a. Blood type (A, B, O, AB)

   b. Actual time when the student's watch says it is 12 noon.

35. Refer to the detailed report labeled as Original Source 13: "2003 CASA National Survey of American Attitudes on Substance Abuse VIII: Teens and Parents" on the companion website. Locate the questions asked of the parents, in Appendix E. For each of the following questions explain whether the response was a nominal variable, an ordinal variable, or a measurement variable.

   a. Question 2

   b. Question 9

   c. Question 12

   d. Question 19

   e. Question 29

   f. Question 30

*36. Explain whether there is likely to be variability in the following measurements if they were to be taken on 10 consecutive days for the same student. If so, explain whether the variability most likely would be due to natural variability across time, natural variability across individuals, measurement error, or some combination of these three causes.

   *a. Systolic blood pressure

   *b. Blood type (A, B, O, AB)

37. Refer to Exercise 36. For each of the following measurements taken on 10 consecutive days for the same student, explain whether there is likely to be variability. If so, explain whether the variability most likely would be due to natural variability across time, natural variability across individuals, measurement error, or some combination of these three causes.

   a. Time on the student's watch when the actual time is 12 noon.

   b. Actual time when the student's watch says it's 12 noon.

38. In Chapter 1, we discussed Lee Salk's experiment in which he exposed one group of infants to the sound of a heartbeat and compared their weight gain to that of a group not exposed. Do you think it would be easier to discover a difference in weight gain between the group exposed to the heartbeat and the "control group" if there were a lot of natural variability among babies, or if there were only a little? Explain.

*39. Give an example of a characteristic that could be measured as either a discrete or a continuous variable, depending on the types of units used.

Exercises 40 to 44 refer to News Story 2, "Research shows women harder hit by hangovers" and Original Source 2, "Development and initial validation of the Hangover Symptoms Scale: Prevalence and correlates of hangover symptoms in college students" on the companion website accompanying the book.

40. The researchers were interested in measuring the severity of hangovers for each person so they developed a "Hangover Symptoms Scale." Read the article and explain what they measured with this scale.

41. Explain whether the Hangover Symptoms Scale for each individual in this study is likely to be a *reliable* measure of hangover severity.

*42. Explain whether the Hangover Symptoms Scale for each individual in this study is likely to be a *valid* measure of hangover severity.

43. To make the conclusion that women are harder hit by hangovers, the researchers measured two variables on each individual. Specify the two variables and explain whether each one is a categorical or a measurement variable.

44. The measurements in this study were self-reported by the participants. Explain the extent to which you think this may systematically have caused the measurements of hangover severity of men or women or both to be biased, and whether that may have affected the conclusions of the study in any way.

Mini-Projects

1. Measure the heights of five males and five females. Draw a line to scale, starting at the lowest height in your group and ending at the highest height, and mark each male with an M and each female with an F. It should look something like this:

F		F	M		FMF		F	MM	M
5′									6′2″

Explain exactly how you measured the heights, and then answer each of the following:

   a. Are your measures valid?

   b. Are your measures reliable?

c. How does the variability in the measurements within each group compare to the difference between the two groups? For example, are all of your men taller than all of your women? Are they completely intermixed?

d. Do you think your measurements would convince an alien being that men are taller, on average, than women? Explain. Use your answer to part c as part of your explanation.

2. Design a survey with three questions to measure attitudes toward something of interest to you. Now design a new version by changing just a few words in each question to make it deliberately biased. Choose 20 people to whom you will administer the survey. Put their names in a hat (or a box or a bag) and draw out 10 names. Administer the first (unbiased) version of the survey to this group and the second (biased) version to the remaining 10 people. Compare the responses and discuss what happened.

3. Find a study that includes an emotion like "depression" or "happiness" as one of the measured variables. Explain how the researchers measured that emotion. Discuss whether the method of measurement is likely to produce valid measurements. Discuss whether the method of measurement is likely to produce reliable measurements.

References

Anastasi, Anne, and Susana Urbina. (1997). *Psychological testing*. 7th ed. New York: Macmillan.

Clark, H. H., and M. F. Schober (1992). "Asking questions and influencing answers," in *Questions about Questions*, J. M. Tanur (ed.), New York: Russell Sage Foundation, pp. 15–48.

Crossen, Cynthia. (1994). *Tainted truth*. New York: Simon and Schuster.

Graeff, Timothy R. (2007). Reducing Uninformed Responses: The Effects of Product-Class Familiarity and Measuring Brand Knowledge on Surveys. *Psychology and Marketing* 24(8), pp. 681–702.

Morin, Richard. (10–16 April 1995). What informed public opinion? *Washington Post*, National Weekly Edition.

Saad, Lydia (2013). Americans' Abortion Views Steady Amid Gosnell Trial. May 10, 2013, http://www.gallup.com/poll/162374/americans-abortion-views-steady-amid-gosnell-trial.aspx.

Schuman, H., and S. Presser. (1981). *Questions and answers in attitude surveys*. New York: Academic Press.

Schuman, H., and J. Scott. (22 May 1987). Problems in the use of survey questions to measure public opinion. *Science* 236, pp. 957–959.

CHAPTER **4**

How to Get a Good Sample

Thought Questions

1. What do you think is the major difference between a *survey* (such as a public opinion poll) and an *experiment* (such as the heartbeat experiment in Case Study 1.1)?

2. Suppose a properly chosen sample of 1600 people across the United States was asked if they regularly watch a certain television program, and 24% said yes. How close do you think that is to the percentage of the entire country who watch the show? Within 30%? 10%? 5%? 1%? Exactly the same?

3. Many television stations conduct polls by asking viewers to call or text one number if they feel one way about an issue and a different number if they feel the opposite. Do you think the results of such a poll represent the feelings of the community? Do you think they represent the feelings of all those watching the TV station at the time or the feelings of some other group? Explain.

4. Suppose you had a "yellow pages" telephone directory listing all the businesses in a county, alphabetized by type of business. If you wanted to phone 100 of them to get a representative sampling of opinion on some issue, how would you select which 100 businesses to phone? Why would it not be a good idea to simply use the first 100 businesses listed?

5. There are many professional polling organizations, such as Gallup and Roper. They often report on surveys they have done, announcing that they have sampled 1243 adults, or some such number. How do you think they select the people to include in their samples?

4.1   Common Research Strategies

In Chapters 1 to 3, we discussed scientific studies in general, without differentiating them by type. In this chapter and the next, we are going to look at proper ways to conduct specific types of studies. When you read the results of a scientific study, the first thing you need to do is determine which research strategy was used. You can then see whether or not the study used the proper methods for that strategy. In this chapter and the next, you will learn about potential difficulties and outright disasters that can befall each type of study, as well as some principles for executing them correctly. First, let's examine the common types of research strategies.

Sample Surveys

You are probably quite familiar with sample surveys, at least in the form of political and opinion polls. In a **sample survey,** a subgroup of a large population is questioned on a set of topics. The results from the subgroup are used as if they were representative of the larger population, which they will be if the sample was chosen correctly. There is no intervention or manipulation of the respondents in this type of research; they are simply asked to answer some questions. We examine sample surveys in more depth later in this chapter.

Randomized Experiments

An **experiment** measures the effect of manipulating the environment in some way. For example, the manipulation may include receiving a drug or medical treatment, going through a training program, following a special diet, and so on. In a **randomized experiment,** the manipulation is assigned to participants on a random basis. In Chapter 5, we will learn more about how this is done.

Most experiments on humans use volunteers because you can't force someone to accept a manipulation. You then measure the result of the feature being manipulated, called the **explanatory variable,** on an outcome, called the **outcome variable** or **response variable.** Examples of outcome variables are cholesterol level (after taking a new drug), amount learned (after a new training program), or weight loss (after a special diet).

As an example, recall Case Study 1.2, a randomized experiment that investigated the relationship between aspirin and heart attacks. The explanatory variable, manipulated by the researchers, was whether a participant took aspirin or a placebo. The variable was then used to help explain the outcome variable, which was whether a participant had a heart attack or not. Notice that the explanatory and outcome variables are both categorical in this case, with two categories each (aspirin/placebo and heart attack/no heart attack).

Randomized experiments are important because, unlike most other studies, they often allow us to determine cause and effect. The participants in an experiment are usually randomly assigned to either receive the manipulation or take part in a control group. The purpose of the random assignment is to make the two groups approximately equal in all respects except for the explanatory variable, which is purposely manipulated. Differences in the outcome variable between the groups, if large enough to rule out natural chance variability, can then be attributed to the manipulation of the explanatory variable.

For example, suppose we flip a coin to assign each of a number of new babies into one of two groups. Without any intervention, we should expect both groups to gain about the same amount of weight, on average. If we then expose one group to the sound of a heartbeat and that group gains significantly more weight than the other group, we can be reasonably certain that the weight gain was due to the sound of the heartbeat. Similar reasoning applies when more than two groups are used.

Observational Studies

As we noted in Chapter 1, an **observational study** resembles an experiment except that the manipulation occurs naturally rather than being imposed by the experimenter. For example, we can observe what happens to people's weight when they quit smoking, but we can't experimentally manipulate them to quit smoking. We must rely on naturally occurring events. This reliance on naturally occurring events leads to problems with establishing a causal connection because we can't arrange to have a similar control group. For instance, people who quit smoking may do so because they are on a "health kick" that also includes better eating habits, a change in coffee consumption, and so on. In this case, if we were to observe a weight loss (or gain) after cessation of smoking, we would not know if it were caused by the changes in diet or the lack of cigarettes. In an observational study, you cannot assume that the explanatory variable of interest to the researchers is the only one that may be responsible for any observed differences in the outcome variable.

A special type of observational study is frequently used in medical research. Called a **case-control study,** it is an attempt to include an appropriate control group. In Chapter 5, we will explore the details of how these and other observational studies are conducted, and in Chapter 6, we will cover some examples in depth.

Observational studies do have one advantage over experiments. Researchers are not required to induce artificial behavior. Participants are simply observed doing what they would do naturally; therefore, the results can be more readily extended to the real world.

Meta-Analyses

A **meta-analysis** is a quantitative review of a collection of studies all done on a similar topic. Combining information from various researchers may result in the emergence of patterns or effects that weren't conclusively available from the individual studies.

It is becoming quite common for results of meta-analyses to appear in the news. For example, a *New York Times* story about a meta-analysis was headlined, "Acupuncture Provides True Pain Relief in Study" (O'Conner, September 11, 2012). The article read, in part:

> *A new study of acupuncture—the most rigorous and detailed analysis of the treatment to date—found that it can ease migraines and arthritis and other forms of chronic pain. The findings provide strong scientific support for*

an age-old therapy used by an estimated three million Americans each year. Though acupuncture has been studied for decades, the body of medical research on it has been mixed and mired to some extent by small and poor-quality studies. Financed by the National Institutes of Health and carried out over about half a decade, the new research was a detailed analysis of earlier research that involved data on nearly 18,000 patients.

Later in the article, the study was explicitly called a meta-analysis, but it should be clear from the last sentence of the above quote that this is the case. Rather than collecting their own data, the study authors compiled and analyzed the results of numerous prior studies. This particular meta-analysis was unusual because the authors actually contacted the original researchers, obtained their data, and reanalyzed it. In most cases of meta-analysis, the researchers simply combine the results given in previously published reports. In Chapter 25, we will examine meta-analysis in more detail.

Case Studies

A **case study** is an in-depth examination of one or a small number of individuals. The researcher observes and interviews that individual and others who know about the topic of interest. For example, to study a purported spiritual healer, a researcher might observe her at work, interview her about techniques, and interview clients who had been treated by the healer. We do not cover case studies of this type because they are descriptive and do not require statistical methods. We will issue one warning, though: Be careful not to assume you can extend the findings of a case study to any person or situation other than the one studied. In fact, case studies may be used to investigate situations precisely because they are rare and unrepresentative.

EXAMPLE 4.1 Two Studies That Compared Diets

There are many claims about the health benefits of various diets, but it is difficult to test them because there are so many related variables. For instance, people who eat a vegetarian diet may be less likely to smoke than people who don't. Most studies that attempt to test claims about diet are observational studies.

For instance, News Story 20 in the Appendix, "Eating organic foods reduces pesticide concentration in children," is based on an observational study in which parents kept a food diary for their children for three days. Concentrations of various pesticides were then measured in the children's urine. The researchers compared the pesticide measurements for children who ate primarily organic produce and those who ate primarily conventional produce. They did find lower pesticide levels in the children who ate organic foods, but there is no way to know if the difference was the result of food choices, or if children who ate organic produce had less pesticide exposure in other ways. (The researchers did attempt to address this issue, as we will see when we revisit this example.) We will learn more about what can be concluded from this type of observational study in Chapter 5.

In contrast, News Story 3, "Rigorous veggie diet found to slash cholesterol," was based on a randomized experiment. The study used volunteers who were willing to be told what to eat during the month-long study. The volunteers were randomly assigned to one of three diet groups and reduction in cholesterol was measured and compared for the three groups at the end of the study. Because the participants were randomly assigned to the three diet groups, other variables that may affect cholesterol, such as weight or smoking behavior, should have been similar across all three groups.

This example illustrates that a choice can sometimes be made between conducting a randomized experiment and an observational study. An advantage of a randomized experiment is that factors other than the one being manipulated should be similar across the groups being compared. An advantage of an observational study is that people do what comes naturally to them. Knowing that a particular diet substantially reduces cholesterol doesn't help much if no one in the real world would follow the diet. We will re-visit these ideas in much greater depth in Chapter 5. ■

4.2 Defining a Common Language

In the remainder of this chapter, we explore the methods used in sample surveys. To make our discussion of sampling methods clear, let's establish a common language. As we have seen before, statisticians borrow words from everyday language and attach specialized meaning to them.

The first thing you need to know is that researchers sometimes speak synonymously of the individuals being measured and the measurements themselves. You can usually figure this out from the context. The relevant definitions cover both meanings.

- A **unit** is a single individual or object to be measured.
- The **population** (or **universe**) is the entire collection of units about which we would like information or the entire collection of measurements we would have if we could measure the whole population.
- The **sample** is the collection of units we actually measure or the collection of measurements we actually obtain.
- The **sampling frame** is a list of units from which the sample is chosen. Ideally, it includes the whole population.
- In a **sample survey,** measurements are taken on a subset, or sample, of units from the population.
- A **census** is a survey in which the entire population is measured.

EXAMPLE 4.2 Determining Monthly Unemployment in the United States

In the United States, the Bureau of Labor Statistics (BLS) is responsible for determining monthly unemployment rates. To do this, the BLS does not collect information on all adults; that is, it does not take a census. Instead, employees visit approximately 60,000 households, chosen from a list of all known households in the country, and obtain information on the approximately 110,000 adults living in them. They classify each person as employed, unemployed, or "not in the labor force." The last category includes the "discouraged workers" discussed in Chapter 3. The unemployment rate is the number of unemployed persons divided by the sum of the employed and unemployed. Those "not in the labor force" are not included at all. (See the U.S. Department of Labor's BLS *Handbook of Methods,* referenced at the end of the chapter, for further details, or visit their website, http://www.bls.gov.)

Before reading any further, try to apply the definitions you have just learned to the way the BLS calculates unemployment. In other words, specify the units, the population, the sampling frame, and the sample. Be sure to include both forms of each definition when appropriate. The *units* of interest to the BLS are adults in the labor force, meaning adults who meet their definitions of employed and unemployed. Those who are "not in the labor force" are not relevant units. The *population of units* consists of all adults who are in the labor force. The *population of measurements,* if we could obtain it, would consist of the employment status (working or not working) of everyone in the labor force. The *sampling frame* is the list of all known households in the country. The people who actually get asked about their employment status by the BLS constitute the *units in the sample,* and their actual employment statuses constitute the *measurements in the sample.*

4.3   The Beauty of Sampling

Here is some information that may astound you. If you use commonly accepted methods to sample 1500 adults from an entire population of millions of adults, you can estimate fairly accurately, to within 3%, the percentage of the entire population who have a certain trait or opinion. (There is nothing magical about 1500 and 3%, as you will soon see.) Even more amazing is the fact that this result doesn't depend on how big the population is (as long as it's much bigger than the sample); it depends only on how many are in the sample. Our sample of 1500 would do equally well at estimating, to within 3%, the percentage of a population of 10 million or 10 billion. Of course, you have to use a proper sampling method—but we address that later.

You can see why researchers are content to rely on public opinion polls rather than trying to ask everyone for their opinion. It is much cheaper to ask 1500 people than several million, especially when you can get an answer that is almost as accurate. It also takes less time to conduct a sample survey than a census, and because fewer interviewers are needed, there is better quality control.

Accuracy of a Sample Survey: Margin of Error

Most sample surveys are used to estimate the proportion or percentage of people who have a certain trait or opinion. For example, the Nielsen ratings, used to determine the percentage of American television sets tuned to a particular show, are based on a sample of a few thousand households. News organizations routinely conduct surveys of a few thousand people to determine public opinion on current topics of interest. As we have said, these surveys, if properly conducted, are amazingly accurate. The measure of accuracy is a number called the **margin of error.** The sample proportion and the population proportion differ by more than the margin of error less than 5% of the time, or in fewer than 1 in 20 surveys.

> *As a general rule, the amount by which the proportion obtained from the sample will differ from the true population proportion rarely exceeds 1 divided by the square root of the number in the sample.* This is expressed by the simple formula $1/\sqrt{n}$, where the letter n represents the number of people in the sample. To express results in terms of percentages instead of proportions, simply multiply everything by 100.

For example, with a sample of 1600 people, we usually get an estimate that is accurate to within $1/40 = 0.025$ or 2.5% of the truth because the square root of 1600 is 40. You might see results such as "Fifty-five percent of respondents support the president's economic plan. The margin of error for this survey is plus or minus 2.5 percentage points." This means that it is almost certain that between 52.5% and 57.5% of the entire population support the plan. In other words, add and subtract the margin of error to the sample value, and the resulting interval almost surely covers the true population value. If you were to follow this method every time you read the results of a properly conducted survey, the interval would only miss covering the truth about 1 in 20 times.

EXAMPLE 4.3   Measuring Teen Drug Use

In News Story 13 in the Appendix, "Three factors key for drug use in kids," the margin of error is provided at the end of the article, as follows:

> *QEV Analytics surveyed 1,987 children ages 12 to 17 and 504 parents. . . . The margin of error was plus or minus two percentage points for children and plus or minus four percentage points for parents.*

Notice that $n = 1987$ children were surveyed, so the margin of error is $1/\sqrt{1987} = 0.0224$, or about 2.2%. There were 504 parents interviewed, so the margin of error for their responses is about $1/\sqrt{504} = 0.0445$, or about 4.45%. These values were rounded off in the news story, to 2% and 4%, respectively. The more accurate values of 2.2% and 4.4% are given on page 30 in Original Source 13, along with an explanation of what they mean.

The margin of error can be applied to any percent reported in the study to estimate the percent of the *population* that would respond the same way. For instance, the news story reported that 20% of the children in the study said they could buy marijuana in an hour or less. Applying the margin of error, we can be fairly confident that somewhere between 18% and 22% of *all* teens in the *population* represented by those in this survey would respond that way if asked. Notice that the news story misinterprets this information when stating that "more than 5 million children ages

12 to17, or 20 percent, said they could buy marijuana in an hour or less." In fact, only a total of 1987 children were even asked! The figure of 5 million is a result of multiplying the percent of the *sample* who responded affirmatively (20%) by the total number of children in the *population* of 12- to 17-year-olds in the United States, presumably about 25 million at the time of the study. ∎

Other Advantages of Sample Surveys

When a Census Isn't Possible

Suppose you needed a laboratory test to see if your blood had too high a concentration of a certain substance. Would you prefer that the lab measure the entire population of your blood, or would you prefer to give a sample? Similarly, suppose a manufacturer of firecrackers wanted to know what percentage of its products were duds. It would not make much of a profit if it tested them all, but it could get a reasonable estimate of the desired percentage by testing a properly selected sample. As these examples illustrate, there are situations where measurements destroy the units being tested and thus a census is not feasible.

Speed

Another advantage of a sample survey over a census is the amount of time it takes to conduct a sample survey. For example, it takes several years to successfully plan and execute a census of the entire population of the United States. Getting monthly unemployment rates would be impossible with a census; the results would be quite out-of-date by the time they were released. It is much faster to collect a sample than a census if the population is large.

Accuracy

A final advantage of a sample survey is that you can devote your resources to getting the most accurate information possible from the sample you have selected. It is easier to train a small group of interviewers than a large one, and it is easier to track down a small group of nonrespondents than the larger one that would inevitably result from trying to conduct a census.

4.4 Simple Random Sampling

The ability of a relatively small sample to accurately reflect the opinions of a huge population does not happen haphazardly. It works only if proper sampling methods are used. Everyone in the population must have a specified chance of making it into the sample. Methods with this characteristic are called **probability sampling plans.**

The simplest way of accomplishing this goal is to use a **simple random sample.** With a simple random sample, every conceivable group of people of the required size has the same chance of being the selected sample.

To actually produce a simple random sample, you need two things. First, you need a list of the units in the population. Second, you need a source of **random numbers.**

Random numbers can be found in tables designed for that purpose, called "tables of random digits," or they can be generated by computers and calculators. If the population isn't too large, physical methods can be used, as illustrated in the next hypothetical example.

EXAMPLE 4.4 How to Sample from Your Class

Suppose you are taking a class with 400 students and are unhappy with the teaching method. To substantiate that a problem exists so that you can complain to higher powers, you decide to collect a simple random sample of 25 students and ask them for their opinions.

Notice that a sample of this size would have a margin of error of about 20% because $1/\sqrt{25} = 1/5 = 0.20$. Thus, the percentage of those 25 people who were dissatisfied would almost surely be within 20% of the percentage of the entire class who were dissatisfied. If 60% of the sample said they were dissatisfied, you could tell the higher powers that somewhere between 40% and 80% of the entire class was probably dissatisfied. Although that's not a very precise statement, it is certainly enough to show major dissatisfaction.

To collect your sample, you would proceed as follows:

Step 1: Obtain a list of the students in the class, numbered from 1 to 400.

Step 2: Obtain 25 random numbers between 1 and 400. One simple way to do this would be to write each of the numbers from 1 to 400 on equally sized slips of paper, put them in a bag, mix them very well, and draw out 25. However, we will instead use a free website, www.randomizer.org (Urbaniak and Plous, 2013), to select the 25 numbers. (In Chapter 15 you will learn the details of how to use the randomizer.org website.) Here are the results of asking for 25 numbers to be randomly drawn from the integers 1 to 400, put into numerical order for ease of reading:

```
6, 17, 24, 52, 81, 88, 116, 145, 153, 181, 193, 207, 216, 236,
255, 256, 268, 296, 301, 344, 345, 351, 364, 368, 384
```

Step 3: Locate and interview the people on your list whose numbers were selected. For instance, the first person in your sample would be person #6 on your list. Notice that it is important to try to locate the actual 25 people resulting from this process. For instance, if you contacted people by e-mail and gave up if they didn't respond to your first e-mail, you would bias your results toward people who are conscientious about answering e-mail. They might not have the same opinion of the class as those who don't readily respond to email. If you collected your sample correctly, as described, you would have legitimate data to present to the higher powers. ■

4.5   Other Sampling Methods

By now you may be asking yourself how polling organizations could possibly get a numbered list of all voters or of all adults in the country. In truth, they don't. Instead, they rely on more complicated sampling methods. Here, we describe a few other sampling methods, all of which are good substitutes for simple random sampling in most situations. In fact, they often have advantages over simple random sampling.

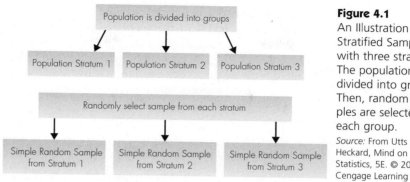

Figure 4.1
An Illustration of Stratified Sampling with three strata: The population is divided into groups. Then, random samples are selected from each group.
Source: From Utts and Heckard, Mind on Statistics, 5E. © 2015 Cengage Learning

Stratified Random Sampling

Sometimes the population of units falls into natural groups, called **strata** (the singular is "**stratum**"). For example, public opinion pollsters often take separate samples from each region of the country so they can spot regional differences as well as measure national trends. Political pollsters may sample separately from each political party to compare opinions by party.

A **stratified random sample** is collected by first dividing the population of units into groups (strata) and then taking a simple random sample from each. Figure 4.1 illustrates this sampling method. For example, the strata might be regions of the country or political parties. You can often recognize this type of sampling when you read the results of a survey because the results will be listed separately for each of the strata. Stratified sampling has other advantages besides the fact that results are available separately by strata. One is that different interviewers may work best with different people. For example, people from separate regions of the country (South, Northeast, and so on) may feel more comfortable with interviewers from the same region. It may also be more convenient to stratify before sampling. If we were interested in opinions of college students across the country, it would probably be easier to train interviewers at each college rather than to send the same interviewer to all campuses.

So far we have been focusing on the collection of categorical variables, such as opinions or traits people might have. Surveys are also used to collect measurement variables, such as age at first intercourse or number of cigarettes smoked per day. We are often interested in the population average for such measurements. The accuracy with which we can estimate the average depends on the natural variability among the measurements. The less variable they are, the more precisely we can assess the population average on the basis of the sample values. For instance, if everyone in a relatively large sample reports that his or her age at first intercourse was between 16 years 3 months and 16 years 4 months, then we can be relatively sure that the average age in the population is close to that. However, if reported ages range from 13 years to 25 years, then we cannot pinpoint the average age for the population nearly as accurately.

Stratified sampling can help to solve the problem of large natural variability. Suppose we could figure out how to stratify in a way that allowed little natural variability in the answers within each of the strata. We could then get an accurate estimate for each stratum and combine estimates to get a much more precise answer for the group than if we measured everyone together. For example, if we wanted to estimate the average weight gain of newborn babies during the first four days of life, we could do so more accurately by dividing the babies into groups based on their initial birth weight. Very heavy newborns actually tend to lose weight during the first few days, whereas very light ones tend to gain more weight.

Stratified sampling is sometimes used instead of simple random sampling for the following reasons:

1. We can find individual estimates for each stratum.
2. If the variable measured gives more consistent values within each of the strata than within the whole population, we can get more accurate estimates of the population values.
3. If strata are geographically separated, it may be cheaper to sample them separately.
4. We may want to use different interviewers or interviewing methods within each of the strata.

Cluster Sampling

Cluster sampling is often confused with stratified sampling, but it is actually a radically different concept and can be much easier to accomplish. Figure 4.2 illustrates how cluster sampling works. The population units are again divided into groups,

Figure 4.2
An Illustration of Cluster Sampling: The population is divided into clusters. Then, a random sample of clusters is selected.
Source: From Utts and Heckard, Mind on Statistics, 5E. © 2015 Cengage Learning

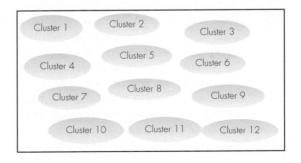

Randomly select clusters

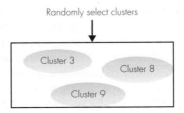

called **clusters,** but rather than sampling within each group, we select a random sample of clusters and measure only those clusters. One obvious advantage of cluster sampling is that you need only a list of clusters, instead of a list of all individual units.

For example, suppose we wanted to sample students living in the dormitories at a college. If the college had 30 dorms and each dorm had six floors, we could consider the 180 floors to be 180 clusters of units. We could then randomly select the desired number of floors and measure everyone on those floors. Doing so would probably be much cheaper and more convenient than obtaining a simple random sample of all dormitory residents.

If cluster sampling is used, the analysis must proceed differently because similarities may exist among the members of the clusters, and these must be taken into account. Numerous books are available that describe proper analysis methods based on which sampling plan was employed. (See, for example, *Sampling Design and Analysis, 2nd edition* by Sharon Lohr, Brooks-Cole/Cengage, 2010.)

Systematic Sampling

Suppose you had a list of 5000 names and e-mail addresses from which you wanted to select a sample of 100. That means you would want to select 1 of every 50 people on the list. The first idea that might occur to you is to simply choose every 50th name on the list. If you did so, you would be using a **systematic sampling plan.** With this plan, you divide the list into as many consecutive segments as you need, randomly choose a starting point in the first segment, then sample at that same point in each segment. In our example, you would randomly choose a starting point in the first 50 names, then sample every 50th name after that. When you were finished, you would have selected one person from each of 100 segments, equally spaced throughout the list.

Systematic sampling is often a good alternative to simple random sampling. In a few instances, however, it can lead to a biased sample, and common sense must be used to avoid those. As an example, suppose you were doing a survey of potential noise problems in a high-rise college dormitory. Further, suppose a list of residents was provided, arranged by room number, with 20 rooms per floor and two people per room. If you were to take a systematic sample of, say, every 40th person on the list, you would get people who lived in the same location on every floor—and thus a biased sampling of opinions about noise problems.

Random Digit Dialing

Most of the national polling organizations in the United States now use a method of sampling called **random digit dialing.** This method results in a sample that approximates a simple random sample of all adults in the United States who have telephones. Until about 2008, this technique consisted of a sophisticated method of randomly sampling landline telephone numbers and calling them. Now, polling organizations include cell phones in random digit dialing because many households do not have a landline. In early 2013, both the Gallup and Pew organizations used telephone samples that were about 60% landlines and 40% cell phones. In May of 2013, the Gallup organization announced that they were increasing the proportion of cell phones in their samples to 50%, after investigating why their polling results for the 2012 presidential election were not as accurate as they had been in the past. This split will probably change again in the near future as more people give up landlines.

Once the phone numbers to be included in the sample have been determined, the pollsters make multiple attempts to reach someone at those numbers. For landline phones, they often ask to speak to a specific type of individual, to make sure that all ages and both sexes are represented in proportion to the population of interest. For instance, they may ask to speak to the youngest adult male (18 or older), or they may ask to speak with the person who has had the most recent birthday. This method ensures that the person in each household who is most likely to answer the phone is not always the person interviewed. Also, younger adults are more likely to live in households that have no landline, so pollsters must take care to ensure that they are adequately represented in both the landline and cell phone samples.

For cell phones, the assumption is that only one person owns that number, so the pollsters only verify that the respondent is at least 18 years old before proceeding with the survey. One of the complications of including cell phones is that geographic data may not be accurate, so geographic information may need to be obtained during the interview. Unlike with landlines, when people move they often keep their cell phone number, so knowing someone's area code no longer ensures that the pollsters know where the person lives.

Polling organizations are constantly monitoring communication patterns to keep up with the trends in how people can be contacted and what methods are likely to get people to respond. It is quite likely that, in the near future, they will need to modify their methods again, to include texting, visits to residences, and other methods to help ensure that they obtain representative samples of the populations of interest.

EXAMPLE 4.5   Finding Teens and Parents Willing to Talk

The survey described in News Story 13 in the Appendix was conducted by telephone and Original Source 13 on the companion website describes in detail how the sample was obtained. The researchers started with an "initial pool of random telephone numbers" consisting of 94,184 numbers, which "represented all 48 continental states in proportion to their population, and were prescreened by computer to eliminate as many unassigned or nonresidential telephone numbers as possible" (p. 29).

Despite the prescreening, the initial pool of 94,184 numbers eventually resulted in only 1987 completed interviews! There is a detailed table of why this is the case on page 31 of the report. For instance, 12,985 of the numbers were "not in service." Another 25,471 were ineligible because there was no resident in the required age group, 12 to 17 years old. Another 27,931 refused to provide the information that was required to know whether the household qualified. Only 8597 were abandoned because of no answer, partly because "at least four call back attempts were made to each telephone number before the telephone number was rejected" (p. 29).

An important question is whether any of the reasons for exclusion were likely to introduce significant bias in the results. The report does address this question with respect to one reason, refusal on the part of a parent to allow the teen to participate:

> *While the refusal rate of parents, having occurred in 544 cases, seems modest, this represents the loss of 11 percent of other eligible households, which is substantial enough to have an impact on the achieved sample. This may be a contributing factor to the understatement of substance use rates, and to the underrepresentation of racial and ethnic populations. (p. 30)*

Multistage Sampling

Many large surveys, especially those that are conducted in person rather than by phone or e-mail, use a combination of the methods we have discussed. They might stratify by region of the country; then stratify by urban, suburban, and rural; and then choose a random sample of communities within those strata. They would then divide those communities into city blocks or fixed areas, as clusters, and sample some of those. Everyone on the block or within the fixed area may then be sampled. This is called a **multistage sampling plan.**

4.6 Difficulties and Disasters in Sampling

Difficulties

1. Using the wrong sampling frame
2. Not reaching the individuals selected
3. Having a low response rate

Disasters

1. Getting a volunteer or self-selected sample
2. Using a convenience or haphazard sample

In theory, designing a good sampling plan is easy and straightforward. However, the real world rarely cooperates with well-designed plans, and trying to collect a proper sample is no exception. Difficulties that can occur in practice need to be considered when you evaluate a study. If a proper sampling plan is never implemented, the conclusions can be misleading and inaccurate.

Difficulties in Sampling

Following are some problems that can occur even when a sampling plan has been well designed.

Using the Wrong Sampling Frame

Remember that the sampling frame is the list of the population of units from which the sample is drawn. Sometimes a sampling frame either will include unwanted units or exclude desired units. For example, using a list of registered voters to predict election outcomes includes those who are not likely to vote as well as those who are likely to do so. Using only landline telephones excludes cell phone users who don't have landlines.

Common sense can often lead to a solution for this problem. In the example of registered voters, interviewers may try to first ascertain the voting history of the person contacted by asking where he or she votes and then continuing the interview only if the person knows the answer. Instead of using only landlines, random digit dialing now includes cell phones. This solution still excludes those without any phone at all and those who don't answer unless they know who is calling, but it is more inclusive than relying on landlines alone.

Not Reaching the Individuals Selected

Even if a proper sample of units is selected, the units may not be reached. For example, *Consumer Reports* magazine mails a lengthy survey to its subscribers to obtain information on the reliability of various products. If you were to receive such a survey, and you had a close friend who had been having trouble with a highly rated automobile, you may very well decide to pass the questionnaire on to your friend to answer. That way, he would get to register his complaints about the car, but *Consumer Reports* would not have reached the intended recipient.

Landline telephone surveys tend to reach a disproportionate number of women because they are more likely to answer the phone. To try to counter that problem, researchers sometimes ask to speak to the oldest adult male at home. Surveys are also likely to have trouble contacting people who work long hours and are rarely home or those who tend to travel extensively.

In recent years, news organizations have been pressured to produce surveys of public opinion quickly. When a controversial story breaks, people want to know how others feel about it. This pressure results in what *Wall Street Journal* reporter Cynthia Crossen calls "quickie polls." As she notes, these are "most likely to be wrong because questions are hastily drawn and poorly pretested, and it is almost impossible to get a random sample in one night" (Crossen, 1994, p. 102). Even with the computer randomly generating phone numbers for the sample, many people are not likely to be available by phone that night—and they may have different opinions from those who are likely to be available. Most responsible reports about polls include information about the dates during which they were conducted. If a poll was done in one night, beware! It is important that once a sample has been selected, those individuals are the ones who are actually measured. It is better to put resources into getting a smaller sample than to get one that has been biased because the survey takers moved on to the next person on the list when a selected individual was initially unavailable.

Having a Low Response Rate

Even the best surveys are not able to contact everyone on their list, and not everyone contacted will respond. The General Social Survey (GSS), run by the prestigious

National Opinion Research Center (NORC) at the University of Chicago, conducts surveys on a wide variety of topics every year or two. The highest response rate they ever achieved was 82.4% in 1993, which they noted was "testimony to the extraordinary skill and dedication of the NORC field staff." The response rate has dropped considerably since then, and in 2006 (the latest year with available data), it was only 71.2%. (*Source*: http://publicdata.norc.org:41000/gss/.%5CDocuments%5CCodebook%5CA.pdf, accessed May 13, 2013.)

Beyond having a dedicated staff, not much can be done about getting everyone in the sample to respond. Response rates should simply be reported in research summaries. As a reader, remember that the lower the response rate, the less the results can be generalized to the population as a whole. Responding to a survey (or not) is voluntary, and those who respond are likely to have stronger opinions than those who do not.

With mail or e-mail surveys, it may be possible to compare those who respond immediately with those who need a second prodding, and in phone surveys you could compare those who answer on the first try with those who require numerous callbacks. If those groups differ on the measurement of interest, then those who were never reached are probably different as well.

In a mail survey, it is best not to rely solely on "**volunteer response**." In other words, don't just accept that those who did not respond the first time can't be cajoled into it. Often, sending a reminder with a brightly colored stamp or following up with a personal phone call will produce the desired effect. Surveys that simply use those who respond voluntarily are sure to be biased in favor of those with strong opinions or with time on their hands.

EXAMPLE 4.6   ## Which Scientists Trashed the Public?

According to a poll taken among scientists and reported in the prestigious journal *Science* (Mervis, 1998), scientists don't have much faith in either the public or the media. The article reported that, based on the results of a "recent survey of 1400 professionals" in science and in journalism, 82% of scientists "strongly or somewhat agree" with the statement, "the U.S. public is gullible and believes in miracle cures or easy solutions," and 80% agreed that "the public doesn't understand the importance of federal funding for research." About the same percentage (82%) also trashed the media, agreeing with the statement "the media do not understand statistics well enough to explain new findings." It isn't until the end of the article that we learn who responded: "The study reported a 34% response rate among scientists, and the typical respondent was a white, male physical scientist over the age of 50 doing basic research." Remember that those who feel strongly about the issues in a survey are the most likely to respond. With only about a third of those contacted responding, it is inappropriate to generalize these findings and conclude that most scientists have so little faith in the public and the media. This is especially true because we were told that the respondents represented only a narrow subset of scientists. ∎

Disasters in Sampling

A few sampling methods are so bad that they don't even warrant a further look at the study or its results.

Getting a Volunteer or Self-Selected Sample

Although relying on volunteer *responses* presents somewhat of a difficulty in determining the extent to which surveys can be generalized, relying on a volunteer *sample* is a complete waste of time. If a magazine, website or television station runs a survey and asks any readers or viewers who are interested to respond, the results reflect only the opinions of those who decide to volunteer. As noted earlier, those who have a strong opinion about the question are more likely to respond than those who do not. Thus, the responding group is simply not representative of any larger group. Most media outlets now acknowledge that such polls are "unscientific" when they report the results, but most readers are not likely to understand how misleading the results can be. The next example illustrates the contradiction that can result between a scientific poll and one relying solely on a volunteer sample.

EXAMPLE 4.7   A Meaningless Poll

On February 18, 1993, shortly after Bill Clinton became president of the United States, a television station in Sacramento, California, asked viewers to respond to the question: "Do you support the president's economic plan?" The next day, the results of a properly conducted study asking the same question were published in the newspaper. Here are the results:

	Television Poll (Volunteer sample)	Survey (Random sample)
Yes (support plan)	42%	75%
No (don't support plan)	58%	18%
Not sure	0%	7%

As you can see, those who were dissatisfied with the president's plan were much more likely to respond to the television poll than those who supported it, and no one who was "Not sure" called the television station because they were not invited to do so. Trying to extend those results to the general population is misleading. It is irresponsible to publicize such studies, especially without a warning that they result from an unscientific survey and are not representative of general public opinion. You should never interpret such polls as anything other than a count of who bothered to go to the telephone and call.   ■

Using a Convenience or Haphazard Sample

Another sampling technique that can produce misleading results for surveys is to use the most convenient group available or to decide on the spot who to sample. In many cases, the group is not likely to represent any larger population for the information measured. In some cases, the respondents may be similar enough to a population of interest that the results can be extended, but extreme caution should be used in deciding whether this is likely to be so.

Many of the research findings in psychology are based on studies using undergraduate psychology students as the participants. In fact, this phenomenon is so common that psychologist Charles Tart (personal communication) once joked that "psychology is the study of current psychology students by former psychology students for the benefit of future psychology students." Data from surveys in introductory statistics classes are often used for examples in statistics textbooks. Students in introductory psychology or statistics classes may be representative of all students at a university, or even all people in their age group in certain ways, but not in others. For instance, they may represent others in their age group on how they respond in experiments on perception, but may not represent them on issues like how many hours they spend on the Internet per day.

EXAMPLE 4.8   ### Haphazard Sampling Finds Ignorant Students

A few years ago, the student newspaper at a California university announced as a front page headline: "Students ignorant, survey says." The article explained that a "random survey" indicated that American students were less aware of current events than international students were. However, the article quoted the undergraduate researchers, who were international students themselves, as saying that "the students were randomly sampled on the quad." The quad is an open-air, grassy area where students relax, eat lunch, and so on. There is simply no proper way to collect a random sample of students by selecting them in an area like that. In such situations, the researchers are likely to approach people who they think will support the results they intended for their survey. Or, they are likely to approach friendly looking people who appear as though they will easily cooperate. This is called a haphazard sample, and it cannot be expected to be representative at all. ■

You have seen the proper way to collect a sample and have been warned about the many difficulties and dangers inherent in the process. We finish the chapter with a famous example that helped researchers learn some of these pitfalls.

CASE STUDY 4.1   ## The Infamous *Literary Digest* Poll of 1936

Before the election of 1936, a contest between Democratic incumbent Franklin Delano Roosevelt and Republican Alf Landon, the magazine *Literary Digest* had been extremely successful in predicting the results in U.S. presidential elections. But 1936 turned out to be the year of its downfall, when it predicted a 3-to-2 victory for Landon. To add insult to injury, young pollster George Gallup, who had just founded the American Institute of Public Opinion in 1935, not only correctly predicted Roosevelt as the winner of the election, he also predicted that the *Literary Digest* would get it wrong. He did this before the magazine even conducted its poll. And Gallup surveyed only 50,000 people, whereas the *Literary Digest* sent questionnaires to 10 million people (Freedman, Pisani, Purves, and Adhikari, 1991, p. 307).

The *Literary Digest* made two classic mistakes. First, the lists of people to whom it mailed the 10 million questionnaires were taken from magazine subscribers, car owners, telephone directories, and, in just a few cases, lists of registered voters.

(Continued)

In 1936, those who owned telephones or cars, or subscribed to magazines, were more likely to be wealthy individuals who were not happy with the Democratic incumbent. The sampling frame did not match the population of interest.

Despite what many accounts of this famous story conclude, the bias produced by the more affluent list was not likely to have been as severe as the second problem (Bryson, 1976). *The main problem was a low response rate.* The magazine received 2.3 million responses, a response rate of only 23%. Those who felt strongly about the outcome of the election were most likely to respond. And that included a majority of those who wanted a change, the Landon supporters. Those who were happy with the incumbent were less likely to bother to respond.

Gallup, however, knew the value of random sampling. He was able not only to predict the election but to predict the results of the *Literary Digest* poll within 1%. How did he do this? According to Freedman and colleagues (1991, p. 308), "he just chose 3000 people at random from the same lists the *Digest* was going to use, and mailed them all a postcard asking them how they planned to vote." This example illustrates the beauty of random sampling and the idiocy of trying to base conclusions on nonrandom and biased samples. The *Literary Digest* went bankrupt the following year, and so never had a chance to revise its methods. The organization founded by George Gallup has flourished, although not without making a few sampling blunders of its own (see, for example, Exercise 28).   ■

Thinking About Key Concepts

- Surveys, randomized experiments, observational studies, and meta-analyses are all common types of statistical studies. To interpret the results of a study properly, you need to know which type of study was conducted.
- For large surveys that measure the proportion of respondents with a particular trait or opinion, we can estimate the proportion of the entire population with that trait or opinion with a reasonable degree of accuracy. Most of the time, the population proportion will be within one "margin of error" of the sample proportion. The margin of error is found as $1/\sqrt{n}$, where n is the number of individuals in the sample. To convert proportions to percents, multiply everything by 100.
- The best way to make sure you get a representative sample from a larger population is to take a simple random sample, which is one in which all groups of the desired size have the same chance of being the sample. Other "probability sampling plans" or methods include stratified sampling, cluster sampling, and systematic sampling. Each of these types has advantages and disadvantages, but all of them are better than haphazard sampling.
- Once individuals have been chosen for the sample, it's very important to try to get those individuals to participate. If you substitute other individuals you might get biased results, because you will have underrepresented the people who are more difficult to reach. They might differ in important ways from people who are easy to reach.

• Polls in which people voluntarily include themselves in the sample are usually worthless, because only people who feel strongly about the issue will volunteer to participate. These "volunteer, self-selected *samples*" are even more of a problem than "volunteer *response*." In the latter case, the sample is properly selected, but not everyone chooses to respond.

Exercises

Exercises with numbers divisible by 3 (3, 6, 9, etc.) are included in the Solutions at the back of the book. They are marked with an asterisk (*).

1. Patients who visit a clinic to help them stop smoking are given a choice of two treatments: undergoing hypnosis or applying nicotine patches. The percentages who quit smoking are compared for the two methods. Is this study a survey, an experiment, an observational study, or a case study? Explain your reasoning.

2. A large company wants to compare two incentive plans for increasing sales. The company randomly assigns a number of its sales staff to receive each kind of incentive and compares the average change in sales of the employees under the two plans. Is this study a survey, an experiment, an observational study, or a case study? Explain your reasoning.

*3. A doctor claims to be able to cure migraine headaches. A researcher administers a questionnaire to each of the patients the doctor claims to have cured. Is this study a survey, an experiment, an observational study, or a case study? Explain your reasoning.

4. The U.S. government uses a multitude of surveys to measure opinions, behaviors, and so on. Yet, every 10 years it takes a census.

What can the government learn from a census that it could not learn from a sample survey?

5. Each of the following quotes is based on the results of an experiment or an observational study. Explain which was used. If an observational study was used, explain whether an experiment could have been used to study the topic instead.

a. "A recent Stanford study of more than 6000 men found that tolerance for exercise (tested on a treadmill) was a stronger predictor of risk of death than high blood pressure, smoking, diabetes, high cholesterol and heart disease" (Kalb, 2003, p. 64).

b. "In a three-month study, researchers randomly assigned 250 black men and women to one of four groups: one received a placebo and the others received 1000, 2000, or 4000 international units of vitamin D daily [to see if vitamin D helps reduce hypertension in African-Americans]. The effect was modest." (*New York Times*, March 19, 2013, p. D4).

*6. Explain whether a survey or a randomized experiment would be most appropriate to find out about each of the following:

*a. Who is likely to win the next presidential election

*b. Whether the use of nicotine gum reduces cigarette smoking

7. Explain whether a survey or a randomized experiment would be most appropriate to find out about each of the following:

a. Whether there is a relationship between height and happiness

b. Whether a public service advertising campaign has been effective in promoting the use of condoms

8. Refer to Case Study 1.1, "Heart or Hypothalamus?" in which Lee Salk exposed newborn infants to the sound of a heartbeat or to silence, and then measured their weight gain at the end of 4 days.

a. What is the sample of units for this study?

b. What two variables did Salk measure for each unit in the sample? (Hint: He imposed one of these two variables and measured the other one.)

c. To what population of units do you think the results apply?

*9. Refer to Case Study 1.1, "Heart or Hypothalamus?" in which Lee Salk exposed newborn infants to the sound of a heartbeat or to silence, and then measured their weight gain at the end of 4 days.

*a. What is the explanatory variable in Salk's study?

*b. What is the outcome (response) variable in Salk's study?

10. Explain the difference between a proportion and a percentage as used to present the results of a sample survey. Include an explanation of how you would convert results from one form to the other.

11. The *Sacramento Bee* (11 Feb. 2001, p. A20) reported on a *Newsweek* poll that was based on interviews with 1000 adults, asking questions about a variety of issues.

a. What is the margin of error for this poll?

b. One of the statements in the news story was *"a margin of error of plus or minus three percentage points means that the 43 percent of Americans for and the 48 percent of Americans against oil exploration in Alaska's Arctic National Wildlife Refuge are in a statistical dead heat."* Explain what is meant by this statement.

*12. In the March 8, 1994, edition of the *Scotsman*, a newspaper published in Edinburgh, Scotland, a headline read, "Reform study finds fear over schools." The article described a survey of 200 parents who had been asked about proposed education reforms and indicated that most parents felt uninformed and thought the reforms would be costly and unnecessary. The report did not clarify whether a random sample was chosen, but make that assumption in answering the following questions.

*a. What is the margin of error for this survey?

*b. It was reported that "about 80 percent added that they were satisfied with the current education set-up in Scotland." What is the range of values that almost certainly covers the percentage of the population of parents who were satisfied?

*c. The article quoted Lord James Douglas-Hamilton, the Scottish education minister, as saying, "If you took a similar poll in two years' time, you would have a different result." Comment on this statement.

13. The student newspaper at a university in California reported a debate between two student council members, revolving around a survey of students (*California Aggie*, 8 November 1994, p. 3). The newspaper reported that "according to an AS [Associated Students] Survey Unit poll, 52 percent of the students surveyed said they opposed a diversity requirement." The report said that one council member "claimed that the roughly 500 people

polled were not enough to guarantee a statistically sound cross section of the student population." Another council member countered by saying that "three percent is an excellent random sampling, so there's no reason to question accuracy." (Note that the 3% figure is based on the fact that there were about 17,000 undergraduate students currently enrolled at that time.)

a. Comment on the remark attributed to the first council member, that the sample size is not large enough to "guarantee a statistically sound cross section of the population." Is the *size* of the sample the relevant issue to address his concern?

b. Comment on the remark by the second council member that "three percent is an excellent random sampling, so there's no reason to question accuracy." Is she correct in her use of terminology and in her conclusion?

c. Assuming a random sample was used, produce an interval that almost certainly covers the true percentage of the population of students who opposed the diversity requirement. Use your result to comment on the debate. In particular, do these results allow a conclusion as to whether the majority of students on campus opposed the requirement?

14. A *Washington Post*/Kaiser Family Foundation poll taken July 25 to August 5, 2012, asked a random sample of 3130 adults: "Do you support or oppose putting a special tax on junk food—that is, things like soda, chips, and candy—and using the money for programs to fight obesity?" They reported that 53% of the respondents said "Support" and that the margin of error for the poll was ± 2 percentage points. (Source: http://www.pollingreport.com/food.htm, accessed May 13, 2013.)

a. Verify that the reported margin of error of ± 2 percentage points is correct (rounded to a whole number of percentage points).

b. What is the numerical interval that is likely to cover the true percentage of the entire population that would have answered "Support" if asked at that time?

c. Refer to the seven pitfalls in Chapter 3 (listed on page 42) that might be encountered when asking questions in a survey. Which one of them might have been a problem for this survey question? Explain.

*15. Refer to Exercise 14. Based on these sample results, are you convinced that a majority of the population (that is, over 50%) at that time supported putting a special tax on junk food? Provide a numerical argument to support your answer.

16. Make a list of 20 people you know. Go to the website www.randomizer.org, and use it to choose a simple random sample of five people from your list.

a. Explain what you did, and give your results by listing the numbers corresponding to the people selected.

b. Now use the randomizer website to draw a systematic sample of five people from your list of 20. Give your results (by listing the numbers).

c. Would the sample you chose in part (a) have been a possible sample in part (b)? Explain.

d. Would the sample you chose in part (b) have been a possible sample in part (a)? Explain.

17. An article in the *Sacramento Bee* (12 January 1998, p. A4) was titled "College freshmen show conservative side" and reported the results of a fall 1997 survey "based on responses from a representative sample of 252,082 full-time freshmen at 464 two- and four-year colleges and universities nationwide." The article did not explain how the schools or students were selected.

a. For this survey, explain what a unit is, what the population is, and what the sample is.

b. Assuming a random sample of students was selected at each of the 464 schools, what type of sample was used in this survey? Explain.

*18. Refer to Exercise 17, in which students were chosen from 464 colleges to participate in a survey.

*a. Suppose that the 464 schools were randomly selected from all eligible colleges and universities and that *all* first-year students at those schools were surveyed. Explain what type of sample was used in the survey.

*b. Why would the sampling method described in part (a) have been simpler to implement than a simple random sample of all first-year college students in the United States?

19. Refer to Exercise 17, in which students were chosen from 464 colleges to participate in a survey. Suppose you were designing this study yourself. Visit the website http://www.utexas.edu/world/univ/alpha/ to see a list of colleges and universities in the United States. There are approximately 2050 schools listed.

 a. Suppose you wanted to select a random sample of 464 of these schools. Explain how you would use the website www.randomizer.org to select your random sample of schools.

 b. Use the method you described in part (a) to select the random sample. List the first three schools in your sample.

 c. Now suppose you wanted to take a systematic sample of 205 schools from the list of 2050 schools. Explain how you would use the website www.randomizer.org to do so.

 d. Use the method you described in part (c) to select the systematic random sample. List the first two schools in your sample.

20. Give an example in which:

 a. A sample would be preferable to a census

 b. A cluster sample would be the easiest method to use

 c. A systematic sample would be the easiest to use and would not be biased

*21. For each of the following situations, state which type of sampling plan was used. Explain whether you think the sampling plan would result in a biased sample.

 *a. To survey the opinions of its customers, an airline company made a list of all its flights and randomly selected 25 flights. All of the passengers on those flights were asked to fill out a survey.

*b. A pollster interested in opinions on gun control divided a city into city blocks, then surveyed the third house to the west of the southeast corner of each block. If the house was divided into apartments, the westernmost ground floor apartment was selected. The pollster conducted the survey during the day, but left a notice for those who were not at home to phone her so she could interview them.

*c. To learn how its employees felt about higher student fees imposed by the legislature, a university divided employees into three categories: staff, faculty, and student employees. A random sample was selected from each group and they were telephoned and asked for their opinions.

*d. A large store wanted to know if consumers would be willing to pay slightly higher prices to have computers available throughout the store to help them locate items. The store posted an interviewer at the door and told her to collect a sample of 100 opinions by asking the next person who came in the door each time she had finished an interview.

22. Refer to the previous exercise. Specify the population and the sample, being sure to include both units and measurements, for the situation described in

 a. Exercise 21a

 b. Exercise 21b

 c. Exercise 21c

 d. Exercise 21d

23. What role does natural variability play when trying to determine the population average of a measurement variable from a sample? (*Hint*: Read the section on stratified sampling.)

*24. Is using a convenience sample an example of a probability sampling plan? Explain why or why not.

25. In early September, 2003, California's Governor Gray Davis approved a controversial law allowing people who were not legal residents to obtain a California state driver's license. That week the

California Field Poll released a survey showing that 59% of registered voters opposed the law and 34% supported it. This part of the survey was based on a random sample of just over 300 people.

a. What is the approximate margin of error for the Field Poll results?

b. Provide an interval that is likely to cover the true percentage of registered California voters who supported the law.

26. Refer to the previous exercise. The same week that the Field Poll was released a Web site called SFGate.com (http://www.sfgate.com/polls/) asked visitors to "Click to vote" on their preferred response to "Agree with new law allowing drivers' licenses for illegal immigrants?" The choices and the percent who chose them were "Yes, gesture of respect, makes roads safe" 19%, "No, thwarts immigration law, poses security risk" 79%, and "Oh, great, another messy ballot battle" 2%. The total number of votes shown was 2900.

a. What type of sample was used for this poll?

b. Explain likely reasons why the percent who supported the law in this poll (19%) differed so much from the percent who supported it in the Field Poll (34%).

***27.** Refer to Exercises 25 and 26. Which result, the one from the SFGate poll or from the Field Poll, do you think was more likely to represent the opinion of the population of registered California voters at that time? Explain.

28. Despite his success in 1936, George Gallup failed miserably in trying to predict the winner of the 1948 U.S. presidential election. His organization, as well as two others, predicted that Thomas Dewey would beat incumbent Harry Truman. All three used what is called "quota sampling." The interviewers were told to find a certain number, or quota, of each of several types of people. For example, they might have been told to interview six women under age 40, one of whom was black and the other five of whom were white. Imagine that you are one of their interviewers trying to follow these instructions. Who would you ask?

Now explain why you think these polls failed to predict the true winner and why quota sampling is not a good method.

29. Explain why the main problem with the *Literary Digest* poll is described as "low response rate" and not "volunteer sample."

***30.** Suppose the administration at your school has hired you to help with a survey to find out how students feel about opening a new recreation center on campus. They provide you with a random sample of 100 students and a phone number for each one. Some of them are landlines, and some are cell phones. You call the 100 numbers and interview the person who answers the phone. Explain which one of the "difficulties and disasters" in sampling you are most likely to encounter and how it could bias your results.

31. Gastwirth (1988, p. 507) describes a court case in which Bristol-Myers was ordered by the Federal Trade Commission to stop advertising that "twice as many dentists use Ipana [toothpaste] as any other dentifrice" and that more dentists recommended it than any other dentifrice. Bristol-Myers had based its claim on a survey of 10,000 randomly selected dentists from a list of 66,000 subscribers to two dental magazines. They received 1983 responses, with 621 saying they used Ipana and only 258 reporting that they used the second most popular brand. As for the recommendations, 461 respondents recommended Ipana, compared with 195 for the second most popular choice.

a. Specify the sampling frame for this survey, and explain whether you think "using the wrong sampling frame" was a difficulty here, based on what Bristol-Myers was trying to conclude.

b. Of the remaining four "difficulties and disasters in sampling" listed in Section 4.6 (other than "using the wrong sampling frame"), which do you think was the most serious in this case? Explain.

c. What could Bristol-Myers have done to improve the validity of the results after it had mailed the 10,000 surveys and received 1983 back? Assume the company kept track of who had responded and who had not.

32. Suppose that a gourmet food magazine wants to know how its readers feel about serving beer with various types of food. The magazine sends surveys to 1000 randomly selected readers. Explain which one of the "difficulties and disasters" in sampling the magazine is most likely to face.

*33. Explain the difference between a low response rate and a volunteer sample. Explain which is worse, and why.

34. Find a news article describing a survey that is obviously biased. Explain why you think it is biased.

35. Construct an example in which a systematic sampling plan would result in a biased sample.

 For Exercises 36 to 38, locate the News Story in the Appendix and Original Source on the companion website. In each case, consult the Original Source and then explain what type of sample was used. Then discuss whether you think the results can be applied to any larger population on the basis of the type of sample used.

*36 Original Source 2: "Development and initial validation of the Hangover Symptoms Scale: Prevalence and correlates of hangover symptoms in college students."

37. Original Source 7: "Auto body repair inspection pilot program: Report to the legislature."

38. "Control subjects" in Original Source 12b: "Night shift work, light at night, and risk of breast cancer."

Mini-Projects

1. One interesting application of statistics is in trying to identify who wrote important historical works that were published using pseudonyms. A classic paper on this topic is "On Sentence-Length as a Statistical Characteristic of Style in Prose: With Application to Two Cases of Disputed Authorship" by G. Udny Yule, published in the journal *Biometrika* (Vol. 30, No. 3/4 Jan., 1939, pp. 363–390). In that paper, Yule identified the length of sentences as a feature that tends to remain consistent across written works by the same author. For this project, you are going to figure out how to estimate the distribution of sentence lengths for a book of your choosing. Find a book that is mainly text and relatively uncluttered with pictures, etc. Choose a random sample of 20 sentences from throughout the book and count how many words they have in them. Use two different sampling methods, chosen from simple random sampling, stratified sampling, cluster sampling, or systematic sampling. In each case, make a table showing how many sentences there were of each length (two words, three words, etc.).

    a. Explain exactly how you chose your samples.

    b. Explain which of your two methods was easier to use.

    c. Of the methods you did not use, explain which of them would have been the most difficult to use.

    d. Do you think either of your methods produced biased results? Explain.

    e. Report your results for each of the two methods. Are they generally in agreement with each other?

f. Do you think that average sentence length would be as good an indicator of authorship as listing all of the different sentence lengths and the proportion of time each one occurred? Explain.

2. Go to a large parking lot or a large area where bicycles are parked. Choose a color or a manufacturer. Design a sampling scheme you can use to estimate the percentage of cars or bicycles of that color or model. In choosing the number to sample, consider the margin of error that will accompany your sample result. Now go through the entire area, *actually taking a census,* and compute the population percentage of cars or bicycles of that type.

a. Explain your sampling method and discuss any problems or biases you encountered in using it.

b. Construct an interval from your sample that almost surely covers the true population percentage with that characteristic. Does your interval cover the true population percentage you found when you took the census?

c. Use your experience with taking the census to name one practical difficulty with taking a census. (*Hint:* Did all the cars or bicycles stay put while you counted?)

References

Bryson, M. C. (1976). The *Literary Digest* poll: Making of a statistical myth. *American Statistician* 30, pp. 184–185.

Crossen, Cynthia. (1994). *Tainted truth.* New York: Simon and Schuster.

Freedman, D., R. Pisani, R. Purves, and A. Adhikari. (1991). *Statistics,* 2d ed. New York: W. W. Norton.

Gastwirth, Joseph L. (1988). *Statistical reasoning in law and public policy.* Vol. 2. *Tort law, evidence and health.* Boston: Academic Press.

Kalb, Claudia (2003). Health for life. *Newsweek,* January 20, 2003, pp. 60–64.

Mervis, Jeffrey (1998). Report deplores science–media gap. *Science* 279, p. 2036.

O'Conner, Anahad (2012). Acupuncture provides true pain relief in study. *New York Times,* online article published September 11, 2012, accessed May 13, 2013; http://well.blogs.nytimes.com/2012/09/11/acupuncture-provides-true-pain-relief-in-study/.

Urbaniak, G.C. and S. Plous (2013). Research Randomizer (Version 4.0) [Computer software.] Retrieved on June 28, 2013 from http://www.randomizer.org.

U.S. Department of Labor, Bureau of Labor Statistics (March 2013). *BLS handbook of methods.*

CHAPTER 5

Experiments and Observational Studies

Thought Questions

1. In conducting a study to relate two conditions (activities, traits, and so on), researchers often define one of them as the *explanatory variable* and the other as the *outcome* or *response variable*. In a study to determine whether surgery or chemotherapy results in higher survival rates for a certain type of cancer, whether the patient survived is one variable, and whether the patient received surgery or chemotherapy is the other. Which is the explanatory variable and which is the response variable?

2. In an experiment, researchers assign "treatments" to participants, whereas in an observational study, they simply observe what the participants do naturally. Give an example of a situation in which an experiment would not be feasible for ethical reasons.

3. Suppose you are interested in determining whether a daily dose of vitamin C helps prevent colds. You recruit 20 volunteers to participate in an experiment. You want half of them to take vitamin C and the other half to agree not to take it. You ask them each which they would prefer, and 10 say they would like to take the vitamin and the other 10 say they would not. You ask them to record how many colds they get during the next 10 weeks. At the end of that time, you compare the results reported from the two groups. Give three reasons why this is not a good experiment.

4. When experimenters want to compare two treatments, such as an old and a new drug, they use *randomization* to assign the participants to the two conditions. If you had 50 people participate in such a study, how would you go about randomizing them? Why do you think randomization is necessary? Why shouldn't the experimenter decide which people should get which treatment?

5. "Graduating is good for your health," according to a headline in the *Boston Globe* (3 April 1998, p. A25). The article noted, "According to the Center for Disease Control, college graduates feel better emotionally and physically than do high school dropouts." Do you think the headline is justified based on this statement? Explain why or why not.

5.1   Defining a Common Language

In this chapter, we focus on studies that attempt to detect relationships between variables. In addition to the examples seen in earlier chapters, some of the connections we examine in this chapter and the next include a relationship between baldness and heart attacks in men, between smoking during pregnancy and subsequent lower IQ in the child, between practicing meditation and scoring higher on exams, and between handedness and age at death. We will see that some of these connections are supported by properly conducted studies, whereas other connections are not as solid.

Explanatory Variables, Response Variables, and Treatments

In most studies, we imagine that if there is a causal relationship, it occurs in a particular direction. For example, if we found that left-handed people die at a younger age than right-handed people, we could envision reasons why their handedness might be responsible for the earlier death, such as accidents resulting from living in a right-handed world. It would be more difficult to argue that they were left-handed because they were going to die at an earlier age.

Explanatory Variables versus Response Variables

We define an **explanatory variable** to be one that attempts to explain or is purported to cause (at least partially) differences in a **response variable** (sometimes called an **outcome variable**). In the previous example, handedness would be the explanatory variable and age at death the response variable. In the Salk experiment described in Chapter 1, whether the baby listened to a heartbeat was the explanatory variable and weight gain (in grams) was the response variable. In a study comparing chemotherapy to surgery for cancer, the medical treatment is the explanatory variable and surviving (usually measured as surviving for 5 years) or not surviving is the response variable. Many studies have more than one explanatory variable for each response variable, and there may be multiple response variables. The goal is to relate one or more explanatory variables to each response variable.

Usually we can distinguish which variable is which, but occasionally we examine relationships in which there is no conceivable direct causal connection in either direction. An example is the apparent relationship between baldness and heart attacks discussed in Case Study 5.4 on page 105. Because the level of baldness was measured at the time of the heart attack, the heart attack could not have caused the baldness. It would be farfetched to assume that baldness results in such stress that men are led to have heart attacks. Instead, a third variable, such as testosterone level, may be causing both the baldness and the heart attack. In such cases, we simply refer to the variables generically and do not assign one to be the explanatory variable and one to be the response variable.

Treatments

Sometimes the explanatory variable takes the form of a manipulation applied by the experimenter, such as when Salk played the sound of a heartbeat for some of the babies. A **treatment** is one or a combination of categories of the explanatory variable(s) assigned by the experimenter. The plural term *treatments* incorporates a collection of conditions, each of which is one treatment. In Salk's experiment, there were two treatments: Some babies received the heartbeat treatment and others received the silent treatment.

For the study described in News Story 1, summarized in the Appendix, some participants were assigned to follow an 8-week meditation regime and the others were not. The two treatments were "meditation routine" and "control," where the control group was measured for the response variables at the same times as the meditation group. Response variables were brain electrical activity and immune system functioning. The goal was to ascertain the effect of meditation on these response variables. This study is explored in Example 5.1 on page 96.

Randomized Experiments versus Observational Studies

Ideally, if we were trying to ascertain the connection between the explanatory and response variables, we would keep everything constant except the explanatory variable. We would then manipulate the explanatory variable and notice what happened to the response variable as a consequence. We rarely reach this ideal, but we can come closer with an experiment than with an observational study.

In a **randomized experiment,** we *create* differences in the explanatory variable and then examine the results. In an **observational study,** we *observe* differences in the explanatory variable and then notice whether these are related to differences in the response variable.

For example, suppose we wanted to detect the effects of the explanatory variable "smoking during pregnancy" on the response variable "child's IQ at 4 years of age." In a randomized experiment, we would randomly assign half of the mothers to smoke during pregnancy and the other half to not smoke. In an observational study, we would merely record smoking behavior. This example demonstrates one of the reasons that we can't always perform an experiment. It would be unethical to randomly assign some mothers to smoke!

Two reasons why we must sometimes use an observational study instead of an experiment:

1. It is unethical or impossible to assign people to receive a specific treatment.
2. Certain explanatory variables, such as handedness, are inherent traits and cannot be randomly assigned.

Confounding Variables and Interacting Variables

Confounding Variables

A **confounding variable** is one that has two properties. First, a confounding variable *is related to the explanatory variable* in the sense that individuals who differ for the explanatory variable are also likely to differ for the confounding variable. Second, a confounding variable *affects the response variable.* Because of these two properties, the effect of a confounding variable on the response variable cannot be separated from the effect of the explanatory variable on the response variable.

For instance, suppose we are interested in the relationship between smoking during pregnancy and child's subsequent IQ a few years after birth. The explanatory variable is whether or not the mother smoked during pregnancy, and the response variable is subsequent IQ of the child. But if we notice that women who smoke during pregnancy have children with lower IQs than the children of women who don't smoke, it could be because women who smoke also have poor nutrition, or lower levels of education, or lower income. In that case, mother's nutrition, education, and income would all be confounding variables. They are likely to differ for smokers and nonsmokers, and they are likely to affect the response—the subsequent IQ of the child. The effect of these variables on the child's IQ cannot be separated from the effect of smoking, which was the explanatory variable of interest.

Confounding variables are a bigger problem in observational studies than in experiments. In fact, one of the major advantages of an experiment over an observational study is that in an experiment, the researcher attempts to control for confounding variables. In an observational study, the best the researcher can hope to do is measure possible confounding variables and see if they are also related to the response variable.

Two variables are *confounded* when it's hard to separate how they affect the response. When an explanatory variable and a response variable have been identified, a *confounding variable* is an additional variable that has these properties:

1. Individuals who differ for the explanatory variable also are likely to differ for the confounding variable.
2. Different values of the confounding variable are likely to result in different values of the response variable.

Interacting Variables and Effect Modifiers

Suppose that a relationship exists between an explanatory and response variable, but that the magnitude of the relationship differs for subgroups, such as men and women or smokers and nonsmokers. Then the subgroup variable is called an **effect modifier** because it modifies the effect of the explanatory variable on the outcome. For instance, suppose that exercise has a beneficial effect on blood pressure, but that the benefit is stronger for nonsmokers than for smokers. Then smoking is an *effect modifier* in the relationship between exercise and blood pressure. The term "effect modifier" is used mainly in health-related studies.

A similar, broader concept is that of interacting variables. An **interaction** between two explanatory variables occurs when the relationship of one of them to the response depends on the other one. For instance, suppose a study is done to compare online and traditional lecture offerings of a statistics course, to see which one is better for student learning. The same final exam is given to both groups at the end of the semester. The explanatory variable is the type of course the student is taking (online or traditional), and the response variable is the student's grade on the final exam. Now suppose that the online course is better for students with high GPAs, but the traditional lecture course is better for students with low GPAs. Then there is an *interaction* between course type and GPA in determining how well a student does in the course. When the results are reported, they should be given separately for the high and low GPA students. That way, it is clear that the better choice for an individual student might depend on that student's GPA.

Experimental Units, Subjects, and Volunteers

In addition to humans, it is common for studies to be performed on plants, animals, machine parts, and so on. To have a generic term for this conglomeration of possibilities, we define **experimental units** to be the smallest basic objects to which we can assign different treatments in a randomized experiment and **observational units** to be the objects or people measured in any study.

The terms **participants** or **subjects** are commonly used when the observational units are people. In most cases, the participants in studies are **volunteers.** Sometimes they are passive volunteers, such as when all patients treated at a particular medical facility are asked to sign a consent form agreeing to participate in a study.

Often, researchers recruit volunteers through the newspaper. For example, a weekly newspaper in a small town near the University of California, Davis ran an article with the headline, "UC Davis seeks volunteers for study on health, internet" (*Winters* (CA) *Express,* 7 March, 2013, p. A-3). The article explained that researchers at UC Davis were conducting a study on how people use the internet to find health information and needed volunteers to participate. Readers were told what they would be asked to do if they volunteered and that they would receive a $20 gift card if they participated.

Notice that by recruiting volunteers for studies, the results cannot necessarily be extended to the larger population. For example, if volunteers are enticed to participate by receiving a small payment or free medical care, as is often the case, those who respond are more likely to be from lower socioeconomic backgrounds. Common sense should enable you to figure out if this is likely to be a problem, but researchers should always report the source of their participants so you can judge this for yourself.

5.2   Designing a Good Experiment

Designing a flawless experiment is extremely difficult, and carrying one out is probably impossible. Nonetheless, there are ideals to strive for, and in this section, we investigate those first. We then explore some of the pitfalls that are still quite prevalent in research today.

Randomization: The Fundamental Feature of Experiments

Experiments are supposed to reduce the effects of confounding variables and other sources of bias that are necessarily present in observational studies. They do so by using a simple principle called **randomization.**

Randomization in experiments is related to the idea of random selection discussed in Chapter 4, when we described how to choose a sample for a survey. There, we were concerned that everyone in the population had a specified probability of making it into the sample. In randomized experiments, we are concerned that each of the experimental units (people, animals, and so on) has a specified probability of receiving any of the potential treatments. For example, Salk should have ensured that each group of babies available for study had an equal chance of being assigned to hear the heartbeat or to go into the silent nursery. Otherwise, he could have chosen the babies who looked healthier to begin with to receive the heartbeat treatment.

In statistics, "random" is not synonymous with "haphazard"—despite what your thesaurus might say. Although random assignments may not be possible or ethical under some circumstances, in situations in which randomization is feasible, it is usually not difficult to accomplish. It can be done easily with a table of random digits, a computer, or even—if done carefully—by physical means such as flipping a coin or drawing numbers from a hat. The important feature, ensured by proper randomization, is that the chances of being assigned to each condition are the same for each participant. Or, if the same participants are measured for all of the treatments, then the order in which they are assigned should be chosen randomly for each participant.

Randomly Assigning the Type of Treatments

In the most basic type of randomized experiment, each participant is assigned to receive one treatment. The decision about which treatment each participant receives should be done using randomization. In addition to preventing the experimenter from selectively choosing the best units to receive the favored treatment, randomly assigning the treatments to the experimental units helps protect against hidden or unknown biases. For example, suppose that in the experiment in Case Study 1.2, approximately the first 11,000 physicians who enrolled were given aspirin and the remaining physicians were given placebos. It could be that the healthier, more energetic physicians enrolled first, thus giving aspirin an unfair advantage.

Randomizing the Order of Treatments

In some experiments, all treatments are applied to each unit. In that case, randomization should be used to determine the *order* in which they are applied. For example, suppose an experiment is conducted to determine the extent to which drinking alcohol or smoking marijuana impairs driving ability. These types of experiments usually are done by having drivers navigate a private course specifically created for the experiment. Because drivers are all so different, it makes sense to test the same drivers under all three conditions (alcohol, marijuana, and sober) rather than using different drivers for each condition. But if everyone were tested under alcohol, then marijuana, then sober, by the time they were traveling the course for the second and third times, their performance would improve just from having learned something about the course.

A better method would be to randomly assign some drivers to each of the possible orderings so the learning effect would average out over the three treatments. Notice that it is important that the assignments be made *randomly*. If we let the experimenter decide which drivers to assign to which ordering, or if we let the drivers decide, assignments could be made that would give an unfair advantage to one of the treatments.

EXAMPLE 5.1 Randomly Assigning Mindfulness Meditation

In the study resulting in News Story 1, summarized in the Appendix, the researchers were interested in knowing if regular practice of meditation would enhance the immune system. If they had allowed participants to choose whether or not to meditate (the explanatory variable), there would have been confounding variables, like how hectic participants' daily schedules were, that may also have influenced the immune system (the response variable). Therefore, as explained in the corresponding Original Source 1, they recruited volunteers who were willing to be assigned to meditate or not. There were 41 volunteers, and they were randomly assigned to one of two conditions. The 25 participants randomly assigned to the "treatment group" completed an 8-week program of meditation training and practice. The 16 participants randomly assigned to the "control group" did not receive this training during the study, but for reasons of fairness, were offered the training when the study was completed. The researchers had decided in advance to assign the volunteers to the two groups as closely as possible to a 3:2 ratio. So, all of the volunteers had the same chance (25/41) of being assigned to receive the meditation training. By using random assignment, possible confounding factors, like daily stress, should have been similar for the two groups. Figure 5.1 illustrates the process used for this experiment. ∎

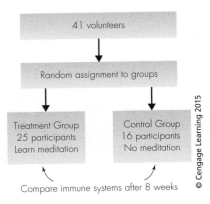

Figure 5.1

The randomized experiment in Example 5.1: Volunteers were randomly assigned to learn and practice meditation or not; immune systems were compared after 8 weeks.

© Cengage Learning 2015

Control Groups, Placebos, and Blinding

Control Groups

To determine whether a drug, heartbeat sound, meditation technique, and so on, has an effect, we need to know what would have happened to the response variable if the treatment had not been applied. To find that out, experimenters create **control groups,** which are handled identically to the treatment group(s) in all respects, except that they don't receive the active treatment.

Placebos

A special kind of control group is usually used in studies of the effectiveness of drugs. A substantial body of research shows that people respond not only to active drugs but also to **placebos.** A placebo looks like the real drug but has no active ingredients. Placebos can be amazingly effective; studies have shown that they can help up to 62% of headache sufferers, 58% of those suffering from seasickness, and 39% of those with postoperative wound pain.

Because the **placebo effect** is so strong, drug research is conducted by randomly assigning some of the volunteers to receive the drug and others to receive a placebo, without telling them which they are receiving. The placebo looks just like the real thing, so the participants will not be able to distinguish between it and the actual drug and thus will not be influenced by belief biases.

Blinding

The patient isn't the only one who can be affected by knowing whether he or she has received an active drug. If the researcher who is measuring the reaction of the patients were to know which group was which, the researcher might take the measurements in a biased fashion. To avoid these biases, good experiments use **double-blind** procedures. A double-blind experiment is one in which neither the participant nor the researcher taking the measurements knows who had which treatment. A **single-blind** experiment is one in which only one of the two, the participant or the researcher taking the measurements, knows which treatment the participant was assigned.

Although double-blind experiments are preferable, they are not always possible. For example, in testing the effect of daily meditation on blood pressure, the subjects would obviously know if they were in the meditation group or the control group. In this case, the experiment could only be single-blind, in which case the person taking the blood pressure measurement would not know who was in which group.

EXAMPLE 5.2

Blindly Lowering Cholesterol

In the study described in News Story 3 in the Appendix, the researchers wanted to compare a special dietary regime with a drug known to lower cholesterol (lovastatin) to see which one would lower cholesterol more. The special dietary regime, called the dietary "portfolio," included elements thought to lower cholesterol, such as soy protein and almonds. The lovastatin group was asked to eat a very low-fat diet in addition to taking the drug, so a "control group" was included that received the same low-fat diet as those taking the lovastatin but was administered a placebo. Thus, there were three treatments—the "portfolio diet," the low-fat diet with lovastatin, and the low-fat diet with placebo.

There were 46 volunteers for the study. Here is a description from the article Original Source 3 on the companion website, illustrating how the researchers addressed random assignment and blinding:

> *Participants were randomized by the statistician using a random number generator . . . . The statistician held the code for the placebo and statin tablets provided with the control and statin diets, respectively. This aspect of the study was therefore double-blind. The dieticians were not blinded to the diet because they were responsible for patients' diets and for checking diet records. The laboratory staff responsible for analyses were blinded to treatment and received samples labeled with name codes and dates. (p. 504)*

In other words, the researchers and participants were both blind as to which drug (lovastatin or placebo) people in those two groups were taking, but the participants and dieticians could not be blind to what the participants were eating. The staff who evaluated cholesterol measurements however, could be and were blind to the treatment. ■

In Example 5.2, the explanatory variable was a combination of diet and drug, and the response variable was how much the person's cholesterol level went down. There were three treatments, all of which included eating special diets, and the participants clearly could not be blind to which diet they ate. Therefore, the drug part of the comparison was double-blind, but the diet part was only single-blind. Case Study 5.1 describes a study that is completely double-blind.

| CASE STUDY 5.1 | **Quitting Smoking with Nicotine Patches** |

SOURCE: Hurt et al. (1994), pp. 595–600.

There is no longer any doubt that smoking cigarettes is hazardous to your health and to those around you. Yet, for someone addicted to smoking, quitting is no simple matter. One promising technique for helping people to quit smoking is to apply a patch to the skin that dispenses nicotine into the blood. These "nicotine patches" have become one of the most frequently prescribed medications in the United States. To test the effectiveness of these patches on the cessation of smoking, Dr. Richard Hurt and his colleagues recruited 240 smokers at Mayo Clinics in Rochester, Minnesota; Jacksonville, Florida; and Scottsdale, Arizona. Volunteers were required to be between the ages of 20 and 65, have an expired carbon monoxide level of 10 ppm or greater (showing that they were indeed smokers), be in good health, have a history of smoking at least 20 cigarettes per day for the past year, and be motivated to quit.

Volunteers were randomly assigned to receive either 22-mg nicotine patches or placebo patches for 8 weeks. They were also provided with an intervention program recommended by the National Cancer Institute, in which they received counseling before, during, and for many months after the 8-week period of wearing the patches.

After the 8-week period of patch use, almost half (46%) of the nicotine group had quit smoking, whereas only one-fifth (20%) of the placebo group had. Having quit was defined as "self-reported abstinence (not even a puff) since the last visit and an expired air carbon monoxide level of 8 ppm or less" (p. 596). After a year, rates in both groups had declined, but the group that had received the nicotine patch still had a higher percentage who had successfully quit than did the placebo group: 27.5% versus 14.2%.

The study was double-blind, so neither the participants nor the nurses taking the measurements knew who had received the active nicotine patches. The study was funded by a grant from Lederle Laboratories and was published in the *Journal of the American Medical Association.* ∎

Matched Pairs, Blocks, and Repeated Measures

It is sometimes easier and more efficient to have each person in a study serve as his or her own control. That way, natural variability in the response variable across individuals doesn't obscure the treatment effects. We encountered this idea when we discussed how to compare driving ability when under the influence of alcohol and marijuana and when sober.

Sometimes, instead of using the same individual for the treatments, researchers will match people on traits that are likely to be related to the outcome, such as age, IQ, or weight. They then randomly assign each of the treatments to one member of each matched pair or grouping. For example, in a study comparing chemotherapy to surgery to treat cancer, patients might be matched by sex, age, and level of severity of the illness. One from each pair would then be randomly chosen to receive the chemotherapy and the other to receive surgery. (Of course, such a study would only be ethically feasible if there were no prior knowledge that one treatment was superior to the other. Patients in such cases are always required to sign an informed consent.)

Matched-Pair Designs

Experimental designs that use either two matched individuals or the same individual to receive each of two treatments are called **matched-pair designs.** For instance, to measure the effect of drinking caffeine on performance on an IQ test, researchers could use the same individuals twice, or they could pair individuals based on initial IQ. If the same people were used, they might drink a caffeinated beverage in one test session, followed by an IQ test, and a noncaffeinated beverage in another test session (on a different day), followed by an IQ test (with different questions). The order in which the two sessions occurred would be decided randomly, separately for each participant. That would eliminate biases, such as learning how to take an IQ test, that would systematically favor one session (first or second) over the other.

If matched pairs of people with similar IQs were used, one person in each pair would be randomly chosen to drink the caffeine and the other to drink the noncaffeinated beverage. The important feature of these designs is that randomization is used to assign the order of the two treatments. Of course, it is still important to try to conduct the experiment in a double-blind fashion so that neither the participant nor the researcher knows which order was used. In the caffeine and IQ example, neither the participants nor the person giving the IQ test would be told which sessions included the caffeine.

Randomized Block Designs and Repeated Measures

An extension of the matched-pair design to three or more treatments is called a **randomized block design,** or sometimes simply a **block design.** In a randomized block design, similar experimental units are first placed together in groups called **blocks.** Then treatments are randomly assigned separately within each block. This is done for every block, thus making sure that all treatments are assigned at least once in every block. The comparison of treatments is done within the blocks first and then summarized across blocks. If the results are very different across blocks, then they are reported separately rather than being combined.

The method described for comparing drivers under three conditions was a randomized block design. Each driver is called a **block.** This somewhat peculiar terminology results from the fact that these ideas were first used in agricultural experiments, in which the experimental units were plots of land that had been subdivided into "blocks." In the social sciences, designs such as these, in which the same participants are measured repeatedly, are referred to as **repeated-measures designs.** Case Study 5.2 (below) illustrates the use of a randomized block design.

Reducing and Controlling Natural Variability and Systematic Bias

Both natural variability and systematic bias can mask differences in the response variable that are due to differences in the explanatory variable. Here are some solutions:

1. **Random assignment to treatments** is used to reduce *unknown* systematic biases due to confounding variables that might otherwise exist between treatment groups.
2. **Matched pairs, repeated measures, and blocks** are used to reduce *known* sources of natural variability in the response variable, so that differences due to the explanatory variable can be detected more easily.

| CASE STUDY 5.2 | **Police Shift and Quality of Life** |

SOURCE: Amendola et al. (2011), pp. 407–442.

Suppose you were in charge of a big city police department and had to decide whether to assign officers to work five 8-hour days, four 10-hour days, or three 12-hour days plus an extra 8-hour day every other week. Which would you choose? Karen Amendola and her colleagues decided to find out which of these would result in better health, sleep, work satisfaction, and other outcomes. They recruited a total of 275 police officers in Detroit, Michigan, and Arlington, Texas, to participate in the experiment. The study ran from January 2007 to June 2009.

Because police work is a 24-hour business, some officers work during the day, some in the evening, and some from midnight to morning. You can imagine that the impact of working a longer shift might be different in the daytime, in the evening, and at night. So a *randomized block design* was used. The officers were grouped into blocks based on their normal work time, creating three blocks (Day, Evening, Night). Then the random assignment of shift length (8, 10, or 12 hours) was done separately within each of these blocks. Once the officers were given their assignments, they were asked to continue that shift length for the full two and a half years of the study. Figure 5.2 displays the process used in assigning the shift lengths.

Here are the results, from the original journal article reporting this study:

The results indicated that those working 10 hour shifts had a significantly higher quality of work life and averaged significantly more sleep than those on 8-hour shifts. Furthermore, those working 8-hour shifts averaged significantly more overtime than did those assigned to 10- and 12-hour shifts. In addition, officers working 12-hour shifts experienced greater levels of sleepiness and reported lower levels of alertness at work than officers on 8-hour shifts (Amendola et al., 2011, p. 407).

The conclusions were similar for all three blocks, so they were reported together. Whether the officers were working during the day, the evening, or at night, the best outcomes were for the 10-hour shifts. As a side note, the two cities were used as blocks as well, but results did not differ for them either, so that part of the study design has been ignored here for simplicity. ■

Figure 5.2
The randomized block design for Case Study 5.2, comparing shift lengths for police officers.

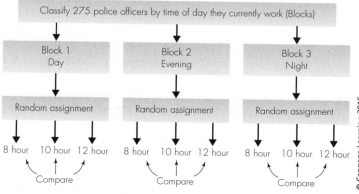

© Cengage Learning 2015

5.3 Difficulties and Disasters in Experiments

We have already introduced some of the problems that can be encountered with experiments, such as biases introduced by lack of randomization. However, many of the complications that result from poorly conducted experiments can be avoided with proper planning and execution.

> *Here are some potential complications:*
>
> **1.** Confounding variables
> **2.** Interacting variables
> **3.** Placebo, Hawthorne, and experimenter effects
> **4.** Lack of ecological validity and consequent generalization

Confounding Variables

The Problem

Variables that are connected with the explanatory variable can distort the results of an experiment because they—and not the explanatory variable—may be the agent actually causing a change in the response variable.

The Solution

Randomization is the solution. If experimental units are randomly assigned to treatments, then the effects of the confounding variables should apply equally to each treatment. Thus, observed differences between treatments should not be attributable to the confounding variables.

EXAMPLE 5.3 Nicotine Patch Therapy

The nicotine patch therapy in Case Study 5.1 was more effective when there were no other smokers in the participant's home. Suppose the researchers had assigned the first 120 volunteers to the placebo group and the last 120 to the nicotine group. Further, suppose that those with no other smokers at home were more eager to volunteer. Then the treatment would have been confounded with whether there were other smokers at home. The observed results showing that the active patches were more effective than the placebo patches could have merely represented a difference between those with other smokers at home and those without. By using randomization, approximately equal numbers in each group should have come from homes with other smokers. Thus, any impact of that variable would be spread equally across the two groups. ■

Interacting Variables

The Problem

Sometimes a second variable interacts with the explanatory variable, but the results are reported without taking that interaction into account. The reader is then misled into thinking the treatment works equally well, no matter what the condition is for the second variable.

The Solution

Researchers should measure and report variables that may interact with the main explanatory variable(s).

EXAMPLE 5.4 Other Smokers at Home

In the experiment described in Case Study 5.1, there was an interaction between the treatment and whether there were other smokers at home. The researchers measured and reported this interaction. After the 8-week patch therapy, the proportion of the nicotine group who had quit smoking was only 31% if there were other smokers at home, whereas it was 58% if there were not. (In the placebo group, 20% had quit whether there were other smokers at home or not.) Therefore, it would be misleading to merely report that 46% of the nicotine recipients had quit, without also providing the information about the interaction.

Placebo, Hawthorne, and Experimenter Effects

The Problem

We have already discussed the strong effect that a placebo can have on experimental outcomes because the power of suggestion is somehow able to affect the result. A related idea is that participants in an experiment respond differently than they otherwise would, just because they *are* in the experiment. This is called the "Hawthorne effect" because it was first detected in 1924 during a study of factory workers at the Hawthorne, Illinois, plant of the Western Electric Company. (The phrase actually was not coined until much later; see French, 1953.) Related to these effects are numerous ways in which the experimenter can bias the results. These include recording the data erroneously to match the desired outcome, treating participants differently based on which condition they are receiving, or subtly letting the participants know the desired outcome.

The Solution

As we have seen, most of these problems can be overcome by using double-blind designs and by including a placebo group or a control group that receives identical handling except for the active part of the treatment. Other problems, such as incorrect data recording, should be addressed by having data entered automatically into a computer as it is collected, if possible. Depending on the experiment, there may still be subtle ways in which experimenter effects can sneak into the results. You should be aware of these possibilities when you read the results of a study.

EXAMPLE 5.5 Dull Rats

In a classic experiment designed to test whether the expectations of the experimenter could really influence the results, Rosenthal and Fode (1963) deliberately conned 12 experimenters. They gave each one five rats that had been taught to run a maze. They told six of the experimenters that the rats had been bred to do well (that is, that they were "maze bright") and told the other six that their rats were "maze dull" and should not be expected to do well. Sure enough, the experimenters who had been told they had bright rats found learning rates far superior to those found by the experimenters who had been told they had dull rats. Hundreds of other studies have since confirmed the "experimenter effect." ■

Lack of Ecological Validity and Consequent Generalization

The Problem

Suppose you wanted to compare three assertiveness training methods to see which was most effective in teaching people how to say no to unwanted requests on their time. Would it be realistic to give them the training, then measure the results by asking them to role-play in situations in which they would have to say no? Probably not, because everyone involved would know it was only role-playing. The usual social pressures to say yes would not be as striking. This is an example of an experiment with little **ecological validity.** In other words, the variables have been removed from their natural setting and are measured in the laboratory or in some other artificial setting. Thus, the results do not accurately reflect the impact of the variables in the real world or in everyday life. Therefore, it might not make sense to generalize the findings to other situations. A related problem is one we have already mentioned—namely, if volunteers are used for a study, can the results be generalized to any larger group?

The Solution

There are no ideal solutions to these problems, other than trying to design experiments that can be performed in a natural setting with a random sample from the population of interest. In most experimental work, these idealistic solutions are impossible. A partial solution is to measure variables for which the volunteers might differ from the general population, such as income, age, or health, and then try to determine the extent to which those variables would make the results less general than desired. In any case, when you read the results of a study, you should question its ecological validity and its generalizability.

EXAMPLE 5.6 **Real Smokers with a Desire to Quit**

The researchers in Case Study 5.1 did many things to help ensure ecological validity and generalizability. First, they used a standard intervention program available from and recommended by the National Cancer Institute instead of inventing their own, so that other physicians could follow the same program. Next, they used participants at three different locations around the country, rather than in one community only, and they involved a wide range of ages (20 to 65). They included individuals who lived in households with other smokers as well as those who did not. Finally, they recorded numerous other variables (sex, race, education, marital status, psychological health, and so on) and checked to make sure these were not related to the response variable or the patch assignment. ■

| CASE STUDY 5.3 | **Exercise Yourself to Sleep** |

SOURCE: King et al. (1997), pp. 32–37.

According to the *UC Davis Health Journal* (November–December 1997, p. 8), older adults constitute only 12% of the population but receive almost 40% of the sedatives prescribed. The purpose of this randomized experiment was to see if regular exercise could help reduce sleep difficulties in older adults. The 43 participants were sedentary volunteers between the ages of 50 and 76 with moderate sleep problems but no heart disease. They were randomly assigned either to participate in a moderate community-based exercise program four times a week for 16 weeks or continue to be sedentary. For ethical reasons, the control group was admitted to the program when the experiment was complete. The results were striking. The exercise group fell asleep an average of 11 minutes faster and slept an average of 42 minutes longer than the control group. Note that this could not be a double-blind experiment because participants obviously knew whether they were exercising. Because sleep patterns were self-reported, there could have been a tendency to err in reporting, in the direction desired by the experimenters. However, this is an example of a well-designed experiment, given the practical constraints, and, as the authors conclude, it does allow the finding that "older adults with moderate sleep complaints can improve self-rated sleep quality by initiating a regular, moderate-intensity exercise program" (p. 32). ∎

5.4   Designing a Good Observational Study

In trying to establish causal links, observational studies start with a distinct disadvantage compared to experiments: The researchers observe, but cannot control, the explanatory variables. However, these researchers do have the advantage that they are more likely to measure participants in their natural setting. Before looking at some complications that can arise, let's look at an example of a well-designed observational study.

| CASE STUDY 5.4 | **Baldness and Heart Attacks** |

SOURCE: Yamada, Hara, Umematsu and Kadowaki (2013), pp. 1-8.

In April, 2013, a new meta-analysis confirmed what individual studies had shown for many years: Men with pattern baldness are at higher risk of heart attacks than men of the same age without it. Pattern baldness affects the crown of the head and is not the same as a receding hairline. About half of men have some balding by age 50, and about 80% by age 70. The analysis found that the higher the amount of baldness, the higher the risk of heart attack, and that the relationship held for younger men as well as older men.

   The meta-analysis combined the results of six studies that examined this relationship, published between 1993 and 2008. All of them were observational studies, because men obviously cannot be randomly assigned to lose hair or not!

Let's look at the earliest of these, an observational study published in 1993 (Lesko, Rosenberg and Shapiro). The study was conducted by researchers at Boston University School of Medicine, in which they compared 665 men who had been admitted to hospitals with their first heart attack to 772 men in the same age group (21- to 54-years-old) who had been admitted to the same hospitals for other reasons. Thirty-five hospitals were involved, all in eastern Massachusetts and Rhode Island.

The study found that the percentage of men who showed some degree of pattern baldness was substantially higher for those who had had a heart attack (42%) than for those who had not (34%). Further, when they used sophisticated statistical tests to ask the question in the reverse direction, they found an increased risk of heart attack for men with any degree of pattern baldness. The analysis methods included adjustments for age and other heart attack risk factors. The increase in risk was more severe with increasing severity of baldness, after adjusting for age and other risk factors.

The authors of both the original study and the meta-analysis speculated that there may be a third variable—perhaps the hormone testosterone, that both increases the risk of heart attacks and leads to a propensity for baldness. With an observational study such as this, scientists can establish a connection, and they can then look for causal mechanisms in future work. ■

Types of Observational Studies

Case-Control Studies

Some terms are used specifically for observational studies. The 1993 study described in Case Study 5.4 is an example of a **case-control study.** In such a study, "cases" who have a particular attribute or condition are compared with "controls" who do not. In this example, those who had been admitted to the hospital with a heart attack were the cases, and those who had been admitted for other reasons were the controls. The cases and controls are compared to see how they differ on the variable of interest, which in Case Study 5.4 was the degree of baldness. Figure 5.3 illustrates the process used for this study.

Sometimes cases are matched with controls on an individual basis. This type of design is similar to a matched-pair experimental design. The analysis proceeds by first comparing the pair, then summarizing over all pairs. Unlike a matched-pair experiment, the researcher does not randomly assign treatments within pairs but is restricted to how they occur naturally. For example, to identify whether left-handed people die at a younger age, researchers might match each left-handed case with a right-handed sibling as a control and compare their ages at death. Handedness could obviously not be randomly assigned to the two individuals, so confounding factors might be responsible for any observed differences.

Figure 5.3

The case-control study for the relationship between baldness and heart attacks.

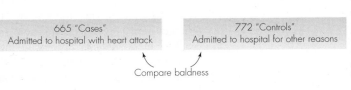

665 "Cases"
Admitted to hospital with heart attack

772 "Controls"
Admitted to hospital for other reasons

Compare baldness

Retrospective or Prospective Studies

Observational studies are also classified according to whether they are **retrospective,** in which participants are asked to recall past events, or **prospective,** in which participants are followed into the future and events are recorded. The latter is a better procedure because people often do not remember past events accurately.

Advantages of Case-Control Studies

Case-control studies have become increasingly popular in medical research, and with good reason. They are much more efficient than experiments, and they do not suffer from the ethical considerations inherent in the random assignment of potentially harmful or beneficial treatments. The purpose of most case-control studies is to find out whether one or more explanatory variables are related to a certain disease. For instance, in an example given later in this book, researchers were interested in whether owning a pet bird is related to incidence of lung cancer.

A case-control study begins with the identification of a suitable number of cases, such as people who have been diagnosed with a disease of interest. Researchers then identify a group of controls, who are as similar as possible to the cases, except that they don't have the disease. To achieve this similarity, researchers often use patients hospitalized for other causes as the controls. For instance, in determining whether owning a pet bird is related to incidence of lung cancer, researchers would identify lung cancer patients as the *cases* and then find people with similar backgrounds who do not have lung cancer as the *controls.* They would then compare the proportions of cases and controls who had owned a pet bird.

Efficiency

The case-control design has some clear advantages over randomized experiments as well as over other observational studies. Case-control studies are very efficient in terms of time, money, and inclusion of enough people with the disease. Imagine trying to design an experiment to find out whether a relationship exists between owning a bird and getting lung cancer. You would randomly assign people to either own a bird or not and then wait to see how many in each group contracted lung cancer. The problem is that you would have to wait a long time, and even then, you would have very few cases of lung cancer in either group. In the end, you may not have enough cases for a valid comparison.

A case-control study, in contrast, would identify a large group of people who had been diagnosed with lung cancer and would then ask them whether they had owned a pet bird. A similar control group would be identified and asked the same question. A comparison would then be made between the proportion of cases (lung cancer patients) who had birds and the proportion of controls who had birds.

Reducing Potential Confounding Variables

Another advantage of case-control studies over other observational studies is that the controls are chosen to try to reduce potential confounding variables. For example, in Case Study 5.4, suppose it were true that bald men were simply less healthy than other men and were therefore more likely to get sick in some way. An observational study that recorded only whether someone had baldness and whether they had had a heart attack would not be able to control for that fact. By using other hospitalized patients as the controls, the researchers were able to at least partially account for general health as a potential confounding factor.

You can see that careful thought is needed to choose controls that reduce potential confounding factors and do not introduce new ones. For example, suppose we wanted to know if heavy exercise induced heart attacks and, as cases, we used people as they were admitted to the hospital with a heart attack. We would certainly not want to use other newly admitted patients as controls. People who were sick enough to enter the hospital (for anything other than sudden emergencies) would probably not have been recently engaging in heavy exercise. When you read the results of a case-control study, you should pay attention to how the controls were selected.

5.5   Difficulties and Disasters in Observational Studies

As with other types of research we have discussed, when you read the results of observational studies you need to watch for problems that could negate the results of the study.

Here are some complications that can arise:

1. Confounding variables and the implications of causation

2. Extending the results inappropriately

3. Using the past as a source of data

Confounding Variables and the Implications of Causation

The Problem
Don't be fooled into thinking that a link between two variables established by an observational study implies that one causes the other. There is simply no way to separate out all potential confounding factors if randomization has not been used.

The Solution
A partial solution is achieved if researchers measure all the potential confounding variables they can imagine and include those in the analysis to see whether they are related to the response variable. Another partial solution can be achieved in case-control studies by choosing the controls to be as similar as possible to the cases. The other part of the solution is up to the reader: Don't be fooled into thinking a causal relationship necessarily exists.

There are some guidelines that can be used to assess whether a collection of observational studies indicates a causal relationship between two variables. These guidelines are discussed in Chapter 11.

EXAMPLE 5.7 Smoking During Pregnancy

In Chapter 1, we introduced a study showing that women who smoked during pregnancy had children whose IQs at age 4 were lower than those of similar women who had not smoked. The difference was as high as nine points before accounting for confounding variables, such as diet and education, but was reduced to just over four points after accounting for those factors. However, other confounding variables could exist that are different for mothers who smoke and that were not measured and analyzed, such as amount of exercise the mother got during pregnancy. Therefore, we should not conclude that smoking during pregnancy necessarily caused the children to have lower IQs. ■

Extending the Results Inappropriately

The Problem

Many observational studies use convenience samples, which often are not representative of any population. Results should be considered with that in mind. Case-control studies often use only hospitalized patients, for example.

In general, results of a study can be extended to a larger population only if the sample is representative of that population for the variables studied. For instance, in News Story 2 in the Appendix, "Research shows women hit harder by hangovers," the research was based on a sample of students in introductory psychology classes at a university in the midwestern United States. The study compared drinking behavior and hangover symptoms in men and women. To whom do you think the results can be extended? It probably isn't reasonable to think they can be extended to all adults because severity of hangover symptoms may change with age. But what about all people in the same age group as the students studied? All college students? College students in the Midwest? Only psychology students? As a reader, you must decide the extent to which you think this sample represents each of these populations on the question of alcohol consumption and hangover symptoms and their differences for men and women.

The Solution

If possible, researchers should sample from all segments of the population of interest rather than just use a convenient sample. In studying the relationship between smoking during pregnancy and child's IQ, described in Example 5.7, the researchers included most of the women in a particular county in upstate New York who were pregnant during the right time period. Had they relied solely on volunteers recruited through the media, their results would not be as extendable.

EXAMPLE 5.8 Baldness and Heart Attacks Revisited

The observational study described in Case Study 5.4, relating baldness and heart attacks, only used men who were hospitalized for some reason. Although that may make sense in terms of providing a more similar control group, you should consider whether the results should be extended to all men. ■

Using the Past as a Source of Data

The Problem

Retrospective observational studies can be particularly unreliable because they ask people to recall past behavior. Some medical studies, in which the response variable is whether someone has died, can be even worse because they rely on the memories of relatives and friends rather than on the actual participants. Retrospective studies also suffer from the fact that variables that confounded things in the past may no longer be similar to those that would currently be confounding variables, and researchers may not think to measure them. Example 5.9 illustrates this problem.

The Solution

If at all possible, prospective studies should be used. That's not always possible. For example, researchers who first considered the potential causes of AIDS or Toxic Shock Syndrome had to start with those who were afflicted and try to find common factors from their pasts. If possible, retrospective studies should use authoritative sources such as medical records rather than relying on memory.

EXAMPLE 5.9 | **Do Left-Handers Die Young?**

A highly publicized study at the end of the 20th century pronounced that left-handed people did not live as long as right-handed people (Coren and Halpern, 1991). In one part of the study, the researchers had sent letters to next of kin for a random sample of recently deceased individuals, asking which hand the deceased had used for writing, drawing, and throwing a ball. They found that the average age of death for those who had been left-handed was 66, whereas for those who had been right-handed, it was 75. What they failed to take into account was that in the early part of the 20th century, many children were forced to write with their right hands, even if their natural inclination was to be left-handed. Therefore, the older people, who died in their 70s and 80s during the time of this study, were more likely to be right-handed than the younger people, who died in their 50s and 60s. The confounding factor of how long ago one learned to write was not taken into account. A better study would be a prospective one, following current left- and right-handers to see which group survived longer. ∎

5.6   Random Sample versus Random Assignment

While random sampling and random assignment to treatments are related ideas, the conclusions that can be made based on each of them are very different. Random sampling is used to get a representative sample from the population of interest. Random assignment is used to control for confounding variables and other possible sources of bias. An ideal study would use both, but for practical reasons, that is rarely done.

Extending Results to a Larger Population: Random Sampling

The main purpose of using a random sample in a study is that the results can be extended to the population from which the sample was drawn. As we will learn in later chapters, there is still some degree of uncertainty, similar to the margin of error introduced in Chapter 4, but it can be stated explicitly. Therefore, it would be ideal to use a random sample from the population of interest in all statistical studies.

Unfortunately, it is almost always impractical to obtain a random sample to participate in a randomized experiment or to be measured in an observational study. For instance, in the study in Example 5.1, it would be impossible to obtain a random sample of people interested in learning meditation from among all adults and then teach some of them how to do so. Instead, the researchers in that study used employees at one company and asked for volunteers.

The extent to which results can be extended to a larger population when a random sample is not used depends on the extent to which the participants in the study are representative of a larger population for the variables being studied. For example, in the nicotine patch experiment described in Case Study 5.1, the participants were patients who came to a clinic seeking help to quit smoking. They probably do represent the population of smokers with a desire to quit. Thus, even though volunteers were used for the study, the results can be extended to other smokers with a desire to quit.

As another example, the experiment in Case Study 1.2, investigating the effect of taking aspirin on heart attack rates, used male physicians. Therefore, the results may not apply to women, or to men who have professions with very different amounts of physical activity than physicians.

As a reader, you must determine the extent to which you think the participants in a study are representative of a larger population for the question of interest. That's why it's important to know the answer to Critical Component 3 in Chapter 2: "The individuals or objects studied and how they were selected."

Establishing Cause and Effect: Random Assignment

The main purpose of random assignment of treatments, or of the order of treatments, is to even out confounding variables across treatments. By doing this, a cause-and-effect conclusion can be inferred that would not be possible in an observational study. With randomization to treatments, the range of values for confounding variables should be similar for each of the treatment groups. For instance, in Case Study 5.1, whether someone is a light or heavy smoker may influence their ability to quit smoking. By randomly assigning participants to wear a nicotine patch or a control patch, about the same proportion of heavy smokers should be in each patch-type group. In Case Study 5.3, caffeine consumption may influence older adults' ability to fall asleep quickly. By randomly assigning them to the exercise program or not, about the same proportion of caffeine drinkers should be in each group.

Without random assignment, naturally occurring confounding variables can result in an apparent relationship between the explanatory and response variables. For instance, in Example 5.1, if the participants had been allowed to choose whether to meditate or not, the two groups probably would have differed in other ways, such as diet, that may affect immune system functioning. If the participants had been assigned in a nonrandom way by the experimenters, they could have chosen those who looked most healthy to participate in the meditation program. Thus, without random assignment, it would not have been possible to conclude that meditation was responsible for the observed difference in immune system functioning.

As a reader, it is important for you to think about Component 6 from Chapter 2: "Differences in the groups being compared, in addition to the factor of interest." If random assignment was used, these differences should be minimized. If random assignment was not used, you must assess the extent to which you think group differences may explain any observed relationships. In Chapter 11, we will learn more about establishing cause and effect when randomization isn't used.

Thinking About Key Concepts

- An *explanatory variable* in a study attempts to explain, at least partially, differences in a *response variable*. But whether that explanation includes cause and effect depends on how the study was done.
- In a *randomized experiment*, the explanatory variable is randomly assigned by the experimenter, whereas in an *observational study,* it is simply observed and recorded.
- A *confounding variable* is one that differs for different values of the explanatory variable and also affects the response. A confounding variable's effect on the response can't be separated from the explanatory variable's effect on the response.
- In randomized experiments, *randomization* (random assignment) is used to assign treatments in the hope that potential confounding variables will be evened out across treatments. That process allows the effect of the explanatory variable on the response to be isolated and, thus, cause and effect conclusions can be made.
- In randomized experiments, *blocks* are sometimes used to control known sources of variation.
- In observational studies, *random sampling* is used to try to get a representative sample that allows the results to be extended to a population beyond just those individuals who are in the study. But confounding variables often are a major problem in observational studies.
- In observational studies, *case-control* studies are sometimes used because they are an efficient way to get enough people who have the disease or condition being studied. However, case-control studies don't use random samples, so the extent to which the cases and controls are representative of their respective populations needs to be considered before the results can be generalized.

Exercises

Exercises with numbers divisible by 3 (3, 6, 9, etc.) are included in the Solutions at the back of the book. They are marked with an asterisk (*).

1. Explain why it may be preferable to conduct a randomized experiment rather than an observational study to determine the relationship between two variables. Support your argument with an example concerning something of interest to you.

2. Suppose a study found that people who drive more than 10 miles to work each day have better knowledge of current events, on average, than people who ride a bicycle to work.

   a. What is the explanatory variable in this study?

   b. What is the response variable in this study?

   c. It was found that people who drive more than 10 miles to work each day also listen to the news on the radio more often than people who ride a bicycle to work. Explain how the variable "how often a person listens to the news on the radio" fits the two properties of a confounding variable (given in the box on page 93) for this study.

*3. A recent study found that people with insomnia are more likely to experience heart problems than people without insomnia.

   *a. What are the explanatory and response variables in this study?

   *b. Explain why amount of caffeine consumed might be a confounding variable in this study by explaining how caffeine consumption fits the two properties of confounding variables given in the box on page 93.

4. A utility company was interested in knowing if agricultural customers would use less electricity during peak hours if their rates were different during those hours. (Agricultural energy use is substantial, for things like irrigation, lighting, wind turbines to reduce frost damage, and so on.) Customers who volunteered for the study were randomly assigned to continue to get standard rates or to receive the time-of-day rate structure. Special meters were attached that recorded usage during peak and off-peak hours, which the customers could read. The technician who read the meter did not know what rate structure each customer had.

   a. What was the explanatory variable in this experiment?

   b. What was the response variable in this experiment?

   c. Was this experiment single-blind, double-blind, or neither? Explain.

   d. Did this experiment use matched pairs, blocks, or neither? Explain.

5. To test the effects of drugs and alcohol use on driving performance, 20 volunteers were each asked to take a driving test under three conditions: sober, after two drinks, and after smoking marijuana. The order in which they drove the three conditions was randomized. An evaluator watched them drive on a test course and rated their accuracy on a scale from 1 to 10, without knowing which condition they were under each time.

   a. What was the explanatory variable in this experiment?

   b. What was the response variable in this experiment?

   c. Was this experiment single-blind, double-blind, or neither? Explain.

   d. Did this experiment use matched pairs, blocks, or neither? Explain.

*6. To compare four brands of tires, one of each brand was randomly assigned to the four tire locations on each of 50 cars. These tires were specially manufactured without any labels identifying the brand. After the tires had been on the cars for 30,000 miles, the researchers removed them and measured the remaining tread. They were not told which brand was which until the experiment was over.

*a. What was the explanatory variable in this experiment?

*b. What was the response variable in this experiment?

*c. Was this experiment single-blind, double-blind, or neither? Explain.

*d. Did this experiment use matched pairs, blocks, or neither? Explain.

7. Suppose an observational study finds that people who use public transportation to get to work have better knowledge of current affairs than those who drive to work, but that the relationship is weaker for well-educated people. What term from this chapter (for example, response variable) applies to each of the following variables?

a. Method of getting to work

b. Knowledge of current affairs

c. Level of education

d. Whether the participant reads a daily newspaper

8. Researchers have found that women who take oral contraceptives (birth control pills) are at higher risk of having a heart attack or stroke than women who do not take them. They also found that the risk is substantially higher for both groups if a woman smokes. Assume that the proportions of women who smoke are similar in both groups. In investigating the relationship between taking oral contraceptives (the explanatory variable) and having a heart attack or stroke (the response variable), explain whether smoking would fit the definition of each of the following:

a. A confounding variable.

b. An effect modifier.

c. An interacting variable.

*9. A study to see whether birds remember color was done by putting birdseed on a piece of red cloth and letting the birds eat the seed. Later, empty pieces of cloth of varying colors (red, purple, white, and blue) were displayed. The birds headed for the red cloth. The researcher concluded that the birds remembered the color.

Using the terminology in this chapter, give an alternative explanation for the birds' behavior.

10. Refer to the previous exercise, about whether birds remember colors. Suppose 20 birds were available for an experiment and they could each be tested separately. Suggest a better method for the study than the one used.

11. Refer to Thought Question 5 at the beginning of this chapter. The headline was based on a study in which a representative sample of over 400,000 adults in the United States were asked a series of questions, including level of education and on how many of the past 30 days they felt physically and emotionally healthy. What were the intended explanatory variable and response variables for this study?

*12. Specify the explanatory and response variables for News Story 10 described in the Appendix, "Churchgoers live longer, study finds."

13. For each of the following observational studies in the news stories in the Appendix, specify the explanatory and response variables.

a. News Story 12: "Working nights may increase breast cancer risk."

b. News Story 16: "More on TV violence."

c. News Story 18: "Heavier babies become smarter adults, study shows."

14. For each of the following news stories in the Appendix, explain whether the study was a randomized experiment or an observational study. If necessary, consult the original source of the study on the website.

a. News Story 4: "Happy people can actually live longer."

b. News Story 6: "Music as brain builder."

*15. For each of the following news stories summarized in the Appendix, explain whether the study was a randomized experiment or an observational study.

*a. News Story 11: "Double trouble behind the wheel."

*b. News Story 15: "Kids' stress, snacking linked."

16. Explain why each of the following is used in experiments. You may refer to an example in the chapter if it makes it easier to explain.

    a. Placebo treatments

    b. Control groups

17. **a.** Explain why blinding is used in experiments.

    b. Explain why it is not always possible to make an experiment double-blind. You may refer to an example in the chapter if it makes it easier to explain.

*18. Refer to each of these news stories printed or summarized in the Appendix, and consult the original source article on the companion website if necessary. In each case, explain whether or not a repeated measures design was used.

    *a. News Story 6: "Music as brain builder."

    *b. News Story 11: "Double trouble behind the wheel."

    *c. News Story 14: "Study: Emotion hones women's memory."

19. Is the "experimenter effect" most likely to be present in a double-blind experiment, a single-blind experiment, or an experiment with no blinding? Explain.

20. Give an example of a randomized experiment that would have poor ecological validity.

*21. Is it possible to conduct a randomized experiment to compare two conditions using volunteers recruited through a local newspaper? If not, explain why not. If so, explain how it would be done and explain any "difficulties and disasters" that would be encountered.

22. Find the story referenced as News Story 5, "Driving while distracted is common, researchers say," and consult the first page of the "Executive Summary" in Original Source 5, "Distractions in Everyday Driving," on the website. Explain the extent to which ecological validity may be a problem in this study and what the researchers did to try to minimize the problem.

23. Refer to Case Study 5.3, "Exercise Yourself to Sleep."

    a. Discuss each of the "difficulties and disasters in experiments" (Section 5.3) as applied to this experiment.

    b. Explain whether the authors can conclude that exercise actually caused improvements in sleep.

    c. Draw a picture similar to Figure 5.1 on page 96 illustrating the process used in this experiment.

*24. To see if eating just before going to bed causes nightmares, volunteers will be recruited to spend the night in a sleep laboratory. They will be randomly assigned to be given a meal before bed or not. Numbers of nightmares will be recorded and compared for the two groups. Explain which of the "difficulties and disasters" is most likely to be a problem in this experiment, and why.

25. Suppose you wanted to know if men or women students spend more money on clothes. You consider two different plans for carrying out an observational study:

    Plan 1: Ask the participants how much they spent on clothes during the last 3 months, and then compare the amounts reported by the men and the women.

    Plan 2: Ask the participants to keep a diary in which they record their clothing expenditures for the next 3 months, and then compare the amounts recorded by the men and the women.

    a. Which of these plans is a retrospective study? What term is used for the other plan?

    b. Give one disadvantage of each plan.

26. Suppose researchers were interested in determining the relationship, if any, between brain cancer and the use of cell phones. Would it be better to use a randomized experiment or a case-control study? Explain.

*27. Read News Story 12 in the Appendix, "Working nights may increase breast cancer risk." The story describes two separate observational studies, one by Scott Davis and co-authors and one by Francine Laden and co-authors. Both studies and an editorial describing them are included on the website, and you may need to consult them. For each of the two studies, explain whether

*a. A case-control study was used. If it was, explain how the "controls" were chosen.

*b. A retrospective or a prospective study was used.

28. A case-control study claimed to have found a relationship between drinking coffee and pancreatic cancer. The cases were people recently hospitalized with pancreatic cancer, and the controls were people hospitalized for other reasons. When asked about their coffee consumption for the past year, it was found that the cancer cases drank more coffee than the controls.

    a. Draw a picture illustrating how this study was done, similar to Figure 5.3 on page 106.

    b. Give a reasonable explanation for this difference in coffee consumption, other than the possibility that drinking coffee causes pancreatic cancer.

29. Explain which of the "difficulties and disasters" in Section 5.5 is most likely to be a problem in each of the following observational studies, and why.

    a. A study measured the number of writing courses taken by college students and their subsequent scores on the quantitative part of the Graduate Record Exam. The students who had taken the largest number of writing courses scored lowest on the exam, so the researchers concluded that students who want to pursue graduate careers in quantitative areas should not take many writing courses.

    b. Successful female social workers and engineers were asked to recall whether they had any female professors in college who were particularly influential in their choice of career. More of the engineers than the social workers recalled a female professor who stood out in their minds.

*30. A company wants to know if placing live green plants in workers' offices will help reduce stress. Employees will be randomly chosen to participate, and plants will be delivered to their offices. One week after they are delivered, all employees will be given a stress questionnaire and those who received plants will be compared with those who did not. Explain which of the "difficulties and disasters" is most likely to be a problem in this experiment, and why.

31. Refer to Thought Question 5 at the beginning of this chapter. The headline was based on a study in which a representative sample of over 400,000 adults in the United States was asked a series of questions, including level of education and on how many of the past 30 days they felt physically and emotionally healthy. Explain how each of the "difficulties and disasters in observational studies" (Section 5.5) applies to this study, if at all.

32. Read the summary of News Story 1 in the Appendix. One of the results reported was that people who participated in the meditation program had better immune system response to a flu vaccine.

    a. Is a cause-and-effect relationship justified in this situation? Explain.

    b. Explain the extent to which you think the results can be extended to a population beyond only those who participated in the study.

*33. Read News Story 13 in the Appendix, and consult the original source article on the website, if necessary. One of the results reported was that teens who had more than $25 a week in spending money were more likely to use drugs than kids with less spending money.

    *a. Is a cause-and-effect relationship justified in this situation? Explain.

    *b. Explain the extent to which you think the results can be extended to a population beyond only those who participated in the study.

34. Find the story referenced as News Story 15 in the Appendix, and consult the original source article on the website, if necessary. One of the results reported was that kids with higher levels of stress in

their lives were more likely to eat high-fat foods and snacks.

a. Is a cause-and-effect relationship justified in this situation? Explain.

b. Explain the extent to which you think the results can be extended to a population beyond only those who participated in the study.

35. Explain why a randomized experiment allows researchers to draw a causal conclusion, whereas an observational study does not.

*36. A headline in the *Sacramento Bee* (11 December 1997, p. A15) read, "Study: Daily drink cuts death," and the article began with the statement, "One drink a day can be good for health, scientists are reporting, confirming earlier research in a new study that is the largest to date of the effects of alcohol consumption in the United States." The article also noted that "most subjects were white, middle-class and married, and more likely than the rest of the U.S. population to be college-educated." Explain why this study could not have been a randomized experiment.

37. Refer to Exercise 36, about a report on the relationship between drinking alcohol and likelihood of death, with the headline "Study: Daily drink cuts death."

a. Explain whether you think the headline is justified for this study.

b. The study was based on recording drinking habits for the 490,000 participants in 1982 and then noting death rates for the next 9 years. Is this a retrospective or a prospective study?

c. Comment on each of the "difficulties and disasters in observational studies" (Section 5.5) as applied to this study.

38. Read News Story 8 in the Appendix, "Education, kids strengthen marriage." Discuss the extent to which each of these problems with observational studies may affect the conclusions based on this study:

a. Confounding variables and the implications of causation

b. Extending the results inappropriately

c. Using the past as a source of data

Each of the situations in Exercises 39 to 41 contains one of the complications listed as "difficulties and disasters" with designed experiments or observational studies (Sections 5.3 and 5.5). Explain the problem and suggest how it could have been either avoided or addressed. If you think more than one complication could have occurred, mention them all, but go into detail about only the most problematic.

*39. (See instructions above.) To study the effectiveness of vitamin C in preventing colds, a researcher recruited 200 volunteers. She randomly assigned 100 of them to take vitamin C for 10 weeks and the remaining 100 to take nothing. The 200 participants recorded how many colds they had during the 10 weeks. The two groups were compared, and the researcher announced that taking vitamin C reduces the frequency of colds.

40. (See instructions before Exercise 39.) A researcher was interested in teaching couples to communicate more effectively. She had 20 volunteer couples, 10 of which were randomly assigned to receive the training program and 10 of which were not. After they had been trained (or not), she presented each of the 20 couples with a hypothetical problem situation and asked them to resolve it while she tape-recorded their conversation. She was blind as to which 10 couples had taken the training program until after she had analyzed the results.

41. (See instructions before Exercise 39.) Researchers ran an advertisement in a campus newspaper asking for sedentary volunteers who were willing to begin an exercise program. The volunteers were allowed to choose which of three programs they preferred: jogging, swimming, or aerobic dance. After 5 weeks on the exercise programs, weight loss was measured. The joggers lost the most weight, and the researchers announced that jogging was better for losing weight than either swimming or aerobic dance.

*42. Refer to Exercise 41, in which volunteers chose their preferred exercise program, and then weight loss was measured after 5 weeks.

   *a. What are the explanatory and response variables in this study?

   b. Draw a picture similar to Figure 5.1 illustrating the process used in this study.

43. For each of the following situations, draw a picture illustrating the process used in the study. The picture should be similar to whichever of Figures 5.1, 5.2, or 5.3 is most appropriate for the situation.

   a. The study described in Example 5.2 on page 97, "Blindly lowering cholesterol."

   b. The study described in Case Study 5.1 on page 98, "Quitting smoking with nicotine patches."

   c. The study described in Exercise 4 on page 113, in which a utility company randomly assigned volunteers to two different rate plans and compared usage.

Mini-Projects

1. Design an experiment to test something of interest to you. Explain how your design addresses each of the four complications listed in Section 5.3, "Difficulties and Disasters in Experiments."

2. Design an observational study to test something of interest to you. Explain how your design addresses each of the three complications listed in Section 5.5, "Difficulties and Disasters in Observational Studies."

3. Go to the library or the Internet, and locate a journal article that describes a randomized experiment. Explain what was done correctly and incorrectly in the experiment and whether you agree with the conclusions drawn by the authors.

4. Go to the library or the Internet, and locate a journal article that describes an observational study. Explain how it was done using the terminology of this chapter and whether you agree with the conclusions drawn by the authors.

5. Design and carry out a single-blind study using 10 participants, as follows. Your goal is to establish whether people write more legibly with their dominant hand. In other words, do right-handed people write more legibly with their right hand, and do left-handed people write more legibly with their left hand? Explain exactly what you did, including how you managed to conduct a single-blind study. Mention things such as whether it was an experiment or an observational study and whether you used matched pairs or not.

 6. Pick one of the news stories in the Appendix that describes a randomized experiment and that has one or more journal articles accompanying it on the companion website. Explain what was done in the experiment using the terminology and concepts in this chapter. Discuss the extent to which you agree with the conclusions drawn by the authors of the study and of the news story. Include a discussion of whether a cause-and-effect conclusion can be drawn for any observed relationships and the extent to which the results can be extended to a larger population.

7. Pick one of the news stories in the Appendix that describes an observational study and that has one or more journal articles accompanying it on the companion website. Explain what was done in the experiment using the terminology and concepts in this chapter. Discuss the extent to which you agree with the conclusions drawn by the authors of the study and of the news story. Include a discussion of whether a cause-and-effect conclusion can be drawn for any observed relationships and the extent to which the results can be extended to a larger population.

References

Amendola, K. L., D. Weisburd, E. E. Hamilton, G. Jones, and M. Slipka (2011). An experimental study of compressed work schedules in policing: advantages and disadvantages of various shift lengths. *Journal of Experimental Criminology,* 7(4) , pp. 407–442.

Coren, S., and D. Halpern. (1991). Left-handedness: A marker for decreased survival fitness. *Psychological Bulletin* 109, no. 1, pp. 90–106.

French, J. R. P (1953). Experiments in field settings. In L. Festinger and D. Katz (eds.), *Research method in the behavioral sciences.* New York: Holt, pp. 98–135.

Hurt, R., L. Dale, P. Fredrickson, C. Caldwell, G. Lee, K. Offord, G. Lauger, Z. Marŭisić, L. Neese, and T. Lundberg. (23 February 1994). Nicotine patch therapy for smoking cessation combined with physician advice and nurse follow-up. *Journal of the American Medical Association* 271, no. 8, pp. 595–600.

King, A. C., R. F. Oman, G. S. Brassington, D. L. Bliwise, and W. L. Haskell. (1 January 1997). Moderate-intensity exercise and self-rated quality of sleep in older adults. A randomized controlled trial. *Journal of the American Medical Association* 277, no. 1, pp. 32–37.

Lesko, S. M., L. Rosenberg, and S. Shapiro. (23 February 1993). A case-control study of baldness in relation to myocardial infarction in men. *Journal of the American Medical Association* 269, no. 8, pp. 998–1003.

Rosenthal, R., and K. L. Fode. (1963). The effect of experimenter bias on the performance of the albino rat. *Behavioral Science* 8, pp. 183–189.

Yamada, T., K. Hara, H. Umematsu, and T. Kadowaki (2013). Male pattern baldness and its association with coronary heart disease: a meta-analysis. *British Medical Journal Open,* 3: e002537. Doi:10.1136/bmjopen-2012-002537.

CHAPTER **6**

Getting the Big Picture

6.1 Final Questions

By now, you should have a fairly clear picture of how data should be acquired in order to be useful. We have examined how to conduct a sample survey, a randomized experiment, and an observational study and how to critically evaluate what others have done. In this chapter, we look at a few examples in depth and determine what conclusions can be drawn from them. The final question you should ask when you read the results of research is whether you will make any changes in your lifestyle or beliefs as a result of the research. To reach that conclusion, you need to answer a series of questions—not all statistical—for yourself.

Here are some guidelines for how to evaluate a study:

Step 1: Determine if the research was a sample survey, a randomized experiment, an observational study, a combination, or based on anecdotes.

Step 2: Consider the Seven Critical Components in Chapter 2 (pp. 20–21) to familiarize yourself with the details of the research.

Step 3: Based on the answer in step 1, review the "difficulties and disasters" inherent in that type of research, and determine if any of them apply.

Step 4: Determine if the information is complete. If necessary, see if you can find the original source of the report or contact the authors for missing information.

Step 5: Ask if the results make sense in the larger scope of things. If they are counter to previously accepted knowledge, see if you can get a possible explanation from the authors.

Step 6: Ask yourself if there is an alternative explanation for the results.

Step 7: Determine if the results are meaningful enough to encourage you to change your lifestyle, attitudes, or beliefs on the basis of the research.

CASE STUDY 6.1   **Can Meditation Improve Test Scores?**

ORIGINAL SOURCE: Mrazek et al. (2013), pp. 776-781.

NEWS SOURCE: "How meditation might boost your test scores"

http://well.blogs.nytimes.com/2013/04/03/how-meditation-might-boost-your-test-scores, accessed May 26, 2013.

Summary

Mindfulness meditation, an ancient technique that helps people focus and concentrate on the present moment, has gained wide popularity in recent years. Studies have shown that it can reduce stress, depression, and chronic pain, and numerous studies have suggested that it even seems to slow aging. Because one of the alleged benefits of meditation is increased concentration and reduced mind-wandering, researchers at the University of California, Santa Barbara, wondered if it might help college students perform better on exams.

To investigate this idea, they recruited 48 undergraduate students and randomly assigned them to take either a mindfulness-based meditation class or a nutrition class. The classes were taught by experts in the respective areas and met 4 days a week for 2 weeks. The students in the meditation class were expected to practice

(Continued)

meditation outside of class for at least 10 minutes a day. To have a requirement taking about the same amount of time, the nutrition students were asked to fill out a food diary every day during the 2 weeks of the class.

Before the experiment started and after it ended, all of the participants took a mock version of the verbal reasoning section of the Graduate Record Exam (GRE), with vocabulary-focused questions removed, thus focusing on reading comprehension. The results were impressive. While the students in the nutrition class did not show a change in GRE scores, the average for the students in the meditation class went from 460 before the training to 520 after the training, a 16-percentile increase. (These scores were based on the scoring method used until 2012, ranging from 200 to 800; the equivalent change using the newer scoring method would be from about 151 to 155.)

Discussion

To evaluate the usefulness of this research, let's analyze it according to the seven steps listed at the beginning of this chapter.

Step 1: Determine if the research was a sample survey, a randomized experiment, an observational study, a combination, or based on anecdotes.

This study was a randomized experiment. The student volunteers were randomly assigned to the two classes.

Step 2: Consider the Seven Critical Components in Chapter 2 (pp. 20–21) to familiarize yourself with the details of the research.

The original journal article gave information necessary for evaluating all of the components. The participants were students at the University of California, Santa Barbara, and the researchers were graduate students and faculty in the Department of Psychological and Brain Sciences there. The study was partially funded by the U.S. Department of Education and the National Science Foundation. The GRE exam was used as the main response variable, and there were a few other tests given to the participants to measure things like how often the mind wandered and how much "working memory capacity" they had. (Working memory capacity is the ability to hold information in your mind and make use of it for the task you are working on.) All of these are described in the journal article.

A few additional details are as follows. The students were not told the purpose of the experiment until it was over. Instead, they were told that the purpose was to compare two equivalent methods for improving cognitive performance. To ensure that the researchers were blind to the conditions, the GRE exams were given to both groups at the same time, and they were computer-graded to avoid any experimenter effects in grading them.

Step 3: Based on the answer in step 1, review the "difficulties and disasters" inherent in that type of research and determine if any of them apply.

The four possible complications listed for an experiment include: (1) confounding variables; (2) interacting variables; (3) placebo, Hawthorne and experimenter effect; and (4) ecological validity. Because students were randomized to the two classes, confounding variables should not be a problem. The article did describe one potential interacting variable, which was the degree to which mind-wandering was present for each student before taking the training. For students with high pre-experiment mind-wandering, meditation seemed to help reduce that as well, which may partially explain why they did better on the GRE exam. Experimenter effects were minimized because the test administrators were blind to which class students had taken. If the Hawthorne effect was a problem, students from both groups would have had improved performance but the nutrition students (who served as a control group) did not. And to test ecological validity the researchers looked at how well the students' GRE scores in the experiment correlated with their SAT scores when they applied to college. They found a high correlation, so they concluded that the "mock" administration of the GRE exam mimicked how well students would have performed on the real thing.

Step 4: Determine if the information is complete. If necessary, see if you can find the original source of the report or contact the authors for missing information.

The news report on this study was lacking some details, but the journal article was quite complete.

Step 5: Ask if the results make sense in the larger scope of things. If they are counter to previously accepted knowledge, see if you can get a possible explanation from the authors.

The journal article gave a reasonable explanation for these results, as follows:

> *The practice of mindfulness encouraged in our intervention entailed promoting a persistent effort to maintain focus on a single aspect of experience, particularly sensations of breathing, despite the frequent interruptions of unrelated perceptions or personal concerns. The present findings suggest that when this ability to concentrate is redirected to a challenging task, it can prevent the displacement of crucial task-relevant information by distractions. At least for people who struggle to maintain focus, our results suggest that the enhanced performance derived from mindfulness training results from a dampening of distracting thoughts. Our findings of reduced mind wandering are consistent with recent accounts that mindfulness training leads to reduced activation of the default network, a collection of brain regions that typically show greater activation at rest than during externally directed cognitive tasks. Both long-term meditators and individuals who have completed 2 weeks of mindfulness training show reduced activation of the default network* (Mrazek et al., 2013, p. 780).

Step 6: Ask yourself if there is an alternative explanation for the results.

The students obviously could not be blind to which class they were taking, so perhaps they guessed the purpose of the experiment. If so, the students in the meditation class may have been motivated to do well on the exam given after the course, but the students in the nutrition class would not have that motivation. However, overall this study was very well done, and it is quite likely that the results represent a real benefit of meditation.

(Continued)

Step 7: Determine if the results are meaningful enough to encourage you to change your lifestyle, attitudes, or beliefs on the basis of the research.

If these results are accurate, they indicate that students can raise their scores substantially on the verbal GRE and similar exams through meditation training. It is not clear that the benefit would extend to more technical material, such as the quantitative portion of the GRE, or to exams in most courses. However, given that meditation training does not seem to have any negative consequences (other than the time and money invested in the training), it seems reasonable to consider the benefits if you are planning to take an exam involving reading comprehension. And, of course, the benefits may extend to reading comprehension in general, which is always useful! ■

CASE STUDY 6.2   **Can Eating Cereal Reduce Obesity?**

Original Source: Frantzen et al. (2013), pp. 511-519.

News Source: "Breakfast cereal tied to lower BMI for kids"

http://www.foxnews.com/health/2013/04/10/breakfast-cereal-tied-to-lower-bmi-for-kids, accessed May 26, 2013.

Summary

The news story began with this enticing information: *"Regularly eating cereal for breakfast is tied to healthy weight for kids, according to a new study that endorses making breakfast cereal accessible to low-income kids to help fight childhood obesity."*

The implication from that quote is that eating cereal for breakfast would help reduce obesity in low-income children. Suppose you were a lawmaker and had to make a decision about whether to fund a program to provide cereal to low-income children. Would you do so based on the results of this study?

The results were based on interviews with 625 children from low-income homes in San Antonio, Texas. The same children were interviewed in grades 4, 5, and 6. For each of three days in each grade (a total of 9 days), the children were asked to list everything they ate. (Only 411 of the 625 children completed all nine interviews.) Other information measured included age, sex, ethnicity, height, and weight. Body mass index (BMI) was calculated as weight (in kilograms) divided by height-squared (in meters). BMI is used to classify people as being underweight, normal weight, overweight, or obese.

The explanatory variable was the number of days (out of 9 possible) the child ate cereal for breakfast. The response variable was the child's BMI percentile. Only 64% of the fourth graders ate breakfast at all, and that dropped to 56% for the fifth graders and only 42% for the sixth graders. Oddly, in the analysis, no differentiation was made between children who did not eat breakfast at all and children who ate something other than cereal. The main focus of the study was the relationship between number of days cereal was eaten and BMI.

On average, the more days the child ate cereal the lower the child's BMI. For instance, according to the news story, *"Kids who ate cereal four out of the nine days tended to be in the 95th percentile for BMI, which is considered overweight, compared to kids who ate cereal all nine days, whose measurements were in the 65th percentile, in the healthy weight range."*

Discussion

Does this study provide evidence that eating cereal would reduce children's BMI? Let's go through the seven steps listed at the beginning of this chapter.

Step 1: Determine if the research was a sample survey, a randomized experiment, an observational study, a combination, or based on anecdotes.

The children simply reported what they ate. They were not randomly assigned to eat breakfast cereal, so this was an observational study.

Step 2: Consider the Seven Critical Components in Chapter 2 (pp. 20–21) to familiarize yourself with the details of the research.

The data were collected from the control group of an experiment designed to study diabetes and consisted of 625 low-income children in San Antonio, Texas. The lead researcher worked for a regional dairy council in Texas, which presumably would benefit from higher consumption of cereal (with milk). The other authors of the journal article were university professors and researchers. Partial funding was provided by The National Institute of Diabetes and Digestive and Kidney Diseases, part of the National Institutes of Health. The journal article explained in detail how the children were interviewed. They were interviewed at school, and there was a separate interview for each of the 3 days, so children only needed to remember what they had eaten the day before.

It was not clear whether the interviewers knew the purpose of the study. Presumably, the children did not know the purpose of the study, but they could have guessed that it was related to food and weight, given that their weights were measured. There may have been other differences in the children who ate cereal and those who did not; most notably the two groups may have had other major dietary differences. Finally, the size of the effect was relatively small. It was reported that, on average, there was a 2% lower BMI percentile for each additional day of eating cereal.

Step 3: Based on the answer in step 1, review the "difficulties and disasters" inherent in that type of research and determine if any of them apply.

The most obvious potential problem in this study is complication 1, "confounding variables and the implication of causation." Both the original journal article and the news source imply that eating cereal actually *causes* children to have lower BMI. But children who eat cereal almost certainly eat differently at other meals than children who do not eat cereal. Therefore, there is no way to separate the effect of eating cereal from the effect of overall diet on BMI. There could be other confounding variables as well, such as amount of exercise. Parents who are more conscientious about the health of their children may be more likely to encourage them to eat a healthy breakfast and also more likely to encourage them to exercise. Therefore, it cannot be concluded that eating cereal actually *caused* the differences in BMI.

Another problem might be "extending the results inappropriately." The children in this study were mostly low income and lived in San Antonio, Texas. It

(Continued)

is not clear whether the results would extend to children in other socioeconomic groups or geographic regions.

Step 4: Determine if the information is complete. If necessary, see if you can find the original source of the report or contact the authors for missing information.

The information in the journal article was relatively complete. However, there was one finding that was not adequately covered. It was clear from the analysis that there was an interaction between sex of the child and relationship of eating cereal to BMI. Therefore, it would have been useful to see the results separately for boys and girls.

Step 5: Ask if the results make sense in the larger scope of things. If they are counter to previously accepted knowledge, see if you can get a possible explanation from the authors.

The results do make some sense. Other studies have shown a relationship between eating breakfast and healthy weight in adults. And some speculate that skipping meals triggers a "starvation mode" reaction in the body, so that it attempts to store fat.

Step 6: Ask yourself if there is an alternative explanation for the results.

In addition to the problem of confounding variables, there are a few other possible explanations for the results. One that was mentioned in the journal article is that obese and overweight children may be inclined to lie about how much food they eat. Therefore, they may be more likely to say that they did not eat breakfast when they actually did. Another possibility is that children with higher metabolism are likely to weigh less and also may need to eat more often. So it could be that the cause and effect relationship is in the opposite direction—that heavier children are able to skip breakfast more easily than underweight or normal weight children.

Step 7: Determine if the results are meaningful enough to encourage you to change your lifestyle, attitudes, or beliefs on the basis of the research.

If it turns out that there really is a causal relationship between eating cereal and obesity and if you have a problem with your weight, you might decide to eat cereal for breakfast. But it would be difficult to determine if there really is a causal relationship. To investigate that, volunteers would need to agree to eat whichever diet they were randomly assigned and would need to stick to it for a long time. It's unlikely that anyone would be willing to do that, especially if one of the "treatments" included skipping breakfast altogether. ■

CASE STUDY 6.3 **Drinking, Driving, and the Supreme Court**

SOURCE: Gastwirth (1988), pp. 524–528.

Summary

This case study doesn't require you to make a personal decision about the results. Rather, it involves a decision that was made by the Supreme Court based on statistical evidence and illustrates how laws can be affected by studies and statistics.

In the early 1970s, a young man between the ages of 18 and 20 challenged an Oklahoma state law that prohibited the sale of 3.2% beer to males under 21 but allowed its sale to females of the same age group. The case (*Craig v. Boren,* 429 U.S. 190, 1976) was ultimately heard by the U.S. Supreme Court, which ruled that the law was discriminatory.

Laws are allowed to use gender-based differences as long as they "serve important governmental objectives" and "are substantially related to the achievement of these objectives" (Gastwirth, 1988, p. 524). The defense argued that traffic safety was an important governmental objective and that data clearly show that young males are more likely to have alcohol-related accidents than young females.

The Court considered two sets of data. The first set, shown in Table 6.1, consisted of the number of arrests for driving under the influence and for drunkenness for most of the state of Oklahoma, from September 1 to December 31, 1973. The Court also obtained population figures for the age groups in Table 6.1. Based on those figures, they determined that the 1393 young males arrested for one of the two offenses in Table 6.1 represented 2% of the entire male population in the 18–21 age group. In contrast, the 126 young females arrested represented only 0.18% of the young female population. Thus, the arrest rate for young males was about 10 times what it was for young females.

The second set of data introduced into the case, partially shown in Table 6.2, came from a "random roadside survey" of cars on the streets and highways around Oklahoma City during August 1972 and August 1973. Surveys like these, despite the name, do not constitute a random sample of drivers. Information is generally collected by stopping some or all of the cars at certain locations, regardless of whether there is a suspicion of wrongdoing.

Discussion

Suppose you are a justice of the Supreme Court. Based on the evidence presented and the rules regarding gender-based differences, do you think the law should be upheld? Let's go through the seven steps introduced in this chapter with a view toward making the decision the Court was required to make.

Step 1: Determine if the research was a sample survey, a randomized experiment, an observational study, a combination, or based on anecdotes.

| TABLE 6.1 | Arrests by Age and Sex in Oklahoma, September–December 1973 |

	Males			Females		
	18–20	21 and Over	Total	18–20	21 and Over	Total
Driving under influence	427	4,973	5,400	24	475	499
Drunkenness	966	13,747	14,713	102	1,176	1,278
Total	1,393	18,720	20,113	126	1,651	1,777

(Continued)

TABLE 6.2 Random Roadside Survey of Driving and Drunkenness in Oklahoma City, August 1972 and August 1973

	Males			Females		
	Under 21	**21 and Over**	**Total**	**Under 21**	**21 and Over**	**Total**
BAC* over .01	55	357	412	13	52	65
Total	481	1926	2407	138	565	703
Percent with BAC over .01	11.4%	18.5%	17.1%	9.4%	9.2%	9.2%

*BAC = Blood alcohol content

The numbers in Table 6.1 showing arrests throughout the state of Oklahoma for a 4-month period are observational in nature. The figures do represent most of the arrests for those crimes, but the people arrested are obviously only a subset of those who committed the crimes. The data in Table 6.2 constitute a sample survey, based on a convenience sample of drivers passing by certain locations.

Step 2: Consider the Seven Critical Components in Chapter 2 (pp. 20–21) to familiarize yourself with the details of the research.

A few details are missing, but you should be able to ascertain answers to most of the components. One missing detail is how the "random roadside survey" was conducted.

Step 3: Based on the answer in step 1, review the "difficulties and disasters" inherent in that type of research and determine if any of them apply.

The arrests in Table 6.1 were used by the defense to show that young males are much more likely to be arrested for incidents related to drinking than are young females. But consider the confounding factors that may be present in the data. For example, perhaps young males are more likely to drive in ways that call attention to themselves, and thus they are more likely to be stopped by the police, whether they have been drinking or not. Thus, young females who were driving while drunk would not be noticed as often. For the data in Table 6.2, because the survey was taken at certain locations, the drivers questioned may not be representative of all drivers. For example, if a sports event had recently ended nearby, there may be more male drivers on the road, and they may have been more likely to have been drinking than normal.

Step 4: Determine if the information is complete. If necessary, see if you can find the original source of the report or contact the authors for missing information.

The information provided is relatively complete, except for the information on how the random roadside survey was conducted. According to Gastwirth (1994, personal communication), this information was not supplied in the original documentation of the court case.

Step 5: Ask if the results make sense in the larger scope of things. If they are counter to previously accepted knowledge, see if you can get a possible explanation from the authors.

Nothing is suspicious about the data in either table. In 1973, when the data were collected, the legal drinking age in Oklahoma was 21, except that females aged 18 to 20 were allowed to drink 3.2% beer. Thus, it makes sense that separate records were kept for those under 21 and those 21 and older.

Step 6: Ask yourself if there is an alternative explanation for the results.

We have discussed one possible source of a confounding variable for the arrest statistics in Table 6.1—namely, that males may be more likely to be stopped for other traffic violations. Let's consider the data in Table 6.2. Notice that almost 80% of the drivers stopped were male. Therefore, at least at that point in time in Oklahoma, males were more likely to be driving than females. That helps explain why 10 times more young men than young women had been arrested for alcohol-related reasons. The important point for the law being challenged in this lawsuit was whether young men were more likely to be driving after drinking than young women. Notice from Table 6.2 that, of those cars with young males driving, 11.4% had blood alcohol levels over 0.01; of those cars with young females driving, 9.4% had blood alcohol levels over 0.01. These rates are statistically indistinguishable.

Step 7: Determine if the results are meaningful enough to encourage you to change your lifestyle, attitudes, or beliefs on the basis of the research.

In this case study, the important question is whether the Supreme Court justices were convinced that the gender-based difference in the law was reasonable. The Supreme Court overturned the law, concluding that the data in Table 6.2 "provides little support for a gender line among teenagers and actually runs counter to the imposition of drinking restrictions based upon age" (Gastwirth, 1988, p. 527). As a consequence of this decision, the legal age for males for drinking 3.2% beer was lowered to 18 in December, 1976. However, the change was short-lived. Along with most other states, Oklahoma raised the drinking age for all alcohol to 21 in the 1980s.   ■

CASE STUDY 6.4   **Smoking During Pregnancy and Child's IQ**

Original Source: Olds, Henderson, and Tatelbaum (February 1994), pp. 221–227.
News Source: Study: Smoking may lower kids' IQs (11 February 1994), p. A-10.

Summary

The news article for this case study is shown in Figure 6.1 (next page).

Discussion

Step 1: Determine if the research was a sample survey, a randomized experiment, an observational study, a combination, or based on anecdotes.

This was an observational study because the researchers could not randomly assign mothers to either smoke or not during pregnancy; they could only observe their smoking behavior.

(Continued)

Figure 6.1

Source: "Study: Smoking May Lower Kids' IQs." Associated Press, February 11, 1994. Reprinted with permission.

Study: Smoking May Lower Kids' IQs

ROCHESTER, N.Y. (AP)—Secondhand smoke has little impact on the intelligence scores of young children, researchers found. But women who light up while pregnant could be dooming their babies to lower IQs, according to a study released Thursday. Children ages 3 and 4 whose mothers smoked 10 or more cigarettes a day during pregnancy scored about 9 points lower on the intelligence tests than the offspring of nonsmokers, researchers at Cornell University and the University of Rochester reported in this month's *Pediatrics* journal. That gap narrowed to 4 points against children of nonsmokers when a wide range of interrelated factors were controlled. The study took into account secondhand smoke as well as diet, education, age, drug use, parents' IQ, quality of parental care and duration of breast feeding. "It is comparable to the effects that moderate levels of lead exposure have on children's IQ scores," said Charles Henderson, senior research associate at Cornell's College of Human Ecology in Ithaca.

Step 2: Consider the Seven Critical Components in Chapter 2 (pp. 20–21) to familiarize yourself with the details of the research.

The brevity of the news report necessarily meant that some details were omitted. Based on the original report, the seven questions can all be answered. Following is some additional information.

The research was supported by a number of grants from sources such as the Bureau of Community Health Services, the National Center for Nursing Research, and the National Institutes of Health. None of the funders seems to represent special interest groups related to tobacco products.

The researchers described the participants as follows:

> *We conducted the study in a semirural county in New York State with a population of about 100,000. Between April 1978 and September 1980, we interviewed 500 primiparous women [those having their first live birth] who registered for prenatal care either through a free antepartum clinic sponsored by the county health department or through the offices of 11 private obstetricians. (All obstetricians in the county participated in the study.) Four hundred women signed informed consent to participate before their 30th week of pregnancy* (Olds et al., 1994, p. 221).

The researchers also noted that "eighty-five percent of the mothers were either teenagers (<19 years at registration), unmarried, or poor. Analysis [was] limited to whites who comprised 89% of the sample" (p. 221).

The explanatory variable, smoking behavior, was measured by averaging the reported number of cigarettes smoked at registration and at the 34th week of pregnancy. For the information included in the news report, the only two groups used were mothers who smoked an average of 10 or more cigarettes per day and those who smoked no cigarettes. Those who smoked between one and nine per day were excluded.

The response variable, IQ, was measured at 12 months with the Bayley Mental Development Index, at 24 months with the Cattell Scales, and at 36 and 48 months with the Stanford-Binet IQ test. In addition to those mentioned in the news source (secondhand smoke, diet, education, age, drug use, parents' IQ, quality of parental care, and duration of breast feeding), other potential confounding variables measured were husband/boyfriend support, marital status, alcohol use, maternal depressive symptoms, father's education, gestational age at initiation of prenatal care, and number of prenatal visits. None of those were found to relate to intellectual functioning. It is not clear if the study was single-blind. In other words, did the researchers who measured the children's IQs know about the mother's smoking status or not?

Step 3: Based on the answer in step 1, review the "difficulties and disasters" inherent in that type of research and determine if any of them apply.

The study was prospective, so memory is not a problem. However, there are problems with potential confounding variables, and there may be a problem with trying to extend these results to other groups, such as older mothers. The fact that the difference in IQ for the two groups was reduced from nine points to four points with the inclusion of several additional variables may indicate that the difference could be even further reduced by the addition of other variables.

The authors noted both of these as potential problems. They commented that "the particular sample used in this study limits the generalizability of the findings. The sample was at considerable risk from the standpoint of its sociodemographic characteristics, so it is possible that the adverse effects of cigarette smoking may not be as strong for less disadvantaged groups" (Olds et al., 1994, p. 225).

The authors also mentioned two potential confounding variables. First, they noted, "We are concerned about the reliability of maternal report of illegal drug and alcohol use" (Olds et al., 1994, p. 225), and, "in addition, we did not assess fully the child's exposure to side-stream smoke during the first four years after delivery" (Olds et al., 1994, p. 225).

Step 4: Determine if the information is complete. If necessary, see if you can find the original source of the report or contact the authors for missing information.

The information in the original report is fairly complete, but the news source left out some details that would have been useful, such as the fact that the mothers were young and of lower socioeconomic status than the general population of mothers.

Step 5: Ask if the results make sense in the larger scope of things. If they are counter to previously accepted knowledge, see if you can get a possible explanation from the authors.

(Continued)

The authors speculate on what the causal relationship might be, if indeed there is one. For example, they speculate that "tobacco smoke could influence the developing fetal nervous system by reducing oxygen and nutrient flow to the fetus" (p. 226). They also speculate that "cigarette smoking may affect maternal/fetal nutrition by increasing iron requirements and decreasing the availability of other nutrients such as vitamins, B12 and C, folate, zinc, and amino acids" (p. 226).

Step 6: Ask yourself if there is an alternative explanation for the results.

As with most observational studies, there could be confounding factors that were not measured and controlled. Also, if the researchers who measured the children's IQs were aware of the mother's smoking status, that could have led to some experimenter bias. You may be able to think of other potential explanations.

Step 7: Determine if the results are meaningful enough to encourage you to change your lifestyle, attitudes, or beliefs on the basis of the research.

If you were pregnant and were concerned about allowing your child to have the highest possible IQ, these results may lead you to decide to quit smoking during the pregnancy. A causal connection cannot be ruled out. ■

CASE STUDY 6.5	**For Class Discussion: Coffee and Longevity**

ORIGINAL SOURCE: Freedman et al. (2012).

NEWS SOURCE: http://www.nih.gov/news/health/may2012/nci-16.htm, accessed May 26, 2013.

Use this Case Study for class discussion or individual practice. The news source read as follows:

NIH study finds that coffee drinkers have lower risk of death

Older adults who drank coffee—caffeinated or decaffeinated—had a lower risk of death overall than others who did not drink coffee, according to a study by researchers from the National Cancer Institute (NCI), part of the National Institutes of Health, and AARP.

Coffee drinkers were less likely to die from heart disease, respiratory disease, stroke, injuries and accidents, diabetes, and infections, although the association was not seen for cancer. These results from a large study of older adults were observed after adjustment for the effects of other risk factors on mortality, such as smoking and alcohol consumption. Researchers caution, however, that they can't be sure whether these associations mean that drinking coffee actually makes people live longer. The results of the study were published in the May 17, 2012 edition of the New England Journal of Medicine.

Neal Freedman, Ph.D., Division of Cancer Epidemiology and Genetics, NCI, and his colleagues examined the association between coffee drinking and risk of death in 400,000 U.S. men and women ages 50 to 71 who participated in the NIH-AARP Diet and Health Study. Information about coffee intake was collected once by questionnaire at study entry in 1995–1996. The participants were followed until the date they died or Dec. 31, 2008, whichever came first.

The researchers found that the association between coffee and reduction in risk of death increased with the amount of coffee consumed. Relative to men and women who did not drink coffee, those who consumed three or more cups of coffee per day had approximately a 10 percent lower risk of death. Coffee drinking was not associated with cancer mortality among women, but there was a slight and only marginally statistically significant association of heavier coffee intake with increased risk of cancer death among men.

"Coffee is one of the most widely consumed beverages in America, but the association between coffee consumption and risk of death has been unclear. We found coffee consumption to be associated with lower risk of death overall and of death from a number of different causes," said Freedman. "Although we cannot infer a causal relationship between coffee drinking and lower risk of death, we believe these results do provide some reassurance that coffee drinking does not adversely affect health."

The investigators caution that coffee intake was assessed by self-report at a single time point and therefore might not reflect long-term patterns of intake. Also, information was not available on how the coffee was prepared (espresso, boiled, filtered, etc.); the researchers consider it possible that preparation methods may affect the levels of any protective components in coffee.

"The mechanism by which coffee protects against risk of death—if indeed the finding reflects a causal relationship—is not clear, because coffee contains more than 1000 compounds that might potentially affect health," said Freedman. "The most studied compound is caffeine, although our findings were similar in those who reported the majority of their coffee intake to be caffeinated or decaffeinated."

Mini-Projects

1. Find a news article about a statistical study. Evaluate it using the seven steps on page 121. If all of the required information is not available in the news article, locate the journal article or other source of the research. As part of your analysis, make sure you discuss step 7 with regard to your own life.

 2. Choose one of the news stories in the Appendix and the accompanying material on the companion website. Evaluate it using the seven steps on page 121. If all of the required information is not available in the news article, locate the journal article or other source of the research. As part of your analysis, make sure you discuss step 7 with regard to your own life.

3. Find the journal article in the *New England Journal of Medicine* on which Case Study 6.5 is based. Evaluate the study using the seven steps on page 121.

References

Frantzen, L. B., R. P. Treviño, R. M. Echon, O. Garcia-Dominic, and N. DiMarco (2013). Association between frequency of ready-to-eat cereal consumption, nutrient intakes, and body mass index in fourth- to sixth-grade low-income minority children. *Journal of the Academy of Nutrition and Dietetics* 113(4), pp. 511–519.

Freedman, N. D., Y Park, C. C. Abnet, A. R. Hollenbeck, and R. Sinha (2012). The association of coffee drinking with total and cause-specific mortality, *New England Journal of Medicine* 366, May 17, 2012, pp. 1896–1904.

Gastwirth, Joseph L. (1988). *Statistical reasoning in law and public policy.* Vol. 2. *Tort law, evidence and health.* New York: Academic Press, pp. 524–528.

Mrazek, M.D., M.S. Franklin, D.T. Phillips, B. Baird, and J.W. Schooler (2013). Mindfulness training improves working memory capacity and GRE performance while reducing mind wandering. *Psychological Science* 24(5), pp. 776–781.

Olds, D. L., C. R. Henderson, Jr., and R. Tatelbaum. (February 1994). Intellectual impairment in children of women who smoke cigarettes during pregnancy. *Pediatrics* 93, no. 2, pp. 221–227.

Study: Smoking may lower kids' IQs. (11 February 1994). *Davis* (CA) *Enterprise,* p. A-10.

PART 2

Finding Life in Data

In Part 1, you learned how data should be collected to be meaningful. In Part 2, you will learn some simple things you can do with data after it has been collected. The goal of the material in this part is to increase your awareness of the usefulness of data and to help you interpret and critically evaluate what you read in the news.

First, you will learn how to take a collection of numbers and summarize them in useful ways. For example, you will learn how to find out more about your own pulse rate by taking repeated measurements and drawing a useful picture.

Second, you will learn to critically evaluate presentations of data made by others. From numerous examples of situations in which the uneducated consumer could be misled, you will learn how to critically read and evaluate graphs, pictures, and data summaries.

CHAPTER 7

Summarizing and Displaying Measurement Data

Thought Questions

1. If you were to read the results of a study showing that daily use of a certain exercise machine for two months resulted in an average 10-pound weight loss, what more would you want to know about the numbers in addition to the average? (*Hint:* Do you think everyone who used the machine lost 10 pounds?)

2. Suppose you are comparing two job offers, and one of your considerations is the cost of living in each area. You record the price of 50 advertised apartments for each community. What summary measures of the rent values for each community would you need in order to make a useful comparison? For instance, would the lowest rent in the list be enough information?

3. In February 2013, the median sales price of a new home in the United States was $264,900, and the average price was $310,000. How do you think these values are computed? Which do you think is more useful to someone considering the purchase of a home, the median or the average? (*Source:* http://www.investmenttools.com/median_and_average_sales_prices_of_houses_sold_in_the_us.htm, June 3, 2013.)

4. The Stanford-Binet IQ test, 5th edition, is designed to have a mean, or average, of 100 for the entire population. It is also said to have a *standard deviation* of 15. What aspect of the population of IQ scores do you think is described by the "standard deviation"? For instance, do you think it describes something about the average? If not, what might it describe?

5. Students in a statistics class at a large state university were given a survey in which one question asked was age (in years). One student was a retired person, and her age was an "outlier." What do you think is meant by an "outlier"? If the students' heights were measured, would this same retired person necessarily have a value that was an "outlier"? Explain.

7.1   Turning Data into Information

How old is the oldest person you know who is currently alive? That question was posed as part of an insurance company commercial during the 2013 Super Bowl and prompted a statistics student to ask the same question of the members of her class as part of a class project. The 31 responses, in the order collected, were as follows*:

75, 90, 60, 95, 85, 84, 76, 74, 92, 62, 83, 80, 90, 65, 72, 79, 36, 78, 65, 98, 70, 88, 99, 60, 82, 65, 79, 76, 80, 52, 75

Suppose the oldest person you know is your great-grandmother Margaret, who is 86, and you are curious about where she falls relative to the people in this class project. It certainly isn't immediately obvious from the list of ages shown above. In fact, looking at a scrambled list of numbers is about as informative as looking at a scrambled set of letters. To get information out of data, the data have to be organized and summarized.

The first thought that may occur to you is to put the ages into increasing order so you could see where Margaret's age is relative to the ages from this class project. Doing that, you find:

36, 52, 60, 60, 62, 65, 65, 65, 70, 72, 74, 75, 75, 76, 76, 78, 79, 79, 80, 80, 82, 83, 84, 85, 88, 90, 90, 92, 95, 98, 99

Now you can see that Margaret would fall quite a bit above the middle, and you can count to see that there are only seven out of the 31 people in the list who are older than she is. But this list still isn't easy to assimilate into a useful picture. It would help if we could summarize the ages. There are many useful summaries of data, and they are generally categorized into four kinds of information. These are the center (mean or median), unusual values called outliers, the variability, and the shape.

Mean and Median as Measures of Center

The first useful concept is the idea of the "center" or "location" of the data. What's a typical or average value? For the ages just given, the numerical **average, or mean,** is 76.3 years. As another measure of "center" consider that there were 31 values in the list of ages, so the **median,** with half of the ages above and half of the ages below it, is 78. There are 15 ages below 78 and 15 ages above it.

To find the mean of a data set, simply add up all of the values and then divide by the number of values. In the age example, the sum of all of the ages is 2365 combined years! But the ages are divided up among 31 people, so the average is 2365/31 = 76.3 years. The median is the middle value after the numbers have been put in order. When the data set has an odd number of values, the median is the one in the middle of the ordered list, as in the age example. There are 31 ages, so the median of 78 has 15 ages above it and 15 ages below it. When the data set has an even number of values, the median is the average of the middle two. Sometimes,

*Hypothetical but realistic data, constructed to illustrate the concepts in this chapter.

there are multiple individuals with the same value, and some of them are tied with the median. So the formal definition is that the median is the value that has half of the ordered list of numbers at or above it and half of the ordered list at or below it.

If one or more values in a data set occur more than once, then the most common one is called the **mode**. The mode is sometimes mentioned as a measure of "center" but in fact it can occur anywhere in the data set, so it doesn't generally represent the center. For the age example, there are many values that occur twice, and the age 65 occurs three times, so the mode is 65. Most of the ages are in fact higher than 65, so the mode is not a very useful representation of "center" in this example. Occasionally, the mode makes sense as a measure of the most "typical" value. For instance, suppose the ages of the students in a kindergarten class were recorded at the start of the school year, and there were three children at each of ages 4 and 6, but 12 children at age 5. Then it would make sense to report that the mode of the ages is 5.

Outliers

You can see that for the oldest ages, the median of 78 is somewhat higher than the mean of 76.3. That's because a very low age, 36, pulled down the mean. It didn't pull down the median because, as long as that very low age was 78 or less, its effect on the median would be the same.

If one or more values are far removed from the rest of the data, they are called **outliers.** There are no hard and fast rules for determining what qualifies as an outlier, but we will learn some guidelines that are often used in identifying them. In this case, most people would agree that the age of 36 is so far removed from the other values that it definitely qualifies as an outlier. The reason for this outlier will be revealed after we look at possible reasons for outliers in general.

Here are three reasons outliers might occur, and what to do about them:

1. The outlier is a legitimate data value and represents natural variation in responses. In that case, the outlier should be retained and included in data summaries.

2. A mistake was made when recording the measurement or a question was misunderstood. In that case, if the correct value can found, replace the outlier. Otherwise, delete it from the dataset before computing numerical summaries. (But always report doing so.)

3. The individual(s) in question belong(s) to a different group than the rest of the individuals. In that case, outliers can be excluded if data summaries are desired for the majority group only. Otherwise, they should be retained.

Which of these three reasons is responsible for the outlier in the oldest ages? It turns out that the student who gave the response of 36 years misunderstood the question. He thought he was supposed to give the age of the oldest person whose *age* he actually *knew*, and that was his father, who was 36 years old. He knew older people, but did not know their exact ages. Whether or not to discard that outlier depends on the question we want to answer. If the question is "What are the oldest ages of the people students know?", then the outlier should be discarded (Reason 2, question was misunderstood). If the question is "What ages would students report when asked this question?", then the value of 36 years is a real response, and should not be discarded (Reason 1). But its value and the reason for it should be noted in any narrative summary of the data.

Variability

The third kind of useful information contained in a set of data is the **variability.** How spread out are the values? Are they all close together? Are most of them together, but a few are outliers? Knowing that the mean is about 76, you might wonder if your great grandmother Margaret's age of 86 is unusually high. It would obviously have a different meaning if the reported ages ranged from 72 to 80 than if they ranged from 50 to 100.

The idea of *natural variability,* introduced in Chapter 3, is particularly important when summarizing a set of measurements. Much of our work in statistics involves comparing an observed difference to what we should expect if the difference is due solely to natural variability. For instance, to determine if global warming is occurring, we need to know how much the temperatures in a given area naturally vary from year to year. To determine if a one-year-old child is growing abnormally slowly, we need to know how much heights of one-year-old children naturally vary.

Minimum, Maximum, and Range

The simplest measure of variability is to find the minimum value and the maximum value and to compute the **range,** which is just the difference between them. In the case of the oldest ages, the reported ages went from 36 to 99, for a range of 63 years. Without the outlier, they covered a 47 year range, from 52 to 99. Margaret's age is not so surprising given that range.

Temperatures over the years on a given date in a certain location may range from a record low of 59 degrees Fahrenheit to a record high of 90 degrees, a 31-degree range. We introduce two more measures of variability, the interquartile range and the standard deviation, later in this chapter.

Shape

The fourth kind of useful information is the **shape,** which can be derived from a certain kind of picture of the data. We can answer questions such as: Are most of the values clumped in the middle with values tailing off at each end? Are there two distinct groupings, with a gap between them? Are most of the values clumped together at one end with a few very high or low values? Even knowing that ages ranged from 36 to 99, Margaret's age of 86 would have a different meaning if the ages were clumped mostly in the 60s and the 90s, for instance, than if they were spread out evenly across that range.

7.2 Picturing Data: Stemplots and Histograms

About Stemplots

A **stemplot** is a quick and easy way to put a list of numbers into order while getting a picture of their shape. The easiest way to describe a stemplot is to construct one. Let's first use the ages we've been discussing, then we will turn to some real data, where each number has an identity. Before reading any further, look at the right-most part of Figure 7.1 so you can see what a completed stemplot looks like. Each of the digits extending to the right represents one data point. The first thing you see is 3|6. That represents the lowest reported age of 36. Each of the digits on the right represents one reported age. For instance, see if you can locate the age of the oldest person, 99. It's the last value to the right of the "stem" value of 9|.

Creating a Stemplot

Stemplots are sometimes called **stem-and-leaf plots** or **stem-and-leaf diagrams.** Only two steps are needed to create a stemplot—creating the stem and attaching the leaves.

Step 1: Create the stems.

The first step is to divide the range of the data into equal units to be used on the **stem.** The goal is to have approximately 6 to 15 stem values, representing equally spaced intervals. In the example shown in Figure 7.1, each of the seven stem values represents a range of 10 years of age. For instance, any age in the 80s, from 80 to 89, would be placed after the 8| on the stem.

Step 2: Attach the leaves.

The second step is to attach a **leaf** to represent each data point. The next digit in the number is used as the leaf, and if there are any remaining digits they are simply dropped. Let's use the unordered list of ages first displayed:

75, 90, 60, 95, 85, 84, 76, 74, 92, 62, 83, 80, 90, 65, 72, 79, 36, 78, 65, 98, 70, 88, 99, 60, 82, 65, 79, 76, 80, 52, 75

Figure 7.1
Building a stemplot of oldest ages

Step 1 Creating the stem	Step 2 Attaching leaves	Step 3 The finished stemplot
3		3		3	6
4		4		4	
5		5		5	2
6		6	0	6	025505
7		7	5	7	5642980965
8		8		8	5430820
9		9	05	9	052089

Example 3|6 = 36

The middle part of Figure 7.1 shows the picture after leaves have been attached for the first four ages, 75, 90, 60, and 95. The finished picture, on the right, has the leaves attached for all 31 ages. Sometimes an additional step is taken and the leaves are ordered numerically on each branch.

Further Details for Creating Stemplots

Suppose you wanted to create a picture of what your own pulse rate is when you are relaxed. You collect 25 values over a few days and find that they range from 54 to 78. If you tried to create a stemplot using the first digit as the stem, you would have only three stem values (5, 6 and 7). If you tried to use both digits for the stem, you could have as many as 25 separate values, and the picture would be meaningless.

The solution to this problem is to reuse each of the digits 5, 6, and 7 in the stem. Because you need to have equally spaced intervals, you could use each of the digits two or five times. If you use them each twice, the first listed would receive leaves from 0 to 4, and the second would receive leaves from 5 to 9. Thus, each stem value would encompass a range of five beats per minute of pulse. If you use each digit five times, each stem value would receive leaves of two possible values. The first stem for each digit would receive leaves of 0 and 1, the second would receive leaves of 2 and 3, and so on. Notice that if you tried to use the initial pulse digits three or four times each, you could not evenly divide the leaves among them because there are always 10 possible values for leaves. Figure 7.2 shows two possible stemplots for the same hypothetical pulse data. Stemplot A shows the digits 5, 6, and 7 used twice; stemplot B shows them used five times. (The first two 5's are not needed and not shown.)

EXAMPLE 7.1   Stemplot of Median Income for Families of Four

Table 7.1 lists the estimated median income for a four-person family in 2014 for each of the 50 states and the District of Columbia, information released by the U.S. government in May 2013 for use in setting aid levels in the Low Income Home

Figure 7.2
Two stemplots for the same pulse rate data

Stemplot A

```
5|4
5|7 8 9
6|0 2 3 3 4 4
6|5 5 5 6 7 7 8 9
7|0 0 1 2 4
7|5 8
```

Stemplot B

```
5|4
5|7
5|8 9
6|0
6|2 3 3
6|4 4 5 5 5
6|6 7 7
6|8 9
7|0 0 1
7|2
7|4 5
7|
7|8
```

TABLE 7.1	Estimated 2014 Median Income for a Family of Four		
Alabama	$64,899	Montana	$68,905
Alaska	$87,726	Nebraska	$74,484
Arizona	$64,434	Nevada	$69,475
Arkansas	$56,994	New Hampshire	$94,838
California	$77,679	New Jersey	$103,852
Colorado	$84,431	New Mexico	$57,353
Connecticut	$103,173	New York	$83,648
Delaware	$83,557	North Carolina	$66,985
District of Columbia	$87,902	North Dakota	$82,605
Florida	$65,406	Ohio	$73,924
Georgia	$67,401	Oklahoma	$63,580
Hawaii	$85,350	Oregon	$69,573
Idaho	$61,724	Pennsylvania	$80,937
Illinois	$81,770	Rhode Island	$87,793
Indiana	$70,504	South Carolina	$62,965
Iowa	$76,905	South Dakota	$71,207
Kansas	$74,073	Tennessee	$64,042
Kentucky	$65,968	Texas	$66,880
Louisiana	$68,964	Utah	$68,017
Maine	$74,481	Vermont	$81,408
Maryland	$105,348	Virginia	$90,109
Massachusetts	$102,773	Washington	$83,238
Michigan	$73,354	West Virginia	$63,863
Minnesota	$87,283	Wisconsin	$79,141
Mississippi	$57,662	Wyoming	$76,868
Missouri	$70,896		

Source: Federal Registry, May 15, 2013, http://www.gpo.gov/fdsys/pkg/FR-2013-05-15/html/2013-11575.htm.

Energy Assistance Program. Scanning the list gives us some information, but it would be easier to get the big picture if it were in some sort of numerical order. We could simply list the states by value instead of alphabetically, but that would not give us a picture of the location, variability, shape, and possible outliers.

Let's create a stemplot for the 51 income levels. The first step is to decide what values to use for the stem. The median family incomes range from a low of $56,994 (for Arkansas) to a high of $105,348 (for Maryland), for a range of $48,354. The goal is to use the first digit or few digits in each number as the stem, in such a way that the stem is divided into about 6 to 15 equally spaced intervals.

We have two reasonable choices for the stem values. If we use the first digit in each income value once, ranging from 5 (representing incomes in the $50,000s) to 10 (representing incomes in the $100,000s), we would have six values on the stem (5, 6, 7, 8, 9, 10). Because we need each part of the stem to represent the same range, our other choice is to divide each group of $10,000 into two intervals of $5000 each. If we divide the incomes into intervals of $5000, we will need to begin the stem with the second half of the $50,000 range (because the lowest value is

```
 5 | 677
 6 | 1233444
 6 | 5566788899
 7 | 00133444
 7 | 6679
 8 | 01123334
 8 | 57777
 9 | 04
 9 |
10 | 233
10 | 5
```

Example: 5|6 = $56,xxx

Figure 7.3
Stemplot of median incomes for families of four

$56,994) and end it with the second half of the $100,000 range, resulting in stem values of 5, 6, 6, 7, 7, 8, 8, 9, 9, 10, 10 for a total of 11 stem values. For a dataset as large as this one, it is better to spread the values out somewhat, so we will divide incomes into intervals of $5000.

Figure 7.3 shows the completed stemplot. Notice that the leaves have been put in order. Notice also that the income values have been *truncated* instead of *rounded*. To truncate a number, simply drop off the unused digits. Thus, the lowest income of $56,994 (for Arkansas) is truncated to $56,000 instead of rounded to $57,000. Rounding could be used instead.

Obtaining Information from the Stemplot

Stemplots help us determine the "shape" of a data set, identify outliers, and locate the center. For instance, the pulse rates in Figure 7.2 have a "bell shape" in which they are centered in the mid-60s and tail off in both directions from there. There are no outliers. The stemplot of ages in Figure 7.1 clearly illustrates the outlier of 36. Aside from that and the age of 52, they are somewhat *uniformly* distributed in the 60s, 70s, 80s, and 90s.

From the stemplot of median income data in Figure 7.3, we can make several observations. First, there is a wide range of values, with the median income in Maryland, the highest, being close to twice that of Arkansas, the lowest. Second, there appear to be four states with unusually high median family incomes, all over $100,000. From Table 7.1, we can see that these are Massachusetts, Connecticut, New Jersey, and Maryland. Then there is a gap before reaching New Hampshire, at $94,838. Other than the four values over $100,000, the incomes tend to be almost "bell-shaped" with a center around the mid $70,000s. There are no obvious outliers.

If we were interested in what factors determine income levels, we could use this information from the stemplot to help us. We would pursue questions like "What is different about the four high-income states?" We might notice that much of their population works in high-income cities. Many New York City employees live in Connecticut and New Jersey, and Washington, D.C. employees live in Maryland. Much of the population of Massachusetts lives and works in the Boston area.

Creating a Histogram

Histograms are pictures related to stemplots. For very large data sets, a histogram is more feasible than a stemplot because it doesn't list every data value.

To create a histogram, divide the range of the data into intervals in much the same way as we did when creating a stemplot. But instead of listing each individual value, simply count how many values fall into each part of the range. Draw a bar with height equal to the count for each part of the range. Or, equivalently, make the height equal to the *proportion* of the total count that falls in that interval.

Figure 7.4

Histogram of estimated 2014 median income for families of four, for states in the United States

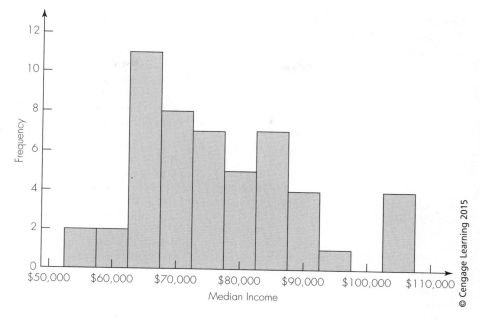

Figure 7.4 shows a histogram for the income data from Table 7.1. Each bar represents an interval of $5000, with the first one centered at $55,000 (ranging from $52,500 to $57,500). Notice that the heights of the bars are represented as frequencies. For example, there are four values in the highest median income range, centered on $105,000. If this histogram had used the proportion in each category on the vertical axis instead, then the height of the bar centered on $105,000 would be 4/51, or about 0.08. In that case, the heights of all of the bars must sum to 1, or 100%. If you wanted to know what proportion of the data fell into a certain interval or range, you would simply sum the heights of the bars for that range. Also, notice that if you were to turn the histogram on its side, it would look very much like a stemplot except that the labels would differ slightly.

EXAMPLE 7.2 Heights of British Males

Figure 7.5 (next page) displays a histogram of the heights, in millimeters, of 199 randomly selected British men. (Marsh, 1988, p. 315; data reproduced in Hand et al., 1994, pp. 179–183). The histogram is rotated sideways from the one in Figure 7.4. Some computer programs display histograms with this orientation. Notice that the heights create a "bell shape" with a center in the mid-1700s (millimeters). There are no outliers. ■

EXAMPLE 7.3 The Old Faithful Geyser

Figure 7.6 (next page) shows a histogram of the times between eruptions of the "Old Faithful" geyser. Notice that the picture appears to have two clusters of values, with one centered around 50 minutes and another, larger cluster centered around 80 minutes. A picture like this may help scientists figure out what causes the geyser to erupt when it does. ■

Figure 7.5
Heights of British males in millimeters ($N = 199$)
Source: Hand et al., 1994.

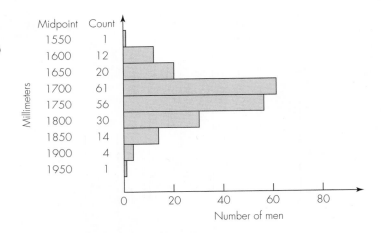

Midpoint	Count
1550	1
1600	12
1650	20
1700	61
1750	56
1800	30
1850	14
1900	4
1950	1

Figure 7.6
Times between eruptions of "Old Faithful" geyser ($N = 299$)
Source: Hand et al., 1994.

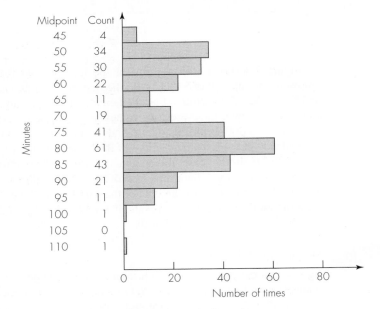

Midpoint	Count
45	4
50	34
55	30
60	22
65	11
70	19
75	41
80	61
85	43
90	21
95	11
100	1
105	0
110	1

EXAMPLE 7.4 How Much Do Students Exercise?

Students in an introductory statistics class were asked a number of questions on the first day of class. Figure 7.7 shows a histogram of 172 responses to the question "How many hours do you exercise per week (to the nearest half hour)?" Notice that the bulk of the responses are in the range from 0 to 10 hours, with a mode of 2 hours. But there are responses trailing out to a maximum of 30 hours a week, with five responses at or above 20 hours a week. ∎

Figure 7.7
Self-reported hours of exercise for 172 college students
Source: The author's students.

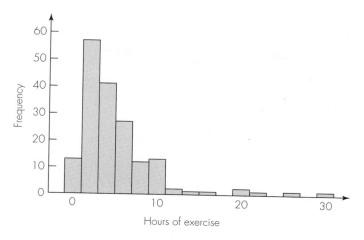

Defining a Common Language about Shape

Symmetric Data Sets

Scientists often talk about the "shape" of data; what they really mean is the shape of the stemplot or histogram resulting from the data. A **symmetric** data set is one in which, if you were to draw a line through the center, the picture on one side would be a mirror image of the picture on the other side. A special case, which will be discussed in detail in Chapter 8, is a **bell-shaped** data set, in which the picture is not only symmetric but also shaped like a bell. The stemplots in Figure 7.2, displaying pulse rates, and the histogram in Figure 7.5, displaying male heights, are approximately symmetric and bell-shaped.

Unimodal or Bimodal

Recall that the mode is the most common value in a set of data. If there is a single prominent peak in a histogram or stemplot, as in Figures 7.2 and 7.5, the shape is called **unimodal,** meaning "one mode." If there are two prominent peaks, the shape is called **bimodal,** meaning "two modes." Figure 7.6, displaying the times between eruptions of the Old Faithful geyser, is bimodal. There is one peak around 50 minutes and a higher peak around 80 minutes.

Skewed Data Sets

In common language, something that is skewed is off-center in some way. In statistics, a **skewed** data set is one that is basically unimodal but is substantially off from being bell-shaped. If it is **skewed to the right,** the higher values are more spread out than the lower values. Figure 7.7, displaying hours of exercise per week for college students, is an example of data skewed to the right. If a data set is **skewed to the left,** then the lower values are more spread out and the higher ones tend to be clumped. This terminology results from the fact that before computers were used, shape pictures were always hand drawn using the horizontal orientation in Figure 7.4. Notice that a picture that is skewed to the right, like Figure 7.7, extends further to the right of the highest peak (the tallest bar) than to the left. Most students think the terminology should be the other way around, so be careful to learn this definition! The direction of the "skew" is the direction with the unusual values, and *not* the direction with the bulk of the data.

7.3   Five Useful Numbers: A Summary

A **five-number summary** is a useful way to summarize a long list of numbers. As the name implies, this is a set of five numbers that provide a good summary of the entire list. Figure 7.8 shows what the five useful numbers are and the order in which they are usually displayed.

Figure 7.8
The five-number summary display

The lowest and highest values are self-explanatory. The median, which we discussed earlier, is the number such that half of the values are at or above it and half are at or below it. If there is an odd number of values in the data set, the median is simply the middle value in the ordered list. If there is an even number of values, the median is the average of the middle two values. For example, the median of the list 70, 75, 85, 86, 87 is 85 because it is the middle value. If the list had an additional value of 90 in it, the median would be 85.5, the average of the middle two numbers, 85 and 86. Make sure you find the middle of the *ordered* list of values.

The median can be found quickly from a stemplot, especially if the leaves have been ordered. Using Figure 7.3, convince yourself that the median of the family income data is the 26th value (51 = 25 + 1 + 25) from either end, which is the lowest of the $74,000 values. Consulting Table 7.1, we can see that the actual value is $74,073, the value for Kansas.

The **quartiles** are simply the medians of the two halves of the ordered list of numbers. The **lower quartile**—because it's halfway into the first half—is one quarter of the way from the bottom of the ordered list. Similarly, the **upper quartile** is one quarter of the way down from the top of the ordered list. Complicated algorithms exist for finding exact quartiles. We can get close enough by simply finding the median of the data first, then finding the medians of all the numbers below it and all the numbers above it. For the family income data, the lower quartile is the median of the 25 values below the median of $74,073. Notice that this would be the 13th value from the bottom because 25 = 12 + 1 + 12. Counting from the low end of the stemplot in Figure 7.3, the 13th value is the first occurrence of $66,000. Consulting Table 7.1, the value is $66,880 (Texas). The upper quartile is the median of the upper 25 values, which is the highest of the values in the $83,000s. Consulting Table 7.1, we see that it is $83,648 (New York). This tells us that three-fourths of the states have median family incomes at or below that for New York, which is $83,648.

The five-number summary for the family income data is thus:

$74,073

$66,880 $83,648

$56,994 $105,348

These five numbers provide a useful summary of the entire set of 51 numbers. We can get some idea of the middle, the spread, and whether or not the values are clumped at one end or the other. The gap between the first quartile and the median ($7193) is somewhat lower than the gap between the median and the third quartile ($9575), indicating that the values in the lower half are somewhat closer together than those in the upper half. The gap between the extremes and the quartiles are larger than between the quartiles and the median, especially at the upper end, indicating that the values are more tightly clumped in the mid-range than at the ends.

Note that, in using stemplots to find five-number summaries, we won't always be able to consult the full set of data values. Remember that we dropped the last three digits on the family incomes when we created the stemplot. If we had used the stemplot only, the family income values in the five-number summary (in thousands) would have been $56, $66, $74, $83, and $105. All of the conclusions we made in the previous paragraph would still be obvious. In fact, they may be more obvious, because the arithmetic to find the gaps would be much simpler. Truncated values from the stemplot are generally close enough to give us the picture we need.

7.4   Boxplots

A visually appealing and useful way to present a five-number summary is through a **boxplot,** sometimes called a **box and whisker plot.** This simple picture also allows easy comparison of the center and spread of data collected for two or more groups.

EXAMPLE 7.5   How Much Do Statistics Students Sleep?

During the spring semester, 190 students in a statistics class at a large university were asked to answer a series of questions in class one day, including how many hours they had slept the night before (a Tuesday night). A five-number summary for the reported number of hours of sleep is

$$
\begin{array}{ccc}
 & 7 & \\
6 & & 8 \\
3 & & 16 \\
\end{array}
$$

Two individuals reported that they slept 16 hours; the maximum for the remaining 188 students was 12 hours. A boxplot for the sleep data is shown in Figure 7.9.   ■

Figure 7.9
Boxplot for hours
of sleep

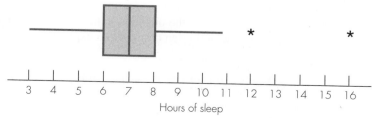

Creating a Boxplot

The boxplot for the hours of sleep is presented in Figure 7.9 and illustrates how a boxplot is constructed. Here are the steps:

1. Draw a horizontal or vertical line, and label it with values from the lowest to the highest values in the data. For the example in Figure 7.9, a horizontal line is used and the labeled values range from 3 to 16 hours.

2. Draw a rectangle, or box, with the ends of the box at the lower and upper quartiles. In Figure 7.9, the ends of the box are at 6 and 8 hours.

3. Draw a line in the box at the value of the median. In Figure 7.9, the median is at 7 hours.

4. Compute the width of the box. This distance is called the **interquartile range** because it's the distance between the lower and upper quartiles. It's abbreviated as "IQR." For the sleep data, the IQR is 2 hours.

5. Compute 1.5 times the IQR. For the sleep data, this is $1.5 \times 2 = 3$ hours. Define an *outlier* to be any value that is more than this distance from the closest end of the box. For the sleep data, the ends of the box are 6 and 8, so any value below $(6 - 3) = 3$, or above $(8 + 3) = 11$, is an outlier.

6. Draw a line or "whisker" at each end of the box that extends from the ends of the box to the farthest data value that isn't an outlier. If there are no outliers, these will be the minimum and maximum values. In Figure 7.9, the whisker on the left extends to the minimum value of 3 hours but the whisker on the right stops at 11 hours.

7. Draw asterisks to indicate data values that are beyond the whiskers and are thus considered to be outliers. In Figure 7.9, we see that there are two outliers, at 12 hours and 16 hours.

If all you have is the information contained in a five-number summary, you can draw a **skeletal boxplot** instead. The only change is that the whiskers don't stop until they reach the minimum and maximum, and thus outliers are not specifically identified. You can still determine if there are any outliers at each end by noting whether the whiskers extend more than $1.5 \times$ IQR. If so, you know that the minimum or maximum value is an outlier, but you don't know if there are any other, less extreme outliers.

Interpreting Boxplots

Boxplots essentially divide the data into fourths. The lowest fourth of the data values is contained in the range of values below the start of the box, the next fourth is contained in the first part of the box (between the lower quartile and the median), the next fourth is in the upper part of the box, and the final fourth is between the box and the upper end of the picture. Outliers are easily identified. Notice that we are now making the definition of an outlier explicit.

> An **outlier** is defined to be any value that is more than $1.5 \times$ IQR beyond the closest quartile.

In the boxplot in Figure 7.9, we can see that one-fourth of the students slept between 3 and 6 hours the previous night, one-fourth slept between 6 and 7 hours, one-fourth slept between 7 and 8 hours, and the final fourth slept between 8 and 16 hours. We can thus immediately see that the data are skewed to the right because the final fourth covers an 8-hour period, whereas the lowest fourth covers only a 3-hour period.

As the next example illustrates, boxplots are particularly useful for comparing two or more groups on the same measurement. Although almost the same information is contained in five-number summaries, the visual display makes similarities and differences much more obvious.

EXAMPLE 7.6 Who Are Those Crazy Drivers?

The survey taken in the statistics class in Example 7.5 also included the question "What's the fastest you have ever driven a car? _____ mph." The boxplots in Figure 7.10 illustrate the comparison of the responses for males and females. Here are the corresponding five-number summaries. (There are only 189 students because one didn't answer this question.)

Males (87 Students)		Females (102 Students)	
	110		89
95	120	80	95
55	150	30	130

Some features are more immediately obvious in the boxplots than in the five-number summaries. For instance, the lower quartile for the men is equal to the upper quartile for the women. In other words, 75% of the men have driven 95 mph or faster, but only 25% of the women have done so. Except for a few outliers (120 and 130), all of the women's maximum driving speeds are close to or below the median for the men. Notice how useful the boxplots are for comparing the maximum driving speeds for the sexes. ■

Figure 7.10
Boxplots for fastest ever driven a car

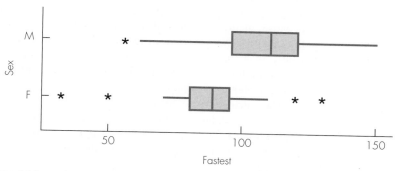

7.5   Traditional Measures: Mean, Variance, and Standard Deviation

The five-number summary has come into use relatively recently. Traditionally, only two numbers have been used to describe a set of numbers: the **mean,** representing the center, and the **standard deviation,** representing the spread or variability in the values. Sometimes the **variance** is given instead of the standard deviation. The standard deviation is simply the square root of the variance, so once you have one you can easily compute the other.

The mean and standard deviation are most useful for symmetric sets of data with no outliers. However, they are very commonly quoted, so it is important to understand what they represent, including their uses and their limitations.

The Mean and When to Use It

As we discussed earlier, the **mean** is the numerical average of a set of numbers. In other words, we add up the values and divide by the number of values. The mean can be distorted by one or more outliers and is thus most useful when there are no extreme values in the data. For example, suppose you are a student taking four classes, and the number of students in each is, respectively, 20, 25, 35, and 200. What is your typical class size? The median is 30 students. The mean, however, is 280/4 or 70 students. The mean is severely affected by the one large class size of 200 students.

As another example, refer to Figure 7.7 from Example 7.4, which displays hours per week students reportedly exercise. The majority of students exercised 10 hours or less, and the median is only 3 hours. But because there were a few very high values, the mean amount is 4.5 hours a week. It would be misleading to say that students exercise an average of 4.5 hours a week. In this case, the median is a better measure of the center of the data.

Data involving incomes or prices of things like houses and cars often are skewed to the right with some large outliers. They are unlikely to have extreme outliers at the lower end because monetary values can't go below 0. Because the mean can be distorted by the high outliers, data involving incomes or prices are usually summarized using the median. For example, the median price of a house in a given area, instead of the mean price, is routinely quoted in the economic news. That's because one house that sold for several million dollars would substantially distort the mean but would have little effect on the median. This is evident in Thought Question 3, in which it is reported that the median price of new homes in the United States in February 2013 was $264,900, but the average price, the mean, was $310,000.

The mean is most useful for symmetric data sets with no outliers. In such cases, the mean and median should be about equal. As an example, notice that the British male heights in Figure 7.5 fit that description. The mean height is 1732.5 millimeters (about 68.25 inches), and the median height is 1725 millimeters (about 68 inches).

The Standard Deviation and Variance

It is not easy to compute the **standard deviation** of a set of numbers, but most calculators and computer programs such as Excel now handle that task for you. For example, in Excel if the data are listed in rows 1 to 20 of Column A, type "=STDEV.S(A1:A20)" into any cell and the standard deviation will be shown. It is more important to know how to interpret the standard deviation, which is a useful measure of how spread out the numbers are. Consider the following two sets of numbers, both with a mean of 100:

Numbers	Mean	Standard Deviation
100, 100, 100, 100, 100	100	0
90, 90, 100, 110, 110	100	10

The first set of numbers has no spread or variability to it at all. It has a standard deviation of 0. The second set has some spread to it; on average, the numbers are about 10 points away from the mean, except for the number that is exactly at the mean. That set has a standard deviation of 10.

Computing the Standard Deviation

Here are the steps necessary to compute the standard deviation:

1. Find the mean.
2. Find the deviation of each value from the mean: value − mean.
3. Square the deviations.
4. Sum the squared deviations.
5. Divide the sum by (the number of values) − 1, resulting in the variance.
6. Take the square root of the variance. The result is the standard deviation.

Let's try this for the set of values 90, 90, 100, 110, 110.

1. The mean is 100.
2. The deviations are −10, −10, 0, 10, 10.
3. The squared deviations are 100, 100, 0, 100, 100.
4. The sum of the squared deviations is 400.
5. The (number of values) − 1 = 5 − 1 = 4, so the variance is 400/4 = 100.
6. The standard deviation is the square root of 100, or 10.

Although it may seem more logical in step 5 to divide by the number of values, rather than by the number of values minus 1, there is a technical reason for subtracting 1. The reason is beyond the level of this discussion but concerns statistical bias, as discussed in Chapter 3.

The easiest interpretation is to recognize that *the standard deviation is roughly the average distance of the observed values from their mean.* Where the data have a bell shape, the standard deviation is quite useful indeed. For example, the Stanford-Binet IQ test (5th edition) is designed to have a mean of 100 and a standard deviation of 15. If we were to produce a histogram of IQs for a large group representative of the whole population, we would find it to be approximately bell-shaped. Its center would be at 100. If we were to determine how far each person's IQ fell from 100, we would find an average distance, on one side or the other, of about 15 points. (In the next chapter, we will see how to use the standard deviation of 15 in a more useful way.) For shapes other than bell shapes, the standard deviation is useful as an intermediate tool for more advanced statistical procedures; it is not very useful on its own, however.

EXAMPLE 7.7 | Putting it All Together with Women's Heights

Let's look at how to combine the information learned in this chapter into one coherent story. Women in a college statistics class were asked to report various measurements, including height, for which 94 women responded. Most of them reported height to the nearest inch, but a few reported it to the nearest half inch.

We probably can consider these women to be representative of all college women for this measurement. What can we learn about college women's heights from these 94 individuals? Figure 7.11 illustrates a boxplot of these measurements, and the five-number summary used to construct it. From these, we learn the following:

- The heights ranged from 59 inches (4 feet, 11 inches) to 70 inches (5 feet, 10 inches).
- The median of 64 means that half of the women reported heights of 64 inches or more, and half reported 64 inches or less.
- The lowest one-fourth of the women ranged from 59 to 63 inches, the next one-fourth from 63 to 64 inches, the next one-fourth from 64 to 66 inches, and the final one-fourth from 66 to 70 inches. Therefore, heights are more spread out in the extremes than in the middle and are slightly more spread out in the upper half than in the lower half.
- There are no outliers.

Figure 7.11
Boxplot of heights of 94 college women, and five-number summary used to create it

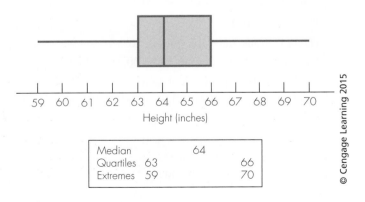

© Cengage Learning 2015

Figure 7.12
Stemplot and histogram for heights of 94 college women for Example 7.7

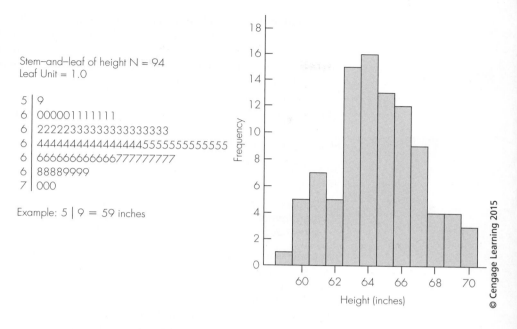

Stem–and–leaf of height N = 94
Leaf Unit = 1.0

```
5 | 9
6 | 000001111111
6 | 22222333333333333333
6 | 44444444444444444455555555555555
6 | 666666666666777777777
6 | 88889999
7 | 000
```

Example: 5 | 9 = 59 inches

© Cengage Learning 2015

Figure 7.11 gives information about location, spread, and outliers but does not tell us much about the shape. For that, we need to use a stemplot or a histogram. Both of these (created by the Minitab Statistical Software program) are shown in Figure 7.12. Note that the stemplot uses seven intervals, allocating 2 inches to each interval, while the histogram uses 12 intervals, one for each inch of height. Both pictures show that the pattern is bell-shaped, although the histogram obscures the smooth shape somewhat because of the level of detail.

The only measures that aren't directly obvious from either Figure 7.11 or Figure 7.12 are the mean and standard deviation. Those can be computed from the original 94 observations. The mean is 64.455 inches, and the standard deviation is 2.51 inches. We will round these off to 64.5 and 2.5 inches. In Chapter 8, we will revisit this example and learn that these two measures are very useful for bell-shaped distributions.

7.6 Caution: Being Average Isn't Normal

By now, you should realize that it takes more than just an average value to describe a set of measurements. Yet, it is a common mistake to confuse "average" with "normal." For instance, if a young boy is tall for his age, people might say something like "He's taller than normal for a three-year-old." In fact, what they mean is that he's taller than the *average* height of three-year-old boys. There is quite a range of possible heights for three-year-old boys, and as we will learn in Chapter 8, any height within a few standard deviations of the mean is quite "normal." Be careful about confusing "average" and "normal" in your everyday speech.

Equating "normal" with average is particularly common in weather data reportage. News stories often confuse these. When reporting rainfall data, this confusion leads to stories about drought and flood years when in fact the rainfall for the year is well within a "normal" range. If you pay attention, you will notice this mistake being made in almost all news reports about the weather.

EXAMPLE 7.8   ## How Much Hotter Than Normal Is Normal?

It's true that the beginning of October, 2001 was hot in Sacramento, California. But how much hotter than "normal" was it? According to the *Sacramento Bee:*

> October came in like a dragon Monday, hitting 101 degrees in Sacramento by late afternoon. That temperature tied the record high for Oct. 1 set in 1980—and was 17 degrees higher than normal for the date. (Korber, 2001)

The article was accompanied by a drawing of a thermometer showing that the "Normal High" for the day was 84 degrees. This is the basis for the statement that the high of 101 degrees was 17 degrees higher than normal. But the high temperature for October 1 is quite variable. October is the time of year when the weather is changing from summer to fall, and it's quite natural for the high temperature to be in the 70s, 80s, or 90s. While 101 was a record high, it was not "17 degrees higher than normal" if "normal" includes the range of possibilities likely to occur on that date.   ∎

CASE STUDY 7.1   ## Detecting Exam Cheating with a Histogram

SOURCE: Boland and Proschan (Summer 1990), pp. 10–14.

A class of 88 students at a university in Florida was taking a 40-question multiple-choice exam when the proctor happened to notice that one student, whom we will call C, was probably copying answers from a student nearby, whom we will call A. Student C was accused of cheating, and the case was brought before the university's supreme court.

At the trial, evidence was introduced showing that of the 16 questions *missed* by both A and C, both had made the same wrong guess on 13 of them. The prosecution argued that a match that close by chance alone was very unlikely, and student C was found guilty of academic dishonesty. The case was challenged, however, partly because in calculating the odds of such a strong match, the prosecution had used an unreasonable assumption. They assumed that any of the four wrong answers on a missed question would be equally likely to be chosen. Common sense, as well as data from the rest of the class, made it clear that certain wrong answers were more attractive choices than others.

A second trial was held, and this time the prosecution used a more reasonable statistical approach. The prosecution created a measurement for each student in the class except A (the one from whom C allegedly copied), resulting in 87 data values. For each student, the prosecution simply counted how many of his or her 40 answers matched the answers on A's paper. For Student C, there were

Figure 7.13
Histogram of the number of matches to A's answers for each student
Source: Data from Boland and Proschan, Summer 1990, p. 14.

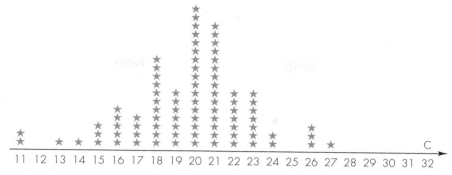

32 matches to Student A, including 19 questions they both got right and 13 out of the 16 questions they both got wrong.

The results are shown in the histogram in Figure 7.13. Student C is coded as a C, and each asterisk represents one other student. Student C is an obvious outlier in an otherwise bell-shaped picture. You can see that it would be quite unusual for that particular student to match A's answers so well without some explanation other than chance.

Unfortunately, the jury managed to forget that the proctor observed Student C looking at Student A's paper. The defense used this oversight to convince them that, based only on the histogram, A could have been copying from C. The guilty verdict was overturned, despite the compelling statistical picture and evidence. ∎

Thinking About Key Concepts

- Knowing the *mean* (average) of a list of numbers is not very information without additional information because *variability* is inherent in almost all measurements. It is useful to know how spread out the numbers are, what *range* they cover, what basic *shape* they have, and whether there are any *outliers*.
- *Outliers* are values that are far removed from the bulk of the data. They distort the mean and standard deviation, but have little effect on the median or the interquartile range. There are three basic reasons outliers occur, and how they should be treated depends on which of these reasons holds.
- The *mean* and *median* for a set of measurements can be very different from each other if there are *outliers* or extreme *skewness* in the numbers. The median is generally a more appropriate representation of a "typical" value than is the mean in that case.
- Useful pictures of data include *stemplots*, *histograms*, and *boxplots*. Shape can be determined from stemplots and histograms, but boxplots are the most useful type of display for comparing two or more groups.
- "Normal" should not be equated with "average." Any number in a range of values that routinely occur should be considered to be normal.

Focus on Formulas

The Data

n = number of observations

x_i = the *ith* observation, $i = 1, 2, \ldots, n$

The Mean

$$\bar{x} = \frac{1}{n}(x_1 + x_2 + \cdots + x_n) = \frac{1}{n}\sum_{i=1}^{n} x_i$$

The Variance

$$s^2 = \frac{1}{(n-1)}\sum_{i=1}^{n}(x_i - \bar{x})^2$$

The Computational Formula for the Variance (easier to compute directly with a calculator)

$$s^2 = \frac{1}{(n-1)}\left(\sum_{i=1}^{n} x_i^2 - \frac{\left(\sum_{i=1}^{n} x_i\right)^2}{n}\right)$$

The Standard Deviation

Use either formula to find s^2, then take the square root to get the standard deviation s.

Exercises

Exercises with numbers divisible by 3 (3, 6, 9, etc.) are included in the Solutions at the back of the book. They are marked with an asterisk (*).

1. At the beginning of this chapter, the following "oldest ages" were listed, and a stemplot was shown for them in Figure 7.1: 75, 90, 60, 95, 85, 84, 76, 74, 92, 62, 83, 80, 90, 65, 72, 79, 36, 78, 65, 98, 70, 88, 99, 60, 82, 65, 79, 76, 80, 52, 75

   a. Create a five-number summary for these ages.

   b. Create a boxplot using the five-number summary from part (a).

2. Refer to the previous exercise, and the stemplot for the "oldest ages" in Figure 7.1.

   a. Create a stemplot for the oldest ages using each 10s value twice instead of once on the stem.

   b. Compare the stemplot created in part (a) with the one in Figure 7.1. Are any features of the data apparent in the new stemplot that were not apparent in Figure 7.1? Are any features of the data apparent in Figure 7.1 that are not apparent in the new stemplot? Explain.

*3. Refer to the pulse rate data displayed in the stemplots in Figure 7.2.

   *a. Find the median.

   *b. Create a five-number summary.

4. Suppose the scores on a recent exam in your statistics class were as follows: 78, 95, 60, 93, 55, 84, 76, 92, 62, 83, 80, 90, 64, 75, 79, 32, 75, 64, 98, 73, 88, 61, 82, 68, 79, 78, 80, 85.

   a. Create a stemplot for these test scores using each 10s value once on the stem.

   b. Create a stemplot for these test scores using each 10s value twice on the stem.

   c. Compare the stemplots created in parts (a) and (b). Are any features of the data apparent in one of them but not the other one? Explain.

5. Refer to the test scores in the Exercise 4.

   a. Create a histogram for the test scores.

   b. Explain how you decided how many intervals to use for the histogram in part (a).

   c. Comment on the shape of the histogram in part (a).

*6. Refer to the test scores in Exercise 4.

   *a. Find the median for the test scores.

   *b. Find the mean for the test scores and compare it to the median. Which one is larger, and why?

   *c. Find the standard deviation for the test scores. If possible, use software or a calculator to do the calculations.

   *d. Do you think the standard deviation would be smaller or larger if the test score of 32 were to be removed from the data? Explain your answer.

7. Refer to the test scores in Exercise 4.

   a. Create a five-number summary.

   b. Create a boxplot.

   c. The test score of 32 should have been identified as an outlier. Which of the three reasons for outliers given on page 139 do you think explains this outlier? Should the outlier be removed when a description of the test scores is presented? Explain your reasoning.

8. Refer to the pulse rates given in the stemplots in Figure 7.2 (page 142).

   a. Create a five-number summary of the pulse rates (also found in Exercise 3b).

   b. Create a boxplot of the pulse rates.

*9. Give an example for which the median would be more useful than the mean as a measure of center. (Do not reuse an example already discussed in this chapter.)

10. Refer to the "oldest ages" given at the beginning of this chapter and in Exercise 1.

   a. Create a histogram for these ages.

   b. Explain how you decided how many intervals to use for the histogram in part (a).

   c. Comment on the shape of the histogram in part (a).

11. Construct an example, and draw a histogram for a measurement that you think would be bell-shaped.

*12. Construct an example, and draw a histogram for a measurement that you think would be skewed to the right.

13. Construct an example, and draw a histogram for a measurement that you think would be bimodal.

14. Give one advantage a stemplot has over a histogram and one advantage a histogram has over a stemplot.

*15. Which set of data is more likely to have a bimodal shape: daily New York City temperatures at noon for the summer months or daily New York City temperatures at noon for an entire year? Explain.

16. Give an example of a set of more than five numbers that has a five-number summary of:

Median	40	
Quartiles	30	70
Extremes	10	80

17. All of the information contained in the five-number summary for a data set is required for constructing a boxplot. What additional information is required?

*18. Give an example of a measurement for which the mode would be more useful than the median or the mean as an indicator of the "typical" value. (Do not reuse the example of kindergarten children's ages given on page 139.)

19. Three types of pictures were presented in this chapter: stemplots, histograms, and boxplots. Explain the features of a data set for which:

    a. Stemplots are most useful.

    b. Histograms are most useful.

    c. Boxplots are most useful.

20. Refer to the data on median family income in Table 7.1; a five-number summary is given in Section 7.3, page 148. What income value would be an outlier at the upper end, using the definition of an outlier on page 150? Determine if there are any outliers, and if so, which values are outliers.

*21. Refer to the data on median family income in Table 7.1; a five-number summary is given in Section 7.3, page 148.

    *a. What is the value of the range?

    *b. What is the value of the interquartile range?

22. Refer to the data on median family income in Table 7.1; a five-number summary is given in Section 7.3, page 148.

    a. Construct a boxplot for this data set.

    b. Discuss which picture is more useful for this data set: the boxplot from part (a), or the histogram in Figure 7.4.

23. The data on hours of sleep discussed in Example 7.5 also included whether each student was male or female. Here are the separate five-number summaries for "hours of sleep" for the two sexes:

Males		Females	
	7		7
6	8	6	8
3	16	3	11

    a. Two males reported sleeping 16 hours and one reported sleeping 12 hours. Using this information and the five-number summaries, draw boxplots that allow you to compare the sexes on number of hours slept the previous night. Use a format similar to Figure 7.10.

    b. Based on the boxplots in part (a), describe the similarities and differences between the sexes for number of hours slept the previous night.

*24. Suppose an advertisement reported that the mean weight loss after using a certain exercise machine for 2 months was 10 pounds. You investigate further and discover that the median weight loss was 3 pounds.

    *a. Explain whether it is most likely that the weight losses were skewed to the right, skewed to the left, or symmetric.

    *b. As a consumer trying to decide whether to buy this exercise machine, would it have been more useful for the company to give you the mean or the median? Explain.

25. Find a set of data of interest to you that includes at least 12 numbers, such as local rents from a website, or scores from a sport of interest. Include the data with your answer.

    a. Create a five-number summary of the data.

    b. Create a boxplot of the data.

c. Describe the data in a paragraph that would be useful to someone with no training in statistics.

26. Draw a boxplot illustrating a data set with each of the following features:

a. Skewed to the right with no outliers.

b. Bell-shaped with the exception of one outlier at the upper end.

c. Values uniformly spread across the range of the data.

*27. Would outliers more heavily influence the range or the quartiles? Explain.

28. Suppose you had a choice of two professors for a class in which your grade was very important. They both assign scores on a percentage scale (0 to 100). You can have access to three summary measures of the last 200 scores each professor assigned. Of the numerical summary measures discussed in this chapter, which three would you choose? Why?

29. The students surveyed for the data on exercising in Example 7.4 were also asked "How many alcoholic beverages do you consume in a typical week?" Five-number summaries for males' and females' responses are

Males		Females	
2		0	
0	10	0	2
0	55	0	17.5

a. Draw side-by-side skeletal boxplots for the data.

b. Are the values within each set skewed to the right, bell-shaped, or skewed to the left? Explain how you know.

*30. Refer to the five-number summaries given in Exercise 29 for amount of alcohol consumed by males and females. In each case, would the mean be higher, lower, or about the same as the median? Explain how you know.

31. Refer to the alcohol consumption data in Exercise 29. Students were also asked if they typically sit in the front, back, or middle of the classroom. Here are the responses to the question about alcohol consumption for the students who responded that they typically sit in the back of the classroom:

Males ($N = 22$): 0, 0, 0, 0, 0, 0, 0, 1, 3, 3, 4, 5, 10, 10, 10, 14, 15, 15, 20, 30, 45, 55

Females ($N = 14$): 0, 0, 0, 0, 0, 1, 2, 2, 4, 4, 10, 12, 15, 17.5

a. Create a five-number summary for the males, and compare it to the one for all of the males in the class, shown in the Exercise 29. What does this say about the relationship between where one sits in the classroom and drinking alcohol for males?

b. Repeat part (a) for the females.

c. Create a stemplot for the males, and comment on its shape.

d. Create a stemplot for the females, and comment on its shape.

32. Refer to the data in Exercise 31 on alcohol consumption for students who typically sit in the back of the classroom. Using the definition of outliers on page 150, identify which value(s) are outliers in each of the two sets of values (males and females).

*33. Refer to the data in Exercise 31 on alcohol consumption for students who typically sit in the back of the classroom. Find the mean and median number of drinks for males. Which one is a better representation of how much a typical male who sits in the back of the room drinks? Explain.

34. a. Give an example of a set of five numbers with a standard deviation of 0.

b. Give an example of a set of four numbers with a mean of 15 and a standard deviation of 0.

c. Is there more than one possible set of numbers that could be used to answer part (a)? Is there more than one possible set of numbers that could be used to answer part (b)? Explain.

35. Find the mean and standard deviation of the following set of numbers: 10, 20, 25, 30, 40.

*36. In each of the following cases, would the mean or the median probably be higher, or would they be about equal?

*a. Prices of all new cars sold in 1 month in a large city.

*b. Heights of all 7-year-old children in a large city.

*c. Shoe sizes of adult women.

37. In each of the following cases, would the mean or the median probably be higher, or would they be about equal?

a. Salaries in a company employing 100 factory workers and two highly paid executives.

b. Ages at which residents of a suburban city die, including everything from infant deaths to the most elderly.

38. Suppose a set of test scores is approximately bell-shaped, with a mean of 70 and a range of 50. Approximately, what would the minimum and maximum test scores be?

*39. What is the variance for the fifth edition of the Stanford-Binet IQ test? (Hint: See Thought Question 4 on page 137.)

40. Explain the following statement in words that someone with no training in statistics would understand: The heights of adult males in the United States are bell-shaped, with a mean of about 70 inches and a standard deviation of about 3 inches.

41. The *Winters* (CA) *Express* on April 4, 2013, reported that the seasonal rainfall (since July 1, 2012) for the year was 16.75 inches, and that the "average to April 3 is 20.09 inches." Does this mean that the area received abnormally low rainfall in the period from July 1, 2012, to April 4, 2013? Explain, and mention any additional information that might help you make this assessment.

*42. Suppose you are interested in genealogy, and want to try to predict your potential longevity by using the ages at death of your ancestors. You find out the ages at death for the eight great-grandparents of your mother and your father. Suppose the ages (in numerical order) are as follows:

Mother's great-grandparents: 78, 80, 80, 81, 81, 82, 82, 84

Father's great-grandparents: 30, 50, 77, 80, 82, 90, 95, 98

*a. Compare the medians for the two sets of ages. How do they compare?

*b. Compare the means for the two sets of ages. How do they compare?

*c. Find the standard deviation for each set of ages.

*d. Which set do you think is more useful for predicting longevity in your family? Explain.

43. According to the National Weather Service, there is about a 10% chance that total annual rainfall for Sacramento, CA, will be less than 11.1 inches and a 20% chance that it will be less than 13.5 inches. At the upper end, there is about a 10% chance that it will exceed 29.8 inches and a 20% chance that it will exceed 25.7 inches. The average amount is about 19 inches. In the 2001 year (July 1, 2000–June 30, 2001), the area received about 14.5 inches of rain. Write two news reports of this fact, one that conveys an accurate comparison to other years and one that does not.

44. Refer to Original Source 5 on the companion website, "Distractions in everyday driving." Notice that on page 86 of the report, when the responses are summarized for the quantitative data in question 8, only the mean is provided. But for questions 7 and 9 the mean and median are provided. Why do you think the median is provided in addition to the mean for these two questions?

*45. Refer to Original Source 20 on the companion website, "Organophosphorus pesticide exposure of urban and suburban preschool children with organic and conventional diets." In Table 4 on page 381, information is presented for estimated dose levels of various pesticides for children who eat organic versus conventional produce. Find the data for malathion. Assume the minimum exposure for both groups is zero. Otherwise, all of the information you need for a five-number summary is provided. (Note that the percentiles are the values with that percent of the data at or below the value. For instance, the median is the 50th percentile.)

*a. Create a five-number summary for the malathion exposure values for each group (organic and conventional).

b. Construct side-by-side skeletal boxplots for the malathion exposure values for the two groups. Write a few sentences comparing them.

c. Notice that in each case the mean is higher than the median. Explain why this is the case.

Mini-Projects

1. Find a set of data that has meaning for you. Internet sites are good sources for data. For instance, find ads for something you would like to buy, sports scores, weather data, and so on. Using the methods given in this chapter, summarize and display the data in whatever ways are most useful. Give a written description of interesting features of the data.

2. Measure your pulse rate 25 times over the next few days, but don't take more than one measurement in any 10-minute period. Record any unusual events related to the measurements, such as if one was taken during exercise or one was taken immediately upon awakening. Create a stemplot and a five-number summary of your measurements. Give a written assessment of your pulse rate based on the data.

References

Boland, Philip J., and Michael Proschan. (Summer 1990). The use of statistical evidence in allegations of exam cheating. *Chance* 3, no. 3, pp. 10–14.

Hand, D. J., F. Daly, A. D. Lunn, K. J. McConway, and E. Ostrowski. (1994). *A handbook of small data sets*. London: Chapman and Hall.

Korber, Dorothy (2001). "Record temperature ushers in October." *Sacramento Bee* (2 Oct. 2001), p. B1.

Marsh, C. (1988). *Exploring data*. Cambridge, England: Policy Press.

CHAPTER **8**

Bell-Shaped Curves
and Other Shapes

Thought Questions

1. The heights of adult women in the United States follow, at least approximately, a bell-shaped curve. What do you think this means?

2. What does it mean to say that a man's weight is in the 30th percentile for all adult males?

3. A "standardized score" is simply the number of standard deviations an individual score falls above or below the mean for the whole group. (Values above the mean have positive standardized scores, whereas those below the mean have negative ones.) Adult male heights in the United States have a mean of 70 inches and a standard deviation of 3 inches. Adult female heights in the United States have a mean of 65 inches and a standard deviation of 2 1/2 inches. Thus, a man who is 73 inches tall has a standardized score of 1. What is the standardized score corresponding to your own height, compared to adults of your sex in the United States?

4. Data sets consisting of physical measurements (heights, weights, lengths of bones, and so on) for adults of the same species and sex tend to follow a similar pattern. The pattern is that most individuals are clumped around the average, with numbers decreasing the farther values are from the average in either direction. Describe what shape a histogram of such measurements would have.

8.1   Populations, Frequency Curves, and Proportions

In Chapter 7, we learned how to draw a picture of a set of data and how to think about its shape. In this chapter, we learn how to extend those ideas to pictures and shapes for populations of measurements. For example, in Figure 7.5 we illustrated that, based on a sample of 199 men, heights of adult British males are reasonably bell-shaped. Because the men were a representative sample, the picture for all of the millions of British men is probably similar. But even if we could measure them all, it would be difficult to construct a histogram with so much data. What is the best way to represent the shape of a large population of measurements?

Frequency Curves

The most common type of picture for a population is a smooth **frequency curve.** Rather than drawing lots of tiny rectangles, the picture is drawn as if the tops of the rectangles were connected with a smooth curve. Figure 8.1 illustrates a frequency curve for the population of British male heights, based on the assumption that the heights shown in the histogram in Figure 7.5 are representative of all British men's heights. Notice that the picture is similar to the histogram in Figure 7.5, except that the curve is smooth and the heights have been converted to inches. The mean and standard deviation for the 199 men in the sample are 68.2 inches and 2.7 inches, respectively, and were used to draw Figure 8.1. Later in this chapter you will learn how to draw a picture like this, based on knowing only the mean and standard deviation.

Notice that the vertical scale is simply labeled "height of curve." This height is determined by sizing the curve so that the area under the entire curve is 1, for reasons that will become clear in the next few pages. Unlike with a histogram, the height of the curve cannot be interpreted as a proportion or frequency, but is chosen simply to satisfy the rule that the entire area under the curve is 1.

The bell shape illustrated in Figure 8.1 is so common that if a population has this shape, the measurements are said to follow a **normal distribution.** Equivalently,

Figure 8.1
A normal frequency curve

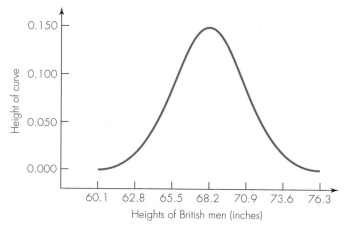

Figure 8.2
A nonnormal frequency curve

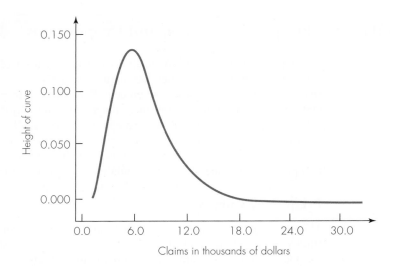

they are said to follow a **bell-shaped curve,** a **normal curve,** or a **Gaussian curve.** This last name comes from the name of Karl Friedrich Gauss (1777–1855), who was one of the first mathematicians to investigate the shape.

A population of measurements that follow a normal distribution is said to be **normally distributed**. For instance, we would say that the heights of British males are (approximately) normally distributed with a mean of 68.2 inches and a standard deviation of 2.7 inches. Because normal distributions are symmetric, the mean and median are equal. For instance, the median height of British males would also be 68.2 inches, so about half of all British males are taller than this height, and about half are shorter.

Not all frequency curves are bell-shaped. Figure 8.2 shows a likely frequency curve for the population of dollar amounts of car insurance damage claims for a Midwestern city in the United States, based on data from 187 claims in that city in the early 1990s (Ott and Mendenhall, 1994). Notice that the curve is skewed to the right. Most of the claims were below $12,000, but occasionally there was an extremely high claim. For the remainder of this chapter, we focus on bell-shaped curves.

EXAMPLE 8.1   College Women's Heights

Based on the heights of 94 college women discussed in Example 7.7 and displayed in Figure 7.12, we can speculate that the population of heights of all college women is approximately normally distributed with a mean of 64.5 inches and a standard deviation of 2.5 inches. Figure 8.3 shows the same histogram as Figure 7.12. It has the normal curve with the same mean (64.5 inches) and standard deviation (2.5 inches) drawn on top of it. Note that the histogram is somewhat "choppy" but if we had thousands of heights it would look much more like the smooth curve. ∎

Figure 8.3

Heights of 94 college women (histogram) and normal curve for heights of all college women (smooth curve)

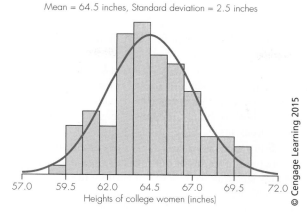

Mean = 64.5 inches, Standard deviation = 2.5 inches

57.0   59.5   62.0   64.5   67.0   69.5   72.0

Heights of college women (inches)

© Cengage Learning 2015

Proportions

Frequency curves are quite useful for determining what **proportion** or percentage of the population of measurements falls into a certain range. If we wanted to find out what proportion of the data fell into any particular range with a stemplot, we would count the number of leaves that were in that range and divide by the total. If we wanted to find the proportion in a certain range using a histogram, we would simply add up the heights of the rectangles for that range, assuming we had used proportions instead of counts for the heights. If not, we would add up the counts for that range and divide by the total number in the sample.

What if we have a frequency curve instead of a stemplot or histogram? Frequency curves are, by definition, drawn to make it easy to represent the proportion of the population falling into a certain range. Recall that they are drawn so the entire area underneath the curve is 1, or 100%. Therefore, to figure out what percentage or proportion of the population falls into a certain range, all you have to do is figure out how much of the area is situated over that range. For example, in Figure 8.1, half of the area is in the range above the mean height of 68.2 inches. In other words, about half of all British men are 68.2 inches or taller.

Although it is easy to visualize what proportion of a population falls into a certain range using a frequency curve, it is not as easy to compute that proportion. For anything but very simple cases, the computation to find the required area involves the use of calculus. However, because bell-shaped curves are so common, tables have been prepared in which the work has already been done (see, for example, Table 8.1 at the end of this chapter), and many calculators and computer applications such as Excel will compute these proportions.

8.2   The Pervasiveness of Normal Curves

Nature provides numerous examples of populations of measurements that, at least approximately, follow a normal curve. If you were to create a picture of the shape of almost any physical measurement within a homogeneous population, you would probably get the familiar bell shape. In addition, many psychological attributes, such as IQ, are normally distributed. Many standard academic tests, such as the Scholastic Assessment Test (SAT), if given to a large group, will result in normally distributed scores.

The fact that so many different kinds of measurements all follow approximately the same shape should not be surprising. The majority of people are somewhere close to average on any attribute, and the farther away you move from the average, either above or below, the fewer people will have those more extreme values for their measurements.

Sometimes a set of data is distorted to make it fit a normal curve. That's what happens when a professor "grades on a bell-shaped curve." Rather than assign the grades students have actually earned, the professor distorts them to make them fit into a normal curve, with a certain percentage of A's, B's, and so on. In other words, grades are assigned *as if* most students were average, with a few strong ones at the top and a few weak ones at the bottom. Unfortunately, this procedure has a tendency to artificially spread out clumps of students who are at the top or bottom of the scale, so that students whose original grades were very close together may receive different letter grades.

8.3 Percentiles and Standardized Scores

Percentiles

Have you ever wondered what percentage of the population of your sex is taller than you are, or what percentage of the population has a lower IQ than you do? Your **percentile** in a population represents the position of your measurement in comparison with everyone else's. It gives the percentage of the population that falls *below* you. If you are in the 50th percentile, it means that exactly half of the population falls below you. If you are in the 98th percentile, 98% of the population falls below you and only 2% is above you.

Your percentile is easy to find if the population of values has an approximate bell shape and if you have just three pieces of information. All you need to know are your own value and the mean and standard deviation for the population. Although there are obviously an unlimited number of potential bell-shaped curves, depending on the magnitude of the particular measurements, each one is completely determined once you know its mean and standard deviation. In addition, each one can be "standardized" in such a way that the same table can be used to find percentiles for any of them.

Standardized Scores

Suppose you knew your IQ was 115, as measured by the 5th edition of the Stanford-Binet IQ test. (Earlier editions had a standard deviation of 16, but the more recent 5th edition has a standard deviation of 15.) Scores from that test have a normal distribution with a mean of 100 and a standard deviation of 15. Therefore, your IQ is exactly 1 standard deviation above the mean of 100. In this case, we would say you have a *standardized score* of 1. In general, a **standardized score** simply represents the number of standard deviations the observed value or score falls from the mean. A positive standardized score indicates an observed value above the mean, whereas a negative standardized score indicates a value below the mean. Someone with an IQ of 85 would have a standardized score of −1 because he or she would be exactly 1 standard deviation below the mean. Sometimes the abbreviated term **standard score** is used instead of "standardized score." The letter z is often used to represent a standardized score, so another synonym is **z-score.**

Once you know the standardized score for an observed value, all you need to find the percentile is the appropriate table, one that gives percentiles for a normal distribution with a mean of 0 and a standard deviation of 1. (Calculators, statistical software, and free websites that provide these percentiles are also available.) A normal curve with a mean of 0 and a standard deviation of 1 is called a **standard normal curve.** It is the curve that results when any normal curve is converted to standardized scores. In other words, the standardized scores resulting from any normal curve will have a mean of 0 and a standard deviation of 1 and will retain the bell shape.

Table 8.1, presented at the end of this chapter, gives percentiles for standardized scores. For example, with an IQ of 115 and a standardized score of +1, you would be at the 84th percentile. In other words, your IQ would be higher than that of 84% of the population. If we are told the percentile for a score but not the value itself, we can also work backward from the table to find the value. Let's review the steps necessary to find a percentile from an observed value, and vice versa.

To find the percentile from an observed value:

1. Find the standardized score: (observed value − mean)/s.d., where s.d. = standard deviation. Don't forget to keep the plus or minus sign.
2. Look up the percentile in Table 8.1 (page 175).

To find an observed value from a percentile:

1. Look up the percentile in Table 8.1, and find the corresponding standardized score.
2. Compute the observed value: mean + (standardized score)(s.d), where s.d. = standard deviation.

EXAMPLE 8.2   Is High Cholesterol Too Common?

According to a publication from the World Health Organization (Lawes et al, 2004), cholesterol levels for women aged 30 to 44 in North American are approximately normally distributed with a mean of about 185 mg/dl and standard deviation of about 35 mg/dl. High cholesterol is defined as anything over 200 mg/dl. What proportion of women in this age group has high cholesterol? In other words, if a woman has an observed level of 200 mg/dl, what is her percentile?

> *Standardized score = (observed value – mean)/(s.d.)*
>
> *Standardized score = (200 – 185)/35*
>
> *Standardized score = 15/35 = 0.43.*

From Table 8.1, we see that a standardized score of 0.43 is between the 66th percentile score of 0.41 and the 67th percentile score of 0.44. Therefore, about 66.7%, or about two-thirds of women in this age group, do not have high cholesterol. That means that about one-third do have high cholesterol. Thus, if you fall into this category, you are not alone! In fact, according to the United States Center for Disease Control, the average total cholesterol level is 200 mg/dl. That means that about half of all adults are categorized as having high cholesterol. (*Source*: http://www.cdc.gov/cholesterol/facts.htm, accessed June 10, 2013) ∎

EXAMPLE 8.3   Tragically Low IQ

In the Edinburgh newspaper the *Scotsman* on March 8, 1994, a headline read, "Jury urges mercy for mother who killed baby" (p. 2). The baby had died from improper care. One of the issues in the case was that "the mother ... had an IQ lower than 98 percent of the population, the jury had heard." From this information, let's compute the mother's IQ. If it was lower than 98% of the population, it was higher than only 2%, so she was in the 2nd percentile. From Table 8.1, we see that her standardized score was −2.05, or 2.05 standard deviations below the mean of 100. We can now compute her IQ:

> *observed value = mean + (standardized score)(s.d.)*
>
> *observed value = 100 + (−2.05)(15)*
>
> *observed value = 100 + (−30.8) = 100 − 30.8*
>
> *observed value = 69.2*

Thus, her IQ was about 69. The jury was convinced that her IQ was, tragically, too low to expect her to be a competent mother. ∎

EXAMPLE 8.4   Calibrating Your GRE Score

The Graduate Record Examination (GRE) is a test taken by college students who intend to pursue a graduate degree in the United States. For people who took the exam between August 2011 and April 2012, the mean for the quantitative reasoning portion of the exam was 151.3 and the standard deviation was 8.7 (Educational Testing Service, 2012). If you had received a score of 163 on that GRE exam, what percentile would you be in, assuming the scores were bell-shaped? We can compute your percentile by first computing your standardized score:

standardized score = (observed value − mean)/(s.d.)
standardized score = (163 − 151.3)/8.7
standardized score = 11.7/8.7 = 1.34

From Table 8.1, we see that a standardized score of 1.34 is at the 91st percentile. In other words, your score was higher than about 91% of the population. Figure 8.4 illustrates the GRE score of 163 for the population of GRE scores and the corresponding standardized score of 1.34 for the standard normal curve. Notice the similarity of the two pictures.

The Educational Testing Service publishes tables showing the exact percentile for various scores on the GRE. For those who took the exam between August 2011 and April 2012, 88% scored below 163 on the quantitative reasoning part. The actual value of 88% is thus very close to the value of 91% that we calculated based only on knowing the mean and standard deviation of the scores. ∎

Figure 8.4
The percentile for a GRE quantitative reasoning score of 163 and corresponding standardized score

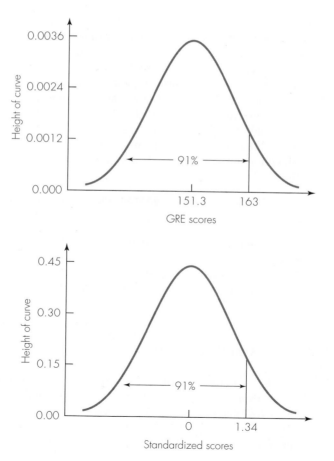

Figure 8.5
Moles inside and outside the legal limits of 68 to 211 grams

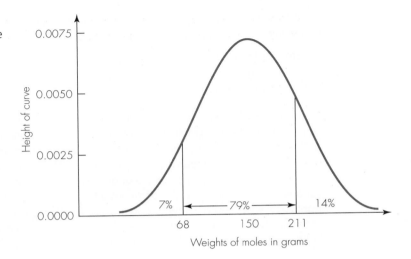

EXAMPLE 8.5 Ian Stewart (17 September 1994, p. 14) reported on a problem posed to a statistician by a British company called Molegon, whose business was to remove unwanted moles from gardens. The company kept records indicating that the population of weights of moles in its region was approximately normal, with a mean of 150 grams and standard deviation of 56 grams. The European Union announced that, starting in 1995, only moles weighing between 68 grams and 211 grams can be legally caught. Molegon wanted to know what percentage of all moles could be legally caught.

To solve this problem, we need to know what percentage of all moles weigh between 68 grams and 211 grams. We need to find two standardized scores, one for each end of the interval, and then find the percentage of the curve that lies between them:

standardized score for 68 grams = (68 − 150)/56 = −1.46
standardized score for 211 grams = (211 − 150)/56 = 1.09

From Table 8.1, we see that about 86% of all moles weigh 211 grams or less. But we also see that about 7% are below the legal limit of 68 grams. Therefore, about 86% − 7% = 79% are within the legal limits. Of the remaining 21%, 14% are too big to be legal and 7% are too small. Figure 8.5 illustrates this situation. ∎

8.4 z-Scores and Familiar Intervals

Any educated consumer of statistics should know a few facts about normal curves. First, as mentioned already, a synonym for a standardized score is a **z-score.** Thus, if you are told that your z-score on an exam is 1.5, it means that your score is 1.5 standard deviations above the mean. You can use that information to find your approximate percentile in the class, assuming the scores are approximately bell-shaped.

Second, some easy-to-remember intervals can give you a picture of where values on any normal curve will fall. This information is known as the **Empirical Rule.**

Empirical Rule

For any normal curve, approximately

68% of the values fall within 1 standard deviation of the mean in either direction;

95% of the values fall within 2 standard deviations of the mean in either direction;

99.7% of the values fall within 3 standard deviations of the mean in either direction.

A measurement would be an extreme outlier if it fell more than 3 standard deviations above or below the mean. You can see why the standard deviation is such an important measure. If you know that a set of measurements is approximately bell-shaped, and you know the mean and standard deviation, then even without a table like Table 8.1, you can say a fair amount about the magnitude of the values.

For example, because adult women in the United States have a mean height of about 65 inches (5 feet 5 inches) with a standard deviation of about 2.5 inches, and heights are bell-shaped, we know that approximately

- 68% of adult women in the United States are between 62.5 inches and 67.5 inches
- 95% of adult women in the United States are between 60 inches and 70 inches
- 99.7% of adult women in the United States are between 57.5 inches and 72.5 inches

Figure 8.6 illustrates the Empirical Rule for the heights of adult women in the United States.

Figure 8.6
The Empirical Rule for heights of adult women

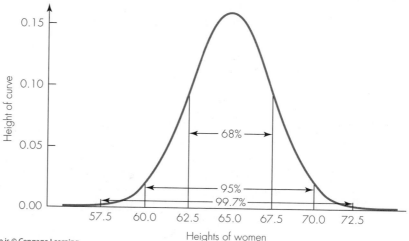

The mean height for adult males in the United States is about 70 inches and the standard deviation is about 3 inches. You can easily compute the ranges into which 68%, 95%, and almost all men's heights should fall.

Using Computers to Find Normal Curve Proportions

There are computer programs and websites that will find the proportion of a normal curve that falls below a specified value, above a value, and between two values. For example, here are two useful Excel functions:

NORMSDIST(value) provides the proportion of the standard normal curve below the value. Example: NORMSDIST(1) = .8413, which rounds to .84, shown in Table 8.1 for $z = 1$.

NORMDIST(value,mean,s.d.,1) provides the proportion of a normal curve with the specified mean and standard deviation (s.d.) that lies below the value given. (If the last number in parentheses is 0 instead of 1, it gives you the height of the curve at that value, which isn't much use to you. The "1" tells it that you want the proportion below the value.) Example: NORMDIST(67.5,65,2.5,1) = .8413, representing the proportion of adult women with heights of 67.5 inches or less.

Thinking About Key Concepts

- A *frequency curve* shows the possible values for a measurement and can be used to find the proportion of the population that falls into various ranges.
- For many types of measurements, the appropriate frequency curve is a *bell-shaped curve* or *normal curve*. Measurements with this shape are clustered around the mean and median in the center, then tail off symmetrically in both directions.
- A *standardized score* or *z-score* for a measurement is the number of standard deviations that measurement falls above the mean (for a positive score) or below the mean (for a negative score).
- For bell-shaped measurements, if the mean and standard deviation are known, you can find the percentile for any measurement with no additional information.
- The *Empirical Rule* provides guidance on where to expect measurements to fall for bell-shaped data. It states that 68% of all measurements should fall within one standard deviation of the mean (in either direction), 95% should fall within two standard deviations of the mean, and almost all (99.7%) should fall within three standard deviations of the mean.

TABLE 8.1   Proportions and Percentiles for Standard Normal Scores

Standard Score, z	Proportion Below z	Percentile	Standard Score, z	Proportion Below z	Percentile
−6.00	0.000000001	0.0000001	0.03	0.51	51
−5.20	0.0000001	0.00001	0.05	0.52	52
−4.26	0.00001	0.001	0.08	0.53	53
−3.00	0.0013	0.13	0.10	0.54	54
−2.576	0.005	0.50	0.13	0.55	55
−2.33	0.01	1	0.15	0.56	56
−2.05	0.02	2	0.18	0.57	57
−1.96	0.025	2.5	0.20	0.58	58
−1.88	0.03	3	0.23	0.59	59
−1.75	0.04	4	0.25	0.60	60
−1.64	0.05	5	0.28	0.61	61
−1.55	0.06	6	0.31	0.62	62
−1.48	0.07	7	0.33	0.63	63
−1.41	0.08	8	0.36	0.64	64
−1.34	0.09	9	0.39	0.65	65
−1.28	0.10	10	0.41	0.66	66
−1.23	0.11	11	0.44	0.67	67
−1.17	0.12	12	0.47	0.68	68
−1.13	0.13	13	0.50	0.69	69
−1.08	0.14	14	0.52	0.70	70
−1.04	0.15	15	0.55	0.71	71
−1.00	0.16	16	0.58	0.72	72
−0.95	0.17	17	0.61	0.73	73
−0.92	0.18	18	0.64	0.74	74
−0.88	0.19	19	0.67	0.75	75
−0.84	0.20	20	0.71	0.76	76
−0.81	0.21	21	0.74	0.77	77
−0.77	0.22	22	0.77	0.78	78
−0.74	0.23	23	0.81	0.79	79
−0.71	0.24	24	0.84	0.80	80
−0.67	0.25	25	0.88	0.81	81
−0.64	0.26	26	0.92	0.82	82
−0.61	0.27	27	0.95	0.83	83
−0.58	0.28	28	1.00	0.84	84
−0.55	0.29	29	1.04	0.85	85
−0.52	0.30	30	1.08	0.86	86
−0.50	0.31	31	1.13	0.87	87
−0.47	0.32	32	1.17	0.88	88
−0.44	0.33	33	1.23	0.89	89
−0.41	0.34	34	1.28	0.90	90
−0.39	0.35	35	1.34	0.91	91
−0.36	0.36	36	1.41	0.92	92
−0.33	0.37	37	1.48	0.93	93
−0.31	0.38	38	1.55	0.94	94
−0.28	0.39	39	1.64	0.95	95
−0.25	0.40	40	1.75	0.96	96
−0.23	0.41	41	1.88	0.97	97
−0.20	0.42	42	1.96	0.975	97.5
−0.18	0.43	43	2.05	0.98	98
−0.15	0.44	44	2.33	0.99	99
−0.13	0.45	45	2.576	0.995	99.5
−0.10	0.46	46	3.00	0.9987	99.87
−0.08	0.47	47	3.75	0.9999	99.99
−0.05	0.48	48	4.26	0.99999	99.999
−0.03	0.49	49	5.20	0.9999999	99.99999
0.00	0.50	50	6.00	0.999999999	99.9999999

Focus on Formulas

Notation for a Population

The lowercase Greek letter "mu" (μ) represents the **population mean.**

The lowercase Greek letter "sigma" (σ) represents the **population standard deviation.**

Therefore, the **population variance** is represented by σ^2.

A **normal distribution** with a mean of μ and variance of σ^2 is denoted by $N(\mu, \sigma^2)$.

For example, the **standard normal distribution** is denoted by $N(0, 1)$.

Standardized Score z for an Observed Value x

$$z = \frac{x - \mu}{\sigma}$$

Observed Value x for a Standardized Score z

$$x = \mu + z\sigma$$

Empirical Rule

If a population of values is $N(\mu, \sigma^2)$, then approximately:

68% of values fall within the interval $\mu \pm \sigma$

95% of values fall within the interval $\mu \pm 2\sigma$

99.7% of values fall within the interval $\mu \pm 3\sigma$

Exercises

Exercises with numbers divisible by 3 (3, 6, 9, etc.) are included in the Solutions at the back of the book. They are marked with an asterisk (*).

1. Using Table 8.1, a computer, or a calculator, determine the percentage of the population falling *below* each of the following standard scores:

    a. -1.00

    b. 1.96

    c. 0.84

2. In each of the following cases, explain how you know that the population of measurements could not be normally distributed.

    a. The population is all families in the world that have exactly four children, and the measurement is the number of boys in the family.

    b. The population is all college students and the measurement is number of hours per week the student exercises. The mean is 4.5 hours, and the standard deviation is also 4.5 hours.

*3. Using Table 8.1, a computer, or a calculator, determine the percentage of the population falling *above* each of the following standard scores:

    *a. 1.28

    *b. -0.25

    *c. 2.33

4. Using Table 8.1, a computer, or a calculator, determine the standard score that has the following percentage of the population below it:

   a. 25%

   b. 75%

   c. 45%

   d. 98%

5. Give an example of a population of measurements that would not have a normal distribution for each of the following reasons.

   a. The measurement can only result in a small number of possible values, instead of the continuum over a substantial range that is required for a normal distribution.

   b. The measurement cannot go below 0, but is likely to be skewed to the right because it is extremely high for a small subset of all individuals.

*6. Using Table 8.1, a computer, or a calculator, determine the standard score that has the following percentage of the population above it:

   *a. 2%

   b. 50%

   *c. 75%

   d. 10%

7. Using Table 8.1, a computer, or a calculator, determine the percentage of the population falling between the two standard scores given:

   a. −1.00 and 1.00

   b. −1.28 and 1.75

   c. 0.0 and 1.00

8. The 84th percentile for the Stanford-Binet IQ test is 115. (Recall that the mean is 100 and the standard deviation is 15.)

   a. Verify that this is true by computing the standardized score and using Table 8.1.

   b. Draw pictures of the original and standardized scores to illustrate this situation, similar to the pictures in Figure 8.4.

*9. Find the percentile for the observed value in the following situations:

   *a. Quantitative reasoning GRE score of 146 (mean = 151.3, s.d. = 8.7).

   *b. Stanford-Binet IQ score of 97 (mean = 100, s.d. = 15).

   c. Woman's height of 68 inches (mean = 65 inches, s.d. = 2.5 inches).

   d. Man's height of 68 inches (mean = 70 inches, s.d. = 3 inches).

10. Draw a picture of a bell-shaped curve with a mean value of 100 and a standard deviation of 10. Mark the mean and the intervals derived from the Empirical Rule in the appropriate places on the horizontal axis. You do not have to mark the vertical axis. Use Figure 8.6 as a guide.

11. Mensa is an organization that allows people to join only if their IQs are in the top 2% of the population.

    a. What is the lowest Stanford-Binet IQ you could have and still be eligible to join Mensa? (Remember that the mean is 100 and the standard deviation is 15.)

    b. Mensa also allows members to qualify on the basis of certain standard tests. If you were to try to qualify on the basis of the quantitative reasoning part of the GRE exam, what score would you need on the exam? (Remember that the mean is 151.3 and the standard deviation is 8.7, and round your answer to the closest whole number.)

*12. Every time you have your cholesterol measured, the measurement may be slightly different due to random fluctuations and measurement error. Suppose that for you, the population of possible cholesterol measurements if you are healthy has a mean of 190 and a standard deviation of 10. Further, suppose you know you should get concerned if your measurement ever gets up to the 97th percentile. What level of cholesterol does that represent?

13. Use Table 8.1 to verify that the Empirical Rule is true. You may need to round off the values slightly.

14. Remember that the whiskers on a boxplot can extend 1.5 interquartile ranges (IQRs) from either end of the box, and values beyond that range are considered to be outliers. For normally distributed data, the mean and median are equal and are both centered in the middle of the box in a boxplot.

    a. Draw the "box" part of a boxplot for normally distributed data. What fraction of an IQR is the distance between the mean/median and the ends of the box? (Hint: How many IQRs is the width of the box?)

    b. For normally distributed data, how many IQRs above or below the mean/median would a data value have to be in order to be considered an outlier? (Hint: Combine the information at the beginning of this exercise and your answer in part (a).)

*15. Recall from Chapter 7 that the interquartile range (IQR) covers the middle 50% of the data.

    *a. What range of standardized scores is covered by the interquartile range for normally distributed populations? In other words, what are the standardized scores for the lower quartile (the 25th percentile) and the upper quartile (the 75th percentile)? (Hint: Draw a standard normal curve and locate the 25th and 75th percentiles using Table 8.1.)

    *b. How many standard deviations are covered by the interquartile range for a normally distributed population? (Hint: Use the standardized scores found in part (a) for the lower and upper quartiles.)

    *c. Data values are outliers for normally distributed data if they are more than two IQRs away from the mean. (See Exercise 14.) At what percentiles (at the upper and lower ends) are data values considered outliers for normally distributed data? Use your answers to parts (a) and (b) to help you determine the answer to this question. (Software required.)

16. Give an example of a population of measurements that you do not think has a normal curve, and draw its frequency curve.

17. A graduate school program in Statistics will admit only students with quantitative reasoning GRE scores in the top 30%. What is the lowest GRE score it will accept? (Recall the mean is 151.3 and the standard deviation is 8.7, and round your answer to the closest whole number.)

*18. Recall that for Stanford-Binet IQ scores the mean is 100 and the standard deviation is 15.

    *a. Use the Empirical Rule to specify the ranges into which 68%, 95%, and 99.7% of Stanford-Binet IQ scores fall.

    b. Draw a picture similar to Figure 8.6 for Stanford-Binet scores, illustrating the ranges from part (a).

19. For every 100 births in the United States, the number of boys follows, approximately, a normal curve with a mean of 51 boys and standard deviation of 5 boys. If the next 100 births in your local hospital resulted in 36 boys (and thus 64 girls), would that be unusual? Explain.

20. Suppose a candidate for public office is favored by only 48% of the voters. If a sample survey randomly selects 2500 voters, the percentage in the sample who favor the candidate can be thought of as a measurement from a normal curve with a mean of 48% and a standard deviation of 1%. Based on this information, how often would such a survey show that 50% or more of the sample favored the candidate?

*21. Suppose you record how long it takes you to get to work or school over many months and discover that the times are approximately bell-shaped with a mean of 15 minutes and a standard deviation of 2 minutes. How much time should you allow to get there to make sure you are on time 90% of the time?

22. Assuming heights for each sex are bell-shaped, with means of 70 inches for men and 65 inches for women, and with standard deviations of 3 inches for men and 2.5 inches for women, what proportion of your sex is shorter than you are? (Be sure to mention your sex and height in your answer!)

23. Math SAT scores for students admitted to a university are bell-shaped with a mean of 520 and a standard deviation of 60.

    a. Draw a picture of these SAT scores, indicating the cutoff points for the middle 68%, 95%, and 99.7% of the scores. See Figure 8.6 for guidance.

    b. A student had a math SAT score of 490. Find the standardized score for this student and draw where her score would fall on your picture in part (a).

*24. According to *Chance* magazine ([1993], 6, no. 3, p. 5), the mean healthy adult temperature is around 98.2° Fahrenheit, not the previously assumed value of 98.6°. Suppose the standard deviation is 0.6 degree and the population of healthy temperatures is bell-shaped.

    *a. What proportion of the population have temperatures at or below the presumed norm of 98.6°?

    *b. Would it be accurate to say that the normal healthy adult temperature is 98.2° Fahrenheit? Explain.

25. Remember from Chapter 7 that the *range* for a data set is found as the difference between the maximum and minimum values. Explain why it makes sense that for a bell-shaped data set of a few hundred values, the range should be about 4 to 6 standard deviations. (Hint: Use the Empirical Rule.)

26. Suppose that you were told that scores on an exam in a large class you are taking ranged from 50 to 100 and that they were approximately bell-shaped.

    a. Estimate the mean for the exam scores.

    b. Refer to the result about the relationship between the range and standard deviation in Exercise 25. Estimate the standard deviation for the exam scores, using that result and the information in this problem.

    c. Suppose your score on the exam was 80. Explain why it is reasonable to assume that your standardized score is about 0.5.

    d. Based on the standardized score in part (c), about what proportion of the class scored higher than you did on the exam?

*27. Over many years, rainfall totals for Sacramento, CA, in January ranged from a low of about 0.05 inch to a high of about 19.5 inches. The median was about 3.1 inches. Based on this information, explain how you can tell that the distribution of rainfall values in Sacramento in January cannot be bell-shaped.

28. Recall that quantitative reasoning GRE scores are approximately bell-shaped with a mean of 151.3 and standard deviation of 8.7. The minimum and maximum possible scores on this part of the GRE exam are 130 and 170, respectively.

    a. What is the range for these GRE scores?

    b. Refer to the result about the relationship between the range and standard deviation in Exercise 25. Does the result make sense for GRE scores? Explain.

29. Refer to Example 8.1 on page 167, and the corresponding Figure 8.3 showing the distribution of heights for college women. The mean of the heights is 64.5 inches, and the standard deviation is 2.5 inches.

    a. Use the Empirical Rule to determine the ranges into which about 68%, 95%, and 99.7% of college women's heights should fall.

    b. Draw a picture illustrating the ranges you found in part (a), indicating the cutoff points for the middle 68%, 95%, and 99.7% of heights of college women.

    c. The table below (top of next page) shows some ranges of heights and the number of women out of the 94 in the sample whose heights fell into these ranges. Compare these 94 women to what would be expected if the Empirical Rule held exactly, using the intervals found in part (a).

Range (inches)	57 to 59.5	59.5 to 62	62 to 64.5	64.5 to 67	67 to 69.5	69.5 to 72
Number of women	1	17	32	33	9	2

References

Educational Testing Service. (2012). http://www.ets.org/s/gre/pdf/gre_guide_table1a.pdf, accessed June 10, 2013. Princeton, NJ: Educational Testing Service.

Lawes, C. M. M., S. Vander Hoorn, M. R. Law and A. Rodgers (2004). High cholesterol. Chapter 7 in *Global and Regional Burden of Diseases Attributable to Selected Major Risk Factors,* Ezzati, M., Lopez, A.D., Rodgers, A., Murray, C.J.L (eds.), World Health Organization; online: http://www.who.int/publications/cra/chapters/volume1/0391-0496.pdf, accessed June 10, 2013.

Ott, R. L., and W. Mendenhall (1994). *Understanding statistics,* 6th ed. Belmont, CA: Duxbury Press.

Stewart, Ian (17 September 1994). Statistical modelling. *New Scientist: Inside Science* 74, p. 14.

CHAPTER 9

Plots, Graphs, and Pictures

Thought Questions

1. You have seen pie charts and bar graphs and should have some rudimentary idea of how to construct them. Suppose you have been keeping track of your living expenses and find that you spend 50% of your money on rent, 25% on food, and 25% on other expenses. Draw a pie chart and a bar graph to depict this information. Discuss which is more visually appealing and useful.

2. Here is an example of a plot that has some problems. Give two reasons why this is not a good plot.

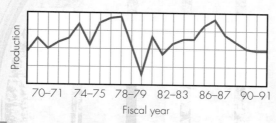

Domestic Water Production
1968–1992

3. Suppose you had a set of data representing two measurement variables—namely, height and weight—for each of 100 people. How could you put that information into a plot, graph, or picture that illustrated the relationship between the two measurements for each person?

4. Suppose you own a company that produces candy bars, and you want to display two graphs. One graph is for customers and shows the price of a candy bar for each of the past 10 years. The other graph is for stockholders and shows the amount the company was worth for each of the past 10 years. You decide to adjust the dollar amounts in one graph for inflation but to use the actual dollar amounts in the other graph. If you were trying to present the most favorable story in each case, which graph would be adjusted for inflation? Explain.

5. What do you think is meant by the term *time series*?

9.1   Well-Designed Statistical Pictures

There are many ways to present data in pictures. The most common are plots and graphs, but sometimes a unique picture is used to fit a particular situation. The purpose of a plot, graph, or picture of data is to give you a visual summary that is more informative than simply looking at a collection of numbers. Done well, a picture can quickly convey a message that would take you longer to find if you had to study the data on your own. Done poorly, a picture can mislead all but the most observant of readers. Here are some basic characteristics that all plots, graphs, and pictures should exhibit:

1. The data should stand out clearly from the background.
2. There should be clear labeling that indicates
   a. the title or purpose of the picture.
   b. what each of the axes, bars, pie segments, and so on, denotes.
   c. the scale of each axis, including starting points.
3. A source should be given for the data.
4. There should be as little "chart junk"—that is, extraneous material—in the picture as possible.

9.2   Pictures of Categorical Data

Categorical data are easy to represent with pictures. The most frequent use of such data is to determine how the whole divides into categories, and pictures are useful in expressing that information. Let's look at three common types of pictures for categorical data, and their uses.

Pie Charts

Pie charts are useful when only one categorical variable is measured. Pie charts show what percentage of the whole falls into each category. They are simple to understand, and they convey information about the relative size of groups more readily than a table. Side-by-side pie charts can be used to compare groups for a categorical variable, while still illustrating how the pie divides up for each group separately. Figure 9.1 shows pie charts representing the percentage of Caucasian American boys and girls who have various eye colors. Note that for both sexes, roughly one-third have blue eyes and another one-third have brown eyes. You can also see that there are slight differences in the sexes, with more boys having blue eyes and more girls having green eyes.

Bar Graphs

Bar graphs also show percentages or frequencies in various categories, but they can be used to represent two or three categorical variables simultaneously. One categorical

Figure 9.1
Pie charts of eye colors
of Caucasian American
boys and girls
Source: Malloy, 2008

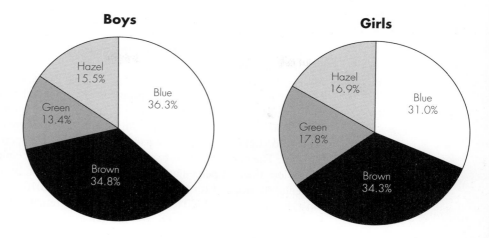

variable is used to label the horizontal axis. Within each of the categories along that axis, a bar is drawn to represent each category of the second variable. Frequencies or percentages are shown on the vertical axis. A third variable can be included if the graph has only two categories by using percentages on the vertical axis. One category is shown, and the other is implied by the fact that the total must be 100%.

For example, Figure 9.2 illustrates employment trends for men and women across decades. The year in which the information was collected is one categorical variable, represented by the horizontal axis. In each year, people were categorized according to two additional variables: whether they were in the labor force and whether they were male or female. Separate bars are drawn for males and females, and the percentage of all adults of that sex who were in the

Figure 9.2
Percentage of males and
females 16 and over in
the labor force
Source: Based on data from
U.S. Dept. of Labor, Bureau of
Labor Statistics, *Current
Population Survey.*

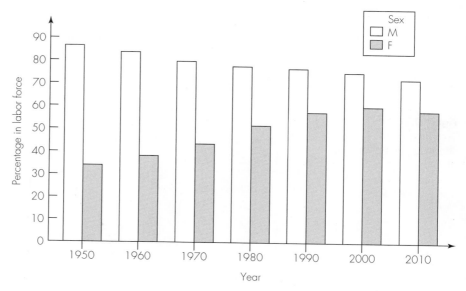

labor force that year determines the heights of the bars. It is implicit that the remainder were not in the labor force. Respondents were part of the Bureau of Labor Statistics' Current Population Survey, the large monthly survey used to determine unemployment rates.

The decision about which variable occupies which position should be made to better convey visually the purpose for the graph. The purpose of the graph in Figure 9.2 is to illustrate that the percentage of women in the labor force increased from 1950 to 2000 (then decreased slightly by 2010), whereas the percentage of men decreased slightly across all years, resulting in the two percentages coming closer together. The gap between the sexes in 1950 was 53 percentage points, but by 2010, it was less than 13 percentage points. If you look closely, you will notice that the percentages dropped for both sexes from 2000 to 2010. For males, it dropped the most, from 74.8% in 2000 to 71.2% in 2010. For females, the drop was from 59.9% in 2000 to 58.6% in 2010. Although it is not evident from the graph, which includes only years divisible by 10, the percentage of women in the labor force peaked at 60% in 1999.

Bar graphs are not always as visually appealing as pie charts, but they are much more versatile. They can also be used to represent actual frequencies instead of percentages and to represent proportions that are not required to sum to 100%.

Pictograms

A **pictogram** is like a bar graph except that it uses pictures related to the topic of the graph. Figure 9.3 shows a pictogram illustrating the proportion of Ph.D.s earned by women in the United States in 2011 in three fields—psychology (72%), biological sciences (52%), and mathematics (29%)—as reported by the National Science Foundation. Notice that in place of bars, the graph uses pictures of diplomas.

It is easy to be misled by pictograms. The pictogram on the left shows the diplomas using realistic dimensions. However, it is misleading because the eye tends to focus on the *area* of the diploma rather than just its height. The heights of the three diplomas reach the correct proportions, with heights of 72%, 52%, and 29%, so the height of the one for psychology Ph.D.s is about two and a half times the height of the one for math Ph.D.s. However, in keeping the proportions realistic, the area of

Figure 9.3
Two pictograms showing percentages of Ph.D.s earned by women
Source: National Science Foundation

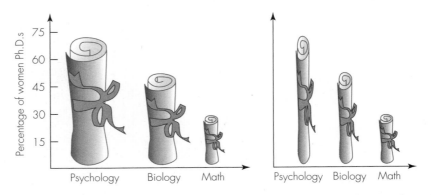

the diploma for psychology is more than six times the area of the one for math, leading the eye to inflate the difference.

The pictogram on the right is drawn by keeping the width of the diplomas the same for each field. The picture is visually more accurate, but it is less appealing because the diplomas are consequently quite distorted in appearance. When you see a pictogram, be careful to interpret the information correctly and not to let your eye mislead you.

9.3   Pictures of Measurement Variables

Measurement variables can be illustrated with graphs in numerous ways. We saw three ways to illustrate a single measurement variable in Chapter 7—namely, stemplots, histograms, and boxplots. Graphs are most useful for displaying the relationship between two measurement variables or for displaying how a measurement variable changes over time. Two common types of displays for measurement variables are illustrated in Figures 9.4 and 9.5.

Line Graphs

Figure 9.4 is an example of a **line graph** displayed over time. It shows the winning times for the men's 500-meter speed skating event in the Winter Olympics from 1924 to 2010. Notice the distinct downward trend, with only a few upturns over the years (including 2006 and 2010). There was a large drop between 1952 and 1956, followed by a period of relative stability. Also notice that no Olympic games were held in 1944 because of World War II. These patterns are much easier to detect with a picture than they would be by scanning a list of winning times.

Figure 9.4

Line graph displaying winning time versus year for men's 500-meter Olympic speed skating

Source: http://www.olympic.org/speed-skating-500m-men

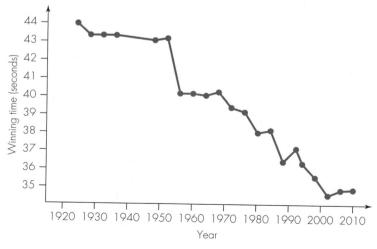

Figure 9.5
Scatterplot of grade point average versus verbal SAT score

Source: Ryan, Joiner, and Ryan, 1985, pp. 309–312.

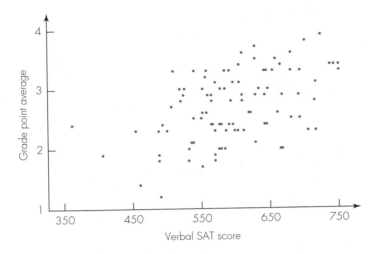

Scatterplots

Figure 9.5 is an example of a **scatterplot.** Scatterplots are useful for displaying the relationship between two measurement variables. Each dot on the plot represents one individual, unless two or more individuals have the same data, in which case only one point is plotted at that location. (An exception is that some software programs replace the dot with the number of points it represents, if it's more than one.) The plot in Figure 9.5 shows the grade point averages (GPAs) and verbal scholastic achievement test (SAT) scores for a sample of 100 students at a university in the northeastern United States.

Although a scatterplot can be more difficult to read than a line graph, it displays more information. It shows outliers, as well as the degree of variability that exists for one variable at each location of the other variable. In Figure 9.5, we can see an increasing trend toward higher GPAs with higher SAT scores, but we can also still see substantial variability in GPAs at each level of verbal SAT scores. A scatterplot is definitely more useful than the raw data. Simply looking at a list of the 100 pairs of GPAs and SAT scores, we would find it difficult to detect the trend that is so obvious in the scatterplot.

9.4   Pictures of Trends across Time

Many measurement variables are recorded across time, and it is of interest to look for trends and patterns from the past to try to predict the future, or simply to learn about the past. Examples include economic data, weather data, and demographics. A **time series** is simply a record of a variable across time, usually measured at equally spaced intervals. For instance, most economic data used by both governments and businesses are measured monthly.

To understand data presented across time, it is important to know how to recognize the various components that can contribute to the ups and downs in a time series. Otherwise, you could mistake a temporary high in a cycle for a permanent increasing trend and make a very unwise economic decision.

Figure 9.6

An example of a time series plot: Jeans sales in the United Kingdom from 1980 to 1984

Source: Hand et al., 1994, p. 314.

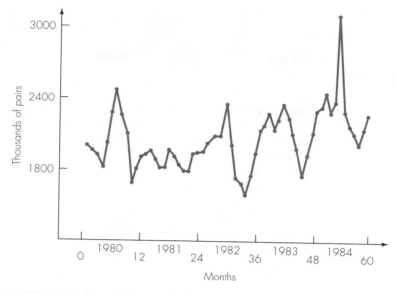

A Time Series Plot

Figure 9.6 illustrates a **time series plot.** The data represent monthly sales of jeans in Britain for the 5-year period from January 1980 to December 1984. Notice that the data points have been connected to make it easier to follow the ups and downs across time. Data are measured in thousands of pairs sold. Month 1 is January 1980, and month 60 is December 1984.

Components of Time Series

Most time series have the same four basic components: *long-term trend, seasonal components, irregular cycles,* and *random fluctuations.* We will examine these in more detail in Chapters 10 and 18, but the basic ideas are discussed here and will help you interpret time series plots.

Long-Term Trend

Many time series measure variables that either increase or decrease steadily across time. This steady change is called a **trend.** If the trend is even moderately large, it should be obvious by looking at a plot of the series. Figure 9.6 clearly shows an increasing trend for jeans sales.

If the long-term trend has a straight line pattern (up or down), we can investigate it using techniques we cover in Chapter 10.

Seasonal Components

Most time series involving economic data or data related to people's behavior have **seasonal components**. In other words, they tend to be high in certain months or seasons and low in others every year. For example, new housing starts are much higher in warmer months. Sales of toys and other standard gifts are much higher just before

Christmas. Unemployment rates in the United States tend to rise in January—when outdoor jobs are minimal and the Christmas season is over—and again in June, when a new graduating class enters the job market. Weddings are more likely to occur in certain months than others.

If you look carefully, you will see that there is indeed a seasonal component to the time series of sales of jeans evident in Figure 9.6. Sales appear to peak during June and July and reach a low in October every year. (You can identify individual months by counting the dots, starting with the dot at 0 representing January, 1980.) Manufacturers need to know about these seasonal components. Otherwise, they might mistake increased sales during June, for example, as a general trend and over-produce their product.

In Chapter 18, we will discuss a variety of economic indicators, almost all of which are subject to seasonal fluctuations. When pictures of these time series are shown or when the latest values are reported in the news, it will be stated that they have been "seasonally adjusted." In other words, the seasonal components have been removed so that more general trends can be seen. Economists have sophisticated methods for seasonally adjusting time series. They use data from the same month or season in prior years to construct a seasonal factor, which is a number either greater than one or less than one by which the current figures are multiplied.

Irregular Cycles and Random Fluctuations

There are two remaining components of time series: the irregular (but smooth) **cycles** that economic systems tend to follow and unexplainable random fluctuations. It is often hard to distinguish between these two components, especially if the cycles are not regular.

Figure 9.7 shows the U.S. unemployment rate, seasonally adjusted, monthly from January, 1990, through May, 2013. Notice the definite irregular cycles during which unemployment rates rise and fall over a number of years. Some of these can

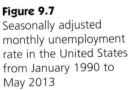

Figure 9.7
Seasonally adjusted monthly unemployment rate in the United States from January 1990 to May 2013
Source: http://data.bls.gov/pdq/ SurveyOutputServlet

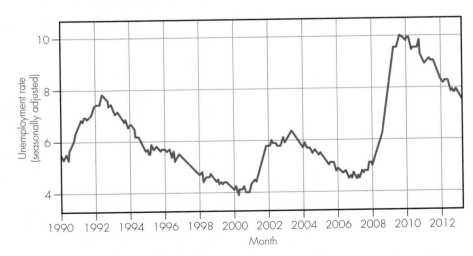

be at least partially explained by social and political factors. For example, the steady decrease in unemployment during the 1990s has been partially attributed to President Clinton's economic policies and partially to the boom in the high tech industry. The rapid increase in unemployment starting in 2008 followed the collapse of the housing market, brought about by lending practices that left many homeowners unable to pay their mortgages.

The final component in a time series, **random fluctuations**, is defined as what's left over when the other three components have been removed. They are part of the natural variability present in all measurements. It is for this reason that the future can never be completely predicted using data from the past.

Improper Presentation of a Time Series

Let's look at one way in which you can be fooled by improper presentation of a time series. In Figure 9.8, a subset of the time series from Figure 9.6, sales of jeans, is displayed.

Suppose an unscrupulous entrepreneur was anxious to have you invest your hard-earned savings into his clothing company. To convince you that sales of jeans can only go up, he presents you with a limited set of data—from October 1982 to June 1984. With only those few months shown, it appears that the basic trend is way up!

A less obvious version of this trick is to present data up to the present time but to start the plot of the series at an advantageous point in the past, rather than providing all of the available data. Be suspicious of time series showing returns on investments that look too good to be true. They probably are. Notice when the time series begins and compare that with your knowledge of recent economic cycles.

Figure 9.8

Distortion caused by displaying only part of a time series: Jeans sales for 21 months

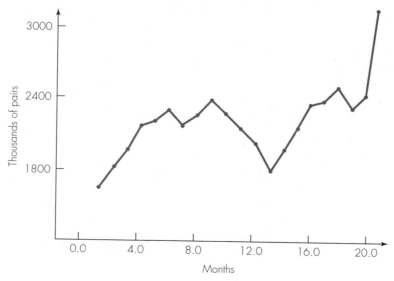

9.5 Difficulties and Disasters in Plots, Graphs, and Pictures

A number of common mistakes appear in plots and graphs that may mislead readers. If you are aware of them and watch for them, you will substantially reduce your chances of misreading a statistical picture.

The most common problems in plots, graphs, and pictures are

1. No labeling on one or more axes
2. Not starting at zero as a way to exaggerate trends
3. Distorting time series plots
4. Change(s) in labeling on one or more axes
5. Misleading units of measurement
6. Using poor information

No Labeling on One or More Axes

You should always look at the axes in a picture to make sure they are labeled. Figure 9.9a gives an example of a plot for which the units were *not* labeled on the vertical axis. The plot appeared in a newspaper insert titled, "May 1993: Water awareness month." When there is no information about the units used on one of the axes, the plot cannot be interpreted. To see this, consider Figure 9.9b and c, displaying two different scenarios that could have produced the actual graph in Figure 9.9a. In Figure 9.9b, the vertical axis starts at zero for the existing plot. In Figure 9.9c, the vertical axis for the original plot starts at 30 and stops at 40, so what appears to be a large drop in 1979 in the other two graphs is only a minor fluctuation. We do not know which of these scenarios is closer to the truth, yet you can see that the two possibilities represent substantially different situations.

Not Starting at Zero

Often, even when the axes are labeled, the scale of one or both of the axes does not start at zero, and the reader may not notice that fact. A common ploy is to present an increasing or decreasing trend over time on a graph that does not start at zero. As we saw for the example in Figure 9.9, what appears to be a substantial change may actually represent quite a modest change. Always make it a habit to check the numbers on the axes to see where they start.

Figure 9.10 shows what the line graph of winning times for the Olympic speed skating data in Figure 9.4 would have looked like if the vertical axis had started at zero. Notice that the drop in winning times over the years does not look nearly as dramatic as it did in Figure 9.4. Be very careful about this form of potential deception if someone is

Figure 9.9

Example of a graph with no labeling (a) and possible interpretations (b and c)

Source: Insert in the *California Aggie* (UC Davis), 30 May 1993.

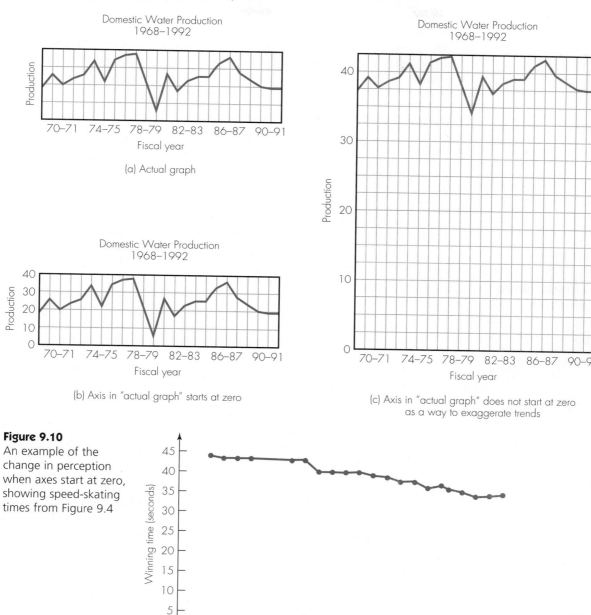

(a) Actual graph

(b) Axis in "actual graph" starts at zero

(c) Axis in "actual graph" does not start at zero as a way to exaggerate trends

Figure 9.10

An example of the change in perception when axes start at zero, showing speed-skating times from Figure 9.4

presenting a graph to display growth in sales of a product, a drop in interest rates, and so on. Be sure to look at the labeling, especially on the vertical axis.

Despite this, be aware that for some graphs it makes sense to start the units on the axes at values different from zero. A good example is the scatterplot of GPAs versus SAT scores in Figure 9.5. It would make no sense to start the horizontal axis (SAT scores) at zero because the range of interest is from about 350 to 800. It is the responsibility of the reader to notice the units. Never assume a graph starts at zero without checking the labeling.

Distorting Time Series Plots

There are several ways a time series plot can be distorted. As demonstrated in the comparison of Figures 9.6 and 9.8, a partial time series can be presented to mask the long-term trends and cycles present in the full series. When you look at data presented across time, make sure the time period is long enough to include likely trends, seasonal components, and irregular cycles.

Time series plots can also be misleading if they are not seasonally adjusted, especially if a short time period is presented. Sometimes, it is useful to forego seasonal adjustments, such as in the jeans sales in Figure 9.6, because picturing seasonal trends is part of the information the picture is designed to convey. But make sure you are told whether or not a time series plot has been seasonally adjusted.

A third method of distorting time series information is to present measurements for an entire population instead of "per person" when the population is increasing over time. If you look at a time series of sales of almost any item over the past few decades, the sales will show a strong increasing trend. That's because the population has been steadily increasing as well. A more appropriate time series would show sales per person or per 100,000 people instead. We will investigate this issue further in Chapter 11.

Changes in Labeling on One or More Axes

Figure 9.11 shows an example of a graph where a cursory look would lead one to think the vertical axis starts at zero. However, notice the white horizontal bar just above the bottom of the graph, in which the vertical bars are broken. That indicates a gap in the vertical axis. In fact, you can see that the bottom of the graph actually corresponds to about 4.0%. It would have been more informative if the graph had simply been labeled as such, without the break.

Misleading Units of Measurement

The units shown on a graph can be different from those that the reader would consider important. For example, Figure 9.12a illustrates a graph similar to one that appeared in *USA Today* on March 7, 1994 with the heading,

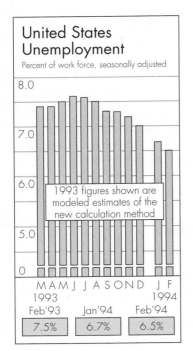

Figure 9.11
A bar graph with gap in labeling
Source: Davis (CA) *Enterprise,* 4 March 1994, p. A-7.

Figure 9.12
Cost of a first class stamp over time, without and with adjustment for inflation
Source: USA Today, 7 March 1994, p. 13A.

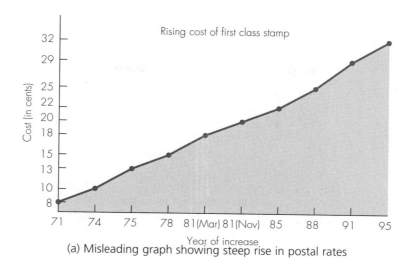

(a) Misleading graph showing steep rise in postal rates

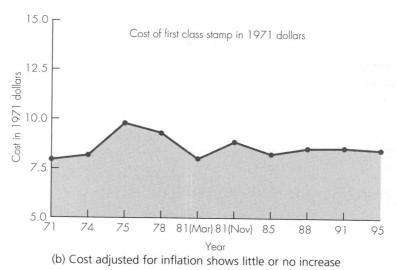

(b) Cost adjusted for inflation shows little or no increase

"Rising Postal Rates." It represents how the cost of a first-class stamp rose from 1971 to 1995. However, the original graph in *USA Today* had a footnote that read "In 1971 dollars, the price of a 32-cent stamp in February 1995 would be 8.4 cents." In other words, the increase from 8 cents in 1971 to 32 cents in 1995 was very little increase at all, when adjusted for inflation. A more truthful picture would show the changing price of a first-class stamp adjusted for inflation, as shown in Figure 9.12b. (You will learn how to adjust for inflation in Chapter 18.) As the footnote implied, such a graph would show little or no rise in postal rates as a function of the worth of a dollar. Both Figures 9.12a and 9.12b have an additional problem, which you will be asked to identify in Exercise 6.

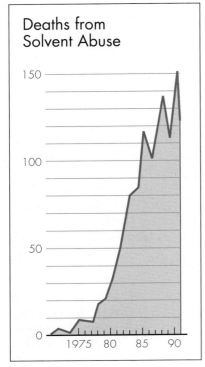

Figure 9.13
A graph based on poor information
Source: The Independent on Sunday (London),
13 March 1994.

Using Poor Information

A picture can only be as accurate as the information that was used to design it. All of the cautions about interpreting the collection of information given in Part 1 of this book apply to graphs and plots as well. You should always be told the source of information presented in a picture, and an accompanying article should give you as much information as necessary to determine the worth of that information.

Figure 9.13 shows a graph that appeared in the London newspaper the *Independent on Sunday* on March 13, 1994. The accompanying article was titled, "Sniffers Quit Glue for More Lethal Solvents." The graph appears to show that very few deaths occurred in Britain from solvent abuse before the late 1970s. However, the accompanying article includes the following quote from a research fellow at the unit where the statistics are kept: "It's only since we have started collecting accurate data since 1982 that we have begun to discover the real scale of the problem" (p. 5). In other words, the article indicates that the information used to create the graph is not at all accurate until at least 1982. Therefore, the apparent sharp increase in deaths linked to solvent abuse around that time period is likely to have been simply a sharp increase in deaths reported and classified.

Don't forget that a statistical picture isn't worth much if the data can't be trusted. Once again, you should familiarize yourself to the extent possible with the Seven Critical Components listed in Chapter 2 (pp. 20–21).

9.6   A Checklist for Statistical Pictures

To summarize, here are 12 questions you should ask when you look at a statistical picture—before you even begin to try to interpret the data displayed.

1. Does the message of interest stand out clearly?
2. Is the purpose or title of the picture evident?
3. Is a source given for the data, either with the picture or in an accompanying article?
4. Did the information in the picture come from a reliable, believable source?
5. Is everything clearly labeled, leaving no ambiguity?
6. Do the axes start at zero or not?
7. For time series data, is a long enough time period shown?
8. Can any observed trends be explained by another variable, such as increasing population?
9. Do the axes maintain a constant scale?

10. Are there any breaks in the numbers on the axes that may be easy to miss?

11. For financial data, have the numbers been adjusted for inflation and/or seasonally adjusted?

12. Is there information cluttering the picture or misleading the eye?

CASE STUDY 9.1   **Time to Panic about Illicit Drug Use?**

The graph illustrated in Figure 9.14 appeared on the website for the U.S. Department of Justice, Drug Enforcement Agency, in Spring, 1998 (http://www.usdoj.gov/dea/drugdata/cp-316.htm). The headline over the graph reads "Emergency Situation among Our Youth." Look quickly at the graph, and describe what you see. Did it lead you to believe that almost 80% of 8th-graders used illicit drugs in 1996, compared with only about 10% in 1992? The graph is constructed so that you might easily draw that conclusion. Notice that careful reading indicates otherwise, and crucial information is missing. The graph tells us only that in 1996 the rate of use was 80% higher, or 1.8 times what it was in 1991. The actual rate of use is *not* provided at all in the graph. Only after searching the remainder of the website does that emerge. The rate of illicit drug use among 8th-graders in 1991 was about 11%, and thus, in 1996, it was about 1.8 times that, or about 19.8%. Additional information elsewhere on the website indicates that about 8% of 8th-graders used marijuana in 1991, and thus, this was the most common illicit drug used. These are still disturbing statistics, but not as disturbing as the graph would lead you to believe. ∎

Figure 9.14
Emergency situation among our youth: 8th-grade drug use
Source: U.S. Dept. of Justice.

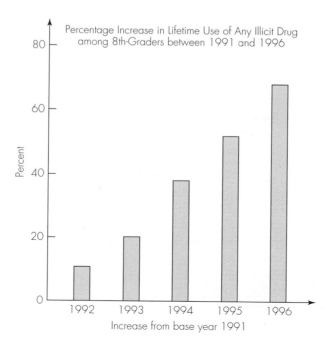

Thinking About Key Concepts

- *Pie charts* and *bar graphs* are useful for picturing categorical data, with the latter more useful for comparing a categorical variable across groups.
- There are numerous ways that pictures of data can convey misleading messages. Make sure you carefully examine graphs and plots and don't be fooled by distorted visual displays of information.
- *Scatter plots* are useful for displaying the relationship between two measurement variables and can reveal outliers that are not apparent from pictures of each variable on its own.
- A perfectly good graph can convey a misleading message if the information used to create it is not thoughtfully presented. For instance, prices across time should be adjusted for inflation, and measurements for groups of individuals across time should be presented as numbers per capita instead of as totals for the whole group.
- *Time series* plots present measurements across regularly spaced intervals. They can show trends, seasonal patterns, and cycles, but can also be distorted to convey a misleading message—for instance, by showing a short time period.

Exercises

Exercises with numbers divisible by 3 (3, 6, 9, etc.) are included in the Solutions at the back of the book. They are marked with an asterisk (*).

1. Use the pie charts in Figure 9.1 to create a bar graph comparing eye colors for Caucasian boys and Caucasian girls.

2. Suppose a real estate company in your area sold 100 houses last month, whereas their two major competitors sold 50 houses and 25 houses, respectively. The top company wants to display its better record with a pictogram using a simple two-dimensional picture of a house. Draw two pictograms displaying this information, one of which is misleading and one of which is not. (The horizontal axis should list the three companies and the vertical axis should list the number of houses sold.)

*3. Give the name of a type of statistical picture that could be used for each of the following kinds of data:

  *a. One categorical variable

  *b. One measurement variable

  *c. Two categorical variables

  *d. Two measurement variables

4. One method used to compare authors or to determine authorship on unsigned writing is to look at the frequency with which words of different lengths appear in a piece of text. For this exercise, you are going to compare your own writing with that of the author of this book.

  a. Using the first full paragraph of this chapter (not the Thought Questions), create a pie chart with three segments, showing the relative frequency of words of 1 to 3 letters, 4 to 5 letters, and 6 or more letters in length. (Do not include the numbered list after the paragraph.)

b. Find a paragraph of your own writing of at least 50 words. Repeat part (a) of this exercise for your own writing.

c. Display the data in parts (a) and (b) of this exercise using a single bar graph that includes the information for both writers.

d. Discuss how your own writing style is similar to or different from that of the author of this book, as evidenced by the pictures in parts (a) to (c).

e. Name one advantage of displaying the information in two pie charts and one advantage of displaying the information in a single bar graph.

5. An article in *Science* (23 January 1998, 279, p. 487) reported on a "telephone survey of 2600 parents, students, teachers, employers, and college professors" in which people were asked the question, "Does a high school diploma mean that a student has at least learned the basics?" Results are shown in Table 9.1.

a. The article noted that "there seems to be a disconnect between the producers [parents, teachers, students] and the consumers [professors, employers] of high school graduates

in the United States. Create a bar graph from this study that emphasizes this feature of the data.

b. Create a bar graph that deemphasizes the issue raised in part (a).

*6. Figure 9.12a, which displays rising postal rates, is an example of a graph with misleading units because the prices are not adjusted for inflation. The graph actually has another problem as well. Use the checklist in Section 9.6 to determine the problem; then redraw the graph correctly (but still use the unadjusted prices). Comment on the difference between Figure 9.12a and your new picture.

7. In its February 24–26, 1995, edition (p. 7), *USA Weekend* gave statistics on the changing status of which parent children live with. As noted in the article, the numbers don't total 100% because they are drawn from two sources: the U.S. Census Bureau and *America's Children: Resources from Family, Government, and the Economy* by Donald Hernandez (New York: Russell Sage Foundation, 1995). Using the data shown in Table 9.2, draw a bar graph presenting the information. Be sure to include all the components of a good statistical picture.

TABLE 9.1

	Professors	Employers	Parents	Teachers	Students
Yes	22%	35%	62%	73%	77%
No	76%	63%	32%	26%	22%

TABLE 9.2

Kids Live With	1960	1980	1990
Father and mother	80.6%	62.3%	57.7%
Mother only	7.7%	18.0%	21.6%
Father only	1.0%	1.7%	3.1%
Father and stepmother	0.8%	1.1%	0.9%
Mother and stepfather	5.9%	8.4%	10.4%
Neither parent	3.9%	5.8%	4.3%

TABLE 9.3	U.S. Population and Violent Crime*								
Year	**1982**	**1983**	**1985**	**1986**	**1987**	**1988**	**1989**	**1990**	**1991**
U.S. population	231	234	239	241	243	246	248	249	252
Violent crime	1.32	1.26	1.33	1.49	1.48	1.57	1.65	1.82	1.91

*Figures for 1984 were not available in the original.

8. Figure 10.4 in Chapter 10 displays the success rate for professional golfers when putting at various distances. Discuss the figure in the context of the material in this chapter. Are there ways in which the picture could be improved?

*9. Table 9.3 indicates the population (in millions) and the number of violent crimes (in millions) in the United States from 1982 to 1991, as reported in the *World Almanac and Book of Facts* (1993, p. 948). (Thankfully, both numbers and rates of violent crime started dropping in 1992 and have continued to do so. Thus, the data for this exercise end in 1991 to illustrate the increasing trend up to that time.)

   a. Draw two line graphs representing the trend in violent crime over time. Draw the first graph to try to convince the reader that the trend is quite ominous. Draw the second graph to try to convince the reader that it is not. Make sure all of the other features of your graph meet the criteria for a good picture.

   *b. Draw a scatterplot of population versus violent crime, making sure it meets all the criteria for a good picture. Comment on the scatterplot. Now explain why drawing a line graph of violent crime versus year, as in part (a) of this exercise, might be misleading.

   c. Rather than using number of violent crimes on the vertical axis, redraw the first line graph (from part [a]) using a measure that adjusts for the increase in population. Comment on the differences between the two graphs.

10. According to the American Red Cross (http://www.redcrossblood.org/learn-about-blood/blood-types), the distribution of blood types in the United States is as shown in Table 9.4.

    a. Draw a pie chart illustrating the blood-type distribution for Caucasians, ignoring the RH factor.

    b. Draw a statistical picture incorporating all of the information given.

11. Find an example of a statistical picture in a newspaper or magazine that has at least one of the problems listed in Section 9.5, "Difficulties and Disasters in Plots, Graphs, and Pictures." Explain the problem. If you think anything should have been done differently, explain what and why. Include the picture with your answer.

TABLE 9.4	Blood Types in the United States			
	Caucasian Americans		**African Americans**	
Blood Type	**Rh+**	**Rh−**	**Rh+**	**Rh−**
A	33%	7%	24%	2%
B	9%	2%	18%	1%
AB	3%	1%	4%	.3%
O	37%	8%	47%	4%

*12. Find a graph that does not start at zero. Redraw the picture to start at zero. Discuss the pros and cons of the two versions.

13. According to an article in *The Seattle Times* (Meckler, 2003), living organ donors are most often related to the organ recipient. Table 9.5 gives the percentages of each type of relationship for all 6613 cases in which an organ was transplanted from a living donor in 2002 in the United States. Create a pie chart displaying the relationship of the donor to the recipient, and write a few sentences describing the data.

14. Table 9.6 provides the total *number* of men and women who were in the labor force in 1971, 1981, 1991, 2001, and 2011 in the United States.

  a. Create a bar graph for the data.

  b. Compare the bar graph to the one in Figure 9.2, which presents the *percent* of men and women who were in the labor force. Discuss what can be learned from each graph that can't be learned from the other.

*15. For each of the following time series, do you think the long-term trend would be positive, negative, or nonexistent?

  *a. The cost of a loaf of bread measured monthly from 1960 to the present month.

  *b. The temperature in Boston measured at noon on the first day of each month from 1960 to the present month.

  c. The price of a basic computer, adjusted for inflation, measured monthly from 1970 to 2012.

  d. The number of personal computers sold in the United States measured monthly from 1970 to 2012.

16. For each of the time series in Exercise 15, explain whether there is likely to be a seasonal component.

17. Global warming is a major concern because it implies that temperatures around the world are going up on a permanent basis. Suppose you were to examine a plot of monthly temperatures in one location for the past 50 years. Explain the role that the three time series components (trend, seasonal, cycles) would play in trying to determine whether global warming was taking place.

*18. If you were to present a time series of the yearly cost of tuition at your local college for the past 30 years, would it be better to first adjust the costs for inflation? Explain.

19. The population of the United States rose from about 179 million people in 1960 to about 314 million people in 2012. Suppose you wanted to examine a time series to see if homicides had become an increasing problem over that time period. Would you simply plot the number of homicides versus time, or is there a better measure to plot against time? Explain.

20. Many statistics related to births, deaths, divorces, and so on across time are reported as rates per 100,000 of population rather than as actual numbers. Explain why those rates may be more meaningful as a measure of change across time than the actual numbers of those events.

TABLE 9.5	Living Donor's Relationship to Organ Transplant Recipient for All Cases in the United States in 2002

Relationship	Percent of Donors
Sibling	30%
Child	19%
Parent	13%
Spouse	11%
Other relative	8%
Not related	19%

TABLE 9.6	Total Number of Men and Women in the U.S. Labor Force (in millions)

Year	Men	Women
1971	51.2	32.2
1981	62.0	46.7
1991	69.2	57.2
2001	76.9	66.8
2011	82.0	71.6

Source: Current Population Survey, Bureau of Labor Statistics, http://www.bls.gov/cps/wlf-databook-2012.pdf, Table 2, accessed June 11, 2013.

*21. Discuss which of the three components of a time series (trend, seasonal, and cycles) are likely to be present in each of the following series, reported monthly for the past 10 years:

*a. Unemployment rates

*b. Hours per day the average child spends watching television

22. Explain which one of the components of an economic time series would be most likely to be influenced by a major war.

23. Find an example of a statistical picture in a newspaper or magazine or on the Internet. Answer the 12 questions in Section 9.6 for the picture. In the process of answering the questions, explain what (if any) features you think should have been added or changed to make it a good picture. Include the picture with your answer.

*24. Which of the three components (trend, seasonal, and cycles) are likely to be present in a time series reporting interest rates paid by a bank each month for the last 10 years?

25. Explain why it is important for economic time series to be seasonally adjusted before they are reported.

26. Refer to Additional News Story 19 on the companion website, a press release from Cornell University entitled "Puppy love's dark side: First study of love-sick teens reveals higher risk of depression, alcohol use and delinquency." The article includes a graph labeled "Adjusted change in depression between interviews." Comment on the graph.

*27. Refer to Figure 1 on page 691 of Original Source 11, linked to the companion website, "Driving impairment due to sleepiness is exacerbated by low alcohol intake."

*a. What type of picture is in the figure?

b. Write a few sentences explaining what you learn from the picture about lane drifting episodes under the different conditions.

28. Refer to Figure 2 on page 691 of Original Source 11, linked to the companion website, "Driving impairment due to sleepiness is exacerbated by low alcohol intake."

a. What type of picture is in the figure?

b. Write a few sentences explaining what you learn from the picture about subjective sleepiness ratings under the different conditions.

Mini-Projects

1. Collect some categorical data on a topic of interest to you, and represent it in a statistical picture. Explain what you have done to make sure the picture is as useful as possible.

2. Collect two measurement variables on each of at least 10 individuals. Represent them in a statistical picture. Describe the picture in terms of possible outliers, variability, and relationship between the two variables.

3. Find some data that represent change over time for a topic of interest to you. Present a line graph of the data in the best possible format. Explain what you have done to make sure the picture is as useful as possible.

References

Hand, D. J., F. Daly, A. D. Lunn, K. J. McConway, and E. Ostrowski. (1994). *A handbook of small data sets*. London: Chapman and Hall.

Krantz, L. (1992). *What the odds are*. New York: Harper Perennial.

Malloy, J. (2008). NLSY blogging: Eye and hair color of Americans, http://www.gnxp.com/blog/2008/12/nlsy-blogging-eye-and-hair-color-of.php, posted December 13, 2008, accessed June 11, 2013.

Meckler, L. (13 August 2003). Giving til it hurts. *The Seattle Times,* p. A3.

Ryan, B. F., B. L. Joiner, and T. A. Ryan, Jr. (1985). *Minitab handbook*. 2d ed. Boston: PWS-Kent.

Source for women and men in labor force for Figure 9.2 and Table 9.6. http://www.bls.gov/cps/wlf-databook-2012.pdf

Source for Figure 9.3 http://www.nsf.gov/statistics/sed/2011/pdf/tab16.pdf

World almanac and book of facts. (1993). Edited by Mark S. Hoffman. New York: Pharos Books.

CHAPTER **10**

Relationships Between Measurement Variables

Thought Questions

1. Judging from the scatterplot in Figure 9.5, there is a *positive correlation* between verbal SAT score and GPA. For used cars, there is a *negative correlation* between the age of the car and the selling price. Explain what it means for two variables to have a positive correlation or a negative correlation.

2. Suppose you were to make a scatterplot of (adult) sons' heights versus fathers' heights by collecting data on both from several of your male friends. You would now like to predict how tall your nephew will be when he grows up, based on his father's height. Could you use your scatterplot to help you make this prediction? Explain.

3. Do you think each of the following pairs of variables would have a positive correlation, a negative correlation, or no correlation?

    a. Calories eaten per day and weight.

    b. Calories eaten per day and IQ.

    c. Amount of alcohol consumed and accuracy on a manual dexterity test.

    d. Number of ministers and number of liquor stores in cities in Pennsylvania.

    e. Height of husband and height of wife.

4. An article in the *Sacramento Bee* (29 May 1998, p. A17) noted, "Americans are just too fat, researchers say, with 54 percent of all adults heavier than is healthy. If the trend continues, experts say that within a few generations virtually every U.S. adult will be overweight." This prediction is based on "extrapolating," which assumes the current rate of increase will continue indefinitely. Is that a reasonable assumption? Do you agree with the prediction? Explain.

10.1   Statistical Relationships

One of the interesting advances made possible by the use of statistical methods is the quantification and potential confirmation of relationships. In the first part of this book, we discussed relationships between aspirin and heart attacks, meditation and test scores, and smoking during pregnancy and child's IQ, to name just a few. In Chapter 9, we saw examples of relationships between two variables illustrated with pictures, such as the scatterplot of verbal SAT scores and college GPAs.

Although we have examined many relationships up to this point, we have not considered how those relationships could be expressed quantitatively. In this chapter, we discuss **correlation,** which measures the *strength* of a certain type of relationship between two measurement variables, and **regression,** which is a numerical method for trying to *predict* the value of one measurement variable from knowing the value of another one.

Statistical Relationships versus Deterministic Relationships

A **statistical relationship** differs from a **deterministic relationship** in that, in the latter case, if we know the value of one variable, we can determine the value of the other exactly. For example, the relationship between volume and weight of water is deterministic. The old saying, "A pint's a pound the world around," isn't quite true, but the deterministic relationship between volume and weight of water does hold. (A pint is actually closer to 1.04 pounds.) We can express the relationship by a formula, and if we know one value, we can solve for the other (weight in pounds = 1.04 × volume in pints).

Natural Variability in Statistical Relationships

In a statistical relationship, natural variability exists in the relationship between the two measurements. For example, we could describe the average relationship between height and weight for adult females, but very few women would fit that exact formula. If we knew a woman's height, we could predict the average weight for all women with that same height, but we could not predict her weight exactly. Similarly, we can say that, on average, taking aspirin every other day reduces one's chance of having a heart attack, but we cannot predict what will happen to one specific individual.

Statistical relationships are useful for describing what happens to a population, or aggregate. The stronger the relationship, the more useful it is for predicting what will happen for an individual. When researchers make claims about statistical relationships, they are not claiming that the relationship will hold for everyone.

10.2 Strength versus Statistical Significance

To find out if a statistical relationship exists between two variables, researchers must usually rely on measurements from only a sample of individuals from a larger population. However, for any particular sample, a relationship may exist even if there is no relationship between the two variables in the population. It may be just the "luck of the draw" that that particular sample exhibited the relationship.

For example, suppose an observational study followed for 5 years a sample of 1000 owners of satellite dishes and a sample of 1000 nonowners and found that four of the satellite dish owners developed brain cancer, whereas only two of the nonowners did. Could the researcher legitimately claim that the rate of cancer among all satellite dish owners is twice that among nonowners? You would probably not be persuaded that the observed relationship was indicative of a problem in the larger population. The numbers are simply too small to be convincing.

Defining Statistical Significance

To overcome this problem, statisticians try to determine whether an observed relationship in a sample is **statistically significant.** To determine this, we ask what the chances are that a relationship that strong or stronger would have been observed in the sample if there really were nothing going on in the population. If those chances are small, we declare that the relationship is statistically significant and was not just a fluke. To be convincing, an observed relationship must also be statistically significant.

Most researchers are willing to declare that a relationship is statistically significant if there is only a small chance of observing the relationship in the sample when actually nothing is going on in the population. A common criterion is to define a "small chance" to be 5%, but sometimes 10%, 1%, or some other value is used. In other words, a relationship observed in sample data is typically considered to be statistically significant if that relationship is stronger than 95% of the relationships we would expect to see just by chance.

Of course, this reasoning carries with it the implication that of all the relationships that do occur by chance alone, 5% of them will erroneously earn the title of statistical significance. However, this is the price we pay for not being able to measure the entire population—while still being able to determine that statistically significant relationships do exist. We will learn how to assess statistical significance in Chapters 13, 22, and 23.

Two Warnings about Statistical Significance

Two important points, which we will study in detail in Chapter 24, often lead people to misinterpret statistical significance. First, it is easier to rule out chance if the observed relationship is based on very large numbers of observations. Even a minor relationship will achieve "statistical significance" if the sample is very large. However, earning that title does not necessarily imply that there is a *strong* relationship or even one of practical importance.

EXAMPLE 10.1 Small but Significant Increase in Risk of Breast Cancer

News Story 12 in the Appendix, "Working nights may increase breast cancer risk," contains the following quote by Francine Laden, one of the co-authors of the study: "The numbers in our study are small, but they are statistically significant." As a reader, what do you think that means? Reading further in the news story reveals the answer:

> The study was based on more than 78,000 nurses from 1988 through 1998. It found that nurses who worked rotating night shifts at least three times a month for one to 29 years were 8 percent more likely to develop breast cancer. For those who worked the shifts for more than 30 years, the relative risk went up by 36 percent.

The "small numbers" Dr. Laden referenced were the small increases in the risk of breast cancer, of 8 percent and 36 percent (especially the 8 percent). Because the study was based on over 78,000 women, even the small relationship observed in the sample probably reflects a real relationship in the population. In other words, the relationship in the sample, while not strong, is "statistically significant." ∎

Second, a very strong relationship won't necessarily achieve "statistical significance" if the sample is very small. If you read about researchers who "failed to find a statistically significant relationship" between two variables, do not be confused into thinking that they have proven that there *isn't* a relationship. It may be that they simply didn't take enough measurements to rule out chance as an explanation.

EXAMPLE 10.2 Do Younger Drivers Eat and Drink More while Driving?

News Story 5 (summarized in the Appendix), "Driving while distracted is common, researchers say" contains the following quote:

> Stutts' team had to reduce the sample size from 144 people to 70 when they ran into budget and time constraints while minutely cataloging hundreds of hours of video. The reduced sample size does not compromise the researchers' findings, Stutts said, although it does make analyzing population subsets difficult.

What does this mean? Consulting Original Source 5 on the companion website, one example explicitly stated is when the researchers tried to compare behavior across age groups. For instance, in Table 7 of the report (p. 36), it is shown that 92.9 percent of 18- to 29-year-old drivers were eating or drinking while driving. Middle-aged drivers weren't as bad, with 71.4 percent of drivers in their 30s and 40s and 78.6 percent of drivers in their 50s eating or drinking. And a mere 42.9 percent of drivers 60 and over were observed eating or drinking while driving. It would seem that these reflect real differences in behavior in the population of all drivers, and not just in the drivers observed in this study. But because there were only 14 drivers observed in each age group, the observed relationship between age and eating behavior is not statistically significant. It is impossible to know whether or not the relationship exists in the population. The authors of the report wrote:

> Compared to older drivers, younger drivers appeared more likely to eat or drink while driving. . . . Sample sizes within age groups, however, were small, prohibiting valid statistical testing. (pp. 61–62)

Notice that in this example, the authors of the original report and the journalist who wrote the news story interpreted the problem correctly. An incorrect, and not uncommon, interpretation would be to say that "no significant difference was found in eating and drinking behavior across age groups." While technically true, this language would lead readers to believe that there is no difference in these behaviors in the population, when in fact the sample was just too small to decide one way or the other. Occasionally a completely misleading statement will be made by saying that "no difference" was found, when the author means that no statistically significant difference was found. ■

10.3 Measuring Strength Through Correlation

A Linear Relationship

It is convenient to have a single number to measure the strength of the relationship between two measurement variables and to have that number be independent of the units used to make the measurements. For instance, if height is reported in inches instead of centimeters (and not rounded off in either case), the strength of the relationship between height and weight should not change.

Many types of relationships can occur between measurement variables, but in this chapter we consider only the most common one. The **correlation** between two measurement variables is an indicator of *how closely their values fall to a straight line* on a scatterplot. Sometimes this measure is called the *Pearson product–moment correlation* or the *correlation coefficient* or is simply represented by the letter *r*.

Notice that the statistical definition of *correlation* is more restricted than its common usage. For example, if the value of one measurement variable is always the square of the value of the other variable, they have a perfect relationship but may still have no statistical correlation. As used in statistics, correlation measures *linear relationships* only; that is, it measures how close the individual points in a scatterplot are to a straight line.

Other Features of Correlations

Here are some other features of correlations:

1. A correlation of $+1$ (or 100%) indicates that there is a perfect linear relationship between the two variables. As one increases, so does the other. All individuals fall on the same straight line, just as when two variables have a deterministic linear relationship.

2. A correlation of -1 also indicates that there is a perfect linear relationship between the two variables. However, as one increases, the other *decreases*. The individuals all fall on a straight line that slopes *downward*.

3. A correlation of zero could indicate that there is no linear relationship between the two variables. It could also indicate that the best straight line through the data on a scatterplot is exactly horizontal.

4. A *positive correlation* indicates that the variables increase together.

5. A *negative correlation* indicates that as one variable increases, the other decreases.

6. Correlations are unaffected if the units of measurement are changed. For example, the correlation between weight and height remains the same regardless of whether height is expressed in inches, feet, or millimeters.

Examples of Positive, Negative, and No Linear Relationship

Following are some examples of both positive and negative relationships. Notice how the closeness of the points to a straight line determines the *magnitude* of the correlation, whereas whether the line slopes up or down determines if the correlation is positive or negative.

EXAMPLE 10.3   Verbal SAT and GPA

In Chapter 9, we saw a scatterplot showing the relationship between the two variables verbal SAT and GPA for a sample of college students. The correlation for the data in the scatterplot is .485, indicating a moderate positive relationship. In other words, students with higher verbal SAT scores tend to have higher GPAs as well, but the relationship is nowhere close to being exact. ∎

EXAMPLE 10.4   Husbands' and Wives' Ages and Heights

Marsh (1988, p. 315) and Hand et al. (1994, pp. 179–183) reported data on the ages and heights of a random sample of 200 married couples in Britain, collected in 1980 by the Office of Population Census and Surveys. Figures 10.1 and 10.2 (see next page) show scatterplots for the ages and the heights, respectively, of the couples. Notice that the ages fall much closer to a straight line than do the heights. In other words, husbands' and wives' ages are likely to be closely related, whereas their heights are less closely related. The correlation between husbands' and wives' ages is .94, whereas the correlation between their heights is only .36. Thus, the values for the correlations confirm what we see from looking at the scatterplots. ∎

Figure 10.1
Scatterplot of British husbands' and wives' ages; correlation = .94
Source: Hand et al., 1994.

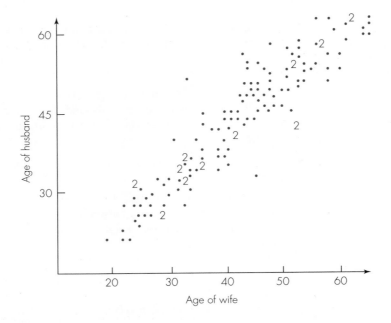

Figure 10.2
Scatterplot of British husbands' and wives' heights (in millimeters); correlation = .36
Source: Hand et al., 1994.

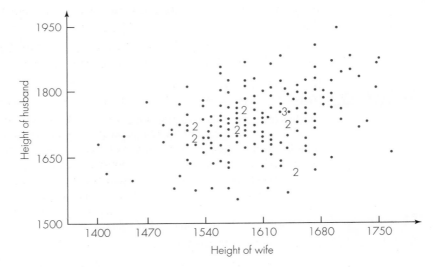

EXAMPLE 10.5 Facebook Friends and Studying for Statistics Class

In early 2011, the social media network Facebook had about 700 million users across the world and was widely used by college students. At that time, was there a relationship between the numbers of Facebook "friends" students had and how much time they spent studying? Were those with more friends likely to spend more time on Facebook and less time studying? Or perhaps those with more friends were more

Figure 10.3

Hours spent studying for a Statistics class and number of Facebook friends; correlation = −.057

Source: The author's students.

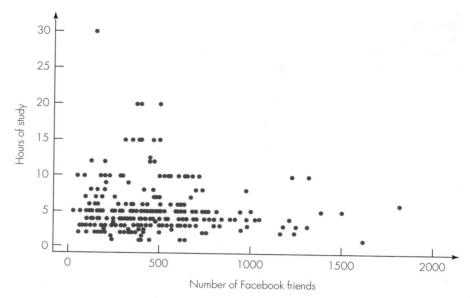

ambitious in general and studied more? Two of the questions on a survey of students in two introductory statistics classes in the 2011 Winter quarter were:

- How many Facebook friends do you have?
- How many hours per week on average do you study for this class? Include time spent studying, doing homework, and in office hours, but not time in class or discussion.

Of the 277 students who responded, 267 were Facebook users. Figure 10.3 displays a scatterplot of their responses to these two questions. (Two outliers with unrealistic values on one or the other of the questions were removed because they were clearly facetious answers.) There does not appear to be much of a relationship between these two variables, and the near-zero correlation of −.057 confirms this fact. The outlier at the top of the plot is a person who studied 30 hours a week and had 150 Facebook friends. That point is somewhat responsible for the negative correlation, and in fact, if that individual is removed the correlation is even closer to 0, at −.032. There does not appear to be a linear relationship between number of Facebook friends and study hours. ■

EXAMPLE 10.6 Professional Golfers' Putting Success

Iman (1994, p. 507) reported on a study conducted by *Sports Illustrated* magazine in which the magazine studied success rates at putting for professional golfers. Using data from 15 tournaments, the researchers determined the percentage of successful putts at distances from 2 feet to 20 feet. We have restricted our attention to the part of the data that follows a linear relationship, which includes putting distances from 5 feet to 15 feet. Figure 10.4 (next page) illustrates this relationship. The correlation between distance and rate of success is −.94. Notice the negative sign, which indicates that as distance goes up, success rate goes down. ■

Figure 10.4
Professional golfers'
putting success rates;
correlation = −.94
Source: Iman, 1994.

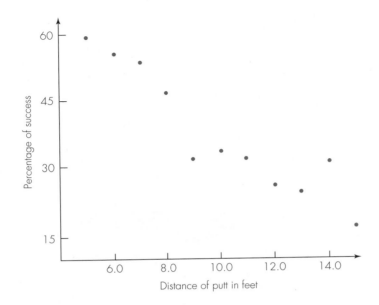

Using Computers to Find Correlation

There are many calculators, computer programs and websites that will calculate correlation for you. Here is how to do it in Excel:

1. List the values of one of the variables in one column. Let's call the cells used "Array 1". For instance, if you use Column A, rows 1 to 10, then Array 1 is A1:A10.
2. List the values of the other variable in the same order in another column. Let's call the cells used "Array 2." Array 1 and Array 2 must be the same length.
3. The Excel function CORREL(Array 1, Array 2) gives the correlation.

10.4 Specifying Linear Relationships with Regression

Sometimes, in addition to knowing the strength of the connection between two variables, we would like a *formula* for the relationship. For example, it might be useful for colleges to have a formula for the connection between verbal SAT score and college GPA. They could use it to predict the potential GPAs of future students. Some colleges do that kind of prediction to decide who to admit, but they use a collection of variables instead of just one. The simplest kind of relationship between two variables is a straight line, and that's the only type we discuss here. Our goal is to find a straight line that comes as close as possible to the points in a scatterplot.

Defining Regression

We call the procedure we use to find a straight line that comes as close as possible to the points in a scatterplot **regression;** the resulting line, the **regression line;** and the formula that describes the line, the **regression equation.** You may wonder why that word is used. Until now, most of the vocabulary borrowed by statisticians had at least some connection to the common usage of the words. The use of the word *regression* dates back to Francis Galton, who studied heredity in the late 1800s. (See Stigler, 1986 or 1989, for a detailed historical account.) One of Galton's interests was whether a man's height as an adult could be predicted by his parents' heights. He discovered that it could, but that the relationship was such that very tall parents tended to have children who were shorter than they were, and very short parents tended to have children taller than themselves. He initially described this phenomenon by saying there was "reversion to mediocrity" but later changed to the terminology "regression to mediocrity." Henceforth, the technique of determining such relationships has been called *regression.*

How are we to find the best straight line relating two variables? We could just take a ruler and try to fit a line through the scatterplot, but each of us would probably get a different answer. Instead, the most common procedure is to find what is called the **least squares line.** In determining the least squares line, priority is given to how close the points fall to the line for the variable represented by the vertical axis. Those distances are squared and added up for all of the points in the sample. For the least squares line, that sum is smaller than it would be for any other line.

The vertical distances are chosen because the equation is often used to predict that variable when the one on the horizontal axis is known. Therefore, we want to minimize how far off the prediction would be in that direction. In other words, the horizontal axis usually represents an explanatory variable, and the vertical axis represents a response variable. We want to predict the value of the response variable from knowing the value of the explanatory variable. The line we use is the one that minimizes the sum of the squared errors resulting from this prediction for the individuals in the sample. The reasoning is that if the line is good at predicting the response for those in the sample, when the response is already known, then it will work well for predicting the response in the future when only the explanatory variable is known.

The Equation for the Line

All straight lines can be expressed by the same formula. Using standard conventions, we call the variable on the vertical axis y and the variable on the horizontal axis x. We can then write the equation for the line relating them as

$$y = a + bx$$

where for any given situation, a and b would be replaced by numbers. We call the number represented by a the **intercept** and the number represented by b the **slope.** The intercept simply tells us at what particular point the line crosses the vertical axis when the horizontal axis is at zero. The slope tells us how much of an increase there is for one variable (the one on the vertical axis) when the other (on the horizontal axis) increases by one unit. A *negative slope* indicates a decrease in one variable as the other increases, just as a negative correlation does.

For example, Figure 10.5 shows the (deterministic) relationship between $y =$ temperature in Fahrenheit and $x =$ temperature in Celsius. The equation for the relationship is

$$y = 32 + 1.8x$$

The intercept, 32, is the temperature in Fahrenheit when the Celsius temperature is zero. The slope, 1.8, is the amount by which Fahrenheit temperature increases when Celsius temperature increases by one unit.

Figure 10.5
A straight line with intercept of 32 and slope of 1.8

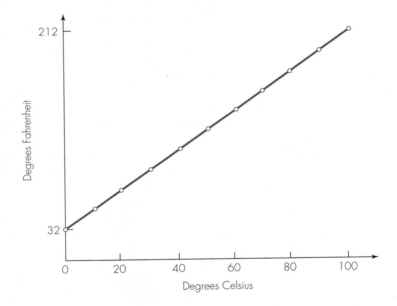

Finding and Using the Regression Equation

There are many calculators, computer programs and websites that will find the intercept and slope of the least squares line for you. Once you have the intercept and slope, you can combine them into the regression equation as follows:

$y = intercept + slope \times (x\text{-}value)$

However, remember that in a statistical relationship knowing the value of x does not give you a precise value for y. Instead, the equation can be used for the following two purposes:

- To predict the value of y in the future, when only the value of x (and not y) is known. In this case, the equation is written as:

  Predicted $y = intercept + slope \times (x\text{-}value)$

- To estimate the average value of y for a particular value of x. In this case, the equation is written as:

$$Estimated\ y = intercept + slope \times (x\text{-}value)$$

For instance, we might want to use x = verbal SAT to predict y = college GPA using the data exhibited in Figure 9.5 on page 186. The regression equation for doing so, found using the data in Figure 9.5, is:

$$Predicted\ GPA = 0.539 + (0.00362)(verbal\ SAT)$$

Using this equation, someone with a verbal SAT score of 600 would be predicted to have a GPA of:

$$Predicted\ GPA = 0.539 + (0.00362)(600) = 2.71$$

This is also what we would estimate as the average (mean) GPA for all individuals with a verbal SAT score of 600. But, remember that it is not going to be the exact GPA for *each* of those individuals, because the relationship is statistical and not deterministic. Some individuals will have higher GPAs, and some will have lower GPAs. But the average will be close to 2.71.

Using Excel to Find the Intercept and Slope of the Least Squares Regression Line

Here is how you find the intercept and slope in Excel:

1. List the y values in one column. Let's call the cells used "Array 1." For instance, if you have 10 individuals and use Column A, rows 1 to 10 for their y values, then Array 1 is A1:A10.

2. List the x values in the same order in another column. Let's call the cells used "Array 2." For instance, Array 2 might be B1:B10, representing the first 10 rows of column B. Array 1 and Array 2 must be the same length.

3. The Excel function INTERCEPT(Array 1, Array 2) gives the value of the intercept.

4. The Excel function SLOPE(Array 1, Array 2) gives the value of the slope.

CAUTION! Note that the first array listed contains the y-values, and the second one contains the x-values. Usually we think of the x values as being listed first, so you have to remember that the order required by Excel is to list the y values first.

EXAMPLE 10.7 Husbands' and Wives' Ages, Revisited

Figure 10.6 shows the same scatterplot as Figure 10.1, relating ages of husbands and wives, except that now we have added the regression line. This line minimizes the sum of the squared vertical distances between the line and the husbands' actual

Figure 10.6
Scatterplot and regression line for British husbands' and wives' ages
Source: Hand et al., 1994.

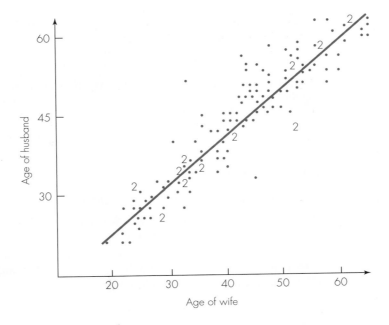

ages. The regression equation for the line shown in Figure 10.6, relating husbands' and wives' ages, is

$$\text{Predicted } y = 3.6 + .97x$$

or, equivalently,

$$\text{Predicted husband's age} = 3.6 + (.97)(\text{wife's age})$$

Notice that the intercept of 3.6 does not have any meaning in this example. It would be the predicted age of the husband of a woman whose age is 0. But obviously that's not a possible wife's age. The slope does have a reasonable interpretation. For every year of difference in two wives' ages, there is a difference of about .97 years in their husbands' ages, close to 1 year. For instance, if two women are 10 years apart in age, their husbands can be expected to be about (.97) × 10 = 9.7 years apart in age.

Let's use the equation to predict husband's age at various wife's ages.

Wife's Age	Predicted Age of Husband
20 years	3.6 + (.97)(20) = 23.0 years
25 years	3.6 + (.97)(25) = 27.9 years
40 years	3.6 + (.97)(40) = 42.4 years
55 years	3.6 + (.97)(55) = 57.0 years

This table shows that for the range of ages in the sample, husbands tend to be 2 to 3 years older than their wives, on average. The older the couple, the smaller the gap in their ages. Remember that with statistical relationships, we are determining what happens to the average and not to any given individual. Thus, although most couples won't fit the pattern given by the regression line exactly, it does show us one way to represent the average relationship for the whole group. ∎

Investigating Long-Term Trends in Time Series

In Chapter 9, we learned that a *time series* is a measurement variable recorded at evenly spaced intervals over time. Recall that one of the four components that help explain data in a time series is *long-term trend*. If the trend is linear, we can estimate the trend by finding a regression line, with time period as the explanatory variable and the measurement in the time series as the response variable. We can then remove the trend to enable us to see what other interesting features exist in the series. When we remove the linear trend in a time series, the result is, aptly enough, called a **detrended time series**. Let's revisit jeans sales in Britain, and see what we can learn about the trend.

EXAMPLE 10.8   Jeans Sales in Britain, Revisited

Figure 9.6 illustrated a time series of sales of jeans in Britain from 1980 to 1984. Figure 10.7 shows the same time series plot, with the regression line for month versus sales superimposed. Month 1 is January 1980, and month 60 is December 1984. The equation for the line is:

$$Sales = 1880 + 6.62 \ (month)$$

If we were to try to forecast sales for January 1985, the first month that is not included in the series, we would use month = 61 and solve the equation for "Sales." The resulting

Figure 10.7
Jeans sales from Figure 9.6 with regression line showing linear trend.

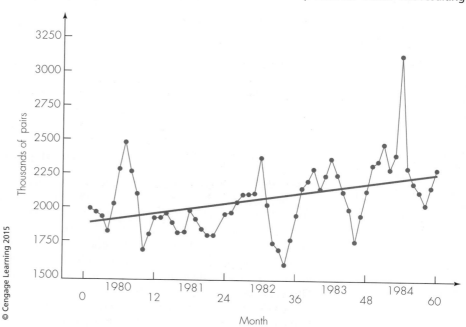

value is $1880 + 6.62(61) = 2284$ thousand pairs of jeans. Actual sales for January 1985 were 2137 thousand pairs. Our prediction is not far off, given that, overall, the data range from about 1600 to 3100 thousand pairs. One reason the actual value may be slightly lower than the predicted value is that sales tend to be lower during the winter months. Remember that seasonal components are another of the factors that affect values in a time series, and in this example, sales tend to be low in January for all years.

The regression line indicates that the trend, on average, is that sales increase by about 6.62 units per month. Because the units represent thousands of pairs, the actual increase is about 6620 pairs per month. Figure 10.8 presents the time series for jeans sales with the trend removed. Compare Figure 10.8 with Figure 10.7. Notice that the fluctuations remaining in Figure 10.8 are similar in character to those in Figure 10.7, but the upward trend is gone. ■

Figure 10.8
Detrended time series of jeans sales (linear trend removed).

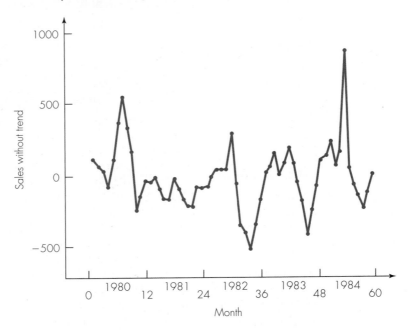

Don't Extrapolate Beyond the Range of Data Values

It is generally not a good idea to use a regression equation to predict values far outside the range where the original data fell. There is no guarantee that the relationship will continue beyond the range for which we have data. For example, using the regression equation illustrated in Figure 10.6, we would predict that women who are 100 years old have husbands whose average age is 100.6 years. But women tend to live to be older than men, so it is more likely that if a woman is married at 100, her husband is younger than she is. The relationship for much older couples would be affected by differing death rates for men and women, and a different equation would most likely apply. It is typically acceptable to use the equation only for a minor extrapolation beyond the range of the original data.

A Final Cautionary Note

It is easy to be misled by inappropriate interpretations and uses of correlation and regression. In the next chapter, we examine how that can happen, and how you can avoid it.

| CASE STUDY 10.1 | Are Attitudes about Love and Romance Hereditary? |

SOURCE: Waller and Shaver (September 1994).

Are you the jealous type? Do you think of love and relationships as a practical matter? Which of the following two statements better describes how you are likely to fall in love?

My lover and I were attracted to each other immediately after we first met.

It is hard for me to say exactly when our friendship turned into love.

If the first statement is more likely to describe you, you would probably score high on what psychologists call the *Eros* dimension of love, characteristic of those who "place considerable value on love and passion, are self-confident, enjoy intimacy and self-disclosure, and fall in love fairly quickly" (Waller and Shaver, 1994, p.268). However, if you identify more with the second statement, you would probably score higher on the *Storge* dimension, characteristic of those who "value close friendship, companionship, and reliable affection" (p. 268). Whatever your beliefs about love and romance, do you think they are partially inherited, or are they completely due to social and environmental influences?

Psychologists Niels Waller and Philip Shaver set out to answer the question of whether feelings about love and romance are partially genetic, as are most other personality traits. Waller and Shaver studied the love styles of 890 adult twins and 172 single twins and their spouses from the California Twin Registry. They compared the similarities between the answers given by monozygotic twins (MZ), who share 100% of their genes, to the similarities between those of dizygotic twins (DZ), who share, on average, 50% of their genes. They also studied the similarities between the answers of twins and those of their spouses. If love styles are genetic, rather than determined by environmental and other factors, then the matches between MZ twins should be substantially higher than those between DZ twins.

Waller and Shaver studied 345 pairs of MZ twins, 100 pairs of DZ twins, and 172 spouse pairs (that is, a twin and his or her spouse). Each person filled out a questionnaire called the "Love Attitudes Scale" (LAS), which asked them to read 42 statements like the two given earlier. For each statement, respondents assigned a ranking from 1 to 5, where 1 meant "strongly agree" and 5 meant "strongly disagree." There were seven questions related to each of six love styles, with a score determined for each person on each love style. Therefore, there were six scores for each person.

In addition to the two styles already described (Eros and Storge), scores were generated for the following four:

- *Ludus* characterizes those who "value the fun and excitement of romantic relationships, especially with multiple alternative partners; they generally are not interested in mutual self-disclosure, intimacy, or 'getting serious' " (p. 268).

(Continued)

- *Pragma* types are "pragmatic, entering a relationship only if it meets certain practical criteria" (p. 269).
- *Mania* types "are desperate and conflicted about love. They yearn intensely for love but then experience it as a source of pain, a cause of jealousy, a reason for insomnia" (p. 269).
- Those who score high on *Agape* "are oriented more toward what they can give to, rather than receive from, a romantic partner. Agape is a selfless, almost spiritual form of love" (p. 269).

For each type of love style, and for each of the three types of pairs (MZ twins, DZ twins, and spouses), the researchers computed a correlation. The results are shown in Table 10.1. (They first removed effects due to age and gender, so the correlations are not due to a relationship between love styles and age or gender.) Notice that the correlations for the MZ twins are lower than they are for the DZ twins for two love styles, and just slightly higher for the other four styles. This is in contrast to most other personality traits. For comparison purposes, three such traits are also shown in Table 10.1. Notice that for those traits, the correlations are much higher for the MZ twins, indicating a substantial hereditary component. Regarding the findings for love styles, Waller and Shaver conclude:

> *This surprising, and very unusual, finding suggests that genes are not important determinants of attitudes toward romantic love. Rather, the common environment appears to play the cardinal role in shaping familial resemblance on these dimensions.* (p. 271) ■

TABLE 10.1 Correlations for Love Styles and for Some Personality Traits

	Correlation		
	Monozygotic Twins	**Dizygotic Twins**	**Spouses**
Love Style			
Eros	.16	.14	.36
Ludus	.18	.30	.08
Storge	.18	.12	.22
Pragma	.40	.32	.29
Mania	.35	.27	−.01
Agape	.30	.37	.28
Personality Trait			
Well-being	.38	.13	.04
Achievement	.43	.16	.08
Social closeness	.38	.01	−.04

Source: Waller and Shaver, September 1994.

| CASE STUDY 10.2 | **A Weighty Issue: Women Want Less, Men Want More** |

Do you like your weight? Let me guess . . . If you're male and under about 175 pounds, you probably want to weigh the same or more than you do. If you're female, no matter what you weigh, you probably want to weigh the same or less. Those were the results uncovered in a large statistics class (119 females and 63 males) when students were asked to give their actual and their ideal weights.

Figure 10.9 shows a scatterplot of ideal versus actual weight for the females, and Figure 10.10 (next page) is the same plot for the males. Each point represents one student, whose ideal weight can be read on the vertical axis and actual weight can be read on the horizontal axis. What is the relationship between ideal and actual weight, on average, for men and for women?

First, notice that if everyone were at their ideal weight, all points would fall on a line with the equation

$$\text{ideal} = \text{actual}$$

That line is drawn in each figure. Most of the women fall below that line, indicating that their ideal weight is below their actual weight. The situation is not as clear for the men, but a pattern is still evident. Most of those weighing under 175 pounds fall on or above the line (would prefer to weigh the same or more than they do), and most of those weighing over 175 pounds fall on or below the line (would prefer to weigh the same or less than they do).

The regression lines are also shown on each scatterplot. The regression equations are:

Women: ideal = 43.9 + 0.6 actual
Men: ideal = 52.5 + 0.7 actual

(Continued)

Figure 10.9
Ideal versus actual weight for females

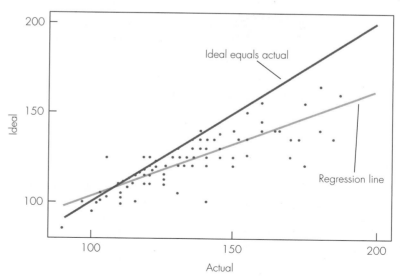

Figure 10.10
Ideal versus actual
weight for males

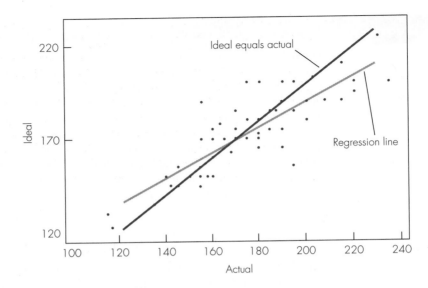

These equations have several interesting features, which, remember, summarize
the relationship between ideal and average weight for the aggregate, not for each
individual:

■ The weight for which *ideal = actual* is about 110 pounds for women and
175 pounds for men. Below those weights, actual weight is less than desired;
above them, actual weight is more than desired.

■ The slopes represent the increase in ideal weight for each 1-pound increase in actual weight. Thus, every 10 pounds of additional weight indicates an increase of
only 6 pounds in ideal weight for women and 7 pounds for men. Another way to
think about the slope is that if two women's actual weights differed by 10 pounds,
their ideal weights would differ by about $(0.6) \times 10 = 6$ pounds.   ■

Thinking About Key Concepts

• In a *deterministic relationship* between two variables, the value of one variable
can be determined exactly when the value of the other one is known. But in a
statistical relationship, there is individual variability, so the value of one variable
can only be approximated from the value of the other one.

• A *statistically significant* relationship is one that is strong enough in the observed
sample that it would have been unlikely to occur if there were no relationship in
the corresponding population. The size of the sample influences the likelihood that
a true population relationship will be statistically significant in the sample.

• When two variables have a *positive correlation,* a scatterplot would show a general
increasing straight line relationship between them.

- When two variables have a *negative correlation,* a scatterplot would show a general decreasing straight line relationship between them.
- A *regression line* is a straight line placed through the points on a scatterplot showing the average relationship between the two variables, and a *regression equation* is the equation for that line. This equation can be used to predict values of y when the value of x is known. It can also be used to estimate the average value of y for a given value of x.
- A regression equation should never be used to *extrapolate* values far beyond the range of the data used to create the equation because the relationship may not stay the same.
- In a *detrended time series* plot, the linear trend has been removed, which allows the other components to be more readily seen.

Focus On Formulas

The Data

n pairs of observations, (x_i, y_i), $i = 1, 2, \ldots, n$, where x_i is plotted on the horizontal axis and y_i on the vertical axis.

Summaries of the Data, Useful for Correlation and Regression

$$\text{SSX} = \sum_{i=1}^{n} (x_i - \bar{x})^2 = \sum_{i=1}^{n} x_i^2 - \frac{\left(\sum_{i=1}^{n} x_i\right)^2}{n}$$

$$\text{SSY} = \sum_{i=1}^{n} (y_i - \bar{y})^2 = \sum_{i=1}^{n} y_i^2 - \frac{\left(\sum_{i=1}^{n} y_i\right)^2}{n}$$

$$\text{SXY} = \sum_{i=1}^{n} (x_i - \bar{x})(y_i - \bar{y}) = \sum_{i=1}^{n} x_i y_i - \frac{\left(\sum_{i=1}^{n} x_i\right)\left(\sum_{i=1}^{n} y_i\right)}{n}$$

Correlation for a Sample of n Pairs

$$r = \frac{\text{SXY}}{\sqrt{\text{SSX}}\sqrt{\text{SSY}}}$$

The Regression Slope and Intercept

$$\text{slope} = b = \frac{\text{SXY}}{\text{SSX}}$$

$$\text{intercept} = a = \bar{y} - b\bar{x}$$

Exercises

Exercises with numbers divisible by 3 (3, 6, 9, etc.) are included in the Solutions at the back of the book. They are marked with an asterisk (*).

1. For each of the following pairs of variables measured for cities in North America, explain whether the relationship between them would be a deterministic one or a statistical one.

   a. Geographic latitude of the city and average temperature in January for the city.

   b. Average temperature in January for the city in degrees Fahrenheit and average temperature in January for the city in degrees Centigrade.

   c. Average temperature in January for the city in degrees Fahrenheit and average temperature in August for the city in degrees Fahrenheit.

2. For each of the following pairs of variables measured on college students, explain whether the relationship between them would be a deterministic one or a statistical one.

   a. Hours per day spent studying, on average, and hours per night spent sleeping, on average.

   b. Height in inches and height in centimeters (neither one rounded off).

   c. Average number of units taken per quarter or semester and GPA.

*3. Suppose 100 different researchers each did a study to see if there was a relationship between daily coffee consumption and height for adults. Suppose there really is no such relationship in the population. Would you expect any of the researchers to find a statistically significant relationship? If so, approximately how many (using the usual criterion for "small chance" of 5%)? Explain your answer.

4. Suppose a weak relationship exists between two variables in a population. Which would be more likely to result in a statistically significant relationship between the two variables: a sample of size 100 or a sample of size 10,000? Explain.

5. The relationship between height and weight is a well-established and obvious fact. Suppose you were to sample heights and weights for a small number of your friends, and you failed to find a statistically significant relationship between the two variables. Would you conclude that the relationship doesn't hold for the population of people like your friends? Explain.

*6. A pint of water weighs 1.04 pounds, so 1 pound of water is 0.96 pint. Suppose a merchant sells water in containers weighing 0.5 pound, but customers can fill them to their liking. It is easier to weigh the filled container than to measure the volume of water the customer is purchasing. Define x to be the combined weight of the container and the water and y to be the volume of the water.

   *a. Write the equation the merchant would use to determine the volume y when the weight x is known.

   *b. Specify the numerical values of the intercept and the slope, and interpret their physical meanings for this example.

   *c. What is the correlation between x and y for this example?

   d. Draw a picture of the relationship between x and y.

7. In Figure 10.2, we observed that the correlation between husbands' and wives' heights, measured in millimeters, was .36. Can you determine what the correlation would be if the heights were converted to inches (and not rounded off)? Explain.

8. Are each of the following pairs of variables likely to have a positive correlation or a negative correlation?

   a. Daily temperatures at noon in New York City and in Boston measured for a year.

   b. Weights of automobiles and their gas mileage in average miles per gallon.

   c. Hours of television watched and GPA for college students.

   d. Years of education and salary.

*9. Which implies a stronger linear relationship, a correlation of $+.4$ or a correlation of $-.6$? Explain.

10. Give an example of a pair of variables that are likely to have a positive correlation and a pair of variables that are likely to have a negative correlation.

11. Explain how two variables can have a perfect curved relationship and yet have zero correlation. Draw a picture of a set of data meeting those criteria.

*12. The table below gives the average June and December temperatures (in Fahrenheit) for eight cities in the United States. (*Source:* http://www.ncdc.noaa.gov/land-based-station-data/climate-normals/1981-2010-normals-data.)

City	June	December
Anchorage	55.0	18.6
Bismarck	64.7	16.2
Boston	67.7	34.7
Chicago	68.9	27.7
Dallas	81.3	47.1
New York	70.4	37.7
Phoenix	90.5	56.1
Portland	63.6	40.4

a. Draw a scatter plot for these temperatures, placing June on the horizontal axis and December on the vertical axis. Comment on whether or not it looks like there is a general linear relationship and, if so, whether it is positive or negative.

*b. Using Excel, a calculator, or other software, find the correlation between average June and December temperatures for these cities. Do the value and the sign (positive or negative) make sense based on the scatterplot from part (a)?

13. The regression line relating verbal SAT scores and GPA for the data exhibited in Figure 9.5 is

$$GPA = 0.539 + (0.00362)(\text{verbal SAT})$$

Predict the average GPA for those with verbal SAT scores of 500.

14. Refer to the previous exercise, giving the regression equation relating verbal SAT and GPA.

a. Explain what the slope of 0.00362 represents.

b. The lowest possible SAT score is 200. Does the intercept of 0.539 have any useful meaning for this example? Explain.

*15. Refer to Case Study 10.2, in which regression equations are given for males and females relating ideal weight to actual weight. The equations are

Women: ideal $= 43.9 + 0.6$ actual

Men: ideal $= 52.5 + 0.7$ actual

Predict the ideal weight for a man who weighs 150 pounds and for a woman who weighs 150 pounds. Compare the results.

16. Refer to the equation for women in the previous exercise, relating ideal and actual weight.

a. Does the intercept of 43.9 have a logical physical interpretation in the context of this example? Explain.

b. Does the slope of 0.6 have a logical interpretation in the context of this example? Explain.

17. Outliers in scatterplots may be within the range of values for each variable individually but lie outside the general pattern when the variables are examined in combination. A few points in Figures 10.9 and 10.10 could be considered as outliers. In the context of this example, explain the characteristics of someone who appears as an outlier.

*18. In Chapter 9, we examined a picture of winning time in men's 500-meter speed skating plotted across time. The data represented in the plot started in 1924 and went through 2010. A regression equation relating winning time and year for 1924 to 2006 is

$$\text{winning time} = 273.06 - (0.11865)(\text{year})$$

*a. Would the correlation between winning time and year be positive or negative? Explain.

*b. In 2010, the actual winning time for the gold medal was 34.91 seconds. Use the regression equation to predict the winning time for 2010, and compare the prediction to what actually happened. Was the actual winning time higher or lower than the predicted time?

*c. Explain what the slope of -0.11865 indicates in terms of how winning times change from year to year.

19. Refer to the equation in the previous exercise, relating year to winning time. The Olympics are held every 4 years. Explain what the slope of -0.11865 indicates in terms of how winning times should change from one Olympics to the next.

20. Explain why we should not use the regression equation we found in Exercise 18 for speed-skating time versus year to predict the winning time for the 2022 Winter Olympics.

*21. In one of the examples in this chapter, we noticed a very strong relationship between husbands' and wives' ages for a sample of 200 British couples, with a correlation of .94. Coincidentally, the relationship between putting distance and success rate for professional golfers had a correlation of $-.94$, based on 11 data points. (See Figure 10.4) This latter correlation was statistically significant, so we can be pretty sure the observed relationship was not just due to chance. Based on this information, do you think the observed relationship between husbands' and wives' ages is statistically significant? Explain.

22. The regression equation relating distance (in feet) and success rate (percent) for professional golfers, based on 11 distances ranging from 5 feet to 15 feet, is

$$\text{success rate} = 76.5 - (3.95)(\text{distance})$$

a. What percent success would you expect for these professional golfers if the putting distance is 6.5 feet?

b. Explain what the slope of -3.95 means in terms of how success changes with distance.

23. The original data for the putting success of professional golfers included values beyond those we used in Exercise 22 (5 feet to 15 feet), in both directions. At a distance of 2 feet, 93.3% of the putts were successful. At a distance of 20 feet, 15.8% of the putts were successful.

a. Use the equation in Exercise 22 to predict success rates for those two distances (2 feet and 20 feet). Compare the predictions to the actual success rates.

b. Use your results from part (a) to explain why it is not a good idea to use a regression equation to predict information beyond the range of values from which the equation was determined.

c. Based on the picture in Figure 10.4 and the additional information in this exercise, draw a picture of what you think the relationship between putting distance and success rate would look like for the entire range from 2 feet to 20 feet.

d. Explain why a regression equation should not be formulated for the entire range from 2 feet to 20 feet.

*24. Suppose a time series across 60 months has a long-term positive trend. Would you expect to find a positive correlation, a negative correlation, or no correlation between the values in the series and the month numbers 1 to 60? Explain.

25. Refer to the temperature data given in the table accompanying Exercise 12.

a. Using Excel, a calculator, or other software, find the intercept and slope for the regression equation with x = average June temperature and y = average December temperature for these cities.

b. Use the equation you found in part (a) to predict the average December temperature for a city with an average June temperature of 70 degrees. What assumption is required about the city in question, in order for this to be a relatively accurate prediction?

c. Do you think the equation you found in part (a) could be used to predict average December temperature from June temperature for Honolulu, Hawaii? Explain.

26. The table below gives the self-reported heights of 10 college women ("Daughter's height"), along with the heights of their mothers.

Daughter's height (y)	Mother's height (x)
60	62
66	67
65	64
66	66
67	65
63	63
69	65
63	61
61	59
65	67

a. Draw a scatter plot for these data, placing Mother's height on the horizontal axis and Daughter's height on the vertical axis. Comment on whether or not it looks like there is a linear relationship and, if so, whether it is positive or negative.

b. Using Excel, a calculator, or other software, find the correlation between the mother's and daughter's heights. Do the value and the sign (positive or negative) make sense based on the scatterplot from part (a)? Explain.

c. Using Excel, a calculator, or other software, find the intercept and slope for the regression equation with x = Mother's height and y = Daughter's height.

d. The equation you found in part (c) might be useful for predicting the height of a female from her mother's height before the daughter is fully grown. Use the equation to predict the height of the daughter of a mother who is 63 inches tall.

*27. Refer to the previous exercise about the relationship between mothers' and daughters' heights. Would the intercept of the regression line relating x = Mother's height to y = Daughter's height have a useful interpretation in this situation? (You do not need to compute the regression line to answer this question.) Explain.

28. Refer to the temperature data given in Exercise 12. A regression equation could be computed to relate x = average June temperature and y = average December temperature across cities. (This was requested in Exercise 25.) Answer the following questions without actually computing the regression equation.

a. Would the intercept of this regression equation have a useful interpretation in this situation? If you think not, explain why not. If you think it would, give the interpretation.

b. Would the slope of this regression equation have a useful interpretation in this situation? If you think not, explain why not. If you think it would, give the interpretation.

29. Refer to the journal article given as Original Source 2 on the companion website, "Development and initial validation of the Hangover Symptoms Scale: Prevalence and correlates of hangover symptoms in college students." On page 1447 it says: "The HSS [Hangover Symptoms Scale] was significantly positively associated with the frequency of drinking ($r = 0.44$)."

a. What two variables were measured for each person to provide this result?

b. Explain what is meant by $r = .44$.

c. What is meant by the word *significantly* as it is used in the quote?

d. The authors did not provide a regression equation relating the two variables. If a regression equation were to be found for the two variables in the quote, which one do you think would be the logical explanatory variable? Explain.

Mini-Projects

1. *(Computer or statistics calculator required.)* Measure the heights and weights of 10 friends of the same sex. Draw a scatterplot of the data, with weight on the vertical axis and height on the horizontal axis. Using a computer or calculator that produces regression equations, find the regression equation for your data. Draw it on your scatter diagram. Use this to predict the average weight for people of that sex who are 67 inches tall.

2. Go to your library or an electronic journal resource and peruse journal articles, looking for examples of scatterplots accompanied by correlations. Find three examples in different journal articles. Present the scatterplots and correlations, and explain in words what you would conclude about the relationship between the two variables in each case.

References

Hand, D. J., F. Daly, A. D. Lunn, K. J. McConway, and E. Ostrowski. (1994). *A handbook of small data sets*. London: Chapman and Hall.

Iman, R. L. (1994). *A data-based approach to statistics*. Belmont, CA: Wadsworth.

Labovitz, S. (1970). The assignment of numbers to rank order categories. *American Sociological Review* 35, pp. 515–524.

Marsh, C. (1988). *Exploring data*. Cambridge, England: Policy Press.

Stigler, S. M. (1986). *The history of statistics: The measurement of uncertainty before 1900*. Cambridge, MA: Belknap Press.

Stigler, S. M. (1989). Francis Galton's account of the invention of correlation. *Statistical Science* 4, pp. 73–79.

Waller, N. G., and P. R. Shaver (September 1994). The importance of nongenetic influences on romantic love styles: A twin-family study. *Psychological Science* 5, no. 5, pp. 268–274.

CHAPTER 11

Relationships Can Be Deceiving

Thought Questions

1. Use the following two pictures to speculate on what influence outliers have on correlation. For each picture, do you think the correlation is higher or lower than it would be without the outlier? (*Hint:* Remember that correlation measures how closely points fall to a straight line.)

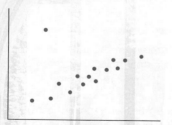 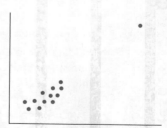

2. A strong correlation has been found in a certain city in the northeastern United States between sales of hot chocolate and sales of facial tissues measured weekly for a year. Would you interpret that to mean that hot chocolate causes people to need facial tissues? Explain.

3. Researchers have shown that there is a positive correlation between the average fat intake and the breast cancer rate across countries. In other words, countries with higher fat intake tend to have higher breast cancer rates. Does this correlation prove that dietary fat is a contributing cause of breast cancer? Explain.

4. If you were to draw a scatterplot of *number of women in the workforce* versus *number of Christmas trees sold* in the United States for each year between 1930 and the present, you would find a very strong correlation. Why do you think this would be true? Does one cause the other?

11.1   Illegitimate Correlations

In Chapter 10, we learned that the correlation between two measurement variables provides information about how closely related they are. A strong correlation implies that the two variables are closely associated or related. With a positive correlation, they increase together, and with a negative correlation, one variable tends to increase as the other decreases.

However, as with any numerical summary, correlation does not provide a complete picture. A number of anomalies can cause misleading correlations. Ideally, all reported correlations would be accompanied by a scatterplot. Without a scatterplot, however, you need to ascertain whether any of the problems discussed in this section may be distorting the correlation between two variables.

Watch out for these problems with correlations:

- Outliers can substantially inflate or deflate correlations.
- Groups combined inappropriately may mask relationships.

The Impact Outliers Have on Correlations

In a manner similar to the effect we saw on means, outliers can have a large impact on correlations. This is especially true for small samples. An outlier that is consistent with the trend of the rest of the data will inflate the correlation. An outlier that is not consistent with the rest of the data can substantially decrease the correlation.

EXAMPLE 11.1   Highway Deaths and Speed Limits

The data in Table 11.1 come from the time when the United States still had a maximum speed limit of 55 miles per hour. The correlation between death rate and speed limit across countries is .55, indicating a moderate relationship. Higher death rates tend to be associated with higher speed limits. A scatterplot of the data is presented in Figure 11.1; the two countries with the highest speed limits are labeled. Notice that Italy has both a much higher speed limit and a much higher death rate than any other country. That fact alone is responsible for the magnitude of the correlation. In fact, if Italy is removed, the correlation drops to .098, a negligible association. Of course, we could now claim that Britain is responsible for the almost zero magnitude of the correlation, and we would be right. If we remove Britain from the plot, the correlation is no longer negligible; it jumps to .70. You can see how much influence outliers have, sometimes inflating correlations and sometimes deflating them. (Of course, the actual relationship between speed limit and death rate is complicated by many other factors, a point we discuss later in this chapter.)   ■

One of the ways in which outliers can occur in a set of data is through erroneous recording of the data. Statisticians speculate that at least 5% of all

TABLE 11.1	Highway Death Rates and Speed Limits	
Country	**Death Rate (Per 100 Million Vehicle Miles)**	**Speed Limit (in Miles Per Hour)**
Norway	3.0	55
United States	3.3	55
Finland	3.4	55
Britain	3.5	70
Denmark	4.1	55
Canada	4.3	60
Japan	4.7	55
Australia	4.9	60
Netherlands	5.1	60
Italy	6.1	75

Source: Rivkin, 1986.

Figure 11.1
An example of how an outlier can inflate correlation
Source: Rivkin, 1986.

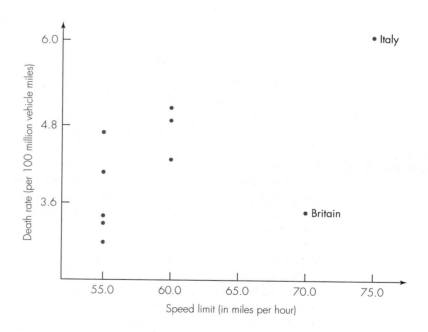

data points are corrupted, either when they are initially recorded or when they are entered into the computer. Good researchers check their data using scatterplots, stemplots, and other methods to ensure that such errors are detected and corrected. However, they do sometimes escape notice, and they can play havoc with numerical measures like correlation.

Figure 11.2
An example of how
an outlier can deflate
correlation
Source: Adapted from
Figure 10.1.

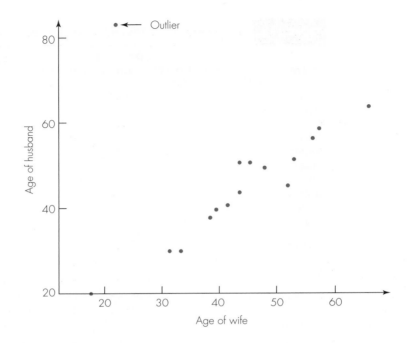

EXAMPLE 11.2 **Ages of Husbands and Wives**

Figure 11.2 shows a subset of the data we examined in Chapter 10, Figure 10.1, relating the ages of husbands and wives in Britain. In addition, an outlier has been added. This outlier could easily have occurred in the data set if someone had erroneously entered one husband's age as 82 when it should have been 28.

The correlation for the picture as shown is .39, indicating a somewhat low correlation between husbands' and wives' ages. However, the low correlation is completely attributable to the outlier. When it is removed, the correlation for the remaining points is .964, indicating a very strong relationship. ∎

Legitimate Outliers, Illegitimate Correlation

Outliers can also occur as legitimate data, as we saw in the example for which both Italy and Britain had much higher speed limits than other countries. However, the theory of correlation was developed with the idea that both measurements were from bell-shaped distributions, so outliers would be unlikely to occur. As we have seen, correlations are quite sensitive to outliers. Be very careful when you are presented with correlations for data in which outliers are likely to occur or when correlations are presented for a small sample, as shown in Example 11.3. Not all researchers or reporters are aware of the havoc outliers can play with correlation, and they may innocently lead you astray by not giving you the full details.

TABLE 11.2	Major Earthquakes in the Continental United States, 1880–2012		
Date	**Location**	**Deaths**	**Magnitude**
August 31, 1886	Charleston, SC	60	6.6
April 18–19,1906	San Francisco, CA	503	8.3
March 10, 1933	Long Beach, CA	115	6.2
February 9, 1971	San Fernando Valley, CA	65	6.6
October 17, 1989	San Francisco area (CA)	62	7.1
June 28, 1992	Yucca Valley, CA	1	7.5
January 17, 1994	Northridge, CA	61	6.8

Source: http://earthquake.usgs.gov/earthquakes/states/historical.php, accessed June 13, 2013

EXAMPLE 11.3

Earthquakes in the Continental United States

Table 11.2 lists major earthquakes that occurred in the continental United States between 1880 and 2012. The correlation between deaths and magnitude for these seven earthquakes is .689, showing a relatively strong association. This relationship implies that, on average, higher death tolls accompany stronger earthquakes.

However, if you examine the scatterplot of the data shown in Figure 11.3 (next page), you will notice that the correlation is entirely due to the famous San Francisco earthquake of 1906. In fact, for the remaining earthquakes, the trend is actually reversed. Without the 1906 quake, the correlation for these six earthquakes is actually strongly negative, at −.92. Higher-magnitude quakes are associated with fewer deaths.

Clearly, trying to interpret the correlation between magnitude and death toll for this small group of earthquakes is a misuse of statistics. The largest earthquake, in 1906, occurred before earthquake building codes were enforced. The next largest quake, with magnitude 7.5, killed only one person but occurred in a very sparsely populated area. ■

The Missing Link: A Third Variable

Another common mistake that can lead to an illegitimate correlation is combining two or more groups when they should be considered separately. The variables for each group may actually fall very close to a straight line, but when the groups are examined together, the individual relationships may be masked. As a result, it will appear that there is very little correlation between the two variables.

This problem is a variation of "Simpson's Paradox" for count data, a phenomenon we will study in the next chapter. However, statisticians do not seem to be as alert to this problem when it occurs with measurement data. When you read that two variables have a very low correlation, ask yourself whether data may have been combined into one correlation when groups should, instead, have been considered separately.

Figure 11.3
A data set for which correlation should not be used.
Source: Data from Table 11.2.

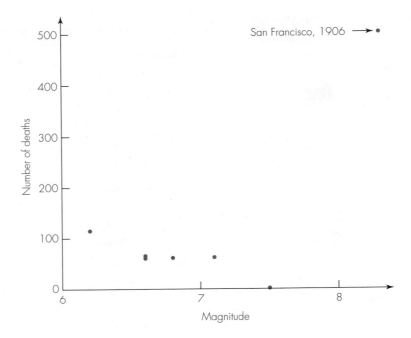

EXAMPLE 11.4   The Fewer the Pages, the More Valuable the Book?

If you peruse the bookshelves of a typical college professor, you will find a variety of books ranging from textbooks to esoteric technical publications to paperback novels. To determine whether the price of a book can be determined by the number of pages it contains, a college professor recorded the number of pages and price for 15 books on one shelf. The numbers are shown in Table 11.3. Is there a relationship between number of pages and the price of the book? The correlation for these figures is −.312. The negative correlation indicates that the more pages a book has, the less it costs, which is certainly a counterintuitive result.

Figure 11.4 illustrates what has gone wrong. It displays the data in a scatterplot, but it also identifies the books by type. The letter H indicates a hardcover book; the letter S indicates a softcover book. The collection of books on the professor's shelf consisted of softcover novels, which tend to be long but inexpensive, and hardcover technical books, which tend to be shorter but very expensive. If the correlations are calculated within each type, we find the result we would expect. The correlation between number of pages and price is .64 for the softcover books alone, and .35 for the hardcover books alone. Combining the two types of books into one collection not only masked the positive association between length and price, but produced an illogical negative association. ∎

Pages	Price	Pages	Price	Pages	Price
104	32.95	342	49.95	436	5.95
188	24.95	378	4.95	458	60.00
220	49.95	385	5.99	466	49.95
264	79.95	417	4.95	469	5.99
336	4.50	417	39.75	585	5.95

TABLE 11.3   Pages versus Price for the Books on a Professor's Shelf

Figure 11.4
Combining groups produces misleading correlations (H = hard-cover; S = softcover).
Source: Data from Table 11.3.

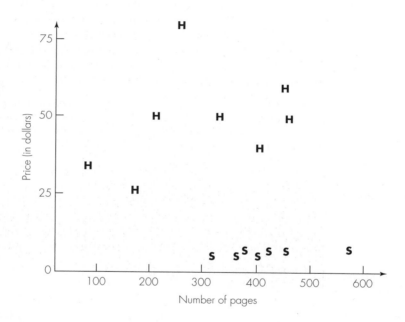

11.2   Legitimate Correlation Does Not Imply Causation

Even if two variables are legitimately related or correlated, do not fall into the trap of believing there is a causal connection between them. Although "correlation does not imply causation" is a very well known saying among researchers, relationships and correlations derived from observational studies are often *reported* as if the connection were causal.

It is easy to construct silly, obvious examples of correlations that do not result from causal connections. For example, a list of weekly tissue sales and weekly hot chocolate sales for a city with extreme seasons would probably exhibit a correlation because both tend to go up in the winter and down in the summer. A list of shoe sizes and vocabulary words mastered by school children would certainly exhibit a correlation because older children tend to have larger feet and to know more words than younger children.

The problem is that sometimes the connections do seem to make sense, and it is tempting to treat the observed association as if there were a causal link. Remember that data from an observational study, in the absence of any other evidence, simply cannot be used to establish causation.

EXAMPLE 11.5

Happiness and Heart Disease

News Story 4 in the Appendix and on the companion website notes that "heart patients who are happy are much more likely to be alive 10 years down the road than unhappy heart patients." Does that mean that if you can somehow force yourself to be happy, it will be good for your heart? Maybe, but this research is clearly an observational study. People cannot be randomly assigned to be happy or not.

The news story provides some possible explanations for the observed relationship between happiness and risk of death:

The experience of joy seems to be a factor. It has physical consequences and also attracts other people, making it easier for the patient to receive emotional support. Unhappy people, besides suffering from the biochemical effects of their sour moods, are also less likely to take their medicines, eat healthy, or to exercise. (p. 9)

Notice there are several possible confounding variables listed in this quote that may help explain why happier people live longer. For instance, it may be the case that whether happy or not, people who don't take their medicine and don't exercise are likely to die sooner, but that unhappy people are more likely to fall into that category. Thus, taking one's medicine and exercising are confounded with the explanatory variable, mood, in determining its relationship with the response variable, length of life. ∎

EXAMPLE 11.6

Does Eating Cereal Reduce Weight?

In Case Study 6.2 we discussed a news story with the heading "Breakfast cereal tied to lower BMI for kids" (http://www.reuters.com/article/2013/04/09/us-health-breakfast-idUSBRE93815320130409), which reported on a dietary study conducted with low income children in Austin, Texas. The study was originally reported in the *Journal of the Academy of Nutrition and Dietetics* (Frantzen et al, 2013). The researchers asked children what they ate for three consecutive days when they were in each of grades 4, 5, and 6. In one part of the study, the explanatory variable was number of days of eating breakfast cereal, and the response variable was the child's percentile for body mass index (BMI). In other words, the response was a measure of how overweight the child was at that time. The study reported that on average, children who ate breakfast cereal had lower BMI than children who did not, and the more cereal they ate the lower their BMI percentile was.

We may be tempted to believe that eating breakfast cereal causes children to weigh less. That's possible, but it is more likely that the observed relationship can be explained by something else. For instance, a possible confounding variable is general dietary habits in the home. Children who are fed a healthy breakfast are more likely to be fed healthy meals at other times of the day and thus are likely to weigh less than those who don't eat as well. Another possibility is that both variables have a common cause, such as high metabolism. People with high metabolism are less likely to be overweight than people who have lower metabolism and are also less able to skip breakfast. (Note that the study did not distinguish between eating no breakfast and eating something other than cereal. The explanatory variable was simply the number of days the child ate cereal.)   ■

In the next section we will examine seven possible reasons for a relationship between two variables. The first reason is that a change in the explanatory variable really is causing a change in the response variable. But there are six other explanations that could account for an observed relationship. Remember the well-worn phrase, "correlation does not imply causation." Always think about other possible explanations.

11.3   Some Reasons for Relationships Between Variables

We have seen numerous examples of variables that are related but for which there is probably not a causal connection. To help us understand this phenomenon, let's examine some of the reasons two variables could be related, including a causal connection.

Some reasons two variables could be related:

1. The explanatory variable is the direct cause of the response variable.
2. The response variable is causing a change in the explanatory variable.
3. The explanatory variable is a contributing but not sole cause of the response variable.
4. Confounding variables may be responsible for the observed relationship.
5. Both variables may result from a common cause.
6. Both variables are changing over time.
7. The association may be nothing more than coincidence.

Reason 1: The explanatory variable is the direct cause of the response variable. Sometimes, a change in the explanatory variable is the direct cause of a change in the response variable. For example, if we were to measure amount of food consumed in the past hour and level of hunger, we would find a relationship. We would probably agree that the differences in the amount of food consumed were responsible for the difference in levels of hunger.

Unfortunately, even if one variable is the direct cause of another, we may not see a strong association. For example, even though intercourse is the direct cause of pregnancy, the relationship between having intercourse and getting pregnant is not strong; most occurrences of intercourse do not result in pregnancy.

Reason 2: The response variable is causing a change in the explanatory variable. Sometimes the causal connection is the opposite of what might be expected. For example, what do you think you would find if you studied hotels and defined the response variable as the hotel's occupancy rate and the explanatory variable as advertising sales (in dollars) per room? You would probably expect that higher advertising expenditures would cause higher occupancy rates. Instead, it turns out that the relationship is negative because, when occupancy rates are low, hotels spend more money on advertising to try to raise them. Thus, although we might expect higher advertising dollars to cause higher occupancy rates, if they are measured at the same point in time, we instead find that low occupancy rates cause higher advertising revenues.

Reason 3: The explanatory variable is a contributing but not sole cause of the response variable. The complex kinds of phenomena most often studied by researchers are likely to have multiple causes. Even if there were a causal connection between diet and a type of cancer, for instance, it would be unlikely that the cancer was caused solely by eating that certain type of diet. It is particularly easy to be misled into thinking you have found a sole cause for a particular outcome, when what you have found is actually a *necessary contributor* to the outcome. For example, scientists generally agree that in order to have AIDS, you must be infected with HIV. In other words, HIV is *necessary* to develop AIDS. But it does not follow that HIV is the *sole* cause of AIDS, and there has been some controversy over whether that is actually the case.

Another possibility, discussed in earlier chapters, is that one variable is a contributory cause of another, but only for a subgroup of the population. If the researchers do not examine separate subgroups, that fact can be masked, as the next example demonstrates.

EXAMPLE 11.7 Delivery Complications, Rejection, and Violent Crime

A study summarized in *Science* (Mann, March 1994) and conducted by scientists at the University of Southern California reported a relationship between violent crime and complications during birth. The researchers found that delivery complications at birth were associated with much higher incidence of violent crime later in life. The data came from an observational study of males born in Copenhagen, Denmark, between 1959 and 1961.

However, the connection held only for those men whose mothers rejected them. Rejection meant that the mother had not wanted the pregnancy, had tried to have the fetus aborted, and had sent the baby to an institution for at least a third of his first year of life. Men who were accepted by their mothers did not exhibit this relationship. Men who were rejected by their mothers but for whom there were no complications at birth did not exhibit the relationship either. In other words, it was the interaction of delivery complications and maternal rejection that was associated with higher levels of violent crime.

This example was based on an observational study, so there may not be a causal link at all. However, even if there is a causal connection between delivery complications and subsequent violent crime, the data suggest that it holds only for a particular subset of the population. If the researchers had not measured the additional variable of maternal rejection, the data would have erroneously been interpreted as suggesting that the connection held for all men.

Reason 4: Confounding variables may be responsible for the observed relationship. We defined confounding variables in Chapter 5, but it is worth reviewing the concept here because it is relevant for explaining relationships. Remember that a confounding variable is one that has two properties. First, a confounding variable is related to the explanatory variable in the sense that individuals who differ for the explanatory variable are also likely to differ for the confounding variable. Second, a confounding variable affects the response variable. Thus, both the explanatory and one or more confounding variables may help cause the change in the response variable, but there is no way to establish how much is due to the explanatory variable and how much is due to the confounding variables. Example 5 in this chapter illustrates the point with several possibilities for confounding variables. For instance, people with differing levels of happiness (the explanatory variable) may have differing levels of emotional support, and emotional support affects one's will to live. Thus, emotional support is a confounding variable for the relationship between happiness and length of life.

Reason 5: Both variables may result from a common cause. We have seen numerous examples in which a change in one variable was thought to be associated with a change in the other, but for which we speculated that a third variable was responsible. For example, a study by Glaser et al. (1992) found that meditators had levels of an enzyme normally associated with people of a younger age. We could speculate that something in the personality of the meditators caused them to want to meditate and also caused them to have lower enzyme levels than others of the same age.

As another example, recall the scatterplot and correlation between verbal SAT scores and college GPAs, exhibited in Chapters 9 and 10. We would certainly not conclude that higher SAT scores caused higher grades in college, except perhaps for a slight benefit of boosted self-esteem. However, we could probably agree that the causes responsible for one variable being high (or low) are the same as those responsible for the other being high (or low). Those causes would include such factors as intelligence, motivation, and ability to perform well on tests.

EXAMPLE 11.8   Do Smarter Parents Have Heavier Babies?

News Story 18 in the Appendix describes a study that found for babies in the normal birth weight range, there was a relationship between birth weight and intelligence in childhood and early adulthood. The study was based on a cohort of about 3900 babies born in Britain in 1946. But there is a genetic component to intelligence, so smarter parents are likely to have smarter offspring. The researchers did include mother's education and father's social class in the analysis, to rule them out as possible confounding variables. However, there are many other variables that may contribute to birth weight, such as mother's diet and alcohol consumption, for which smarter parents may have provided more favorable conditions. Thus, it's possible that heavier birth weight and higher intelligence in the child both result from a common cause, such as parents' intelligence.

Reason 6: Both variables are changing over time.   Some of the most nonsensical associations result from correlating two variables that have both changed over time. If each one is steadily increasing or decreasing across time, you will indeed see a strong correlation, but it may not have any causal link. For example, you would certainly see a correlation between winning times in two different Olympic events because winning times have all decreased over the years.

Sociological variables are the ones most likely to be manipulated in this way, as demonstrated by the next example, relating decreasing marriage rates and increasing life expectancy. Watch out for reports of a strong association between two such variables, especially when you know that both variables are likely to have consistently changed over time.

EXAMPLE 11.9   **Marriage Rates and Life Expectancy**

Table 11.4 shows the marriage rate (marriages per 1000 people) and the life expectancy in the United States for the years 2000 to 2011, and Figure 11.5 (next page) shows a scatterplot of these numbers with the least squares regression line superimposed. There is a very strong negative correlation between these two sets of numbers, at –0.97. Does that mean that avoiding marriage causes people to live longer? No! The explanation is that marriage rates have been declining across time for various social, political, and religious reasons, and life expectancy has been increasing across time due to advanced health care, safety programs, and lifestyle decisions. In fact, both variables are more highly correlated with year than they are with each other, with a correlation of 0.98 for year with life expectancy and –0.984 for year with marriage rate. Any two variables that both change across time will display a correlation with each other.   ∎

TABLE 11.4	Marriage Rates and Life Expectancy in the United States, 2000 to 2011	
Year	**Marriage Rate (per 1000)**	**Life Expectancy**
2000	8.2	76.8
2001	8.2	76.9
2002	8.0	76.9
2003	7.7	77.1
2004	7.8	77.5
2005	7.6	77.4
2006	7.5	77.7
2007	7.3	77.9
2008	7.1	78.0
2009	6.8	78.5
2010	6.8	78.3
2011	6.8	78.7

Source: Life expectancy, http://www.census.gov/compendia/statab/2012/tables/12s0104.pdf Marriage rates, http://www.cdc.gov/nchs/nvss/marriage_divorce_tables.htm

Figure 11.5
Marriage rate versus life expectancy in the United States, 2000 to 2011

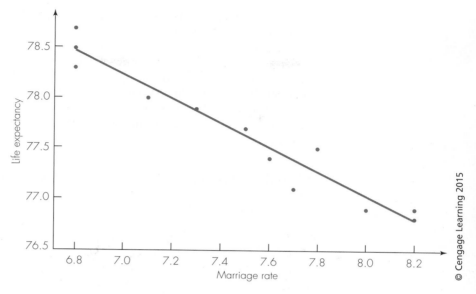

© Cengage Learning 2015

Reason 7: The association may be nothing more than coincidence.   Sometimes an association between two variables is nothing more than coincidence, even though the odds of it happening appear to be very small. For example, suppose a new office building opened, and within a year, there was an unusually high rate of brain cancer among workers in the building. Suppose someone calculated that the odds of having that many cases in one building were only 1 in 10,000. We might immediately suspect that something wrong in the environment was causing people to develop brain cancer.

The problem with this reasoning is that it focuses on the odds of seeing such a rare event occurring in that particular building in that particular city. It fails to take into account the fact that there are thousands of new office buildings. If the odds really were only 1 in 10,000, we should expect to see this phenomenon just by chance in about 1 of every 10,000 buildings. And that would just be for this particular type of cancer. What about clusters of other types of cancer or other diseases? It would be unusual if we did not occasionally see clusters of diseases as chance occurrences.

We will study this phenomenon in more detail in Part 3. For now, be aware that a connection of this sort should be expected to occur relatively often, even though each individual case has low probability.

11.4   Confirming Causation

Given the number of possible explanations for the relationship between two variables, how do we ever establish that there actually is a causal connection? It isn't easy. Ideally, in establishing a causal connection, we would change nothing in the environment except the suspected causal variable and then measure the result on the suspected outcome variable.

The only legitimate way to establish a causal connection statistically is *through the use of randomized experiments.* As we have discussed earlier, in randomized experiments we try to rule out confounding variables through random assignment. If we have a large sample and if we use proper randomization, we can assume that the levels of confounding variables will be about equal in the different treatment groups. This reduces the chances that an observed association is due to confounding variables, even those that we have neglected to measure.

Assessing Possible Causation from Observational Studies

Evidence of a possible causal connection exists when

1. There is a reasonable explanation of cause and effect.
2. The connection happens under varying conditions.
3. Potential confounding variables are ruled out.
4. There is a "dose-response" relationship.

If a randomized experiment cannot be done, then nonstatistical considerations must be used to determine whether a causal link is reasonable. Following are some features that lend evidence to a causal connection:

1. *There is a reasonable explanation of cause and effect.* A potential causal connection will be more believable if an explanation exists for how the cause and effect occur. For instance, in Example 11.4 in this chapter, we established that for hardcover books, the number of pages is correlated with the price. We would probably not contend that higher prices result in more pages, but we could reasonably argue that more pages result in higher prices. We can imagine that publishers set the price of a book based on the cost of producing it and that the more pages there are, the higher the cost of production. Thus, we have a reasonable explanation for how an increase in the length of a book could cause an increase in the price.

2. *The connection happens under varying conditions.* If many observational studies conducted under different conditions all find the same link between two variables, the evidence for a causal connection is strengthened. This is especially true if the studies are not likely to have the same confounding variables. The evidence is also strengthened if the same type of relationship holds when the explanatory variable falls into different ranges.

For example, numerous observational studies have related cigarette smoking and lung cancer. Further, the studies have shown that the higher the number of cigarettes smoked, the greater the chances of developing lung cancer; similarly, a connection has been established between lung cancer and the age at which smoking began. These facts make it more plausible that smoking actually causes lung cancer.

3. *Potential confounding variables are ruled out.* When a relationship first appears in an observational study, potential confounding variables may immediately come to mind. For example, as the news story in Figure 6.1 illustrates, the researchers in Case Study 6.4 relating mother's smoking and child's IQ did take into account possible confounding variables such as mother's education and IQ. The greater the number of confounding factors that can be ruled out, the more convincing the evidence for a causal connection.

4. *There is a "dose-response" relationship.* When the explanatory variable in a study is a measurement variable, it is useful to see if the response variable changes systematically as the explanatory variable changes. For example, the study by Freedman et al (2012) explained in Case Study 6.5 asked people aged 50 to 71 years old how much coffee they drank, and then kept track of them for the next 12 years to see if they died during that time period. They found that the more coffee people drank (up to 5 cups a day), the less likely they were to die during the 12 year follow-up. The "dose" of amount of coffee was related to the "response" of dying or not. The relationship between coffee drinking and likelihood of death during the 12 years held for both men and women. This "dose-response" relationship strengthens the likelihood that there is a causal connection, although there are still potential confounding variables that could explain the relationship. For instance, people seem to have a genetic predisposition to metabolize caffeine at differing rates, and perhaps that is related to both amount of coffee they drink and age of death.

A Final Note

As you should realize by now, it is very difficult to establish a causal connection between two variables by using anything except randomized experiments. Because it is virtually impossible to conduct a flawless experiment, potential problems crop up even with a well-designed experiment. This means that you should look with skepticism on claims of causal connections. Having read this chapter, you should have the tools necessary for making intelligent decisions and for discovering when an erroneous claim is being made.

Thinking About Key Concepts

- An *outlier* in a scatterplot that fits the pattern of the rest of the data will increase the correlation. An outlier that does not fit the pattern of the rest of the data can substantially decrease the correlation, depending on where it falls.
- *Combining two or more groups* in a scatterplot can distort the relationship between the explanatory and response variable. It is better to use a separate symbol in the plot to identify group membership.
- There are many reasons that two variables could be related other than a direct cause-and-effect connection. The most common reason is that there are *confounding variables* that are related to the explanatory variable and affect the response variable.

- If two variables are both *changing consistently over time,* they may show a relationship with each other, even though there is no direct causal association.
- Although a direct causal link between two variables cannot be made based on an observational study, there are features of the relationship that can help support a possible causal connection. These include having a reasonable explanation, observing the relationship under varying conditions, ruling out potential confounding variables, and finding a dose-response relationship.

Exercises

Exercises with numbers divisible by 3 (3, 6, 9, etc.) are included in the Solutions at the back of the book. They are marked with an asterisk (*).

1. Explain why a strong correlation would be found between weekly sales of firewood and weekly sales of cough drops over a 1-year period. Would it imply that fires cause coughs?

2. Each of the following headlines reported on an observational study. In each case, explain whether or not you think the headline is warranted. If not, write a more suitable headline.

   a. "Morning people are happier than night owls, study suggests." (http://todayhealth.today.com/_news/2012/06/11/12169581-morning-people-are-happier-than-night-owls-study-suggests?lite)

   b. "Breakfast cereals prevent overweight in children." (http://worldhealthme.blogspot.com/2013/04/breakfast-cereals-prevent-overweight-in.html)

   c. "Yogurt reduces high blood pressure, says a study." (http://www.ihealthcast.com/?p52418 or the original seems to be from http://www.globalpost.com/dispatch/news/health/120920/yogurt-reduces-high-blood-pressure-says-new-study)

   d. "Aspirin tied to lower melanoma risk." (http://www.reuters.com/article/2013/03/14/us-aspirin-melanoma-idUS-BRE92D0UR20130314)

*3. Suppose a study of employees at a large company found a negative correlation between weight and distance walked on an average day. In other words, people who walked more weighed less. Would you conclude that walking causes lower weight? Can you think of another potential explanation?

4. An article in *Science News* (1 June 1996, 149, p. 345) claimed that "evidence suggests that regular consumption of milk may reduce a person's risk of stroke, the third leading cause of death in the United States." The claim was based on an observational study of 3150 men, and the article noted that the researchers "report strong evidence that men who eschew milk have more than twice the stroke risk of those who drink 1 pint or more daily." The article concluded by noting that "those who consumed the most milk tended to be the leanest and the most physically active." Go through the list of seven "reasons

TABLE 11.5 The Eight Greatest Risks

Activity or Technology	Experts' Rank	Students' Rank
Motor vehicles	1	5
Smoking	2	3
Alcoholic beverages	3	7
Handguns	4	2
Surgery	5	11
Motorcycles	6	6
X rays	7	17
Pesticides	8	4

Source: Iman, 1994, p. 505.

two variables may be related," and discuss each one in the context of this study.

5. Iman (1994, p. 505) presents data on how college students and experts perceive risks for 30 activities or technologies. Each group ranked the 30 activities. The rankings for the eight greatest risks, as perceived by the experts, are shown in Table 11.5.

   a. Prepare a scatterplot of the data, with students' ranks on the vertical axis and experts' ranks on the horizontal axis.

   b. The correlation between the two sets of ranks is .407. Based on your scatterplot in part (a), do you think the correlation would increase or decrease if X rays were deleted? Explain. What if pesticides were deleted instead?

   c. Another technology listed was nuclear power, ranked first by the students and 20th by the experts. If nuclear power was added to the list, do you think the correlation between the two sets of rankings would increase or decrease? Explain.

*6. Give an example of two variables that are likely to be correlated because they are both changing over time.

7. Which one of the seven reasons for relationships listed in Section 11.3 is supposed to be ruled out by designed experiments?

8. Refer to Case Study 10.2, in which students reported their ideal and actual weights. When males and females are not separated, the regression equation is

$$ideal = 8.0 + 0.9 \text{ actual}$$

   a. Draw the line for this equation and the line for the equation *ideal = actual* on the same graph. Comment on the graph as compared to those shown in Figures 10.9 and 10.10, in terms of how the regression line differs from the line where ideal and actual weights are the same.

   b. Calculate the ideal weight based on the combined regression equation and the ideal weight based on separate equations, for individuals whose actual weight is 150 pounds. Recall that the separate equations were

   For women: $ideal = 43.9 + 0.6 \text{ actual}$

   For men: $ideal = 52.5 + 0.7 \text{ actual}$

   c. Comment on the conclusion you would make about individuals weighing 150 pounds if you used the combined equation compared with the conclusion you would make if you used the separate equations.

*9. Refer to the previous exercise. (You do not need to do that exercise to answer this one; just read through it.) Explain which of the problems identified in this chapter has been uncovered with this example.

10. Suppose a study measured total beer sales and number of highway deaths for 1 month in various cities. Explain why it would make sense to divide both variables by the population of the city before determining whether a relationship exists between them.

11. Construct an example of a situation in which an outlier inflates the correlation between two variables. Draw a scatterplot.

*12. Construct an example of a situation in which an outlier deflates the correlation between two variables. Draw a scatterplot.

13. According to *The Wellness Encyclopedia* (University of California, 1991, p.17): "Alcohol consumed to excess increases the risk of cancer of the mouth, pharynx, esophagus, and larynx. These risks increase dramatically when alcohol is used in conjunction with tobacco." It is obviously not possible to conduct a designed experiment on humans to test this claim, so the causal conclusion must be based on observational studies. Explain three potential additional pieces of information that the authors may have used to lead them to make a causal conclusion.

The following information applies to Exercises 14 to 16. Suppose a positive relationship had been found between each of the sets of variables given in Exercises 14 to 16. In Section 11.3, seven potential reasons for such relationships are given. Explain which of the seven reasons is most likely to account for the relationship in each case. If you think more than one reason might apply, mention them all but elaborate on only the one you think is most likely.

14. **a.** Number of deaths from automobiles and beer sales for each year from 1950 to 1990.

    b. Number of ski accidents and average wait time for the ski lift for each day during one winter at a ski resort.

*15. *a. Stomach cancer and consumption of barbecued foods, which are known to contain carcinogenic (cancer-causing) substances.

    *b. Self-reported level of stress and blood pressure.

16. **a.** Amount of dietary fat consumed and heart disease.

    b. Twice as many cases of leukemia in a new high school built near a power plant than at the old high school.

17. Explain why it would probably be misleading to use correlation to express the relationship between number of acres burned and number of deaths for major fires in the United States.

*18. It is said that a higher proportion of drivers of red cars are given tickets for traffic violations than the drivers of any other color car. Does this mean that if you drove a red car rather than a white car, you would be more likely to receive a ticket for a traffic violation? Explain.

19. Construct an example for which correlation between two variables is masked by grouping over a third variable.

20. An article in the *Davis* (CA) *Enterprise* (5 April 1994) had the headline "Study: Fathers key to child's success." The article described the study as follows: "The research, published in the March issue of the *Journal of Family Psychology,* found that mothers still do a disproportionate share of child care. But surprisingly, it also found that children who gain the 'acceptance' of their fathers make better grades than those not close to their dads." The article implies a causal link, with gaining father's acceptance (the explanatory variable) resulting in better grades (the response variable). Choosing from the remaining six possibilities in Section 11.3 (reasons 2 through 7), give three other potential explanations for the observed connection.

*21. Lave (1990) discussed studies that had been done to test the usefulness of seat belts before and after their use became mandatory. One possible method of testing the usefulness of mandatory seat belt laws is to measure the number of fatalities in a particular region for the year before and the year after the law went into effect and to compare them. If such a study were to find substantially reduced fatalities during the year after the law went into effect, could it be claimed that the mandatory seat belt law was completely responsible? Explain. (*Hint:* Consider factors such as weather and the anticipatory effect of the law.)

22. In Case Study 10.1, we learned how psychologists relied on twins to measure the contributions of heredity to various traits. Suppose a study were to find that identical (monozygotic) twins had highly correlated scores on a certain trait but that pairs of adult friends did not. Why would that not be sufficient evidence to conclude that genetic factors were responsible for the trait?

23. An article in *The Wichita Eagle* (24 June 2003, p. 4A) read as follows:

    Scientists have analyzed autopsy brain tissue from members of a religious order who had an average of 18 years of formal education and found that the more years of schooling, the less likely they were to exhibit Alzheimer's symptoms of dementia. The study provides the first biological proof of education's possible protective effect.

    Do you agree with the last sentence, that this study provides proof that education has a protective effect? Explain.

*24. One of the features that may suggest a cause-and-effect relationship from an observational study is a "dose-response" relationship.

    *a. Explain what is meant by a dose-response relationship.

    *b. Give an example of a possible dose-response relationship.

*c. Can a dose-response relationship exist if the explanatory variable is categorical (and not ordinal)? Explain.

 For Exercises 25 to 29, refer to the news stories in the Appendix and/or corresponding original source on the companion website. In each case, identify which of the "Reasons for Relationships between Variables" described in Section 11.3 are likely to apply.

25. News Story 8: "Education, kids strengthen marriage."

26. News Story 12: "Working nights may increase breast cancer risk."

*27. Summary of News Story 15: "Kids' stress, snacking linked."

28. News Story 16: "More on TV Violence."

29. News Story 20: "Eating Organic Foods Reduces Pesticide Concentrations in Children."

Exercises 30 and 31 list the titles of some of the news stories printed or summarized in the Appendix. In each case, determine whether the study was a randomized experiment or an observational study, then discuss whether the title is justified based on the way the study was done.

*30. *a. News Story 3: "Rigorous veggie diet found to slash cholesterol."

     *b. News Story 8: "Education, kids strengthen marriage."

31. a. News Story 2: "Research shows women harder hit by hangovers."

    b. News Story 15: "Kids' stress, snacking linked."

    c. News Story 20: "Eating Organic Foods Reduces Pesticide Concentrations in Children."

Mini-Projects

1. Find a news story or journal article that describes an observational study in which the author's actual goal is to try to establish a causal connection. Read the article, and then discuss how well the author has made a case for a causal connection. Consider the factors discussed in Section 11.4, and discuss whether

they have been addressed by the author. Finally, discuss the extent to which the author has convinced you that there is a causal connection.

2. Peruse journal articles, and find two examples of scatterplots for which the authors have computed a correlation that you think is misleading. For each case, explain why you think it is misleading.

3. (Computer software required.) Find two variables that are both changing over time. A good source is government data available at the website http://fedstats.gov. Record the values of the two variables for at least eight time periods (such as years).

   a. Create a scatterplot of the data, and comment on the relationship.

   b. Find the correlation between the two variables. Then find the correlation between each variable and year (or whatever time period you are using). Comment on whether you think there is an important relationship between the two variables, or if they are related only because they have both changed over time.

References

Freedman, N.D., Y. Park, C. C. Abnet, A. R. Hollenbeck, and R. Sinha, Ph.D. (2012). Association of coffee drinking with total and cause-specific mortality. *The New England Journal of Medicine*, 366 No. 20, pp. 1891–1904.

Glaser, J. L., J. L. Brind, J. H. Vogelman, M. J. Eisner, M. C. Dillbeck, R. K. Wallace, D. Chopra, and N. Orentreich. (1992). Elevated serum dehydroepiandrosterone sulfate levels in practitioners of the Transcendental Meditation (TM) and TM-Sidhi programs. *Journal of Behavioral Medicine* 15, no. 4, pp. 327–341.

Iman, R. L. (1994). *A data-based approach to statistics.* Belmont, CA: Duxbury.

Lave, L. B. (1990). Does the surgeon-general need a statistics advisor? *Chance* 3, no. 4, pp. 33–40.

Mann, C. C. (March 1994). War of words continues in violence research. *Science* 263, no. 11, p. 1375.

Rivkin, D. J. (25 November 1986). Fifty-five mph speed limit is no safety guarantee. *New York Times,* letter to the editor, p. 26.

University of California at Berkeley. (1991). *The wellness encyclopedia.* Boston: Houghton Mifflin.

CHAPTER **12**

Relationships Between Categorical Variables

Thought Questions

1. Students in a statistics class were asked whether they preferred an in-class or a take-home final exam and were then categorized as to whether they had received an A on the midterm. Of the 25 A students, 10 preferred a take-home exam, whereas of the 50 non-A students, 30 preferred a take-home exam. How would you display these data in a table?

2. Suppose a news article claimed that drinking diet soda doubled your risk of developing a certain disease. Assume the statistic was based on legitimate, well-conducted research. What additional information would you want about the risk before deciding whether to quit drinking diet soda? (*Hint:* Does this statistic provide any information on your actual risk?)

3. A study to be discussed in detail in this chapter classified pregnant women according to whether they smoked and whether they were able to get pregnant during the first cycle in which they tried to do so. What do you think is the question of interest? Attempt to answer it. Here are the results:

	Pregnancy Occurred After		
	First Cycle	**Two or More Cycles**	**Total**
Smoker	29	71	100
Nonsmoker	198	288	486
Total	227	359	586

4. A news report noted that the "relative risk" of a woman developing lung cancer if she smoked was 27.9. What do you think is meant by the term *relative risk*?

12.1   Displaying Relationships Between Categorical Variables: Contingency Tables

Remember that *categorical variables* are measurements that place individuals into categories. Examples are male/female, smoker/non-smoker, or disease/no disease. Summarizing and displaying data resulting from the measurement of two categorical variables is easy to do: Simply count the number of individuals who fall into each combination of categories, and present those counts in a table. Such displays are often called **contingency tables** because they cover all contingencies for combinations of the two variables. Each row and column combination in the table is called a **cell**.

In some cases, one variable can be designated as the explanatory variable and the other as the response variable. In these cases, it is conventional to place the explanatory variables down along the side of the table (as labels for the rows) and the response variables along the top of the table (as labels for the columns). This makes it easier to display the percentages of interest.

EXAMPLE 12.1   Aspirin and Heart Attacks

In Case Study 1.2, we discussed an experiment in which there were two categorical variables:

variable *A* = explanatory variable = aspirin or placebo
variable *B* = response variable = heart attack or no heart attack

Table 12.1 illustrates the contingency table for the results of this study. Notice that the explanatory variable (whether the individual took aspirin or placebo) is the row variable, whereas the response variable (whether the person had a heart attack or not) is the column variable. There are four cells, one representing each combination of treatment and outcome.   ■

Conditional Percentages and Rates

It's difficult to make useful comparisons from a contingency table without doing further calculations (unless the number of individuals under each condition is the same). Usually, the question of interest is whether the percentages in each category of the response variable change when the explanatory variable changes.

In Example 12.1, the question of interest is whether the percentage of heart attack sufferers differs for the people who took aspirin and the people who took a placebo. In other words, is the percentage of people who fall into the first column (heart attack) the

TABLE 12.1   Heart Attack Rates After Taking Aspirin or Placebo

	Heart Attack	No Heart Attack	Total
Aspirin	104	10,933	11,037
Placebo	189	10,845	11,034
Total	293	21,778	22,071

TABLE 12.2   Data for Example 12.1 with Percentage and Rate Added

	Heart Attack	No Heart Attack	Total	Heart Attacks (%)	Rate per 1000
Aspirin	104	10,933	11,037	0.94	9.4
Placebo	189	10,845	11,034	1.71	17.1
Total	293	21,778	22,071		

same for the two rows? We can calculate the *conditional percentages* for the response variable by looking separately at each category of the explanatory variable. Thus, in our example, we have two conditional percentages:

Aspirin group: The percentage who had heart attacks was 104/11,037 = 0.0094 = 0.94%.

Placebo group: The percentage who had heart attacks was 189/11,034 = 0.0171 = 1.71%.

Sometimes, for rare events like these heart attack numbers, percentages are so small that it is easier to interpret a rate. The **rate** is simply stated as the number of individuals per 1000 or per 10,000 or per 100,000, depending on what's easiest to interpret. Percentage is equivalent to a rate per 100.

Table 12.2 presents the data from Example 12.1, but also includes the conditional percentages and the rates of heart attacks per 1000 individuals for the two groups. Notice that the rate per 1000 is easier to understand than the percentages.

EXAMPLE 12.2   Young Drivers, Gender, and Driving Under the Influence of Alcohol

In Case Study 6.3, we learned about a court case challenging an Oklahoma law that differentiated the ages at which young men and women could buy 3.2% beer. The Supreme Court had examined evidence from a "random roadside survey" that measured information on age, gender, and drinking behavior. In addition to the data presented in Case Study 6.3, the roadside survey measured whether the driver had been drinking alcohol in the previous 2 hours. Table 12.3 gives the results for the drivers under 20 years of age.

TABLE 12.3   Results of Roadside Survey for Young Drivers

	Drank Alcohol in Last 2 Hours?			
	Yes	No	Total	Percentage Who Drank
Males	77	404	481	16.0%
Females	16	122	138	11.6%
Total	93	526	619	15.0%

Source: Gastwirth, 1988, p. 526.

After looking at the data in Table 12.3, the Supreme Court concluded that "the showing offered by the appellees does not satisfy us that sex represents a legitimate, accurate proxy for the regulation of drinking and driving" (Gastwirth, 1988, p. 527). How did they reach this conclusion? From Table 12.3, we see that 16% of the young male drivers and 11.6% of the young female drivers had been drinking in the past 2 hours. The difference in the percentages is 4.4%. Is that enough to say that young male drivers are more likely to be drinking than young female drivers for the entire population? The Supreme Court did not think so. In the next chapter, we will see that we cannot rule out chance as a reasonable explanation for this 4.4% difference. In other words, if there really is no difference among the percentages of young male and female drivers in the *population* who drink and drive, we still could reasonably expect to see a difference as large as the one observed in a *sample* of this size. Using the language introduced in Chapter 10, the observed difference in percentages is not statistically significant. Therefore, the Supreme Court decided that the law that treated young males and females differently was not justified. ■

EXAMPLE 12.3

Ease of Pregnancy for Smokers and Nonsmokers

In a retrospective observational study, researchers asked women who were pregnant with planned pregnancies how long it took them to get pregnant (Baird and Wilcox, 1985; see also Weiden and Gladen, 1986). Length of time to pregnancy was measured according to the number of cycles between stopping birth control and getting pregnant. Women were also categorized on whether they smoked, with smoking defined as having at least one cigarette per day for at least the first cycle during which they were trying to get pregnant.

For our purposes, we will classify the women on two categorical variables:

variable A = explanatory variable = smoker or nonsmoker

variable B = response variable = pregnant in first cycle or not

The question of interest is whether the same percentages of smokers and nonsmokers were able to get pregnant during the first cycle. We present the contingency table and the percentages in Table 12.4.

As you can see, a much higher percentage of nonsmokers than smokers were able to get pregnant during the first cycle. Because this is an observational study, we cannot conclude that smoking *caused* a delay in getting pregnant. We

TABLE 12.4 Time to Pregnancy for Smokers and Nonsmokers

| | Pregnancy Occurred After | | | |
	First Cycle	Two or More Cycles	Total	Percentage in First Cycle
Smoker	29	71	100	29%
Nonsmoker	198	288	486	41%
Total	227	359	586	39%

merely notice that there is a relationship between smoking status and time to pregnancy, at least for this sample. It is not difficult to think of potential confounding variables. ∎

12.2 Relative Risk, Increased Risk, and Odds

Various measures are used to report the chances of a particular outcome and how the chances increase or decrease with changes in an explanatory variable. Here are some quotes that use different measures to report chance:

- "What they found was that women who smoked had a risk [of getting lung cancer] 27.9 times as great as nonsmoking women; in contrast, the risk for men who smoked regularly was only 9.6 times greater than that for male nonsmokers." (Taubes, 26 November 1993, p. 1375)
- "Use of aspirin for five years or longer was tied to a 30-percent reduction in skin cancer risk [for women aged 50 to 79], according to findings published in the journal *Cancer*." (Pittman, 14 March 2013)
- "On average, the odds against a high school player playing NCAA football are 25 to 1. But even if he's made his college team, his odds are a slim 30 to 1 against being chosen in the NFL draft." (Krantz, 1992, p. 107)

Risk, Probability, and Odds

There are just two basic ways to express the chances that a randomly selected individual will fall into a particular category for a categorical variable. The first of the two methods involves expressing one category as a **proportion** of the *total;* the other involves comparing one category to *another category* in the form of relative **odds.**

Suppose a population contains 1000 individuals, of which 400 carry the gene for a disease. The following are all equivalent ways to express this proportion:

Forty *percent* (40%) of all individuals carry the gene.

The *proportion* who carry the gene is 0.40.

The *probability* that someone carries the gene is .40.

The *risk* of carrying the gene is 0.40.

However, to express this in odds requires a different calculation. The equivalent statement represented in odds would be:

The *odds* of carrying the gene are 4 to 6 (or 2 to 3, or 2/3 to 1).

The odds are usually expressed by reducing the numbers with and without the trait to the smallest whole numbers possible. Thus, we would say that the odds are 2 to 3, rather than saying they are 2/3 to 1. Both formulations would be correct.

The general forms of these expressions are as follows:

$$\text{Percentage with the trait} = \frac{\text{number with trait}}{\text{total}} \times 100\%$$

$$\text{Proportion with the trait} = \frac{\text{number with trait}}{\text{total}}$$

$$\text{Probability of having the trait} = \frac{\text{number with trait}}{\text{total}}$$

$$\text{Risk of having the trait} = \frac{\text{number with trait}}{\text{total}}$$

$$\text{Odds of having the trait} = \frac{\text{number with trait}}{\text{number without trait}} \text{ to } 1$$

Calculating the odds from the proportion and vice versa is a simple operation.

If p is the proportion who have the trait, then the odds of having it are $p/(1 - p)$ to 1.

If the odds of having the trait are a to b, then the proportion who have it is $a/(a + b)$.

For example, if the proportion carrying a certain gene is 0.4, then the proportion not carrying it is 0.6. So the odds of having it are (0.4/0.6) to 1, or 2/3 to 1, or 2 to 3. Going in the other direction, if the odds of having it are 2 to 3, then the proportion who have it is $2/(2 + 3) = 2/5 = 4/10 = 0.40$.

Baseline Risk

When there is a treatment or behavior for which researchers want to study risk, they often compare it to the **baseline risk,** which is the risk without the treatment or behavior. For instance, in determining whether aspirin helps prevent heart attacks, the baseline risk is the risk of having a heart attack without taking aspirin. In studying the risk of smoking and getting lung cancer, the baseline risk is the risk of getting lung cancer without smoking.

In practice, the baseline risk can be difficult to find. When researchers include a placebo as a treatment, the risk for the group taking the placebo is utilized as the baseline risk. Of course, the baseline risk depends on the population being studied as well. For instance, the risk of having a heart attack without taking daily aspirin differs for men and women, for people in different age groups, for differing levels of exercise, and so on. That's why it's important to include a placebo or a control group as similar as possible to the treatment group in studies assessing the risk of traits or behaviors.

Relative Risk

The **relative risk** of an outcome for two categories of an explanatory variable is simply the ratio of the risks for each category. The relative risk is often expressed as a multiple. For example, a relative risk of 3 may be reported by saying that the risk of developing a disease for one group is three times what it is for another group. Notice that a relative risk of 1 would mean that the risk is the same for both categories of the explanatory variable.

It is often of interest to compare the risk of disease for those with a certain trait or behavior to the baseline risk of that disease. In that case, the relative risk usually is a ratio, with the risk for the trait of interest in the numerator and the baseline risk in the denominator. However, there is no hard-and-fast rule, and if the trait or behavior decreases the risk, as taking aspirin appears to do for heart attacks, the baseline risk is often used as the numerator. In general, relative risks greater than 1 are easier to interpret than relative risks between 0 and 1.

EXAMPLE 12.4 Relative Risk of Divorce for Australian Couples

A report from the Australian government (Butterworth et al., 2008) gave the results from a study of 1498 married couples with children who were surveyed yearly in 2001, 2002, and 2003. At the end of the study period, 114 (7.6%) of the couples were separated or divorced and the remaining 1384 were intact. The researchers wondered if couples whose parents had divorced were more likely to get separated or divorced themselves.

Table 12.5 shows the outcome based on whether the wife's parents had divorced. (Results were similar when the husband's parents had divorced; see Exercises 21 and 22.) Let's compute the relative risk of separation (or divorce) for women whose parents divorced:

- Risk of separation for women whose parents were divorced = 42/334 = 0.126 (12.6%)
- Risk of separation for women whose parents were not divorced = 72/1164 = 0.062 (6.2%)
- Relative risk = 0.126/0.062 = 2.03

A relative risk of 2.03 implies that couples in which the wife's parents were divorced were slightly more than twice as likely to separate or divorce during the time of the study as couples for whom the wife's parents had not divorced. ∎

Notice the direction for which the relative risk was calculated in Example 12.4, which was to put the group with lower risk in the denominator. As noted, this is common practice because it's easier for most people to interpret the results in that

TABLE 12.5 Separation Status of Couples based on Wife's Parents' Status

Wife's Parents Divorced?	Couple Separated	Couple Intact	Total
Yes	42	292	334
No	72	1,092	1,164
Total	114	1,384	1,498

Source: Butterworth et al., 2008.

direction. For Example 12.4, the relative risk in the other direction would be 0.49. In other words, the risk of separating is 0.49 times as much for women whose parents were not divorced as it is for women whose parents were divorced. You can see that this relative risk statistic is more difficult to read than the relative risk of 2.03 presented in the example. In this case, the risk of separating for women whose parents were not divorced can also be considered as the baseline risk. Having parents who divorced increases the risk from that baseline, and it makes sense to present the relative risk in that direction.

EXAMPLE 12.5

Relative Risk of Heart Attack with and without Aspirin

When the risk of a negative outcome is *decreased* through a certain behavior, food, and so on, it is common to put the larger, baseline risk, in the numerator of the relative risk calculation. Let's use the data in Table 12.2 to compute the relative risk of having a heart attack when taking a placebo (baseline) with the risk when taking aspirin.

- Risk of heart attack when taking placebo = 189/11034 = 0.0171
- Risk of heart attack when taking aspirin = 104/11037 = 0.0094
- Relative risk = 0.0171/0.0094 = 1.81

The risk of having a heart attack for those who took placebo was 1.81 times the risk of having a heart attack for those who took aspirin.

If we had reported the relative risk by putting the baseline (placebo) risk in the denominator, it would have been 0.0094/0.0171 = 0.55. It is awkward to interpret a relative risk less than one, and in this case, we would say that the risk of heart attack when taking aspirin is only 0.55 of what the risk is when taking a placebo. You can see that it is easier to interpret relative risk when it is a number greater than one. You just need to keep track of which risk is in the numerator and which is in the denominator, and interpret the result with that in mind. ■

Increased Risk

Sometimes the change in risk is presented as a percentage increase instead of a multiple. In situations for which the baseline risk is the smaller of the two risks, the percent increase in risk is calculated as follows:

$$\text{increased risk} = \frac{\text{change in risk}}{\text{baseline risk}} \times 100\%$$

An equivalent way to compute the increased risk is

$$\text{increased risk} = (\text{relative risk} - 1.0) \times 100\%$$

If there is no obvious baseline risk or if the baseline risk is larger than the risk of interest, then the denominator for the increased risk is whatever was used as the denominator for the corresponding relative risk. As you can see from the second version of the formula, if the relative risk is less than 1.0 the "increased risk" will be negative. In that case, we say there is a *decreased risk* instead of an increased risk.

EXAMPLE 12.6 Increased Risk of Separation for Women whose Parents Divorced

In the Australian study discussed in Example 12.4, the *difference* in risk of separation for women whose parents were and were not divorced was 0.126 − 0.062 = 0.064, or 6.4%. Because the baseline risk was 0.062, this difference represents an *increased risk* of 0.064/0.062 = 1.03, or 103%. This fact would be reported by saying that there is a 103% increase in the risk of a couple separating if the wife's parents had been divorced. Notice that this is also (relative risk − 1.0) × 100% = (2.03 − 1.0) × 100% = 103%. ∎

EXAMPLE 12.7 Increased Risk of Heart Attack with Placebo or Decreased Risk with Aspirin

In Example 12.5, we computed the relative risk of having a heart attack when taking a placebo to be 1.81 times the risk when taking aspirin. Therefore, we can compute the *increased risk* to be

Increased risk = (relative risk − 1.0) × 100% = (1.81 − 1.0) × 100% = 81%

We would say that there is an 81% higher risk of a heart attack when taking placebo than when taking aspirin.

However, when an intervention or behavior is beneficial, it may be informative to talk about decreased risk instead of increased risk. To find the decrease in risk of heart attack when taking aspirin compared with placebo, we use the version of the relative risk that has the aspirin risk in the numerator and the placebo risk in the denominator. In Example 12.5, we computed that relative risk to be 0.55. Therefore, the computation is

Increased risk = (relative risk − 1.0) × 100% = (0.55 − 1.0) × 100% = −45%

The negative increased risk tells us that the risk is actually decreased. Therefore, we say that there is a 45% *decrease* in risk of a heart attack when taking aspirin, compared with taking a placebo. As this example illustrates, it is important to include information about which risk is being used as the denominator when you interpret increased or decreased risk, just as it is when you interpret relative risk. ∎

Odds Ratio

Epidemiologists, who study the causes and progression of diseases and other health risks, and others who study risk often represent comparative risks using the *odds ratio* instead of the relative risk. If the risk of a disease or problem is small, these two measures will be about the same. The relative risk is easier to understand, but the odds ratio is easier to work with statistically because it can be computed without knowing the baseline risk. Therefore, you will often find the odds ratio reported in journal articles about health-related and other issues.

To compute the **odds ratio** for a disease, you first compute the odds of getting the disease to not getting the disease for each of the two categories of the explanatory variable. You then take the ratio of those odds. Let's compute the odds ratio for the data in Table 12.2, comparing heart attack rates when taking aspirin versus a placebo:

- Odds of a heart attack to no heart attack when taking a placebo = 189/10845 = 0.0174
- Odds of a heart attack to no heart attack when taking aspirin = 104/10933 = 0.0095
- Odds ratio = 0.0174/0.0095 = 1.83

You can see that the odds ratio of 1.83 is very similar to the relative risk of 1.81. As we have noted, this will be the case as long as the risk of disease in each category is small.

EXAMPLE 12.8   Odds of Separating Based on Wife's Parents' Status

In Example 12.4 and Table 12.5, we presented data from Australia showing that the likelihood of couples separating depended on whether the wife's parents had been divorced. The relative risk of separation was 2.03. Here is the calculation of the odds ratio:

- Odds of separating to not separating, wife's parents divorced = 42/292 = 0.144
- Odds of separating to not separating, wife's parents not divorced = 72/1092 = 0.066
- Odds ratio = 0.144/0.066 = 2.18

In this example, the odds ratio of 2.18 is somewhat different than the relative risk of 2.03, as will always be the case unless the overall risk in one of the categories is quite small. We interpret the odds ratio by saying that "the odds of separating were 2.18 times higher for couples in which the wife's parents had divorced." The comparison of the odds of separating to the odds of not separating is implicit. We could make it explicit by stating that that "the odds of separating to not separating were 2.18 times higher for couples in which the wife's parents had divorced." But the shorter version is generally used unless there is confusion about what the other category is.   ■

There is an easier way to compute the odds ratio, but if you were to simply see the formula you might not understand that it was a ratio of the two odds. The formula proceeds as follows:

1. Multiply the two numbers in the upper left and lower right cells of the table.
2. Divide the result by the numbers in the upper right and lower left cells of the table.

For the Australian marriage data in Table 12.5, the computation would be as follows:

$$\text{odds ratio} = \frac{42 \times 1092}{292 \times 72} = 2.18$$

Depending on how your table is constructed, you might have to reverse the numerator and denominator. As with relative risk, it is conventional to construct the odds ratio so that it is greater than 1. The only difference is which category of the explanatory variable gets counted as the numerator of the ratio and which gets counted as the denominator.

Summary of risk, relative risk, odds, and odds ratio computations

Generally, risk and odds are used to compare the responses for a category of interest with the responses for a baseline category. Let's represent the numbers as follows.

	Response 1	Response 2	Total
Category of Interest	A_1	A_2	T_A
Baseline	B_1	B_2	T_B

The only difference between risk and odds is whether Response 2 or Total is used for the comparison. *Risk* compares Response 1 to the Total. *Odds* compares Response 1 to Response 2.

- *Risk* of Response 1 for the Category of Interest $= A_1/T_A$
- *Odds* of Response 1 to Response 2 for the Category of Interest $= A_1/A_2$
- *Relative risk* $= \dfrac{A_1/T_A}{B_1/T_B}$
- *Odds ratio* $= \dfrac{A_1/A_2}{B_1/B_2}$

Relative Risk and Odds Ratios in Journal Articles

Journal articles often report relative risk and odds ratios, but rarely in the simple form described here. In most studies, researchers measure a number of potential confounding variables. When they compute the relative risk or odds ratio, they "adjust" it to account for these confounding variables. For instance, they might report the relative risk for getting a certain type of cancer if you eat a high-fat diet, after taking into account age, amount of exercise, and whether or not people smoke.

The statistical methods used for these adjustments are beyond the level of this book, but interpreting the results is not. An adjusted relative risk or odds ratio has almost the same interpretation as the straightforward versions we have learned. The only difference is that you can think of them as applying to two groups for which the other variables are held approximately constant. For instance, suppose the relative risk for getting a certain type of cancer for those with a high-fat and low-fat diet is reported to be 1.3, adjusted for age and smoking status. That means that the relative risk applies (approximately) to two groups of individuals of the same age and smoking status, where one group has a high-fat diet and the other has a low-fat diet.

EXAMPLE 12.9

Night Shift Work and Odds for Breast Cancer

News Story 12, "Working nights may increase breast cancer risk" reported that "women who regularly worked night shifts for three years or less were about 40 percent more likely to have breast cancer than women who did not work such shifts." Consulting the journal article in which this statistic is found reveals that it is not a simple increased risk as defined in this chapter. The result is found in Table 3 of the journal article "Night Shift Work, Light at Night, and Risk of Breast Cancer," in a column headed "Odds ratio." It's actually an adjusted odds ratio, for women who worked at least one night shift a week for up to 3 years. The footnote to the table explains that "odds ratios were adjusted for parity [number of pregnancies], family history of breast cancer (mother or sister), oral contraceptive use (ever), and recent (<5 years) discontinued use of hormone replacement therapy" (p. 1561). Also, notice that the news story reports an increased risk of 40 percent, which would be correct if the relative risk were 1.4. Remember that for most diseases the odds ratio and relative risk are very similar. The news report is based on the assumption that this is the case for this study. ■

12.3   Misleading Statistics about Risk

You can be misled in a number of ways by statistics presenting risks. Unfortunately, statistics are often presented in the way that produces the best story rather than in the way that is most informative. Often, you cannot derive the information you need from news reports.

Common ways the media misrepresent statistics about risk:

1. The baseline risk is missing.
2. The time period of the risk is not identified.
3. The reported risk is not necessarily your risk.

Missing Baseline Risk

A news report of a study done at McGill University in Montreal had bad news for men who drink beer. (*Source:* http://www.reuters.com/article/2009/08/23/us-drinking-idUSTRE57M2AZ20090823, accessed June 19, 2013.) Headlined "A beer a day may raise risk of several cancers," the story gave this example: "When it came to esophageal cancer, for instance, men who drank one to six times per week had an 83 percent higher risk than teetotalers and less-frequent drinkers, while daily drinkers had a three-fold higher risk." In other words, the relative risk of esophageal cancer is about 3.0 for men who are daily drinkers compared to men who don't drink. The original journal article on which the report was based (Benedetti et al., 2009) separated the risks for those drinking beer, wine, and liquor, and reported that the odds ratio for esophageal cancer for daily beer drinkers (men only) was 2.78. If you were a beer drinker reading about this study, would it encourage you to reduce your beer consumption?

Although a relative risk of *three times* sounds ominous, it is not much help in making lifestyle decisions without also having information about what the risk is *without* drinking beer, the baseline risk. If a threefold risk increase means that your chances of developing this cancer go from 1 in 100,000 to 3 in 100,000, you are much less likely to be concerned than if it means your risk jumps from 1 in 10 to 3 in 10. When a study reports relative risk, it should always give you a baseline risk as well.

The National Cancer Institute provides baseline rates for various types of cancer. For esophageal cancer, the annual incidence for men is about 7.7 cases per 100,000.

The lifetime risk is estimated to be 1 in 198. (*Source:* http://seer.cancer.gov/statfacts/html/esoph.html.) However, these numbers include men who drink beer and men who don't, so they do not provide the baseline risk for men who do not drink. Also remember that because these results were based on an observational study, we cannot conclude that drinking beer actually *caused* the greater observed risk. For instance, there may be confounding dietary factors.

Risk over What Time Period?

"Italian scientists report that a diet rich in animal protein and fat—cheeseburgers, french fries, and ice cream, for example—increases a woman's risk of breast cancer threefold," according to *Prevention Magazine's Giant Book of Health Facts* (1991, p. 122). Couple this with the fact that the National Cancer Institute estimates that 1 in 8 women in the United States will get breast cancer. Does that mean that if a woman eats a diet rich in animal protein and fat, her chances of developing breast cancer are more than 1 in 3?

There are two problems with this line of reasoning. First, the statement attributed to the Italian scientists was woefully incomplete. It did not specify anything about how the study was conducted. It also did not specify the ages of the women studied or what the baseline rate of breast cancer was for the study. Why would we need to know the baseline rate for the study when we already know that 1 in 8 women will develop breast cancer? The answer is that age is a critical factor. The baseline rate of 1 in 8 is a lifetime risk. As with most diseases, accumulated risk increases with age.

According to the National Cancer Institute (http://www.cancer.gov/cancertopics/factsheet/detection/probability-breast-cancer), the risk of a woman being diagnosed with breast cancer in the next 10 years depends on her current age:

age 40: 1 in 68

age 50: 1 in 42

age 60: 1 in 28

age 70: 1 in 26

age 80: 1 in 11

The annual risk of developing breast cancer is only about 1 in 3700 for women in their early 30s but is 1 in 235 for women in their early 70s (Fletcher, Black, Harris, Rimer, and Shapiro, 20 October 1993, p. 1644). If the Italian study had been done on very young women, the threefold increase in risk could represent a small increase. Unfortunately, *Prevention Magazine's Giant Book of Health Facts* did not give even enough information to lead us to the original report of the work. Therefore, it is impossible to intelligently evaluate the claim.

Reported Risk versus Your Risk

When you hear about the risk of something, remember that it may not apply to you. Most risk assessments come from observational studies, and confounding variables mean that your risk may differ substantially from the reported risk.

A classic example involves the risk of various years and models of cars being stolen. The following headline was enough to make you want to go out and buy a new car: "Older cars stolen more often than new ones" [*Davis* (CA) *Enterprise*, 15 April 1994, p. C3]. The article reported that "among the 20 most popular auto models stolen [in California] last year, 17 were at least 10 years old."

Suppose you own two cars; one is 15 years old and the other is new. You park them both on the street outside of your home. Are you at greater risk of having the old one stolen? Perhaps, but the information quoted in the article gives you no information about that question.

Numerous factors determine which cars are stolen. We can easily speculate that many of those factors are strongly related to the age of cars as well. Certain neighborhoods are more likely to be targeted than others, and those same neighborhoods are probably more likely to have older cars parked in them. Cars parked in locked garages are less likely to be stolen and are more likely to be newer cars. Cars with easily opened doors are more likely to be stolen and more likely to be old. Cars that are not locked and/or don't have alarm systems are more likely to be stolen and are more likely to be old. Cars with high value for used parts are more likely to be stolen and are more likely to be old, discontinued models.

You can see that the real question of interest to a consumer is, "If I were to buy a new car, would my chances of having it stolen increase or decrease over those of the car I own now?" That question can't be answered based only on information about which cars have been stolen most often. Simply too many confounding variables are related to both the age of the car and its risk of being stolen.

EXAMPLE 12.10   Your Risk of Problems with Distracted Driving

A report posted by the American Automobile Association (AAA) on June 10, 2013, summarized research to date on the risks of distracted driving. (*Source:* http://news-room.aaa.com/wp-content/uploads/2013/06/Cognitive-Distraction_AAAFTS-Research-Compendium.pdf.) Here are two quotes given on the third page of the report:

- "With regard to visual distractions, NHTSA [National Highway Traffic Safety Administration], using naturalistic data, found that glances away from the forward roadway lasting more than two seconds increased the risk of a crash or near-crash to over two times that of "normal" driving."
- "In examining cell phone use and texting, studies have found that these activities increase risk to drivers. [Two studies] for example, each found a fourfold increase in crash risk for drivers using cell phones, and research at the Virginia Tech Transportation Institute (VTTI) showed that commercial truck drivers who were texting were 23 times as likely to have a safety critical event as those who were not distracted."

There is no question that being distracted while driving increases the risk of an accident. The first quote says that the relative risk of an accident when glancing away for more than two seconds is more then 2.0, compared to not glancing away for that long. The second quote says that the relative risk of a crash when using a cell phone is about 4.0, compared to not using one. It also says that the relative risk of a "safety critical event" for truck drivers when texting compared to not texting is 23, which is obviously an extremely large relative risk.

But let's examine these risks in the context of your own risk of an accident. Baseline risk of an accident is very difficult to assess because it depends on so many factors, such as speed, weather conditions, traffic density, and so on. Therefore, a two-fold or four-fold risk may mean a very high actual risk under certain conditions (such as high speed in bad weather with high traffic density), but a very low actual risk under other conditions (such as driving slowly on a straight road with no other cars around).

It is not possible to randomly assign drivers to be distracted or not, so clearly the quoted results are based on observational studies. (However, studies have been done using driving simulators to mimic distracted driving, and results are similar.) Confounding variables are almost surely an issue. It is quite likely that drivers who are willing to text or talk on a cell phone while driving also engage in other risky driving behaviors. Therefore, they are likely to have a higher risk of accidents even when they are not talking or texting. This means that the reported risks in these studies may not be your risk. If you are a careful driver, your risk of an accident may be lower than that reported, even if you were to talk on a cell phone. If you are a driver who takes risks in other ways, your relative risk may be higher than that reported. ∎

12.4   Simpson's Paradox: The Missing Third Variable

In Chapter 11, we saw an example in which omitting a third variable masked the positive correlation between number of pages and price of books. A similar phenomenon can happen with categorical variables, and it goes by the name of **Simpson's Paradox.** It is a paradox because the relationship appears to be in one direction if the third variable is not considered and in the other direction if it is.

EXAMPLE 12.11   Simpson's Paradox for Hospital Patients

We illustrate Simpson's Paradox with a hypothetical example of a new treatment for a disease. Suppose two hospitals are willing to participate in an experiment to test the new treatment. Hospital A is a major research facility, famous for its treatment of advanced cases of the disease. Hospital B is a local area hospital in an urban area.

Both hospitals agree to include 1100 patients in the study. Because the researchers conducting the experiment are on the staff of Hospital A, they decide to perform the majority of cases with the new procedure in-house. They randomly assign 1000 patients to the new treatment, with the remaining 100 receiving the standard treatment. Hospital B, which is a bit reluctant to try something new on too many patients, agrees to randomly assign 100 patients to the new treatment, leaving 1000 to receive the standard. The numbers who survived and died with each treatment in each hospital are shown in Table 12.6.

Table 12.7 shows how well the new procedure worked compared with the standard. It looks as though the new treatment is a success. The risk of dying from the standard procedure is higher than that for the new procedure in both hospitals. In fact, the

TABLE 12.6 Survival Rates for Standard and New Treatments at Two Hospitals

	Hospital A			Hospital B		
	Survive	Die	Total	Survive	Die	Total
Standard	5	95	100	500	500	1000
New	100	900	1000	95	5	100
Total	105	995	1100	595	505	1100

TABLE 12.7 Risk Compared for Standard and New Treatments

	Hospital A	Hospital B
Risk of dying with the standard treatment	$95/100 = 0.95$	$500/1000 = 0.50$
Risk of dying with the new treatment	$900/1000 = 0.90$	$5/100 = 0.05$
Relative risk	$0.95/0.90 = 1.06$	$0.50/0.05 = 10.0$

risk of dying when given the standard treatment is an overwhelming 10 times higher than it is for the new treatment in Hospital B. In Hospital A the risk of dying with the standard treatment is only 1.06 times higher than with the new treatment, but it is nonetheless higher.

The researchers would now like to estimate the overall reduction in risk for the new treatment, so they combine all of the data (Table 12.8).

What has gone wrong? It now looks as though the standard treatment is superior to the new one! In fact, the relative risk, taken in the same direction as before, is $0.54/0.82 = 0.66$. The death rate for the standard treatment is only 66% of what it is for the new treatment. How can that be true, when the death rate for the standard treatment was higher than for the new treatment in both hospitals?

The problem is that the more serious cases of the disease presumably were treated by the famous research hospital, Hospital A. Because they were more serious cases, they were more likely to die. But because they went to Hospital A, they were also more likely to receive the new treatment. When the results from both hospitals are combined, we lose the information that the patients in Hospital A had both a higher overall death rate and a higher likelihood of receiving the new treatment. The combined information is quite misleading. ∎

TABLE 12.8 Estimating the Overall Reduction in Risk

	Survive	Die	Total	Risk of Death
Standard	505	595	1100	$595/1100 = 0.54$
New	195	905	1100	$905/1100 = 0.82$
Total	700	1500	2200	

Simpson's Paradox makes it clear that it is dangerous to summarize information over groups, especially if patients (or experimental units) were not randomized into the groups. Notice that if patients had been randomly assigned to the two hospitals, this phenomenon probably would not have occurred. It would have been unethical, however, to do such a random assignment.

If someone else has already summarized data for you by collapsing three variables into two, you cannot retrieve the information to see whether Simpson's Paradox has occurred. Common sense should help you detect this problem in some cases. When you read about a relationship between two categorical variables, try to find out if the data have been collapsed over a third variable. If so, think about whether separating results for the different categories of the third variable could change the direction of the relationship between the first two. Exercise 29 at the end of this chapter presents an example of this.

| CASE STUDY 12.1 | **Assessing Discrimination in Hiring and Firing** |

The term *relative risk* is obviously not applicable to all types of data. It was developed for use with medical data where risk of disease or injury are of concern. An equivalent measure used in discussions of employment is the **selection ratio,** which is the ratio of the proportion of successful applicants for a job from one group (sex, race, and so on) compared with another group. For example, suppose a company hires 10% of the men who apply and 15% of the women. Then the selection ratio for women compared with men is $15/10 = 1.50$. Comparing this with our discussion of relative risk, it says that women are 1.5 times as likely to be hired as men. The ratio is often used in the reverse direction when arguing that discrimination has occurred. For instance, in this case, it might be argued that men are only $10/15 = 0.67$ times as likely to be hired as women. Gastwirth (1988, p. 209) explains that government agencies in the United States have set a standard for determining whether there is potential discrimination in practices used for hiring: "If the minority pass (hire) rate is *less* than four-fifths (or 0.8) of the majority rate, then the practice is said to have a disparate or disproportionate impact on the minority group, and employers are required to justify its job relevance." In the case where 10% of men and 15% of women who apply are hired, the men would be the minority group. The selection ratio of men to women would be $10/15 = 0.67$, so the hiring practice could be examined for potential discrimination.

Unfortunately, as Gastwirth and Greenhouse (1995) argue, this rule may not be as clear as it needs to be. They present a court case contesting the fairness of layoffs by the U.S. Labor Department, in which both sides in the case tried to interpret the rule in their favor. The layoffs were concentrated in Labor Department offices in the Chicago area, and the numbers are shown in Table 12.9.

If we consider the selection ratio based on people who were laid off, it should be clear that the four-fifths rule was violated. The percentage of whites who were laid off compared to the percentage of African Americans who were laid off is $3.0/8.6 = 0.35$, clearly less than the 0.80 required for fairness. However, the defense argued that the selection ratio should have been computed using those

TABLE 12.9	Layoffs by Ethnic Group for Labor Department Employees			
	Laid Off?			
Ethnic Group	**Yes**	**No**	**Total**	**% Laid Off**
African American	130	1382	1512	8.6
White	87	2813	2900	3.0
Total	217	4195	4412	

Data Source: Gastwirth and Greenhouse, 1995.

who were *retained* rather than those who were laid off. Because 91.4% of African Americans and 97% of whites were retained, the selection ratio is 91.4/97 = 0.94, well above the ratio of 0.80 required to be within acceptable practice. As for which claim was supported by the court, Gastwirth and Greenhouse (1995, p. 1642) report: "The lower court accepted the defendant's claim [using the selection ratio for those retained] but the appellate opinion, by Judge Cudahy, remanded the case for reconsideration." The judge also asked for further statistical information to rule out chance as an explanation for the difference.

The issue of whether or not the result is statistically significant, meaning that chance is not a reasonable explanation for the difference, will be considered in Chapter 13. However, notice that we must be careful in its interpretation. The people in this study are not a random sample from a larger population; they are the only employees of concern. Therefore, it may not make sense to talk about whether the results represent a real difference in some hypothetical larger population.

Gastwirth and Greenhouse point out that the discrepancy between the selection ratio for those laid off versus those retained could have been avoided if the *odds ratio* had been used instead of the selection ratio. The **odds ratio** compares the odds of being laid off to the odds of being retained for each group. Therefore, the plaintiffs and defendants could not manipulate the statistics to get two different answers. The odds ratio for this example can be computed using the simple formula

$$\text{odds ratio} = \frac{130 \times 2813}{1382 \times 87} = 3.04$$

This number tells us that the odds of being laid off compared with being retained are three times higher for African Americans than for whites. Equivalently, the odds ratio in the other direction is 1/3.04, or about 0.33. It is this figure that should be assessed using the four-fifths rule. Gastwirth and Greenhouse argue that "should courts accept the odds ratio measure, they might wish to change the 80% rule to about 67% or 70% since some cases that we studied and classified as close, had ORs [odds ratios] in that neighborhood" (1995, p. 1642).

Finally, Gastwirth and Greenhouse argue that the selection ratio may still be appropriate in some cases. For example, some employment practices require applicants to meet certain requirements (such as having a high school diploma) before they can be considered for a job. If 98% of the majority group meets the criteria but only 96% of the minority group does, then the odds of meeting the criteria versus not meeting them would only be about half as high for the minority as for the majority. To see this, consider a hypothetical set of 100 people from each group, with 98 of the 100 in the majority group and 96 of the 100 in the minority group meeting the criteria. The odds ratio for meeting the criteria to not meeting them for the minority compared to the majority group would be $(96 \times 2)/(4 \times 98) = 0.49$, so even if all qualified candidates were hired, the odds of being hired for the minority group would only be about half of those for the majority group. But the selection ratio would be $96/98 = 98\%$, which should certainly be legally acceptable. As always, statistics must be combined with common sense to be useful. ∎

Thinking About Key Concepts

- A *contingency table* displays the counts of how many individuals fall into the possible combinations of categories for two categorical variables. The categories of the explanatory variable form the rows of the table, and the categories of the response variable form the columns.
- The *risk* of being in a particular category of the response variable is found as the number of people in that category compared to the total number of people. The risk is generally calculated separately for each of the categories of the explanatory variable.
- The *odds* of being in a particular category of the response variable are found as the number of people in that category compared to the number of people not in that category.
- *Relative risk* is a comparison of risk for two categories of the explanatory variable. *Odds ratio* is a comparison of odds for two categories of the explanatory variable.
- When deciding whether to change a behavior based on a quoted risk, find out the *baseline* risk, the *time period* over which the quoted risk applies, and the extent to which the risk would apply to your *personal* circumstances.
- *Simpson's Paradox* occurs when the relationship between an explanatory variable and a response variable changes direction when a third variable is taken into account.

Focus On Formulas

An $r \times c$ contingency table is one with r categories for the row variable and c categories for the column variable. To represent the *observed numbers* in a 2×2 contingency table, we use the notation

	Variable 2		
Variable 1	**Yes**	**No**	**Total**
Yes	a	b	$a + b$
No	c	d	$c + d$
Total	$a + c$	$b + d$	n

Relative Risk and Odds Ratio

Using the notation for the observed numbers, if variable 1 is the explanatory variable and variable 2 is the response variable, then we can compute

$$\text{relative risk} = \frac{a(c + d)}{c(a + b)}$$

$$\text{odds ratio} = \frac{a \times d}{b \times c}$$

Exercises

Exercises with numbers divisible by 3 (3, 6, 9, etc.) are included in the Solutions at the back of the book. They are marked with an asterisk (*).

1. Suppose a study on the relationship between gender and political party included 200 men and 200 women and found 180 Democrats and 220 Republicans. Is that information sufficient for you to construct a contingency table for the study? If so, construct the table. If not, explain why not.

2. In a survey for a statistics class project, students in the class were asked whether they commute to school by car (Yes or No) and whether they had ever received a parking ticket (Yes or No). Of the 23 students who commuted to school by car, 19 had received a parking ticket. Of the 25 students who did not commute to school by car, seven had received a parking ticket.

   a. What is the explanatory variable, and what is the response variable in this survey?

   b. Create a contingency table for these data.

   c. Find the risk of having ever received a parking ticket for students who commuted by car

   and the risk for those who did not commute by car.

   d. Find the relative risk of receiving a parking ticket for those who commuted by car and those who did not. Write a sentence giving this relative risk in words that would be understood by someone with no training in statistics.

   e. Find the odds of having received a ticket for students who commuted by car, and write the result in a sentence that would be understood by someone with no training in statistics.

*3. In a survey for a statistics class project, students in the class were asked if they had ever been in a traffic accident, including a minor "fender-bender." Of the 23 males in the class, 16 reported having been in an accident. Of the 34 females in the class, 18 reported having been in an accident.

   *a. What is the explanatory variable, and what is the response variable in this survey?

   *b. Create a contingency table for these data.

*c. Find the risk of having been in an accident for males and the risk for females.

*d. Find the relative risk of having been in an accident for males compared to females. Write a sentence giving this relative risk in words that would be understood by someone with no training in statistics.

*e. Find the odds of having been in an accident for males, and write the result in a sentence that would be understood by someone with no training in statistics.

4. According to the United States Centers for Disease Control, from 2005 to 2009 "the drowning death rate among males (2.07 per 100,000 population) was approximately four times that for females (0.54 per 100,000 population)." (*Source:* http://www.cdc.gov/mmwr/preview/mmwrhtml/mm6119a4.htm, accessed June 21, 2013.)

   a. Express the rate for males as a percentage of the population; then explain why it is better expressed as a rate per 100,000 than as a percentage.

   b. Find the relative risk of drowning for males compared to females. Write your answer in a sentence that would be understood by someone with no training in statistics.

5. According to the *University of California at Berkeley Wellness Letter* (February 1994, p. 1), only 40% of all surgical operations at that time required an overnight stay at a hospital. Rewrite this fact as a proportion, as a risk, and as the odds of an overnight stay. In each case, express the result as a full sentence.

*6. *Science News* (25 February 1995, p. 124) reported a study of 232 people, aged 55 or over, who had heart surgery. The patients were asked whether their religious beliefs give them feelings of strength and comfort and whether they regularly participate in social activities. Of those who said yes to both, about 1 in 50 died within 6 months after their operation. Of those who said no to both, about 1 in 5 died within 6 months after their operation. What is the relative risk of death (within 6 months) for the two groups? Write your answer in a sentence or two that would be understood by someone with no training in statistics.

7. Raloff (1995) reported on a study conducted by Dimitrios Trichopolous of the Harvard School of Public Health in which researchers "compared the diets of 820 Greek women with breast cancer and 1548 others admitted to Athens-area hospitals for reasons other than cancer." One of the results had to do with consumption of olive oil, a staple in many Greek diets. The article reported that "women who eat olive oil only once a day face a 25 percent higher risk of breast cancer than women who consume it twice or more daily."

   a. The article states that the increased risk of breast cancer for those who consume olive oil only once a day is 25%. What is the relative risk of breast cancer for those who consume olive oil only once a day, compared to those who consume it twice or more?

   b. What information is missing from this article that would help individuals assess the importance of the result in their own lives?

8. The headline in an article in the *Sacramento Bee* read "Firing someone? Risk of heart attack doubles" (Haney, 1998). The article explained that "between 1989 and 1994, doctors interviewed 791 working people who had just undergone heart attacks about what they had done recently. The researchers concluded that firing someone or having a high-stakes deadline doubled the usual risk of a heart attack during the following week. . . . For a healthy 50-year-old man or a healthy 60-year-old woman, the risk of a heart attack in any given hour without any trigger is about 1 in a million." Assuming the relationship is indeed as stated in the article, write sentences that could be understood by someone with no training in statistics, giving each of the following for this example:

   a. The odds ratio

   b. Increased risk

   c. Relative risk

*9. The previous exercise described the relationship between recently firing someone and having a heart attack.

*a. Refer to the types of studies described in Chapter 5. What type of study is described in the previous exercise?

*b. Refer to the reasons for relationships listed in Section 11.3. Which do you think is the most likely explanation for the relationship found between firing someone and having a heart attack? Do you think the headline for this article was appropriate? Explain.

10. In a test for extrasensory perception (ESP), described in Case Study 13.1 in the next chapter, people were asked to try to use psychic abilities to describe a hidden photo or video segment being viewed by a "sender." They were then shown four choices and asked which one they thought was the real answer, based on what they had described. By chance alone, 25% of the guesses would be expected to be successful. The researchers tested 354 people and 122 (about 34.5%) of the guesses were successful. In both parts (a) and (b), express your answer in a full sentence.

a. What are the *odds* of a successful guess by chance alone?

b. What were the *odds* of a successful guess in the experiment?

11. A newspaper story released by the Associated Press noted that "a study by the Bureau of Justice Statistics shows that a motorist has about the same chance of being a carjacking victim as being killed in a traffic accident, 1 in 5000" [*Davis* (CA) *Enterprise,* 3 April 1994, p. A9]. Discuss this statement with regard to your *own chances* of each event happening to you.

*12. The Roper Organization (1992) conducted a study as part of a larger survey to ascertain the number of American adults who had experienced phenomena such as seeing a ghost, "feeling as if you left your body," and seeing a UFO. A representative sample of adults (18 and over) in the continental United States were interviewed in their homes during July, August, and September, 1991. The results when respondents were asked about seeing a ghost are shown in Table 12.10. Find numbers for each of the following:

*a. The percentage of the younger group who reported seeing a ghost

*b. The proportion of the older group who reported seeing a ghost

*c. The risk of reportedly seeing a ghost in the younger group

*d. The odds of reportedly seeing a ghost to not seeing one in the older group

13. Refer to the previous exercise and Table 12.10.

a. What is the *relative risk* of reportedly seeing a ghost for one group compared to the other? Write your answer in the form of a sentence that could be understood by someone who knows nothing about statistics.

b. Repeat part (a) using *increased risk* instead of relative risk.

TABLE 12.10 Age and Ghost Sitings			
	Reportedly Has Seen a Ghost		
	Yes	**No**	**Total**
Aged 18 to 29	212	1313	1525
Aged 30 or over	465	3912	4377
Total	677	5225	5902

Data Source: The Roper Organization, 1992, p. 35.

14. Using the terminology of this chapter, what name (for example, odds, risk, relative risk) applies to each of the boldface numbers in the following quotes?

    a. "Fontham found increased risks of lung cancer with increasing exposure to secondhand smoke, whether it took place at home, at work, or in a social setting. A spouse's smoking alone produced an overall **30** percent increase in lung-cancer risk" (*Consumer Reports,* January 1995, p. 28).

    b. "What they found was that women who smoked had a risk [of getting lung cancer] **27.9** times as great as nonsmoking women; in contrast, the risk for men who smoked regularly was only **9.6** times greater than that for male nonsmokers" (Taubes, 26 November 1993, p. 1375).

*15. Using the terminology of this chapter, what name (for example, odds, risk, relative risk) applies to the following quote? "**One student in five** reports abandoning safe-sex practices when drunk" (*Newsweek,* 19 December 1994, p. 73).

16. *Newsweek* (18 April 1994, p. 48) reported that "clinically depressed people are at a 50 percent greater risk of killing themselves." This means that when comparing people who are clinically depressed to those who are not, the former have an *increased risk* of killing themselves of 50%. What is the *relative risk* of suicide for those who are clinically depressed compared with those who are not?

17. According to *Consumer Reports* (1995 January, p. 29), "among nonsmokers who are exposed to their spouses' smoke, the chance of death from heart disease increases by about 30%." Rewrite this statement in terms of relative risk, using language that would be understood by someone who does not know anything about statistics.

*18. Drug use among college students in the United States was particularly heavy during the 1970s. Reporting on a study of drinking and drug use among college students in the United States in 1994, a *Newsweek* reporter wrote:

    Why should college students be so impervious to the lesson of the morning after? Efforts to discourage them from using drugs actually did work. The proportion of college students who smoked marijuana at least once in 30 days went from one in three in 1980 to one in seven last year [1993]; cocaine users dropped from 7 percent to 0.7 percent over the same period. (19 December 1994, p. 72)

    Do you agree with the statement that "efforts to discourage them from using drugs actually did work"? Explain your reasoning.

19. Refer to the quote in the previous exercise, about marijuana and cocaine use in 1980 and 1993.

    a. What was the relative risk of cocaine use for college students in 1980 compared with college students in 1993? Write your answer as a statement that could be understood by someone who does not know anything about statistics.

    b. Are the figures given for marijuana use (for example, "one in three") presented as proportions or as odds? Whichever they are, rewrite them as the other.

20. A case-control study in Berlin, reported by Kohlmeier, Arminger, Bartolomeycik, Bellach, Rehm, and Thamm (1992) and by Hand and colleagues (1994), asked 239 lung cancer patients and 429 controls (matched to the cases by age and sex) whether they had kept a pet bird during adulthood. Of the 239 lung cancer cases, 98 said yes. Of the 429 controls, 101 said yes.

    a. Construct a contingency table for the data.

    b. Compute the risk of lung cancer for bird and non-bird owners for this study.

    c. Can the risks of lung cancer for the two groups, computed in part (b), be used as baseline risks for the populations of bird and non-bird owners? Explain.

    d. How much more likely is lung cancer for bird owners than for non-bird owners in this study; that is, what is the increased risk?

TABLE 12.11   Separation Status based on Husband's Parents' Status (for Exercises 21 and 22)

Husband's Parents Divorced?	Couple Separated	Couple Intact	Total
Yes	40	245	285
No	74	1139	1213
Total	114	1384	1498

Source: Butterworth et al., 2008.

e. What information about risk would you want, in addition to the information on increased risk in part (d) of this problem, before you made a decision about whether to own a pet bird?

*21. In Example 12.4 and Table 12.5, we presented data from Australia showing that the likelihood of couples separating depended on whether the wife's parents had been divorced. For this exercise, we consider the likelihood of the couple separating based on whether the husband's parents had been divorced. Table 12.11 shows the results for this situation.

   *a. What is the risk of a couple separating if the husband's parents had been divorced?

   *b. What is the risk of a couple separating if the husband's parents had not been divorced?

   *c. Find the relative risk of a couple separating under the two conditions. Write your answer in a sentence that would be understood by someone with no training in statistics.

   *d. Find the increased risk of a couple separating if the husband's parents had been divorced. Write your answer in a sentence that would be understood by someone with no training in statistics.

22. Refer to Exercise 21 and Table 12.11, investigating the risk of separation for Australian couples based on whether the husband's parents had been divorced.

   a. What are the odds of a couple separating to not separating when the husband's parents had been divorced?

   b. What are the odds of a couple separating to not separating when the husband's parents had not been divorced?

   c. What is the odds ratio for the two conditions? Write your answer in a sentence that would be understood by someone with no training in statistics.

23. Explain the difference between risk and odds. Use the summary on page 252 for guidance.

*24. A statement quoted in this chapter was, "Use of aspirin for five years or longer was tied to a 30-percent reduction in skin cancer risk [for women aged 50 to 79], according to findings published in the journal *Cancer*" (Pittman, 14 March 2013).

   *a. What term from this chapter applies to the phrase "30 percent reduction in skin cancer risk?"

   *b. Based on the information that there was a 30% reduction in skin cancer risk when using aspirin compared to not using aspirin, find the relative risk of skin cancer when using aspirin compared to not using aspirin.

   *c. Is it possible to calculate the baseline risk of skin cancer from the information provided in the quote? If so, calculate it. If not, briefly explain why not.

25. The data in Table 12.12 are reproduced from Case Study 12.1 and represent employees laid off by the U.S. Department of Labor.

   a. Compute the odds of being retained to being laid off for each ethnic group.

| **TABLE 12.12** | Layoffs by Ethnic Group for Labor Department Employees | | | |

	Laid Off?			
Ethnic Group	Yes	No	Total	% Laid Off
African American	130	1382	1512	8.6
White	87	2813	2900	3.0
Total	217	4195	4412	

Data Source: Gastwirth and Greenhouse, 1995.

b. Use your results in part (a) to compute the odds ratio and confirm that it is about 3.0, as computed in Case Study 12.1 (where the shortcut method was used).

26. One of your teachers tells your class that skipping class often can lead to dire consequences. He says that data from his previous classes show that students who often skip class have twice the risk of failing compared to students who attend regularly, even though class attendance and participation are not part of the grade.

a. What term from this chapter applies to the phrase "twice the risk" (for example, odds ratio, increased risk, etc.)?

b. Give two reasons why the information provided by the teacher does *not* imply that if you skip class regularly you will probably fail the class.

*27 Kohler (1994, p. 427) reports data on the approval rates and ethnicity for mortgage applicants in Los Angeles in 1990. Of the 4096 African American applicants, 3117 were approved. Of the 84,947 white applicants, 71,950 were approved.

a. Construct a contingency table for the data.

*b.** Compute the proportion of each ethnic group that was approved for a mortgage.

28. Refer to the previous exercise.

a. Compute the ratio of the proportions of each ethnic group that were approved for a mortgage. Would that ratio be more appropriately called a relative risk or a selection ratio? Explain.

b. Would the data pass the four-fifths rule used in employment and described in Case Study 12.1? Explain.

29. A well-known example of Simpson's Paradox, published by Bickel, Hammel, and O'Connell (1975), examined admission rates for men and women who had applied to graduate programs at the University of California at Berkeley. The actual breakdown of data for specific programs is confidential, but the point can be made with similar, hypothetical numbers. For simplicity, we will assume there are only two graduate programs. The figures for acceptance to each program are shown in Table 12.13.

a. Combine the data for the two programs into one aggregate table. What percentage of all

| **TABLE 12.13** | An Example of Simpson's Paradox | | | | | |

	Program A			**Program B**		
	Admit	Deny	Total	Admit	Deny	Total
Men	400	250	650	50	300	350
Women	50	25	75	125	300	425
Total	450	275	725	175	600	775

men who applied were admitted? What percentage of all women who applied were admitted? Which sex was more successful?

b. What percentage of the men who applied did Program A admit? What percentage of the women who applied did Program A admit? Repeat the question for Program B. Which sex was more successful in getting admitted to Program A? Program B?

c. Explain how this problem is an example of Simpson's Paradox. Provide a potential explanation for the observed figures by guessing what type of programs A and B might have been. (*Hint:* Which program was easier to get admitted to overall? Which program interested men more? Which program interested women more?)

For Exercises 30 to 32: Read News Story 13 in the Appendix and on the companion website, "3 factors key for drug use in kids."

*30. What is the statistical term for "twice" in the following quote? (For example, relative risk, odds, etc.) "And kids with $25 or more a week in spending money are nearly **twice** as likely to smoke, drink or use drugs as children with less money."

31. Identify the statistical term for the number(s) in bold in each of the following quotes (for example, relative risk).

a. "High stress was experienced more among girls than boys, with nearly **one in three** saying they were highly stressed compared with fewer than **one in four** boys."

b. "Kids at schools with more than 1200 students are **twice** as likely as those attending schools with fewer than 800 students to be at high risk for substance abuse."

c. "Children ages 12 to 17 who are frequently bored are **50 percent** more likely to smoke, drink, get drunk, or use illegal drugs."

32. Identify or calculate a numerical value for each of the following from the information in the news story:

a. The increased risk of smoking, drinking, getting drunk, or using illegal drugs for teens

who are frequently bored, compared with those who are not.

b. The relative risk of smoking, drinking, getting drunk, or using illegal drugs for teens who are frequently bored, compared with those who are not.

c. The *proportion* of teens in the survey who said they have no friends who regularly drink.

d. The *percent* of teens in the survey who *do* have friends who use marijuana.

For Exercises 33 and 34: Refer to the Additional News Story 10 in the Appendix, " 'Keeping the faith' UC Berkeley researchers links weekly church attendance to longer, healthier life." Based on the information in the story, identify or calculate a numerical value for each of the following:

*33. The increased risk of dying from circulatory diseases for people who attended religious services less than once a week or never, compared to those who attended at least weekly.

34. a. The relative risk of dying from circulatory diseases for people who attended religious services less than once a week or never, compared to those who attended at least weekly.

b. The increased risk of dying from digestive diseases for people who attended religious services less than once a week or never, compared to those who attended at least weekly.

c. The relative risk of dying from digestive diseases for people who attended religious services less than once a week or never, compared to those who attended at least weekly.

35. News Story 10, described in the Appendix, is titled "Churchgoers live longer, study finds." One of the statements in the news story is "women who attend religious services regularly are about 80 percent as likely to die as those not regularly attending." Discuss the extent to which each of the three "common ways the media misrepresent statistics about risk" from Section 12.3, listed as parts a–c, apply to this quote.

a. The baseline risk is missing.

b. The time period of the risk is not identified.

c. The reported risk is not necessarily your risk.

 For Exercises 36 and 37: In Original Source 10, "Religious attendance and cause of death over 31 years" (not available on the companion website), the researchers used a complicated statistical method to assess relative risk by adjusting for factors such as education and income. The resulting numbers are called "relative hazards" instead of relative risks (abbreviated RH), but have the same interpretation as relative risk. The relative risks and increased risks in Exercises 36 and 37 are from Table 4 of the article. Write a sentence or two that someone with no training in statistics would understand presenting each of these in context.

***36.** The relative risk of dying from all causes for women under age 70, for those who do not attend weekly religious services compared with those who do, is 1.22.

37. For those who do not attend weekly religious services compared with those who do:

a. The increased risk of dying from all causes for men under age 70 is 0.27.

b. The relative risk of dying from cancer for men age 70+ is 1.00.

c. The relative risk of dying from cancer for men under age 70 is 0.74.

Mini-Projects

1. Carefully collect data cross-classified by two categorical variables for which you are interested in determining whether there is a relationship. Do not get the data from a book, website or journal; collect it yourself. Be sure to get counts of at least 5 in each cell and be sure the individuals you use are not related to each other in ways that would influence their data.

a. Create a contingency table for the data.

b. Compute and discuss the risks and relative risks. Are those terms appropriate for your situation? Explain.

c. Write a summary of your findings, including whether a cause-and-effect conclusion could be made if you observed a relationship.

2. Find a news story that discusses a study showing increased (or decreased) risk of one variable based on another. Write a report evaluating the information given in the article and discussing what conclusions you would reach based on the information in the article. Discuss whether any of the features in Section 12.3, "Misleading Statistics about Risk," apply to the situation.

 3. Refer to News Story 12, "Working nights may increase breast cancer risk" in the Appendix and on the companion website and the accompanying three journal articles on the companion website. Write a two- to four-page report describing how the studies were done and what they found in terms of relative risks and odds ratios. (Note that complicated statistical methods were used that adjusted for things like reproductive history, but you should still be able to interpret the odds ratios and relative risks reported in the articles.) Discuss shortcomings you think might apply to the results, if any.

References

Baird, D. D., and A. J. Wilcox. (1985). Cigarette smoking associated with delayed conception. *Journal of the American Medical Association* 253, pp. 2979–2983.

Benedetti, A., M. E. Parent, and J. Siemiatycki (2009). Lifetime consumption of alcoholic beverages and risk of 13 types of cancer in men: Results from a case-control study in Montreal. Cancer Detection and Prevention, 32, pp. 352–362.

Bickel, P. J., E. A. Hammel, and J. W. O'Connell. (1975). Sex bias in graduate admissions: Data from Berkeley. *Science* 187, pp. 298–304.

Butterworth, P., T. Oz, B. Rodgers, and H. Berry. (2008). Factors associated with relationship dissolution of Australian families with children. Social Policy Research Paper No. 37, Australian Government, Department of Families, Housing, Community Services and Indigenous Affairs. http://www.fahcsia.gov.au/about-fahcsia/publications-articles/research-publications/social-policy-research-paper-series/number-37-factors-associated-with-relationship-dissolution-of-australian-families-with-children, accessed June 19, 2013.

Fletcher, S. W., B. Black, R. Harris, B. K. Rimer, and S. Shapiro. (20 October 1993). Report of the international workshop on screening for breast cancer. *Journal of the National Cancer Institute* 85, no. 20, pp. 1644–1656.

Gastwirth, J. L. (1988). *Statistical reasoning in law and public policy.* New York: Academic Press.

Gastwirth, J. L., and S. W. Greenhouse. (1995). Biostatistical concepts and methods in the legal setting. *Statistics in Medicine* 14, no. 15, pp. 1641–1653.

Hand, D. J., F. Daly, A. D. Lunn, K. J. McConway, and E. Ostrowski. (1994). *A handbook of small data sets.* London: Chapman and Hall.

Haney, Daniel Q. (20 March 1998). Firing someone? Risk of heart attack doubles. *Sacramento Bee*, p. E1-2.

Kohler, H. (1994). *Statistics for business and economics.* 3d ed. New York: Harper-Collins College.

Kohlmeier, L., G. Arminger, S. Bartolomeycik, B. Bellach, J. Rehm, and M. Thamm. (1992). Pet birds as an independent risk factor for lung cancer: Case-control study. *British Medical Journal* 305, pp. 986–989.

Krantz, L. (1992). *What the odds are.* New York: HarperPerennial.

Pittman, G. (2013). Aspirin tied to lower melanoma risk. http://www.reuters.com/article/2013/03/14/us-aspirin-melanoma-idUSBRE92D0UR20130314, accessed June 19, 2013.

Prevention magazine's giant book of health facts. (1991). Edited by John Feltman. New York: Wings Books.

Raloff, J. (1995). Obesity, diet linked to deadly cancers. *Science News* 147, no. 3, p. 39.

The Roper Organization. (1992). *Unusual personal experiences: An analysis of the data from three national surveys.* Las Vegas: Bigelow Holding Corp.

Taubes, G. (26 November 1993). Claim of higher risk for women smokers attacked. *Science* 262, p. 1375.

Weiden, C. R., and B. C. Gladen. (1986). The beta-geometric distribution applied to comparative fecundability studies. *Biometrics* 42, pp. 547–560.

CHAPTER 13

Statistical Significance for 2 × 2 Tables

Thought Questions

1. Suppose that a sample of 400 people included 100 under age 30 and 300 aged 30 and over. Each person was asked whether or not they supported requiring public school children to wear uniforms. Fill in the number of people who would be expected to fall into the cells in the following table, *if* there is no relationship between age and opinion on this question. Explain your reasoning. (*Hint:* Notice that overall, 30% favor uniforms.)

	Yes, Favor Uniforms	No, Don't Favor Uniforms	Total
Under 30			100
30 and over			300
Total	120	280	400

2. Suppose that in a random sample of 10 males and 10 females, seven of the males (70%) and four of the females (40%) admitted that they have fallen asleep at least once while driving. Would these numbers convince you that there is a difference in the proportions of males and females in the *population* who have fallen asleep while driving? Now suppose the sample consisted of 1000 of each sex but the proportions remained the same, with 700 males and 400 females admitting that they had fallen asleep at least once while driving. Would these numbers convince you that there is a difference in the population proportions who have fallen asleep? Explain the difference in the two scenarios.

3. Based on the data from Example 12.1, we can conclude that there is a statistically significant relationship between taking aspirin or not and having a heart attack or not. What do you think it means to say that the relationship is "statistically significant"?

4. Refer to the previous question. Do you think that a statistically significant relationship is the same thing as an important and sizeable relationship? Explain.

13.1   Measuring the Strength of the Relationship

The Meaning of Statistical Significance

The purpose of this chapter is to help you understand what researchers mean when they say that a relationship between two categorical variables is *statistically significant*. In plain language, it means that a relationship the researchers observed in a sample was unlikely to have occurred unless there really is a relationship in the population. In other words, the relationship in the sample is probably not just a statistical fluke. However, it does not mean that the relationship between the two variables is significant in the common English definition of the word. The relationship in the population may be real, but so small as to be of little practical importance.

Suppose researchers want to know if there is a relationship between two categorical variables. One example in Chapter 12 is whether there is a relationship between taking aspirin (or not) and having a heart attack (or not). Another example is whether there is a relationship between smoking (or not) and getting pregnant easily when trying. In most cases, it would be impossible to measure the two variables on everyone in the population. So, researchers measure the two categorical variables on a *sample* of individuals from a population, and they are interested in whether or not there is a relationship between the two variables in the *population*. It is easy to see whether or not there is a relationship in the sample. In fact, there almost always is. The percentage responding in a particular way is unlikely to be *exactly* the same for all categories of an explanatory variable. Researchers are interested in assessing whether the differences in observed percentages in the sample are just chance differences, or if they represent a real difference for the population. If a relationship as strong as the one observed in the sample (or stronger) would be unlikely without a real relationship in the population, then the relationship in the sample is said to be **statistically significant.** The notion that it could have happened just by chance is deemed to be implausible.

Measuring the Relationship in a 2 × 2 Contingency Table

We discussed the concept of statistical significance briefly in Chapter 10. Recall that the term can be applied if an observed relationship in a sample is stronger than what is expected to happen by chance when there is no relationship in the population. The standard rule for concluding that a relationship is statistically significant is that it is larger than 95% of those that would be observed just by chance. (Sometimes a number other then 95% is used, depending on the seriousness of being wrong about the conclusion.) Let's see how that rule can be applied to relationships between categorical variables.

We will consider only the simplest case, that of 2 × 2 contingency tables. In other words, we will consider only the situation in which each of two variables has two categories. The same principles and interpretation apply to tables of two variables with more than two categories each, but the details are more cumbersome.

In Chapter 12, we saw that relative risk and odds ratios were both useful for measuring the relationship between outcomes in a 2 × 2 table. Another way to measure the strength of the relationship is by the *difference* in the percentages of outcomes for the two categories of the explanatory variable. In many cases, this measure will be easier to interpret than the relative risk or odds ratio, and it provides a very general method for measuring the strength of the relationship between the two variables. However, before we assess statistical significance, we need to incorporate information about the size of the sample as well. Our examples will illustrate why this feature is necessary. Let's revisit some examples from Chapter 12 and use them to illustrate why the size of the sample is important.

EXAMPLE 13.1   Aspirin and Heart Attacks

In Case Study 1.2 and Example 12.1, we learned about an experiment in which physicians were randomly assigned to take aspirin or a placebo. They were observed for 5 years and the response variable for each man was whether or not he had a heart attack. As shown in Table 12.1 on page 248, 104 of the 11,037 aspirin takers had heart attacks, whereas 189 of the 11,034 placebo takers had them. Notice that the difference in percentage of heart attacks between aspirin and placebo takers is only 1.71% − 0.94% = 0.77%—less than 1%. Based on this small difference in percents, can we be convinced by the data in this sample that there is a real relationship in the population between taking aspirin and risk of heart attack? Or would 293 men have had heart attacks anyway, and slightly more of them just happened to be assigned to the placebo group? Assessing whether or not the relationship is statistically significant will allow us to answer that question. ∎

EXAMPLE 13.2   Young Drivers, Gender, and Driving Under the Influence of Alcohol

In Case Study 6.3 and Example 12.2, data were presented for a roadside survey of young drivers. Of the 481 males in the survey, 77, or 16%, had been drinking in the past 2 hours. Of the 138 females in the survey, 16 of them, or 11.6%, had been drinking. The difference between the percentages of males and females who had been drinking is 16% − 11.6% = 4.4%. Is this difference large enough to provide convincing evidence that there is a difference in the percent of young males and females in the *population* who drink and drive? If in fact the population percents are equal, how likely would we be to observe a sample with a difference as large as 4.4% or larger? We will determine the answer to that question later in this chapter. ∎

EXAMPLE 13.3   Ease of Pregnancy for Smokers and Nonsmokers

In Example 12.3, the explanatory variable is whether or not a woman smoked while trying to get pregnant and the response variable is whether she was able to achieve pregnancy during the first cycle of trying. The difference between the percentage of nonsmokers and smokers who achieved pregnancy during the first cycle is 41% − 29% = 12%. There were 486 nonsmokers and 100 smokers in the study. Is a difference of 12% large enough to rule out chance, or could it be that this particular sample just happened to have more smokers in the group that had trouble getting pregnant? For the population of all smokers and nonsmokers who are trying to get pregnant, is it really true that smokers are less likely to get pregnant in the first cycle? We will determine the answer to that question in this chapter. ■

Strength of the Relationship versus Size of the Study

Can we conclude that the relationships observed for the samples in these examples also hold for the populations from which the samples were drawn? The difference of less than 1% (0.77%) having heart attacks after taking aspirin and placebo seems rather small, and in fact if it had occurred in a study with only a few hundred men, it would probably not be convincing. But the experiment included over 22,000 men, so perhaps such a small difference should convince us that aspirin really does work for the population of all men represented by those in the study.

The difference of 4.4% between male and female drinkers is also rather small and was not convincing to the Supreme Court. The difference of 12% in Example 3 is much larger, but is it large enough to be convincing based on fewer than 600 women? Perhaps another study, on a different 600 women, would yield exactly the opposite result.

At this point, you should be able to see that whether we can rule out chance depends on both the strength of the relationship and on how many people were involved in the study. An observed relationship is much more believable if it is based on 22,000 people (as in Example 13.1) than if it is based on about 600 people (as in Examples 13.2 and 13.3).

13.2   Steps for Assessing Statistical Significance

In Chapter 22 we will learn about assessing statistical significance for a variety of situations using a method called *hypothesis testing*. There are four basic steps required for any situation in which this method is used.

The basic steps for hypothesis testing are:

1. Determine the null hypothesis and the alternative hypothesis.
2. Collect the data, and summarize them with a single number called a *test statistic*.
3. Determine how unlikely the test statistic would be if the null hypothesis were true.
4. Make a decision.

In this chapter, we discuss how to carry out these steps when the question of interest is whether there is a relationship between two categorical variables.

Step 1: Determine the null hypothesis and the alternative hypothesis.

In general, the **alternative hypothesis** is what the researchers are interested in showing to be true, so it is sometimes called the **research hypothesis.** The **null hypothesis** is usually some form of "nothing interesting happening." In the situation in this chapter, in which the question of interest is whether two categorical variables are related, the hypotheses can be stated in the following general form:

Null hypothesis: There is no relationship between the two variables in the population.

Alternative hypothesis: There is a relationship between the two variables in the population.

In specific situations, these hypotheses are worded to fit the context. But in general, the null hypothesis is that there is no relationship in the population and the alternative hypothesis is that there is a relationship in the population.

Notice that the alternative hypothesis does not specify a direction for the relationship. In Chapter 22, we will learn how to add that feature. Also, remember that a cause-and-effect conclusion cannot be made unless the data are from a randomized experiment. The hypotheses are stated in terms of whether or not there is a relationship and do not mention that one variable may cause a change in the other variable. When the data are from a randomized experiment, often the alternative hypothesis can be interpreted to mean that the explanatory variable *caused* a change in the response variable.

Here are the hypotheses for the three examples from the previous section:

| EXAMPLE 13.1 CONTINUED | Aspirin and Heart Attacks |

The participants in this experiment were all male physicians. The population to which the results apply depends on what larger group they are likely to represent. That isn't a statistical question; it's one of common sense. It may be all males in white-collar occupations, or all males with somewhat active jobs. You need to decide for yourself what population you think is represented. We will state the hypotheses by simply using the word *population* without reference to what it is.

Null hypothesis: There is no relationship between taking aspirin and risk of heart attack in the population.

Alternative hypothesis: There is a relationship between taking aspirin and risk of heart attack in the population.

Because the data in this case came from a randomized experiment, if the alternative hypothesis is the one chosen, it is reasonable to state in the conclusion that aspirin actually *causes* a change in the risk of heart attack in the population. ∎

| EXAMPLE 13.2 CONTINUED | Drinking and Driving |

For this example, the sample was drawn from young drivers (under 20 years of age) in the Oklahoma City area in August of 1972 and 1973. Again, the population is defined as the larger group represented by these drivers. We could consider that to be only young drivers in that area at that time period or young drivers in general.

> *Null hypothesis:* Males and females in the population are equally likely to drive within 2 hours of drinking alcohol.
>
> *Alternative hypothesis:* Males and females in the population are not equally likely to drive within 2 hours of drinking alcohol.

Notice that the alternative hypothesis does not specify whether males or females are more likely to drink and drive, it simply states that they are not equally likely to do so. ∎

EXAMPLE 13.3

CONTINUED

Smoking and Pregnancy

> *Null hypothesis:* Smokers and nonsmokers are equally likely to get pregnant during the first cycle in the population of women trying to get pregnant.
>
> *Alternative hypothesis:* Smokers and nonsmokers are not equally likely to get pregnant during the first cycle in the population of women trying to get pregnant.

As in the previous two examples, notice that the alternative hypothesis does not specify which group is likely to get pregnant more easily. It simply states that there is a difference. In later chapters, we will learn how to test for a difference of a particular type.

The data for this example were obviously not based on a randomized experiment; it would be unethical to randomly assign women to smoke or not. Therefore, even if the data allow us to conclude that there is a relationship between smoking and time to pregnancy, it isn't appropriate to conclude that smoking *causes* a change in ease of getting pregnant. There could be confounding variables, such as diet or alcohol consumption, that are related to smoking behavior and also to ease of getting pregnant. ∎

13.3 The Chi-Square Test

The hypotheses are established before collecting any data. In fact, it is not acceptable to examine the data *before* specifying the hypotheses. One sacred rule in statistics is that *it is not acceptable to use the same data to determine and test hypotheses*. That would be cheating, because one could collect data for a variety of potential relationships, then test only those for which the data appear to show a statistically significant difference.

However, the remaining steps are carried out after the data have been collected. In the scenario of this chapter, in which we are trying to determine if there is a relationship between two categorical variables, the procedure is called a **chi-square test.** Steps 2 through 4 proceed as follows:

Step 2: Collect the data, and summarize it with a single number called a *test statistic*.

The "test statistic" in this case is called a *chi-square statistic*. It compares the data in the sample to what would be expected if there were no relationship between the two variables in the population. Details are presented after a summary of the remaining steps.

Step 3: Determine how unlikely the test statistic would be if the null hypothesis were true.

This step is the same for any hypothesis test, and the resulting number is called the *p-value* because it's a probability. (We will learn the technical definition of probability in Chapter 14.) Specifically, the **p-value** is the probability of observing a test statistic as extreme as the one observed or more so if the null hypothesis is really true. In the scenario in this chapter, *extreme* simply means "large." So, the *p*-value is the probability that the chi-square statistic found in step 2 would be as large as it is or larger if in fact the two variables are not related in the population. The details of computing this probability are beyond the level of this book, but it is simple to find using computer software such as Microsoft Excel. We will see how to do this later in the chapter.

A large value of the chi-square statistic will result in a small *p*-value, and vice versa. The more the sample results deviate from what would be expected when there is no real relationship, the larger the chi-square statistic will be. The larger the chi-square statistic is, the smaller the *p*-value will be, because large chi-square values are simply implausible when there is no real relationship. So a large chi-square statistic and a small *p*-value are clues that tell us that there might really be a relationship in the population. We will utilize this information to make a decision in step 4.

Step 4: Make a decision

The relationship between the two variables is said to be *statistically significant* if the *p*-value is small enough. The usual criterion used for "small enough" is 0.05 (5%) or less. This is an arbitrary criterion, but it is well established in almost all areas of research. Occasionally, researchers will use the more stringent criterion of 0.01 (1%) or less stringent criterion of 0.10 (10%), and if that's the case, the value used will be stated explicitly. Whatever number is used as the *p*-value cutoff for statistical significance is called the **level of the test** or **level of significance**, sometimes shortened to **level**.

Remember that the larger the chi-square statistic, the smaller the *p*-value. In general, a computer or calculator can be used to find the *p*-value for a given chi-square statistic. For a chi-square test based on a 2 × 2 table (i.e., only two categories for each variable), three examples of the correspondence are as follows:

- The *p*-value is 0.10 or less if the chi-square statistic is 2.71 or more.
- The *p*-value is 0.05 or less if the chi-square statistic is 3.84 or more.
- The *p*-value is 0.01 or less if the chi-square statistic is 6.63 or more.

The reason for declaring statistical significance only when the *p*-value is small is that the observed sample relationship in that case would be implausible unless there really is a relationship in the population. For example, if the two variables are not related in the population, then the chi-square statistic will be 3.84 or larger only 5% of the time just by chance. Since we know that the chi-square statistic will not often be that large (3.84 or larger) if the two variables are not related, researchers conclude that if it is that large, then the two variables probably are related.

Here is a summary of how to state the decision that's made for a 2 × 2 contingency table using a level of 0.05 for the test:

- If the chi-square statistic is at least 3.84, the *p*-value is 0.05 or less, so conclude that the relationship in the population is real. Equivalent ways to state this result are

  The relationship is statistically significant.

  Reject the null hypothesis (that there is no relationship in the population).

  Accept the alternative hypothesis (that there is a relationship in the population).

  The evidence supports the alternative hypothesis.

- If the chi-square statistic is less than 3.84, the *p*-value is greater than 0.05, so there isn't enough evidence to conclude that the relationship in the population is real. Equivalent ways to state this result are

  The relationship is not statistically significant.

  Do not reject the null hypothesis (that there is no relationship in the population).

  The relationship in the sample could have occurred by chance.

  We do not have enough evidence to reject the null hypothesis.

  We do not have enough evidence to support the alternative hypothesis.

Notice that we do not *accept* the null hypothesis; we simply conclude that the evidence isn't strong enough to reject it. The reason for this will become clear later.

Computing the Chi-Square Statistic

To assess whether a relationship in a 2 × 2 table achieves statistical significance, we need to know the value of the **chi-square statistic** for the table. This statistic is a measure that combines the strength of the relationship with information about the size of the sample to give one summary number. For a 2 × 2 table, if that summary number is larger than 3.84 the *p*-value will be less than 0.05, so the relationship in the table is considered to be statistically significant at the 0.05 level. If it is larger than 6.63, the relationship is statistically significant at the 0.01 level.

The actual computation of the chi-square statistic and assessment of statistical significance is tedious but not difficult. There are different ways to represent the

Computing a chi-square statistic:

1. Compute the *expected counts,* assuming the null hypothesis is true.
2. Compare the observed and expected counts.
3. Compute the chi-square statistic.

Note: This method is valid only if there are no empty cells in the table and if all expected counts are at least 5.

TABLE 13.1 Time to Pregnancy for Smokers and Nonsmokers

	Pregnancy Occurred After			
	First Cycle	Two or More Cycles	Total	Percentage in First Cycle
Smoker	29	71	100	29.00%
Nonsmoker	198	288	486	40.74%
Total	227	359	586	38.74%

necessary formula, some of them useful only for 2 × 2 tables. Here we present only one method, and this method can be used for tables with any number of rows and columns. As we list the necessary steps, we will demonstrate the computation using the data from Example 13.3, shown in Table 12.4 and again in Table 13.1.

Compute the Expected Counts, Assuming the Null Hypothesis Is True

Compute the number of individuals that would be expected to fall into each of the cells of the table if there were no relationship. The formula for finding the expected count in any row and column combination is

$$\text{expected count} = \text{(row total)(column total)/(table total)}$$

The expected counts in each row and column must sum to the same totals as the observed numbers, so for a 2 × 2 table, we need only compute one of these using the formula. We can obtain the rest by subtraction.

It is easy to see why this formula would give the number to be expected if there were no relationship. Consider the first column. The *proportion* who fall into the first column overall is (column 1 total)/(table total). For instance, in Table 13.1, 227/586 = .3874, or 38.74%, of all of the women got pregnant in the first cycle. If there is no relationship between the two variables, then that proportion should be the same for both rows. In the example, we would expect the same proportion of smokers and nonsmokers to get pregnant in the first cycle if indeed there is no effect of smoking. Therefore, to find how many of the smokers (in row 1) would be expected to be in column 1, simply take the overall proportion who are in column 1 (.3874) and multiply it by the number of smokers (the row 1 total). In other words, use

$$\text{expected count in first row and first column} = \frac{\text{(row 1 total)(column 1 total)}}{\text{(table total)}}$$

EXAMPLE 13.3
CONTINUED

Expected Counts if Smoking and Ease of Pregnancy Are Not Related

Let's begin by computing the expected number of smokers achieving pregnancy in the first cycle, assuming smoking does not affect ease of pregnancy:

$$\text{expected count for row 1 and column 1} = (100)(227)/586 = 38.74$$

TABLE 13.2 Computing the Expected Counts for Table 13.1

	Pregnancy Occurred After		
	First Cycle	**Two or More Cycles**	**Total**
Smoker	(100)(227)/586 = 38.74	100 − 38.74 = 61.26	100
Nonsmoker	227 − 38.74 = 188.26	486 − 188.26 = 297.74	486
Total	227	359	586

It's very important that the numbers not be rounded off at this stage. Now that we have the expected count for the first row and column, we can fill in the rest of the expected counts (see Table 13.2), making sure the row and column totals remain the same as they were in the original table. (In this example we had the unusual situation that there were 100 people in Row 1. Since 38.74% of 100 is also 38.74, the expected count just happens to match the overall column percent. That won't happen in general.) ■

Compare the Observed and Expected Counts

For this step, we compute the difference between what we would expect by chance (the "expected counts") and what we have actually observed. To remove negative signs and to standardize these differences based on the number in each combination, we compute the following for each of the cells of the table:

$$(\text{observed count} - \text{expected count})^2 / (\text{expected count})$$

The resulting number is called the chi-square contribution for that cell because it's the amount the cell contributes to the final chi-square statistic. In a 2×2 table, the numerator will actually be the same for each cell. In contingency tables with more than two rows or columns (or both), this would not be the case.

EXAMPLE 13.3
CONTINUED

Comparing Observed and Expected Numbers of Pregnancies

The denominator for the first cell is 38.74 and the numerator is

$$(\text{observed count} - \text{expected count})^2 = (29 - 38.74)^2 = (-9.74)^2 = 94.87$$

Convince yourself that this same numerator applies for the other three cells. The chi-square contribution for each cell is shown in Table 13.3. ■

TABLE 13.3 Comparing the Observed and Expected Counts for Table 13.1

	First Cycle	**Two or More Cycles**
Smoker	94.87/38.74 = 2.45	94.87/61.26 = 1.55
Nonsmoker	94.87/188.26 = 0.50	94.87/297.74 = 0.32

Compute the Chi-Square Statistic

To compute the chi-square statistic, simply add the numbers (chi-square contributions) in all of the cells from step 2. The result is the chi-square statistic.

EXAMPLE 13.3 **CONTINUED**	The Chi-Square Statistic for Comparing Smokers and Nonsmokers

$$\text{chi-square statistic} = 2.45 + 1.55 + 0.50 + 0.32 = 4.82$$ ∎

Making the Decision

Let's revisit the rationale for the decision. Remember that for a 2 × 2 table, the relationship earns the title of "statistically significant" at the 0.05 level if the chi-square statistic is at least 3.84, because in that case the p-value will be less than 0.05. The origin of the "magic" number 3.84 is too technical to describe here. It comes from a table of percentiles representing what should happen by chance, similar to the percentile table for z-scores that was included in Chapter 8. For larger contingency tables, you would need to find the appropriate cut-off number or the p-value using a computer or calculator.

The interpretation of the value 3.84 is straightforward. Of all 2 × 2 tables for sample data from populations in which there is no relationship, 95% of the tables will have a chi-square statistic of 3.84 or less. The remaining 5% will have a chi-square statistics larger than 3.84. Relationships in the sample that reflect a real relationship in the population are likely to produce larger chi-square statistics. Therefore, if we observe a relationship that has a chi-square statistic larger than 3.84, we can assume that the relationship in the sample did not occur by chance. In that case, we say that the relationship is statistically significant. Of all relationships that have occurred just by chance, 5% of them will erroneously earn the title of statistically significant.

It is also possible to miss a real relationship. In other words, it's possible that a sample from a population with a real relationship will result in a chi-square statistic that's less than the magic 3.84. As will be described later in this chapter, this is most likely to happen if the size of the sample is too small. In that case, a real relationship may not be detected as statistically significant. Remember that the chi-square statistic depends on both the strength of the relationship and the size of the sample.

EXAMPLE 13.3 **CONTINUED**	Deciding if Smoking Affects Ease of Pregnancy

The chi-square statistic computed for the example is 4.82, which is larger than 3.84. Using a computer, we find that the corresponding p-value is 0.028. Thus, we can say there is a statistically significant relationship between smoking and time to pregnancy. In other words, we can conclude that the difference we observed in time to pregnancy between smokers and nonsmokers in the sample indicates a real difference for the population of all similar women. It was not just the luck of the draw for this sample. This result is based on the assumption that the women studied can be considered to be a random sample from that population. ∎

Figure 13.1
Minitab (Version 16)
Results for Example 13.3
on Smoking and
Pregnancy

```
Expected counts are printed below observed counts
Chi-Square contributions are printed below expected counts


                   Two or
          First     More
          Cycle    Cycles    Total
    1        29        71       100
          38.74     61.26
          2.448     1.548

    2       198       288       486
         188.26    297.74
          0.504     0.318


 Total      227       359       586

 Chi-Sq = 4.817, DF = 1, P-Value = 0.028
```

Note: Minitab does not provide row labels.
        Row 1 is Smoker, Row 2 is Nonsmoker

Computers, Calculators, and Chi–Square Tests

Many simple computer programs and graphing calculators will compute the chi-square statistic and the p-value for you. Figure 13.1 shows the results of using a statistical computing program called Minitab to carry out the example we have just computed by hand. Notice that the computer has done all the work for us and presented it in summary form. For each cell, three numbers are provided. The first number is the observed count, the number below that is the expected count for the cell, and the final number is the contribution of that cell to the chi-square statistic. The chi-square statistic (Chi-Sq) and p-value are given below the table. In this example, the chi-square statistic is found as $2.448 + 1.548 + 0.504 + 0.318 = 4.818$. (The reported value of 4.817 differs slightly due to rounding off the contributions of each cell.)

The only thing the computer did not supply is the decision. But it tells us that the chi-square statistic is 4.817 and the p-value is 0.028. Remember what the p-value of 0.028 tells us: If there is no relationship between smoking and ease of pregnancy in the *population*, we would only see a *sample* relationship this larger or larger about 2.8% of the time. Based on that information, we can reach our own conclusion that the relationship is statistically significant.

Microsoft Excel will provide the p-value for you once you know the chi-square statistic. The function is CHIDIST(x,df) where "x" is the value of the chi-square statistic and "df" is short for "degrees of freedom." In general, for a chi-square test based on a table with r rows and c columns (not counting the "totals" row and column), the degrees of freedom $= (r - 1)(c - 1)$. When there are 2 rows and 2 columns, df $= 1$.

EXAMPLE 13.3
CONTINUED

Using Excel to find the p-value for Smoking and Ease of Pregnancy

Let's illustrate the use of Excel by finding the p-value for the relationship between smoking and ease of getting pregnant. The chi-square statistic is 4.817, and because there are two rows and two columns, df = (2 − 1)(2 − 1) = 1. Entering =CHIDIST(4.817,1) into any cell in Excel returns the value of 0.028180352, or about 0.028, the same value provided by Minitab. This tells us that in about 2.8% of all samples from populations in which there is no relationship, the chi-square statistic will be 4.817 or larger just by chance. Again, for our example, we use this as evidence that the sample didn't come from a population with no relationship—it came from one in which the two variables of interest (smoking and ease of pregnancy) probably are related.   ∎

Excel can also be used to find the chi-square statistic, but you first need to compute the expected counts by hand, and then enter the observed and expected counts into the spreadsheet. The function is CHISQ.TEST(ARRAY1,ARRAY2) where ARRAY1 gives the range of cells containing the observed counts, and ARRAY2 gives the range of cells containing the expected counts. The process is somewhat complicated, and you should consult the Excel Help function for an example if you wish to use this feature.

There also are numerous free statistical calculators on the web. The website http://statpages.org has links to hundreds of free resources. In particular, links to calculators for finding the chi-square statistic and the corresponding p-value can be found at http://statpages.org/#CrossTabs.

Chi-Square Tests for Tables Larger than 2 × 2

Although we have focused on 2 × 2 contingency tables for simplicity, everything covered in this chapter carries over to hypothesis tests for contingency tables with more than two rows and/or columns. The hypotheses are stated in the same way, the expected counts are found in the same way, and the chi-square statistic is found in the same way, by adding the contributions from the individual cells. The only difference is that the value of the chi-square statistic corresponding to statistical significance changes and a calculator or computer is required to find it. In other words, a p-value of 0.05 no longer corresponds to a chi-square statistic of 3.84, a p-value of 0.01 no longer corresponds to a chi-square statistic of 6.63, and so on.

To actually find the p-value when the table is larger than 2 × 2, first compute a number called *degrees of freedom,* abbreviated as df. If there are r rows and c columns in the table, then df = $(r − 1)(c − 1)$. Find the value of the chi-square statistic using the usual procedure, then use Excel, a calculator, or other computer software to find the p-value. For instance, for a 2 × 3 table, df = (2 − 1)(3 − 1) = (1)(2) = 2. Suppose the chi-square statistic is computed to be 7.00. Then the Excel function CHIDIST(7.00,2) yields the p-value of 0.030197. Using a test with level 0.05, the null hypothesis would be rejected, because the p-value of 0.03 is less than 0.05.

EXAMPLE 13.4   ### Age at Birth of First Child and Breast Cancer

Pagano and Gauvreau (1993, p. 133) reported data for women participating in the first National Health and Nutrition Examination Survey (Carter, Jones, Schatzkin, and Brinton, January–February, 1989). The explanatory variable was whether the age at which a woman gave birth to her first child was 25 or older, and the response variable was whether she developed breast cancer. The results are shown in Table 13.4. The relative risk of breast cancer for these women was 1.33, with women having their first

TABLE 13.4 Age at Birth of First Child and Breast Cancer

First Child at Age 25 or Older?	Breast Cancer	No Breast Cancer	Total
Yes	31	1597	1628
No	65	4475	4540
Total	96	6072	6168

Source: Pagano and Gauvreau (1993).

child at age 25 or older having greater risk. The study was based on a sample of over 6000 women. Is the relationship statistically significant? Let's go through the four steps of testing this hypothesis.

Step 1: Determine the null hypothesis and the alternative hypothesis.

Null hypothesis: There is no relationship between age at birth of first child and breast cancer in the population of women who have had children.

Alternative hypothesis: There is a relationship between age at birth of first child and breast cancer in the population of women who have had children.

Step 2: Collect the data, and summarize it with a single number called a *test statistic.*
Expected count for "Yes and Breast Cancer" = (1628)(96)/6168 = 25.34. By subtraction, the other expected counts can be found as shown in Table 13.5. Therefore, the chi-square statistic is

$$\frac{(31 - 25.34)^2}{25.34} + \frac{(1597 - 1602.66)^2}{1602.66} + \frac{(65 - 70.66)^2}{70.66} + \frac{(4475 - 4469.34)^2}{4469.34} = 1.75$$

Step 3: Determine how unlikely the test statistic would be if the null hypothesis were true.
Using Excel, the *p*-value can be found as CHIDIST(1.75, 1) = 0.186.

This *p*-value tells us that if there is no relationship between age of bearing first child and breast cancer, we would still see a relationship as larger as or larger than the one here about 18.6% of the time, just by chance.

Step 4: Make a decision.
Because the chi-square statistic is less than 3.84 and the *p*-value of 0.186 is greater than .05, the relationship is *not* statistically significant, and we cannot conclude that the increased risk observed in the *sample* would hold for the *population* of women. The

TABLE 13.5 Expected Counts for Age at Birth of First Child and Breast Cancer

First Child at Age 25 or Older?	Breast Cancer	No Breast Cancer	Total
Yes	(1628)(96)/6168 = 25.34	1628 − 25.34 = 1602.66	1628
No	96 − 25.34 = 70.66	4540 − 70.66 = 4469.34	4540
Total	96	6072	6168

relationship could simply be due to the luck of the draw for this sample. Even though the relative risk in the *sample* was 1.33, the relative risk in the population may be 1.0, meaning that both groups are at equal risk for developing breast cancer.

Remember that even if the null hypothesis had been rejected, we would not have been able to conclude that delaying childbirth *causes* breast cancer. Obviously, the data are from an observational study, because women cannot be randomly assigned to have children at a certain age. Therefore, there are possible confounding variables, such as use of oral contraceptives at a young age, that may be related to age at birth of first child and may have an effect on likelihood of breast cancer. ■

13.4 Practical versus Statistical Significance

You should be aware that "statistical significance" does not mean the two variables have a relationship that you would necessarily consider to be of practical importance. A table based on a very large number of observations will have little trouble achieving statistical significance, even if the relationship between the two variables is only minor. Conversely, an interesting relationship in a population may fail to achieve statistical significance in the sample if there are too few observations. It is difficult to rule out chance unless you have either a very strong relationship or a sufficiently large sample.

To see this, consider the relationship in Case Study 1.2, between taking aspirin instead of a placebo and having a heart attack or not. The chi-square statistic, based on the result from the 22,071 participants in the study, is 25.01, so the relationship is clearly statistically significant. Now suppose there were only one-tenth as many participants, or 2207—still a fair-sized sample. Further suppose that the heart attack rates remained about the same, at 9.4 per thousand for the aspirin group and 17.1 per thousand for the placebo group. What would happen then?

If you look at the method for computing the chi-square statistic, you will realize that if all numbers in the study are divided by 10, the resulting chi-square statistic is also divided by 10. This is because the numerator of the contribution for each cell is squared, but the denominator is not. Therefore, if the aspirin study had only 2207 participants instead of 22,071, the chi-square statistic would have been only about 2.501 (25.01/10), with a *p*-value of 0.114. It would not have been large enough to conclude that the relationship between heart attacks and aspirin consumption was statistically significant, even though the rates of heart attacks per thousand people were still 9.4 for the aspirin group and 17.1 for the placebo group.

No Relationship versus No Statistically Significant Relationship

Some researchers report the lack of a statistically significant result erroneously, by implying that a relationship must therefore not exist. When you hear the claim that a study "failed to find a relationship" or that "no relationship was found" between two variables, it does not mean that a relationship was not observed in the sample. It means that whatever relationship was observed did not achieve statistical significance. When you hear of such a result, always check to make sure the study was not based on a small number of individuals. If it was, remember that with a small sample, it takes a very strong relationship for it to earn the title of "statistical significance."

Be particularly wary if researchers report that *no relationship* was found, or that the proportions with a certain response are *equal* for the categories of the explanatory variable in the *population*. In general, they really mean that no statistically significant relationship was found. It is impossible to conclude, based on a sample, anything *exact* about the population. This is why we don't say that we can *accept* the null hypothesis of no relationship in the population.

EXAMPLE 13.2

CONTINUED

Drinking and Driving

Let's examine in more detail the evidence in Example 13.2 that was presented to the Supreme Court to see if we can rule out chance as an explanation for the higher percentage of male drivers who had been drinking. In Figure 13.2, we present the results of asking the Minitab program to carry out the chi-square test. The chi-square statistic is only 1.637—which is not large enough to find a statistically significant difference in percentages of males and females who had been drinking. The *p*-value is 0.201. You can see why the Supreme Court was reluctant to conclude that the difference in the sample represented sufficient evidence for a real difference in the population.

This example provides a good illustration of the distinction between statistical and practical significance and how it relates to the size of the sample. You might think that a real relationship for the population is indicated by the fact that 16% of the males but only 11.6% of the females in the sample had been drinking. But the chi-square test tells us that a difference of that magnitude in a *sample* of this size would not be at all surprising, if in fact equal percentages of males and females in the *population* had been drinking. If those same percents were found in a much larger sample, then the evidence would be convincing. Notice that if the sample were three times as large, but the percents drinking remained at 16% and 11.6%, then the chi-square statistic would be about $(3)(1.637) = 4.91$, and the difference would indeed be statistically significant. ∎

Figure 13.2
Minitab Results for Example 13.2 on Drinking and Driving

```
Expected counts are printed below observed counts
Chi-Square contributions are printed below expected counts

          Yes       No    Total
     1     77      404      481
         72.27   408.73
         0.310    0.055

     2     16      122      138
         20.73   117.27
         1.081    0.191

Total     93      526      619

Chi-Sq = 1.637, DF = 1, P-Value = 0.201
```

Note: Minitab does not provide row labels.
       Row 1 is Male, Row 2 is Female.

CASE STUDY 13.1   **Extrasensory Perception Works Best with Movies**

Extrasensory perception (ESP) is the apparent ability to obtain information in ways that exclude ordinary sensory channels. Early laboratory research studying ESP focused on having people try to guess at simple targets, such as symbols on cards, to see if the subjects could guess at a better rate than would be expected by chance. In recent years, experimenters have used more interesting targets, such as photographs, outdoor scenes, or short movie segments.

In a study of ESP reported by Bem and Honorton (January 1994), subjects (called *receivers*) were asked to describe what another person (the *sender*) had just watched on a television screen in another room. The receivers were shown four possible choices and asked to pick which one they thought had been viewed by the sender in the other room. Because the actual target was randomly selected from among the four choices, the guesses should have been successful by chance 25% of the time. Surprisingly, they were actually successful 34% of the time.

For this case study, we are going to examine a categorical variable that was involved and ask whether the results were affected by it. The researchers had hypothesized that moving pictures might be received with better success than ordinary photographs. To test that theory, they had the sender sometimes look at a single, "static" image on the television screen and sometimes look at a "dynamic" short video clip, played repeatedly. The additional three choices shown to the receiver for judging (the "decoys") were always of the same type (static or dynamic) as the actual target, to eliminate biases due to a preference for one over the other. The question of interest was whether the success rate changed based on the type of picture. The results are shown in Table 13.6.

Figure 13.3 shows the results from the Minitab program. (The format is slightly different from the format shown in Figures 13.1 and 13.2 because a different version of Minitab was used.) Notice that the chi-square statistic is 6.675, and the p-value is 0.010. The p-value tells us that if there is no relationship between type of picture and successful ESP guess for the population, then an outcome as extreme or more extreme than what was observed in this sample would occur only about 1% of the time by chance. Therefore, it does appear that

TABLE 13.6   Results of ESP Study

	Successful ESP Guess?			
	Yes	**No**	**Total**	**% Success**
Static picture	45	119	164	27%
Dynamic picture	77	113	190	41%
Total	122	232	354	34%

Source: Bem and Honorton, 1994.

Figure 13.3
Minitab results for Case Study 13.1

```
Expected counts are printed below observed counts

             Yes      No     Total
Static        45      119      164
            56.52   107.48

Dynamic       77      113      190
            65.48   124.52

Total        122      232      354

Chi-Sq = 2.348 + 1.235 +
         2.027 + 1.066 = 6.675
DF = 1,  P-Value = 0.010
```

success in ESP guessing depends on the type of picture used as the target. You can see that guesses for the static pictures were almost at chance (27% compared to 25% expected by chance), whereas the guesses for the dynamic videos far exceeded what was expected by chance (41% compared to 25%). More recent studies have confirmed this effect for dynamic videos. See, for example, Storm et. al (2010) and Storm et. al (2013). ■

Thinking About Key Concepts

- *Hypothesis tests* are used to decide whether an observed relationship in a sample provides evidence of a real relationship in the population represented by the sample.
- A *statistically significant* relationship is one in which the evidence provided by the sample is strong enough to conclude that there is a real relationship in the population.
- *Null* and *alternative hypotheses* are statements about whether two variables are related in the population. The null hypothesis states that they are not related, and the alternative hypothesis states that they are related.
- A *chi-square test* is used to determine whether the evidence provided in a contingency table is strong enough to conclude that the alternative hypothesis (of a real relationship) is true.
- The *expected counts* for a chi-square test are the counts that would be expected, on average, if there really is no relationship between the two variables (that is, if the null hypothesis really is true).
- The *chi-square statistic* for a contingency table measures how far the observed counts are from the expected counts. It is a measure of how different the observed results are from what would be expected just by chance if there is no real relationship.

- The *p-value* for a chi-square test is the probability that the chi-square statistic would be as large as the value observed or even larger, if there is no real relationship in the population.
- The *level of a test* is the threshold for declaring statistical significance. If the *p*-value is less than or equal to the desired level of the test, then the relationship is statistically significant. Typically, a level of 0.05 is used.
- The *p*-value, and thus the conclusion, depends on both the size of the relationship and the size of the sample. With a large sample, a very minor relationship may be statistically significant. With a small sample, an important relationship may not be detected as statistically significant.

Focus on Formulas

To represent the *observed counts* in a 2 × 2 contingency table, we use the notation

Variable 1	Variable 2 Yes	No	Total
Yes	a	b	$a + b$
No	c	d	$c + d$
Total	$a + c$	$b + d$	n

Therefore, the *expected counts* are computed as follows:

Variable 1	Variable 2 Yes	No	Total
Yes	$(a + b)(a + c)/n$	$(a + b)(b + d)/n$	$a + b$
No	$(c + d)(a + c)/n$	$(c + d)(b + d)/n$	$c + d$
Total	$a + c$	$b + d$	n

Computing the Chi-Square Statistic, χ^2, for an $r \times c$ Contingency Table

Let O_i = observed count in cell i and E_i = expected count in cell i, where $i = 1, 2, \ldots, r \times c$. Then,

$$\chi^2 = \sum_{i=1}^{r \times c} \frac{(O_i - E_i)^2}{E_i}$$

Exercises

Exercises with numbers divisible by 3 (3, 6, 9, etc.) are included in the Solutions at the back of the book. They are marked with an asterisk (*).

1. The table below shows the partial results of a survey of 500 middle-aged drivers who were asked whether they had ever had a speeding ticket. Specify the number of people who would fall into each cell if there is no relationship between sex and getting a speeding ticket. Explain your reasoning.

Ever had speeding ticket?

	Yes	No	Total
Male			200
Female			300
Total	400	100	500

2. If there is a relationship between two variables in a population, which is more likely to result in a statistically significant relationship in a sample from that population—a small sample, a large sample, or are they equivalent? Explain.

*3. If there is *no* relationship between two variables in a population, which is more likely to result in a statistically significant relationship in a sample—a small sample, a large sample, or are they equivalent? Explain. (*Hint:* If there is no relationship in the population, how often will the *p*-value be less than 0.05? Does it depend on the size of the sample?)

4. Suppose a relationship between two variables is found to be statistically significant. Explain whether each of the following is true in that case:

   a. There is definitely a relationship between the two variables in the *sample.*

   b. There is definitely a relationship between the two variables in the *population.*

   c. It is likely that there is a relationship between the two variables in the *population.*

5. Are null and alternative hypotheses statements about samples, about populations, or does it depend on the situation? Explain.

*6. Explain what "expected counts" represent. In other words, under what condition are they "expected"?

7. For each of the following possible conclusions, state whether it would follow when the *p*-value is less than 0.05 (assuming a level of 0.05 is desired for the test).

   a. Reject the null hypothesis.

   b. Reject the alternative hypothesis.

   c. Accept the null hypothesis.

   d. Accept the alternative hypothesis.

   e. The relationship is not statistically significant.

   f. The relationship is statistically significant.

   g. We do not have enough evidence to reject the null hypothesis.

8. For each of the following possible conclusions, state whether it would follow when the *p*-value is greater than 0.05 (assuming a level of 0.05 is desired for the test).

   a. Reject the null hypothesis.

   b. Reject the alternative hypothesis.

   c. Accept the null hypothesis.

   d. Accept the alternative hypothesis.

   e. The relationship is not statistically significant.

   f. The relationship is statistically significant.

   g. We do not have enough evidence to reject the null hypothesis.

*9. For each of the following situations, would a chi-square test based on a 2×2 table using a level of 0.05 be statistically significant? Justify your answer.

   *a. chi-square statistic $= 1.42$

   *b. chi-square statistic $= 14.2$

   *c. *p*-value $= 0.02$

   *d. *p*-value $= 0.15$

10. For each of the following situations, would a chi-square test based on a 2 × 2 table using a level of 0.01 be statistically significant? Justify your answer.

  a. chi-square statistic = 1.42

  b. chi-square statistic = 14.2

  c. p-value = 0.02

  d. p-value = 0.15

11. Use software (such as Excel), a calculator, or a website to find the p-value for each of the following chi-square statistics calculated from a 2 × 2 table. You may round off your answer to three decimal places.

  a. chi-square statistic = 3.17

  b. chi-square statistic = 5.02

  c. chi-square statistic = 7.88

  d. chi-square statistic = 10.81

*12. The chi-square test described in this chapter can be used for tables with more than two rows and/or columns. Use software (such as Excel), a calculator, or a website to find the p-value for each of the following chi-square statistics calculated from a table of the specified number of rows and columns. You may round off your answer to three decimal places. In each case, specify the degrees of freedom you used to find the p-value. (*Hint*: Remember that in general, degrees of freedom for a table with r rows and c columns are df = $(r - 1)$ $(c - 1)$.) Then make a conclusion about whether you would reject the null hypothesis for a test with level 0.05.

  *a. Two rows and three columns; chi-square statistic = 7.4

  *b. Three rows and three columns; chi-square statistic = 11.15

  *c. Four rows and three columns; chi-square statistic = 7.88

  *d. Four rows and two columns; chi-square statistic = 12.20

13. In each of the following situations, specify the population. Also, state the two categorical variables that would be measured for each unit in the sample and the two categories for each variable.

  a. Researchers want to know if there is a relationship between having graduated from college or not and voting in the last presidential election, for all registered voters over age 25.

  b. Researchers want to know if there is a relationship between smoking (either partner smokes) and divorce for people who were married between 1990 and 2000.

  c. Researchers classify midsize cities in the United States according to whether the city's median family income is higher or lower than the median family income for the state in which the city is located. They want to know if there is a relationship between that classification and airport availability, defined as whether or not one of the 30 busiest airports in the country is within 50 miles of the city.

14. Refer to the previous exercise. In each case, state the null and alternative hypotheses.

*15. A political poll based on a random sample of 1000 likely voters classified them by sex and asked them if they planned to vote for Candidate A or Candidate B in the upcoming election. Results are shown in the accompanying table.

  *a. State the null and alternative hypotheses in this situation.

  *b. Calculate the expected counts.

  *c. Explain in words the rationale for the expected counts in the context of this example.

  *d. Calculate the value of the chi-square statistic.

  *e. Make a conclusion using a level of 0.05. State the conclusion in the context of this situation.

	Candidate A	Candidate B	Total
Male	200	250	450
Female	300	250	550
Total	500	500	1000

16. Refer to Example 13.1, investigating the relationship between taking aspirin and risk of heart attack. As shown in Table 12.1 on page 248, 104 of the 11,037 aspirin takers had heart attacks, whereas 189 of the 11,034 placebo takers had them. The hypotheses for this example are already given in this chapter.

    a. Calculate the expected counts.

    b. Calculate the value of the chi-square statistic.

    c. Make a conclusion using a level of 0.01. State the conclusion in the context of this situation.

17. In Exercise 12 of Chapter 12 results were given for a Roper Poll in which people were classified according to age and were asked if they had ever seen a ghost. The results from asking Minitab to compute the chi-square statistic are shown in Figure 13.4. (The format is slightly different from the format shown in Figures 13.1 and 13.2 because a different version of Minitab was used here.) What can you conclude about the relationship between age group and reportedly seeing a ghost?

*18. In a national survey, 1500 randomly selected adults will be asked if they favor or oppose a ban on texting while driving and if they have personally texted while driving during the previous month. Write null and alternative hypotheses about the relationship between the two variables in this situation. Make your hypotheses specific to this situation.

19. This is a continuation of Exercise 20 in Chapter 12. A case-control study in Berlin, reported by Kohlmeier et al. (1992) and by Hand et al. (1994), asked 239 lung cancer patients and 429 controls (matched to the cases by age and sex) whether they had kept a pet bird during adulthood. Of the 239 lung cancer cases, 98 said yes. Of the 429 controls, 101 said yes.

    a. State the null and alternative hypotheses for this situation.

    b. Construct a contingency table for the data.

    c. Calculate the expected counts.

    d. Calculate the value of the chi-square statistic.

    e. Make a conclusion about statistical significance using a level of 0.05. State the conclusion in the context of this situation.

20. If a relationship has practical significance, does it guarantee that statistical significance will be achieved in every study that examines it? Explain.

*21. Howell (1992, p. 153) reports on a study by Latané and Dabbs (1975) in which a researcher entered an elevator and dropped a

Figure 13.4
Minitab results for the Roper poll on seeing a ghost

```
Expected counts are printed below observed counts

            Yes       No      Total
18 to 29    212      1313      1525
          174.93   1350.07

over 29     465      3912      4377
          502.07   3874.93

Total       677      5225      5902

Chi-Sq =  7.857 +  1.018 +
          2.737 +  0.355 = 11.967
DF = 1, P-Value = 0.001
```

handful of pencils, with the appearance that it was an accident. The question was whether the males or females who observed this mishap would be more likely to help pick up the pencils. The results are shown in the table below.

*a. Compute and compare the proportions of males and females who helped pick up the pencils.

*b. Compute the chi-square statistic, and use it to determine whether there is a statistically significant relationship between the two variables in the table using a level of 0.05. Explain your result in a way that could be understood by someone who knows nothing about statistics.

*c. Would the conclusion in part (b) have been the same if only 262 people had been observed but the pattern of results was the same? Explain how you reached your answer and what it implies about research of this type.

Helped Pick Up Pencils?

	Yes	No	Total
Male observer	370	950	1320
Female observer	300	1003	1303
Total	670	1953	2623

Data Source: Howell, 1992, p. 154, from a study by Latané and Dabbs, 1975.

22. This is a continuation of Exercise 25 in Chapter 12. The data (shown in the accompanying table) are reproduced from Case Study 12.1 and represent employees laid off by the U.S. Department of Labor.

Ethnic Group	Laid Off	Not Laid Off	Total
African American	130	1382	1512
White	87	2813	2900
Total	217	4195	4412

Data Source: Gastwirth and Greenhouse, 1995.

Minitab computed the chi-square statistic as 66.595. Explain what this means about the relationship between the two variables. Include an explanation that could be understood by someone with no knowledge of statistics. Make the assumption that these employees are representative of a larger population of employees.

23. This is a continuation of Exercise 27 in Chapter 12. Kohler (1994, p. 427) reported data on the approval rates and ethnicity for mortgage applicants in Los Angeles in 1990. Of the 4096 African American applicants, 3117 were approved. Of the 84,947 white applicants, 71,950 were approved. The chi-square statistic for these data is about 220, so the difference observed in the approval rates is clearly statistically significant. Now suppose that a random sample of 890 applicants had been examined, a sample size 100 times smaller than the one reported. Further, suppose the pattern of results had been almost identical, resulting in 40 African American applicants with 30 approved, and 850 white applicants with 720 approved.

a. Construct a contingency table for these numbers.

b. Compute the chi-square statistic for the table.

c. Make a conclusion based on your result in part (b), and compare it with the conclusion that would have been made using the full data set. Explain any discrepancies, and discuss their implications for this type of problem.

*24. Example 12.4 and Table 12.5 provided the results of a study in Australia showing that the couples in the study had a higher risk of separation if the wife's parents had been divorced. The numbers are shown again in the table below. Write appropriate null and alternative hypotheses for this situation.

Wife's Parents Divorced?	Couple Separated	Couple Intact	Total
Yes	42	292	334
No	72	1092	1164
Total	114	1384	1498

Source: Butterworth et al., 2008.

25. Refer to the previous exercise about the separation status of Australian couples.

  a. Find the expected counts for the table.

  b. Calculate the chi-square test statistic.

  c. Make a conclusion about whether there is a statistically significant relationship between the two variables using a level of 0.05. Explain your result in a way that would be understood by someone who knows nothing about statistics.

26. Is it harder to find statistical significance using a test with level 0.05 or a test with level 0.01? In other words, would a test that is statistically significant using 0.05 always be statistically significant using 0.01, would it be the other way around, or does it depend on the situation? Explain your answer.

*27. Explain whether each of the following is possible.

  *a. A relationship exists in the observed sample but not in the population from which the sample was drawn.

  *b. A relationship does not exist in the observed sample but does exist in the population from which the sample was drawn.

  *c. A relationship does not exist in the observed sample, but an analysis of the sample shows that there is a statistically significant relationship, so it is inferred that there is a relationship in the population.

Exercises 28 to 33 are based on News Story 2, "Research shows women harder hit by hangovers" and the accompanying Original Source 2. In the study, 472 men and 758 women, all of whom were college students and alcohol drinkers, were asked about whether they had experienced each of 13 hangover symptoms in the previous year.

28. What population do you think is represented by the sample for this study? Explain.

29. One of the statements in the Original Source was "men and women were equally likely to experience at least one of the hangover symptoms in the past year (men: 89%; women: 87%; chi-square statistic = 1.2, $p = 0.282$)" (Slutske, Piasecki, and Hunt-Carter, 2003, p. 1445).

  a. State the null and alternative hypotheses for this result.

  b. Given what you have learned in this chapter about how to state conclusions, do you agree with the wording of the conclusion, that men and women were *equally* likely to experience at least one of the hangover symptoms in the past year? If so, explain how you reached that conclusion. If not, rewrite the conclusion using acceptable wording.

*30. One of the results in the Original Source was "there were only two symptoms that men experienced more often than women: vomiting (men: 50%; women: 44%; chi-square statistic = 4.7, $p = 0.031$) and sweating more than usual (men: 34%; women: 23%; chi-square statistic = 18.9, $p < 0.001$)" (Slutske, Piasecki, and Hunt-Carter, 2003, p. 1446). State the null and alternative hypotheses for each of these two results.

31. Refer to the two results given in the previous exercise. State the conclusion that would be made for each of these two results, both in statistical terms and in the context of the situation. Use a level of 0.05

32. Participants were asked how many times in the past year they had experienced at least one of the 13 hangover symptoms listed. Responses were categorized as 0 times, 1–2 times, 3–11 times, 12–51 times, and ≥ 52 times. For the purposes of this exercise, responses have been categorized as less than an average of once a month (0–11 times) versus 12 or more

```
Expected counts are printed below observed counts

                  ≤11      ≥12      Total
        Male      326      140       466
                343.27   122.73

        Female    569      180       749
                551.73   197.27

        Total     895      320      1215

        Chi-Sq =   0.869 +   2.429 +
                   0.540 +   1.511 = 5.350
        DF = 1, P-Value = 0.021
```

times. The Minitab output at the top of the page shows the frequency of symptoms categorized in this way versus the categorical variable male, female. (The format is slightly different from the format shown in Figures 13.1 and 13.2 because a different version of Minitab was used.) In this exercise and the next, we will determine if there is convincing evidence that one of the two sexes is more likely than the other to experience hangover symptoms at least once a month, on average. State the null and alternative hypotheses being tested. Make your answer specific to this situation.

*33. Refer to the previous exercise about hangover symptoms. Use the Minitab output at the top of the page for this exercise.

   *a. Show how the expected count of 343.27 for the "Male, ≤ 11" category was computed.

   *b. Give the value of the chi-square statistic and the *p*-value, and make a conclusion. State the conclusion in statistical terms and in the context of the situation. Use a level of 0.05.

Exercises 34 to 37 are based on News Story 5 (summarized in the Appendix), "Driving while distracted is common, researchers say," and the accompanying Original Source 5, "Distractions in Everyday Driving."

34. Refer to Table 8 on page 37 of Original Source 5 on the companion website. Notice that there is a footnote to the table that reads: "*p<.05 and **p<.01, based on chi-square test of association with sex." The footnote applies to "Grooming**" and "External distraction*."

   a. Explain what null and alternative hypotheses are being tested for "Grooming**." (Notice that the definition of "grooming" is given on page 41 of the report.)

   b. Explain what the footnote means.

35. The figure at the top of the next page provides Minitab output for testing for a relationship between sex and "External distraction", but the expected counts have been removed for you to fill in.

   a. Fill in the expected counts.

   b. State the null and alternative hypotheses being tested.

   c. Give the value of the chi-square statistic and the *p*-value, and make a conclusion. State the conclusion in statistical terms and in the context of the situation.

   d. Explain how this result confirms the footnote given for the table in connection with "External distraction."

```
Expected counts are printed below observed counts

              Yes        No     Total
   Male        27          8       35

   Female      33          2       35

   Total       60         10       70

Chi-Sq =   0.300 +   1.800 +
           0.300 +   1.800 = 4.200
DF = 1,  P-Value = 0.040
```

*36. The table below shows participants categorized by sex and by whether they were observed conversing (yes, no). What is the expected count for "Male, Yes?" Find the remaining expected counts by subtraction.

Conversing?

	Yes	No	Total
Male	28	7	35
Female	26	9	35
Total	54	16	70

37. Refer to the previous exercise and the accompanying table, categorizing people by sex and whether they were conversing.

a. State the null and alternative hypotheses that can be tested with this table.

b. Compute the chi-square statistic.

c. Make a conclusion in statistical terms and in the context of the situation.

Mini-Projects

1. Carefully collect data cross-classified by two categorical variables for which you are interested in determining whether there is a relationship. Do not get the data from a book, website, or journal; collect it yourself. Be sure to get counts of at least five in each cell and be sure the individuals you use are not related to each other in ways that would influence their data.

   a. Create a contingency table for the data.

   b. Compute and discuss the risks and relative risks. Are those terms appropriate for your situation? Explain.

   c. Determine whether there is a statistically significant relationship between the two variables.

   d. Discuss the role of sample size in making the determination in part (c).

   e. Write a summary of your findings.

2. Find a poll that has been conducted at two different time periods or by two different sources. For instance, many polling organizations ask opinions about certain issues

on an annual or other regular basis. (A good source for recent polls is http://www.pollingreport.com.)

a. Create a 2 × 2 table in which "time period" is one categorical variable and "response to poll" is the other, categorized into just two choices, such as "favor" and "do not favor" for an opinion question.

b. State the null and alternative hypotheses for comparing responses across the two time periods. Be sure you differentiate between samples and populations.

c. Carry out a chi-square test to see if opinions have changed over the two time periods.

d. Write a few sentences giving your conclusion in words that someone with no training in statistics could understand.

3. Find a journal article that uses a chi-square test based on a contingency table.

a. State the hypotheses being tested.

b. Write the contingency table.

c. Give the value of the chi-square statistic and the *p*-value as reported in the article.

d. Write a paragraph or more (as needed) explaining what was tested and what was concluded, as if you were writing for a newspaper.

References

Bem, D., and C. Honorton. (January 1994). Does psi exist? Replicable evidence for an anomalous process of information transfer. *Psychological Bulletin* 115, no. 1, pp. 4–18.

Butterworth, P., T. Oz, B. Rodgers, and H. Berry. (2008). Factors associated with relationship dissolution of Australian families with children. Social Policy Research Paper No. 37, Australian Government, Department of Families, Housing, Community Services and Indigenous Affairs. http://www.fahcsia.gov.au/about-fahcsia/publications-articles/research-publications/social-policy-research-paper-series/number-37-factors-associated-with-relationship-dissolution-of-australian-families-with-children, accessed June 19, 2013.

Carter, C. L., D. Y. Jones, A. Schatzkin, and L. A. Brinton. (January–February 1989). A prospective study of reproductive, familial, and socioeconomic risk factors for breast cancer using NHANES I data. *Public Health Reports* 104, pp. 45–49.

Gastwirth, J. L., and S. W. Greenhouse. (1995). Biostatistical concepts and methods in the legal setting. *Statistics in Medicine* 14, no. 15, pp. 1641–1653.

Hand, D. J., F. Daly, A. D. Lunn, K. J. McConway, and E. Ostrowski. (1994). *A handbook of small data sets.* London: Chapman and Hall.

Howell, D. C. (1992). *Statistical methods for psychology.* 3d ed. Belmont, CA: Duxbury Press.

Kohler, H. (1994). *Statistics for business and economics.* 3d ed. New York: Harper-Collins College.

Kohlmeier, L., G. Arminger, S. Bartolomeycik, B. Bellach, J. Rehm, and M. Thamm. (1992). Pet birds as an independent risk factor for lung cancer: Case-control study. *British Medical Journal* 305, pp. 986–989.

Latané, B., and J. M. Dabbs, Jr. (1975). Sex, group size and helping in three cities. *Sociometry* 38, pp. 180–194.

Pagano M., and K. Gauvreau. (1993). *Principles of biostatistics.* Belmont, CA: Duxbury Press.

The Roper Organization. (1992). *Unusual personal experiences: An analysis of the data from three national surveys.* Las Vegas: Bigelow Holding Corp.

Slutske, W. S., T. M. Piasecki and E. E. Hunt-Carter. (2003). Development and initial validation of the Hangover Symptoms Scale: Prevalence and correlates of hangover symptoms in college students. *Alcoholism: Clinical and Experimental Research* 27, 1442–1450.

Storm, L., P. E. Tressoldi and L. DiRisio. (2010). Meta-analysis of free-response studies, 1992–2008: Assessing the noise reduction model in parapsychology. *Psychological Bulletin*, 136, pp. 471–485. doi:10.1037/a0019457

Storm, L., P. E. Tressoldi and J. Utts. (2013). Testing the Storm et al. (2010) meta-analysis using Bayesian and frequentist approaches: Reply to Rouder et al.. *Psychological Bulletin*, 139(1), Jan 2013, pp. 248–254.

Understanding Uncertainty in Life

In Parts 1 and 2 of this book, you learned how data should be collected and summarized. Some simple ideas about chance were introduced in the context of whether chance could be ruled out as an explanation for a relationship observed in a sample.

The purpose of the material in Part 3 is to acquaint you with some simple ideas about probability in ways that can be applied to your daily life. In Chapter 14, you will learn how to determine and interpret probabilities for simple events. You will also see that it is sometimes possible to make long-term predictions, even when specific events can't be predicted well. In Chapter 15 you will learn how to use simulation to solve problems about probability, and in Chapters 16 and 17, you will learn how psychological factors can influence judgments involving uncertainty. As a consequence, you will learn some hints that will help you make better decisions in your own life.

CHAPTER **14**

Understanding Probability and Long-Term Expectations

Thought Questions

1. Here are two very different queries about probability:

    a. If you flip a coin and do it fairly, what is the probability that it will land heads up?

    b. What is the probability that you will eventually own a home; that is, how likely do you think it is? (If you already own a home, what is the probability that you will own a different home within the next 5 years?)

    For which question was it easier to provide a precise answer? Why?

2. Explain what it means for someone to say that the probability of his or her eventually owning a home is 70%.

3. Explain what's wrong with the following statement, given by a student as a partial answer to Thought Question 1b: "The probability that I will eventually own a home, or of any other particular event happening, is 1/2 because either it will happen or it won't."

4. Why do you think insurance companies charge young men more than they do older men for automobile insurance, but charge older men more for life insurance?

5. How much would you be willing to pay for a ticket to a contest in which there was a 1% chance that you would win $500 and a 99% chance that you would win nothing? Explain your answer.

14.1   Probability

The word *probability* is so common that in all probability you will run across it today in everyday language. But we rarely stop to think about what the word means. For instance, when we speak of the probability of winning a lottery based on buying a single ticket, are we using the word in the same way as when we speak of the probability that we will eventually buy a home? In the first case, we can quantify the chances exactly. In the second case, we are basing our assessment on personal beliefs about how life will evolve for us. The conceptual difference illustrated by these two examples leads to two distinct interpretations of what is meant by the term *probability*.

14.2   The Relative-Frequency Interpretation

The **relative-frequency interpretation** of probability applies to situations in which we can envision observing results over and over again. For example, it is easy to envision flipping a coin over and over again and observing whether it lands heads or tails. It then makes sense to discuss the probability that the coin lands heads up. It is simply the relative frequency, over the long run, with which the coin lands heads up.

Here are some more interesting situations to which this interpretation of probability can be applied:

- Buying a weekly lottery ticket and observing whether it is a winner.
- Commuting to work daily and observing whether a certain traffic signal is red when we encounter it.
- Testing individuals in a population and observing whether they carry a gene for a certain disease.
- Observing births and noting if the baby is male or female.

The Idea of Long-Run Relative Frequency

If we have a situation such as those just described, we can define the **probability** of any specific outcome as *the proportion of time it occurs over the long run*. This is also called the **relative frequency** of that particular outcome.

Notice the emphasis on what happens *in the long run*. We cannot assess the probability of a particular outcome by observing it only a few times. For example, consider a family with five children, in which only one child is a boy. We would not take that as evidence that the probability of having a boy is only 1/5. However, if we noticed that out of thousands of births only one in five of the babies were boys, then it would be reasonable to conclude that the probability of having a boy is only 1/5.

According to the Centers for Disease Control (http://www.cdc.gov/nchs/pressroom/05facts/moreboys.htm), the long-run relative frequency of males born in the United States from 1940 to 2002 is about .512. In other words, over the long run, out of every 1000 babies born, 512 are male and 488 are female. Suppose we were to record

TABLE 14.1	Relative Frequency of Male Births					
Weeks of Watching	**1**	**4**	**12**	**24**	**36**	**52**
Number of boys	12	47	160	310	450	618
Number of babies	30	100	300	590	880	1200
Proportion of boys	.400	.470	.533	.525	.511	.515

births in a certain city for the next year. Table 14.1 shows what we might observe. Notice how the proportion, or relative frequency, of male births jumps around at first but starts to settle down to something just above .51 in the long run. If we had tried to determine the true proportion after just 1 week, we would have been seriously misled.

Determining the Probability of an Outcome

Method 1: Make an Assumption about the Physical World

Two methods for determining the probability of a particular outcome fit the relative-frequency interpretation. The first method is to *make an assumption about the physical world* and use it to determine the probability of an outcome. For example, we generally assume that coins are manufactured in such a way that they are equally likely to land with heads up or tails up when flipped. Therefore, we conclude that the probability of a flipped coin showing heads up is 1/2. (This probability is based on the assumption that the physics of the situation allows the coin to flip around enough to become unpredictable. With practice, you can learn to toss a coin to come out the way you would like more often than not.)

As a second example, we can determine the probability of winning the lottery by assuming that the physical mechanism used to draw the winning numbers gives each number an equal chance. For instance, many state-run lotteries in the United States have participants choose three digits, each from the set 0 to 9. If the winning set is drawn fairly, each of the 1000 possible combinations should be equally likely. (The 1000 possibilities are 000, 001, 002, . . . , 999.) Therefore, each time you play, your probability of winning is 1/1000. You win only on those rare occasions when the set of numbers you chose is actually drawn. In the long run, that should happen about 1 out of 1000 times. Notice that this does not mean it will happen exactly once in every thousand draws.

Method 2: Observe the Relative Frequency

The other way to determine the probability of a particular outcome is by *observing the relative frequency over many, many repetitions of the situation*. We used that method when we observed the relative frequency of male births in a given city over the course of a year. By using this method, we can get a very accurate figure for the probability that a birth will be a male. As mentioned, the relative frequency of male births in the United States has been consistently close to .512. For example, in 2002, there were a total of 4,021,726 live births in the United States, of which 2,057,979 were males. Therefore, in 2002, the probability that a live birth would result in a male was 2,057,979/4,021,726 = .5117, or about .512.

Sometimes relative-frequency probabilities are reported on the basis of sample surveys. For example, on June 13, 2013, the Gallup Organization reported "Fewer Americans are now optimistic about their future personal financial situations, with 57% saying they will be better off in a year, down from 66% who said so last October" (Brown, 2013). In such cases, a margin of error should be included. The Gallup Organization reports the margin of error in a technical detail box at the end of their news stories. For the example just presented, the margin of error was reported to be ±3 percentage points. Therefore, the statement that 57% of Americans said they would be better off in a year really means that 57% of the *sample* replied that way.

Summary of the Relative-Frequency Interpretation of Probability

- The relative-frequency interpretation of probability can be applied when a situation can be repeated numerous times, at least conceptually, and the outcome can be observed each time.
- In scenarios for which this interpretation applies, the relative frequency with which a particular outcome occurs should settle down to a constant value over the long run. That value is then defined to be the probability of that outcome.
- The interpretation does not apply to situations in which the outcome one time is influenced by or influences the outcome the next time because the probability would not remain the same from one time to the next. We cannot determine a number that is always changing.
- Probability cannot be used to determine whether the outcome will occur on a single occasion but can be used to predict the long-term proportion of the times the outcome will occur.

Relative-frequency probabilities are quite useful in making daily decisions. For example, suppose you have a choice between two flights to reach your destination. All other factors are equivalent, but the airline's website tells you that one has a probability of .90 of being on time, whereas the other has only a probability of .70 of being on time. Even though you can't predict the outcome for your particular flight, you would be likely to choose the one that has the better performance in the long run.

14.3   The Personal-Probability Interpretation

The relative-frequency interpretation of probability is clearly limited to repeatable conditions. Yet, uncertainty is a characteristic of most events, whether they are repeatable under similar conditions or not. We need an interpretation of probability that can be applied to situations even if they will never happen again.

Will you fare better by taking calculus than by taking statistics? If you decide to drive downtown this Saturday afternoon, will you be able to find a good parking space? Should a movie studio release a potential new hit movie before Christmas, when many others are released, or wait until January, when it might have a better chance of being the top box-office attraction? Would a trade alliance with a new country cause problems in relations with a third country?

These are unique situations, not likely to be repeated. They require people to make decisions based on an assessment of how the future will evolve. We could each assign a personal probability to these events, based on our own knowledge and experiences, and we could use that probability to help us with our decisions. We may not agree on what the probabilities of differing outcomes are, but none of us would be considered wrong.

Defining Personal Probability

We define the **personal probability** of an event to be the degree to which a given individual believes the event will happen. There are very few restrictions on personal probabilities. They must fall between 0 and 1 (or, if expressed as a percentage, between 0 and 100%). They must also fit together in certain ways if they are to be *coherent.* By **coherent,** we mean that your personal probability of one event doesn't contradict your personal probability of another. For example, if you thought that the probability of finding a parking space downtown Saturday afternoon was .20, then to be coherent, you must also believe that the probability of not finding one is .80. We explore some of these logical rules later in this chapter.

How We Use Personal Probabilities

People routinely base decisions on personal probabilities. This is why committee decisions are often so difficult. For example, suppose a committee is trying to decide which candidate to hire for a job. Each member of the committee has a different assessment of the candidates, and each may disagree with the others on the probability that a particular candidate would fit the job best. We are all familiar with the problem juries sometimes have when trying to agree on someone's guilt or innocence. Each member of the jury has his or her own personal probability of guilt and innocence. One of the benefits of committee or jury deliberations is that such deliberations may help members reach some consensus in their personal probabilities.

Personal probabilities often take relative frequencies of similar events into account. For example, the late astronomer Carl Sagan believed that the probability of a major asteroid hitting the Earth soon is high enough to be of concern: "The probability that the Earth will be hit by a civilization-threatening small world in the next century is a little less than one in a thousand" (Arraf, 14 December, 1994, p. 4). To arrive at that probability, Sagan obviously could not use the long-run frequency definition of probability. He would have to use his own knowledge of astronomy, combined with past asteroid behavior. (See Exercise 10 for an updated probability.)

14.4 Applying Some Simple Probability Rules

Situations often arise in which we already know probabilities associated with simple events, such as the probability that a birth will result in a girl, and we would like to find probabilities of more complicated events, such as the probability that we will eventually have at least one girl if we ultimately have four children. Some simple, logical rules about probability allow us to do this.

These rules apply naturally to relative-frequency probabilities, and they must apply to personal probabilities if those probabilities are to be coherent. For example, we can never have a probability below 0 or above 1. An impossible event has a probability of 0 and a sure thing has a probability of 1. Here are four additional useful rules:

> *Rule 1:* If there are only two possible outcomes in an uncertain situation, then their probabilities must add to 1.

EXAMPLE 14.1 If the probability of a single birth resulting in a boy is .51, then the probability of it resulting in a girl is .49. ∎

EXAMPLE 14.2 If you estimate the chances that you will eventually own a home to be 70%, in order to be coherent (consistent with yourself), you are also estimating that there is a 30% chance that you will never own one. ∎

EXAMPLE 14.3 According to the *New York Times* (Stellin, 2013), if figures from 2012 are used, the probability that a piece of checked luggage will be lost, delayed, damaged, or pilfered on a flight with a U.S. airline is 3/1000. Thankfully, that means the probability of finding the luggage waiting intact at the end of a trip is 997/1000. ∎

> *Rule 2:* If two outcomes cannot happen simultaneously, they are said to be **mutually exclusive.** The probability of one or the other of two mutually exclusive outcomes happening is the sum of their individual probabilities.

EXAMPLE 14.4 The two most common primary causes of death in the United States are heart disease, which killed about 24% of the Americans who died in the year 2010, and various cancers, which killed about 23%. Therefore, if this year is like the year 2010, the probability that a randomly selected American who dies will die of either heart disease or cancer is the sum of these two probabilities, or about 0.47 (47%). Notice that this is based on death rates for the year 2010 and could well change long before you have to worry about it. This calculation also assumes that one cannot die simultaneously of both causes—in other words, the two causes of death are mutually exclusive. Given the way deaths are recorded, this fact is actually guaranteed because only one primary cause of death may be entered on a death certificate. (*Source*: National Center for Health Statistics) ∎

EXAMPLE 14.5 If you estimate your chances of getting an A in your statistics class to be 50% and your chances of getting a B to be 30%, then you are estimating your chances of getting either an A or a B to be 80%. Notice that you are therefore estimating your chances of getting a C or less to be 20% by Rule 1. ∎

EXAMPLE 14.6 If you estimate your chances of getting an A in your statistics class to be 50% and your chances of getting an A in your history class to be 60%, are you estimating your chances of getting one or the other, or both, to be 110%? Obviously not, because probabilities cannot exceed 100%. The problem here is that Rule 2 stated explicitly that the events under consideration couldn't happen simultaneously. Because it is possible for you to get an A in both courses simultaneously, Rule 2 does not apply here. In case you are curious, Rule 2 could be modified to apply. You would have to subtract the probability that both events happen, which would require you to estimate that probability as well. We see one way to do that using Rule 3. ∎

> *Rule 3:* If two events do not influence each other, and if knowledge about one doesn't help with knowledge of the probability of the other, the events are said to be **independent** of each other. If two events are independent, the probability that they both happen is found by multiplying their individual probabilities.

EXAMPLE 14.7 Suppose a woman has two children. Assume that the outcome of the second birth is independent of what happened the first time and that the probability that each birth results in a boy is .51, as observed earlier. Then the probability that she has a boy *followed* by a girl is $(.51) \times (.49) = .2499$. In other words, there is about a 25% chance that a woman having two children will have a boy and then a girl. ∎

EXAMPLE 14.8 From Example 14.6, suppose you continue to believe that your probability of getting an A in statistics is .5 and an A in history is .6. Further, suppose you believe that the grade you receive in one is independent of the grade you receive in the other. Then you must also believe that the probability that you will receive an A in both is $(.5) \times (.6) = .3$. Notice that we can now complete the calculation we started at the end of Example 14.6. The probability that you will receive at least one A is found by taking $.5 + .6 - .3 = .8$, or 80%. Note that, by Rule 1, you must also believe that the probability of not receiving an A in either class is 20%. ∎

Rule 3 is sometimes difficult for people to understand, but if you think of it in terms of the relative-frequency interpretation of probability, it's really quite simple. Consider women who have had two children. If about half of the women had a boy for their first child, and only about half of those women had a girl the second time around, it makes sense that we are left with only about 25%, or one-fourth, of the women. In other words, one half *of* one half is the same as $(1/2) \times (1/2) = 1/4$ or .25 (25%).

EXAMPLE 14.9   Let's try one more example of Rule 3, using the logic just outlined. Suppose you encounter a red light on 30% of your commutes and get behind a bus on half of your commutes. The two are unrelated because whether you have the bad luck to get behind a bus presumably has nothing to do with the red light. The probability of having a really bad day and having both happen is 15%. This is logical because you get behind a bus half of the time. Therefore, you get behind a bus half of the 30% of the time you encounter the red light, resulting in total misery only 15% of the time. Using Rule 3 directly, this is equivalent to $(.30) \times (.50) = .15$, or 15% of the time both events happen. ■

One more rule is such common sense that it almost doesn't warrant writing down. However, as we will see in Chapter 16, in certain situations this rule will actually seem counterintuitive. Here is the rule:

> *Rule 4:* If the ways in which one event can occur are a subset of those in which another event can occur, then the probability of the subset event *cannot* be higher than the probability of the one for which it is a subset.

EXAMPLE 14.10   Suppose you are 18 years old and speculating about your future. You decide that the probability that you will eventually get married and have children is 75%. By Rule 4, you must then assume that the probability that you will eventually get married is *at least* 75%. The possible futures in which you get married *and* have children are a subset of the possible futures in which you get married. ■

Miscalculated Probabilities

A common mistake when calculating probabilities is to use Rule 3 to find the probability of two events both happening when the events are *not* independent. Remember that Rule 3 only applies to independent events. The following two examples illustrate that how far off such calculations can be.

EXAMPLE 14.11   **A Tragic Miscalculation**

In 1999, Sally Clark, a lawyer and mother in England, was convicted of murder because two of her infants died of sudden infant death syndrome (SIDS). Part of the evidence against her was given by a physician who testified that "the risk of two infants in the same family dying of unexplained natural causes—SIDS—was 1 in 73 million" (Marshall, 2005). How was that risk calculated? According to the expert, the probability of one SIDS death in an affluent nonsmoking family like the Clarks was 1 in 8543. Therefore, he reasoned, the probability of two SIDS death would be

(1/8543)(1/8543) = 1/72,982,849, about 1 in 73 million. Unfortunately, there are a few tragic flaws in this reasoning. First, it assumes that the probability of a SIDS death is constant across all families. Second, it assumes that the deaths are independent, even within the same family. Both of those assumptions are problematic if in fact there is a genetic component to SIDS, as has been speculated. It is quite possible that a genetic component increases the likelihood of SIDS for some families and that the probability of a second SIDS death, given that one has occurred already, is much higher than the initial probability estimate. In other words, the two deaths are not statistically independent. Clark's conviction was overturned in 2003 after she spent three years in jail; but the incident took a devastating toll, and she died of alcohol poisoning in 2007 at the age of 42.

EXAMPLE 14.12   ## The Probability of Twin Sisters Living to be 100

On January 14, 2011, the *Los Angeles Times* reported the happy event of identical twin sisters, Inez and Venice, celebrating their 100th birthday together. The article quoted an official at their celebration as describing "the odds of identical twins living to be 100 as about 1 in 700 million" (Pool, 2011). Further investigation revealed that this reported chance originated with a spokesperson from the Guinness Book of World Records. But how was that probability calculated? It appears to be a wild exaggeration. The logical probability question is "For all sets of identical twins born, what is the proportion for which both twins are still alive 100 years later?" There is no definitive way to answer that question, but some guidance comes from a *New York Times* story titled "Live Long? Die Young? Answer Isn't Just in Genes" (Kolata, 2006). According to the story, the probability of someone born in 1910 living to be at least 100 was about .02 (2%). Further, if a woman lives to be 100, the probability that her sister will live to be 100 is about .04 (4%). The probability for a twin sister may be even higher and certainly would not be lower. So, let's suppose that when Inez was born the probability of her living to 100 was .02, since she was born close to 1910, on January 15, 1911. Then, given that Inez lived to be 100, the probability that Venice would also live to be 100 is at least .04. So the probability that both would happen is at least (.02)(.04) = .0008, or about 1 in 1250. Notice that the events are not independent, but the multiplication rule still applies when the probability given for the second event is based on the first event happening.

A more detailed analysis would take into account whether the twins were male or female, perhaps whether they were both born alive and so on. Other estimates put the probability of living to 100 at about .01 (1%), a bit lower than the .02 value used in this calculation, which would result in a probability of .0004 instead of .0008, or about 1 in 2500. A calculation of a different probability question—the probability of two people being born as identical twins and then living to 100—resulted in an estimate of 1 in 600,000 (Hayes, 2011). But no matter how the calculation is done, thankfully for Inez and Venice (and the other living twin pairs who are over 100), the probability is many times higher than the one in 700 million given in the news story.

14.5   When Will It Happen?

Often, we would like an event to occur and will keep trying to make it happen until it does, such as when a couple keeps having children until they have one of the desired sex. Also, we often gamble that something won't go wrong, even though we know it could, such as when people have unprotected sex and hope they won't get infected with HIV, the virus that causes AIDS.

A simple application of our probability rules allows us to determine the chances of waiting one, two, three, or any given number of repetitions for such events to occur. Suppose (1) we know the probability of each possible outcome on any given occasion, (2) those probabilities remain the same for each occasion, and (3) the outcome each time is *independent* of the outcome all of the other times.

Let's use some shorthand. Define the probability that the outcome of interest will occur on any given occasion to be p so that the probability that it will not occur is $(1 - p)$ by Rule 1. For instance, if we are interested in giving birth to a girl, p is .49 and $(1 - p)$ is .51.

We already know the probability that the outcome occurs on the first try is p. By Rule 3, the probability that it *doesn't* occur on the first try but *does* occur on the *second* try is found by multiplying two probabilities. Namely, it doesn't happen at first $(1 - p)$ and then it does happen (p). Thus, the probability that it happens for the *first* time on the second try is $(1 - p)p$. We can continue this logic. We multiply $(1 - p)$ for each time it *doesn't* happen, followed by p for when it finally *does* happen. We can represent these probabilities as shown in Table 14.2, and you can see the emerging pattern.

EXAMPLE 14.13   **Number of Births to First Girl**

The probability of a birth resulting in a boy is about .51, and the probability of a birth resulting in a girl is about .49. Suppose a couple would like to continue having children until they have a girl. Assuming the outcomes of births are independent of each other, the probabilities of having the first girl on the first, second, third, fifth, and seventh tries are shown in Table 14.3. ∎

TABLE 14.2   Calculating Probabilities

Try on Which the Outcome First Happens	Probability
1	p
2	$(1 - p)p$
3	$(1 - p)(1 - p)p = (1 - p)^2 p$
4	$(1 - p)(1 - p)(1 - p)p = (1 - p)^3 p$
5	$(1 - p)(1 - p)(1 - p)(1 - p)p = (1 - p)^4 p$

TABLE 14.3	Probability of a Birth Resulting in a First Girl
Number of Births to First Girl	**Probability**
1	.49
2	$(.51)(.49) = .2499$
3	$(.51)(.51)(.49) = .1274$
5	$(.51)(.51)(.51)(.51)(.49) = .0331$
7	$(.51)(.51)(.51)(.51)(.51)(.51)(.49) = .0086$

Accumulated Probability

We are often more interested in the *cumulative probability* of something happening by a certain time than just the specific occasion on which it will occur. For example, we would probably be more interested in knowing the probability that we would have had the first girl *by the time of* the fifth child, rather than the probability that it would happen at that specific birth. It is easy to use the probability rules to find this accumulated probability. Notice that the probability of the first occurrence *not happening* by occasion n is $(1 - p)^n$. Therefore, the probability that the first occurrence *has happened* by occasion n is $[1 - (1 - p)^n]$ from Rule 1. For instance, the probability that a girl will *not* have been born by the third birth is $(1 - .49)^3 = (.51)^3 = .1327$. Thus, the probability that a girl will have been born by the third birth is $1 - .1327 = .8673$. This is equivalent to adding the probabilities that the first girl occurs on the first, second, or third tries: $.49 + .2499 + .1274 = .8673$.

EXAMPLE 14.14 Getting Infected with HIV

According to a study published in the *Journal of Infectious Diseases* (Hughes et al, 2012), the probability of a female becoming infected with HIV from a single sexual encounter with an infected male is about 0.0019. We will round this to 0.002 (1 in 500) for simplicity. Therefore, on such a single encounter, the probability of not getting infected is $499/500 = .998$. However, the risk of getting infected goes up with multiple encounters, and, using the strategy we have just outlined, we can calculate the probabilities associated with the number of encounters it would take to become infected. Of course, the real interest is in whether infection will have occurred after a certain number of encounters, and not just on the exact encounter during which it occurs. In Table 14.4 (next page), we show this accumulated probability as well. It is found by adding the probabilities to that point, using Rule 2. Equivalently, we could use the general form we found just prior to this example. For instance, the probability of HIV infection by the second encounter is $[1 - (1 - .002)^2]$ or $[1 - .998^2]$ or .003996.

Table 14.4 tells us that although the risk after a single encounter is only 1 in 500, after 10 encounters the accumulated risk has risen to almost .02, or almost 1 in 50. This means that out of all those people who have 10 encounters, about 1 in 50 of them is likely to get infected with HIV.

Note that these calculations are based on the assumption that the probability remains the same from one encounter to the next. In fact, the situation may be more complex

TABLE 14.4   The Probability of Getting Infected with HIV from Unprotected Sex

Number of Encounters	Probability of First Infection	Accumulated Probability of HIV
1	.002	.002
2	(.998)(.002) = .001996	.003996
4	$(.998)^3(.002) = .001988$	.007976
10	$(.998)^9(.002) = .001964$	.019821

than that. Hughes et al (2012) found that the probability changes based on many factors. For instance, it is lower if the male has been circumcised than if he has not.   ■

EXAMPLE 14.15   **Winning the Lottery**

To play the New Jersey Pick Six game, a player picks six numbers from the choices 1 to 49. Six winning numbers are selected. If the player has matched at least three of the winning numbers, the ticket is a winner. Matching three numbers results in a prize of $3.00; matching four or more results in a prize determined by the number of other successful entries. The probability of winning anything at all is 1/54. How many times would you have to play before winning anything? See Table 14.5.

If you do win, your most likely prize is only $3.00. Notice that even after purchasing five tickets, which cost one dollar each, your chance of having won anything is still under 10%; in fact, it is about 9%. After 20 tries, your probability of having won anything is .3119 or just over 31%. In the next section, we learn how to determine the average expected payoff from playing games like this.   ■

TABLE 14.5   Probabilities of Winning Pick Six

Number of Plays	Probability of First Win	Accumulated Probability of Win
1	1/54 = .0185	.0185
2	(53/54)(1/54) = .0182	.0367
5	$(53/54)^4(1/54) = .0172$	.0892
10	$(53/54)^9(1/54) = .0157$	.1705
20	$(53/54)^{19}(1/54) = .0130$	.3119

14.6   Long-Term Gains, Losses, and Expectations

The concept of the long-run relative frequency of various outcomes can be used to predict long-term gains and losses. Although it is impossible to predict the result of one random happening, we can be remarkably successful in predicting aggregate or long-term results. For example, we noted that the probability of winning anything at all in the New Jersey Pick Six game with a single ticket is 1 in 54. Among the millions of people who play that game regularly, some will be winners and some will be losers. We cannot predict who will win and who will lose. However, we can predict that in the long run, about 1 in every 54 tickets sold will be a winner.

Long-Term Outcomes Can Be Predicted

It is because aggregate or long-term outcomes can be accurately predicted that lottery agencies, casinos, and insurance companies are able to stay in business. Because they can closely predict the amount they will have to pay out over the long run, they can determine how much to charge and still make a profit.

EXAMPLE 14.16 Insurance Policies

Suppose an insurance company has thousands of customers, and each customer is charged $500 a year. The company knows that about 10% of them will submit a claim in any given year and that claims will always be for $1500. How much can the company expect to make per customer?

Notice that there are two possibilities. With probability .90 (or for about 90% of the customers), the amount gained by the company is $500, the cost of the policy. With probability .10 (or for about 10% of the customers), the "amount gained" by the company is the $500 cost of the policy minus the $1500 payoff, for a loss of $1000. We represent the loss by saying that the "amount gained" is $-$1000, or negative one thousand dollars. Here are the possible amounts "gained" and their probabilities:

Claim Paid?	Probability	Amount Gained
Yes	.10	$-$1000
No	.90	$+$ 500

What is the average amount gained, per customer, by the company? Because the company gains $500 from 90% of its customers and loses $1000 from the remaining 10%, its average "gain" per customer is:

$$\text{average gain} = .90\ (\$500) - .10\ (\$1000) = \$350$$

In other words, the company makes an average of $350 per customer. Of course, to succeed this way, it must have a large volume of business. If it had only a few customers, the company could easily lose money in any given year. As we have seen, long-run frequencies apply only to a large aggregate. For example, if the company had only two customers, we could use Rule 3 to find that the probability of the company's having to pay both of them during a given year is $.1 \times .1 = .01 = 1/100$. This calculation assumes that the probability of paying for one individual is independent of that for the other individual, which is a reasonable assumption unless the customers are somehow related to each other. ∎

Expected Value

Statisticians use the phrase **expected value (EV)** to represent the average value of any measurement over the long run. The average gain of $350 per customer for our hypothetical insurance company is called the *expected value* for the amount the company earns per customer. Notice that the expected value does not have to be one of the possible values. For our insurance company, the two possible values were $500 and –$1000. Thus, the expected value of $350 was not even a possible value for any one customer. In that sense, "expected value" is a real misnomer. It doesn't have to be a value that's ever expected in a single outcome.

To compute the expected value for any situation, we need only be able to specify the possible *amounts*—call them $A_1, A_2, A_3, \ldots, A_k$—and the associated *probabilities*, which can be denoted by $p_1, p_2, p_3, \ldots, p_k$. Then the expected value can be found by multiplying each possible amount by its probability and adding them up. Remember, the expected value is the average value per measurement over the long run and not necessarily a typical value for any one occasion or person.

Computing the Expected Value

$$EV = \text{expected value} = A_1 p_1 + A_2 p_2 + A_3 p_3 + \cdots + A_k p_k$$

EXAMPLE 14.17 | **California Decco Lottery Game**

The California lottery has offered a number of games over the years. One such game in the 1990s was Decco, in which players chose one card from each of the four suits in a regular deck of playing cards. For example, the player might choose the 4 of hearts, 3 of clubs, 10 of diamonds, and jack of spades. A winning card was then drawn from each suit. If even one of the choices matched the winning cards drawn, a prize was awarded. It cost one dollar for each play, so the net gain for any prize is one dollar less than the prize. Table 14.6 shows the prizes and the probability of winning each prize, taken from the back of a game card.

We can thus compute the expected value for this Decco game:

$$EV = (\$4999 \times 1/28{,}561) + (\$49 \times 1/595) + (\$4 \times .0303) + (-\$1 \times .726)$$
$$= -\$0.35$$

TABLE 14.6 Probability of Winning the Decco Game

Number of Matches	Prize	Net Gain	Probability
4	$5000	$4999	1/28,561 = .000035
3	$50	$49	1/595 = .00168
2	$5	$4	1/33 = .0303
1	Free ticket	0	.2420
0	None	–$1	.7260

Notice that we count the free ticket as an even trade because it is worth $1, the same amount it cost to play the game. This result tells us that over many repetitions of the game, you will *lose* an *average* of 35 cents each time you play. From the perspective of the Lottery Commission, about 65 cents was paid out for each one dollar ticket sold for this game. (Astute readers will realize that this is an underestimate for the Lottery Commission and an overestimate for the player. The true cost of giving the free ticket as a prize is the expected payout per ticket, not the $1.00 purchase price of the ticket.) ■

Expected Value as Mean Number

If the measurement in question is one taken over a large group of individuals, rather than across time, the expected value can be interpreted as the mean value per individual. For example, according to the Centers for Disease Control, in 2011, 19% of adults in the United States smoked cigarettes. (Source: http://www.cdc.gov/tobacco/data_statistics/fact_sheets/adult_data/cig_smoking/) Let's suppose that they each smoked a pack of cigarettes a day (20 cigarettes) and the remaining 81% of adults smoked none. Then the expected value for the number of cigarettes smoked per day by one person would be

$$EV = (.19 \times 20 \text{ cigarettes}) + (.81 \times 0 \text{ cigarettes}) = 3.8 \text{ cigarettes}$$

In other words, on average, about four cigarettes were smoked per person per day. If we were to measure each person in the population by asking them how many cigarettes they smoked per day (and they answered truthfully), then the arithmetic average would be 3.8, or about four. This example further illustrates the fact that the expected value is not a value we actually expect to measure on any one individual.

CASE STUDY 14.1   **Birthdays and Death Days—Is There a Connection?**

SOURCE: Phillips, Van Voorhies, and Ruth (1992).

Is the timing of death random, or does it depend on significant events in one's life? That's the question University of California at San Diego sociologist David Phillips and his colleagues attempted to answer. Previous research had shown a possible connection between the timing of death and holidays and other special occasions. This study focused on the connection between birthday and death day.

The researchers studied death certificates of all Californians who had died between 1969 and 1990. Because of incomplete information before 1978, we report only on the part of their study that included the years 1979 to 1990. They limited their study to adults (over 18) who had died of natural causes. They eliminated anyone for whom surgery had been a contributing factor to death because there is some choice as to when to schedule surgery. They also omitted those born on February 29 because there was no way to know on which date these people celebrated their birthday in non–leap years.

Because there is a seasonal component to birthdays and death days, the researchers adjusted the numbers to account for those as well. They determined the number of deaths that would be expected on each day of the year if date of birth and date of death were independent of each other. Each death was then classified as to how many weeks after the birthday it occurred. For example, someone who died from 0 to 6 days after his or her birthday was classified as dying in "Week 0," whereas someone who died from 7 to 13 days after the birthday was classified in "Week 1," and so on. Thus, people who died in Week 51 died within a few days before their birthdays.

Finally, the researchers compared the actual numbers of deaths during each week with what would be expected based on the seasonally adjusted data. Here is what they found. For women, the biggest peak was in Week 0. For men, the biggest peak was in Week 51. In other words, the week during which the highest number of women died was the week *after* their birthdays. The week during which the highest number of men died was the week *before* their birthdays.

Perhaps this observation is due only to chance. Each of the 52 weeks is equally likely to show the biggest peak. What is the probability that the biggest peak for the women would be Week 0 *and* the biggest peak for the men would be Week 51? Using Rule 3, the probability of both events occurring is $(1/52) \times (1/52) = 1/2704 = .0004$.

As we will learn in Chapter 17, unusual events often do happen just by chance. Many facts given in the original report, however, add credence to the idea that this is not a chance result. For example, the peak for women in Week 0 remained even when the deaths were separated by age group, by race, and by cause of death. It was also present in the sample of deaths from 1969 to 1977. Further, earlier studies from various cultures have shown that people tend to die just after holidays important to that culture.

A more recent study (Ajdacic-Gross et al, 2012) looked at the exact day of death (rather than just the week) and found that there was an excess of almost 14% over what would be expected for deaths on one's own birthday. The higher than expected birthday death rates held for men and women, and for deaths from various causes, including cancer and cardiovascular diseases. ■

Thinking About Key Concepts

- The *relatively frequency interpretation* of probability says that the probability of any specific outcome is the proportion of time it occurs over the long run.
- The *personal probability* of an event is the degree to which an individual believes it will happen and can be different for each individual without either one being wrong. However, your own personal probabilities must be coherent, which means that they don't contradict each other.
- Rule 1 states that the probability of an outcome happening and the probability of that outcome not happening must sum to one.
- Rule 2 states that the probability of one *or* the other of two non-overlapping outcomes is the sum of their individual probabilities. Such outcomes are called *mutually exclusive events*.

- Rule 3 states that the probability of independent events *both* happening is the product of their probabilities. *Independent events* are events that do not influence each other. Knowing the probability that one of them will or has happened does not change the probability of the other one happening.
- Rule 4 states that the probability of a set of outcomes cannot be higher than the probability of a subset of those same outcomes.
- The probability rules can be combined to find more complicated probabilities. For example, if the probability of an outcome is p on each try and is independent from one try to the next, the probability that the outcome will happen for the first time on the nth try is $p(1 - p)^{n-1}$.
- The *expected value* for a measurement is the mean or average value over the long run or over a large number of individuals. It may or may not be an actual possible value for the measurement. For monetary outcomes, a negative expected value means there is a loss, on average, over the long run.

Focus On Formulas

Notation

Denote "events" or "outcomes" with capital letters *A, B, C,* and so on.

If A is one outcome, all other possible outcomes are part of "A complement" $= A^C$.

$P(A)$ is the probability that the event or outcome A occurs. For any event A, $0 \leq P(A) \leq 1$.

Rule 1

$$P(A) + P(A^C) = 1$$

A useful formula that results from this is

$$P(A^C) = 1 - P(A)$$

Rule 2

If events A and B are *mutually exclusive,* then

$$P(A \text{ or } B) = P(A) + P(B)$$

Rule 3

If events A and B are *independent,* then

$$P(A \text{ and } B) = P(A)\, P(B)$$

Rule 4

If the ways in which an event B can occur are a subset of those for event A, then

$$P(B) \leq P(A)$$

Exercises

Exercises with numbers divisible by 3 (3, 6, 9, etc.) are included in the Solutions at the back of the book. They are marked with an asterisk (*).

1. Recall that there are two interpretations of probability: relative frequency and personal probability.

   a. Which interpretation applies to this statement: "The probability that I will get the flu this winter is 30%"? Explain.

   b. Which interpretation applies to this statement: "The probability that a randomly selected adult in America will get the flu this winter is 30%"? Explain. (Assume it is known that the proportion of adults who get the flu each winter remains at about 30%.)

2. Explain which of the following more closely describes what it means to say that the probability of a tossed coin landing with heads up is 1/2:

   Explanation 1: After more and more tosses, the fraction of heads will get closer and closer to 1/2.

   Explanation 2: The number of heads will always be about half the number of tosses.

*3. Explain why probabilities cannot always be interpreted using the relative-frequency interpretation. Give an example of when that interpretation would not apply.

4. Suppose you wanted to test your extrasensory perception (ESP) ability using an ordinary deck of 52 cards, which has 26 red and 26 black cards. You have a friend shuffle the deck and draw cards at random, replacing the card and reshuffling after each guess. You attempt to guess the color of each card.

   a. What is the probability that you guess the color correctly by chance?

   b. Is the answer in part (a) based on the relative-frequency interpretation of probability, or is it a personal probability?

   c. Suppose another friend has never tried the experiment but believes he has ESP and can guess correctly with probability .60. Is the value of .60 a relative-frequency probability or a personal probability? Explain.

   d. Suppose another friend guessed the color of 1000 cards and got 600 correct. The friend claims she has ESP and has a .60 probability of guessing correctly. Is the value of .60 a relative-frequency probability or a personal probability? Explain.

5. Suppose your are able to obtain a list of the names of everyone in your school and you want to determine the probability that someone randomly selected from your school has the same first name as you.

   a. Assuming you had the time and energy to do it, how would you go about determining that probability? (Assume all names listed are fully spelled out.)

   b. Using the method you described in part (a), would your result be a relative-frequency probability or a personal probability? Explain.

*6. In Section 14.2 (page 307), you learned two ways in which relative-frequency probabilities can be determined. Explain which method you think was used to determine the following statement: The probability that a particular flight from New York to San Francisco will be on time is .78.

7. Which of the two methods for determining relative-frequency probabilities (given in Section 14.2, page 307) was used to determine each of the following?

   a. On any given day, the probability that a randomly selected American adult will read a book for pleasure is .33.

   b. The probability that a five-card poker hand contains "four of a kind" is .00024.

8. Use your own particular expertise to assign a personal probability to something, such as the probability that a certain sports team will win

next week. Now assign a personal probability to another *related* event. Explain how you determined each probability, and explain how your assignments are *coherent*.

*9. For each of the following situations, state whether the relative-frequency interpretation or the personal probability interpretation is appropriate. If it is the relative-frequency interpretation, specify which of the two methods for finding such probabilities (given on page 307) would apply.

    *a. If a spoon is tossed 10,000 times and lands with the rounded head face up on 3000 of those times, we would say that the probability of the rounded head landing face up for that spoon is about .30.

    *b. In a debate with you, a friend says that she thinks there is a 50/50 chance that God exists.

10. On February 13, 2013, the *Wall Street Journal* reported, "The odds that…an asteroid impact would make us the last generation of human civilization are no lower than the odds of an average American dying in an earthquake (about 0.001%)" (Lu and Rees, 2013). Do you think this probability is based on relative frequency? Explain.

11. Use the probability rules in this chapter to solve each of the following:

    a. According to the U.S. Census Bureau, in 2012, the probability that a randomly selected child in the United States was living with his or her mother as the sole parent was .244 and with his or her father as the sole parent was .040. What was the probability that a child was living with just one parent? (*Source:* http://www.census.gov/hhes/families/data/cps2012.html.)

    b. In 2010 in the United States, the probability that a birth would result in twins was .0331, and the probability that a birth would result in triplets or more was .0014. What was the probability that a birth in 2010 resulted in a single child? (*Source:* http://www.cdc.gov/nchs/fastats/multiple.htm.)

*12. Figure 9.1 (page 183) illustrates that 17.8% of Caucasian girls have green eyes and 16.9% of them have hazel eyes.

    *a. What is the probability that a randomly selected Caucasian girl will have green eyes?

    *b. What is the probability that a randomly selected Caucasian girl will have hazel eyes?

    *c. What is the probability that a randomly selected Caucasian girl will have either green or hazel eyes?

    *d. What is the probability that a randomly selected Caucasian girl will *not* have either green or hazel eyes?

13. Example 14.3 states that "the probability that a piece of checked luggage will be lost, delayed, damaged, or pilfered on a flight with a U.S. airline is 3/1000." Interpret that statement, using the appropriate interpretation of probability.

14. There is something wrong in each of the following statements. Explain what is wrong.

    a. The probability that a randomly selected driver will be wearing a seat belt is .75, whereas the probability that he or she will not be wearing one is .30.

    b. The probability that a randomly selected car is red is 1.20.

    c. The probability that a randomly selected car is red is .20, whereas the probability that a randomly selected car is a red sports car is .25.

*15. A small business performs a service and then bills its customers. From past experience, 90% of the customers pay their bills within a week.

    *a. What is the probability that a randomly selected customer will not pay within a week?

    *b. The business has billed two customers this week. What is the probability that neither of them will pay within a week? What assumption did you make to compute that probability? Is it a reasonable assumption?

16. According to Krantz (1992, p. 111), the probability of being born on a Friday the 13th is about 1/214.

    a. What is the probability of not being born on a Friday the 13th?

    b. In any particular year, Friday the 13th can occur once, twice, or three times. Is the probability of being born on Friday the 13th the same every year? Explain.

c. Explain what it means to say that the probability of being born on Friday the 13th is 1/214.

17. Suppose the probability that you get an interesting piece of mail on any given weekday is 1/20. Is the probability that you get at least one interesting piece of mail during the week (Monday to Friday) equal to 5/20? Why or why not?

*18. You cross a train track on your drive to work or school. If you get stopped by a train you are late.

  *a. Are the events "stopped by train" and "late for work or school" independent events? Explain.

  *b. Are the events "stopped by train" and "late for work or school" mutually exclusive events? Explain.

19. On any given day, the probability that a randomly selected adult male in the United States drinks coffee is .51 (51%), and the probability that he drinks alcohol is .31 (31%). (Source: http://www.ars.usda.gov/SP2UserFiles/Place/12355000/pdf/DBrief/6_beverage_choices_adults_0708.pdf) What assumption would we have to make in order to use Rule 3 to conclude that the probability that a person drinks both is (.51) × (.31) = .158? Do you think that assumption holds in this case? Explain.

20. A classic study by Kahneman and Tversky (1982, p. 496) asked people the following question: "Linda is 31 years old, single, outspoken, and very bright. She majored in philosophy. As a student, she was deeply concerned with issues of discrimination and social justice, and also participated in antinuclear demonstrations. Please check off the most likely alternative:

A. Linda is a bank teller.

B. Linda is a bank teller and is active in the feminist movement."

Nearly 90% of the 86 respondents chose alternative B. Explain why alternative B *cannot* have a higher probability than alternative A.

*21. Read the definition of "independent events" given in Rule 3. Explain whether each of the following pairs of events is likely to be independent:

  *a. A married couple will vote in an upcoming presidential election. Event A is that the husband votes for the Republican candidate; event B is that the wife votes for the Republican candidate.

  *b. Event A is that a major earthquake will occur somewhere in the world in the next month; event B is that the Dow Jones Industrial Average will be higher in one month than it is now.

22. Read the definition of "independent events" given in Rule 3. Explain whether each of the following pairs of events is likely to be independent:

  a. Event A is that it snows tomorrow; event B is that the high temperature tomorrow is at least 60 degrees Fahrenheit.

  b. You buy a lottery ticket, betting the same numbers two weeks in a row. Event A is that you win in the first week; event B is that you win in the second week.

23. People are surprised to find that it is not all that uncommon for two people in a group of 20 to 30 people to have the same birthday. We will learn how to find that probability in a later chapter. For now, consider the probability of finding two people who have birthdays in the same month. Make the simplifying assumption that the probability that a randomly selected person will have a birthday in any given month is 1/12. Suppose there are three people in a room and you consecutively ask them their birthdays. Your goal, following parts (a–d), is to determine the probability that at least two of them were born in the same calendar month.

  a. What is the probability that the second person you ask will not have the same birth month as the first person? (*Hint:* Use Rule 1.)

  b. Assuming the first and second persons have different birth months, what is the probability that the third person will have yet a different birth month? (*Hint:* Suppose January and

February have been taken. What proportion of all people will have birth months from March to December?)

c. Explain what it would mean about overlap among the three birth months if the outcomes in part (a) and part (b) both happened. What is the probability that the outcomes in part (a) and part (b) will both happen?

d. Explain what it would mean about overlap among the three birth months if the outcomes in part (a) and part (b) did *not* both happen. What is the probability of that occurring?

*24. A restaurant server knows that the probability that a customer will order coffee is .30, the probability that a customer will order a diet soda is .40, and the probability that a customer will request a glass of water is .70. Explain what is wrong with his reasoning in each of the following.

*a. Using Rule 2, he concludes that the probability that a customer will order either coffee or request a glass of water is .30 + .70 = 1.0.

*b. Using Rule 3, he concludes that the probability that a customer will order coffee and a diet soda is (.30)(.40) = .12.

25. Suppose you routinely check coin-return slots in vending machines to see if they have any money in them. You have found that about 10% of the time you find money.

a. What is the probability that you do not find money the next time you check?

b. What is the probability that the next time you will find money is on the third try?

c. What is the probability that you will have found money by the third try?

26. Lyme disease is a disease carried by ticks, which can be transmitted to humans by tick bites. Suppose the probability of contracting the disease is 1/100 for each tick bite.

a. What is the probability that you will not get the disease when bitten once?

b. What is the probability that you will not get the disease from your first tick bite and will get it from your second tick bite?

*27. Suppose you play a carnival game that requires you to toss a ball to hit a target. The probability that you will hit the target on each play is .2 and is independent from one try to the next. You win a prize if you hit the target by the third try.

*a. What is the probability that you hit the target on the first try?

*b. What is the probability that you miss the target on the first try but hit it on the second try?

*c. What is the probability that you miss the target on the first and second tries but hit it on the third try?

*d. What is the probability that you win a prize?

28. According to Krantz (1992, p. 161), the probability of being injured by lightning in any given year is 1/685,000. Assume that the probability remains the same from year to year and that avoiding a strike in one year doesn't change your probability in the next year.

a. What is the probability that someone who lives 80 years will never be struck by lightning? You do not need to compute the answer, but write down how it would be computed.

b. According to Krantz, the probability of being injured by lightning over the average lifetime is 1/9100. Show how that probability should relate to your answer in part (a), assuming that average lifetime is about 80 years.

c. Do the probabilities given in this exercise apply specifically to you? Explain.

d. Over 300 million people live in the United States. In a typical year, assuming Krantz's figure is accurate, about how many people out of 300 million would be expected to be struck by lightning?

29. Suppose you have to cross a train track on your commute. The probability that you will have to wait for a train is 1/5, or .20. If you don't have to wait, the commute takes 15 minutes, but if you have to wait, it takes 20 minutes.

a. What is the expected value of the time it takes you to commute?

b. Is the expected value ever the actual commute time? Explain.

*30. Suppose the probability that you get an A in any class you take is .3, and the probability that you get a B is .7. To construct a GPA, an A is worth 4.0, and a B is worth 3.0. What is the expected value for your GPA? Would you expect to have this GPA separately for each quarter or semester? Explain.

31. Remember that the probability that a birth results in a boy is about .51. You offer a bet to an unsuspecting friend. Each day you will call the local hospital and find out how many boys and how many girls were born the previous day. For each girl, you will give your friend $1 and for each boy your friend will give you $1.

   a. Suppose that on a given day there are three births. What is the probability that you lose $3 on that day? What is the probability that your friend loses $3?

   b. Notice that your net profit is $1 if a boy is born and –$1 if a girl is born. What is the expected value of your profit for each birth?

   c. Using your answer in part (b), how much can you expect to make after 1000 births?

32. In the "3 Spot" version of the former California Keno lottery game, the player picked three numbers from 1 to 40. Ten possible winning numbers were then randomly selected. It cost $1 to play. The accompanying table shows the possible outcomes. Compute the expected value, the net "gain," for this game. Interpret what it means.

Number of Matches	Amount Won	Probability
3	$20	.012
2	$2	.137
0 or 1	$0	.851

*33. We have seen many examples for which the term *expected value* seems to be a misnomer. Construct an example of a situation in which the term *expected value* would *not* seem to be a misnomer for what it represents.

34. According to the U.S. Census Bureau, in 2012, about 68% (.68) of children in the United States were living with both parents, 24.4% (.244) were living with mother only, 4% (.04) were living with father only, and 3.6% (.036) were not living with either parent. What is the expected value for the number of parents a randomly selected child was living with? Does the concept of expected value have a meaningful interpretation for this example? Explain.

35. Find out your yearly car insurance cost. If you don't have a car, find out the yearly cost for a friend or relative. Now assume you will either have an accident or not, and if you do, it will cost the insurance company $5000 more than the premium you pay. Calculate what yearly accident probability would result in a "break-even" expected value for you and the insurance company. Comment on whether you think your answer is an accurate representation of your yearly probability of having an accident.

Mini-Projects

1. Refer to Exercise 20. Present the question to 10 people, and note the proportion who answer with alternative B. Explain to the participants why it cannot be the right answer, and report on their reactions.

2. Flip a coin 100 times. Stop each time you have done 10 flips (that is, stop after 10 flips, 20 flips, 30 flips, and so on), and compute the proportion of heads using *all* of the flips up to that point. Plot that proportion versus the number of flips. Comment on how the plot relates to the relative-frequency interpretation of probability.

3. Pick an event that will result in the same outcome for everyone, such as whether it will rain next Saturday. Ask 10 people to assess the probability of that event,

and note the variability in their responses. (Don't let them hear each other's answers, and make sure you don't pick something that would have 0 or 1 as a common response.) At the same time, ask them the probability of getting a heart when a card is randomly chosen from a fair deck of cards. Compare the variability in responses for the two questions, and explain why one is more variable than the other.

4. Find two lottery or casino games that have fixed payoffs and for which the probabilities of each payoff are available. (Some lottery tickets list them on the back of the ticket or on the lottery's website. Some books about gambling give the payoffs and probabilities for various casino games.)

    a. Compute the expected value for each game. Discuss what they mean.

    b. Using both the expected values and the list of payoffs and probabilities, explain which game you would rather play and why.

References

Ajdacic-Gross, V., D. Knöpfli, K. Landolt, M. Gostynski, S. T. Engelter, P. A. Lyrer, F. Gutzwiller, W. Rössler. (2012). Death has a preference for birthdays—An analysis of death time series. *Annals of Epidemiology* 22, pp. 603–606.

Arraf, Jane. (14 December 1994). Leave Earth or perish: Sagan. *China Post (Taiwan)*, p. 4.

Brown, Alyssa. (2013). Fewer Americans expect to be better off in a year. http://www.gallup.com/poll/163058/fewer-americans-expect-better-off-year.aspx, June 13, 2013, accessed June 24, 2013.

Hayes, B. (2011). The prime twin conjecture. http://bit-player.org/2011/the-prime-twins-conjecture, posted January 18, 2011, accessed June 25, 2013.

Hughes, J. P., J.M. Baeten, J.R. Lingappa, A.S. Margaret, A. Wald, G. de Bruyn, J. Kiarie, M. Inambao, W. Kilembe, C. Farguhar, C. Celum, and the Partners in Prevention HSV/HIV Transmission Study Team. (2012). Determinants of per-coital-act HIV-1 infectivity among African HIV-1-serodiscordant couples. *Journal of Infectious Diseases*, 205(3), pp. 358-365.

Kahneman, D., and A. Tversky. (1982). On the study of statistical intuitions. In D. Kahneman, P. Slovic, and A. Tversky (eds.), *Judgment under uncertainty: Heuristics and biases* (Chapter 34). Cambridge, England: Cambridge University Press.

Kolata, G. (2006). Live long? Die Young? Answer isn't just in genes. *New York Times* online, http://www.nytimes.com/2006/08/31/health/31age.html, August 31, 2006, accessed June 25, 2013.

Krantz, Les. (1992). *What the odds are.* New York: Harper Perennial.

Lu, E. and M. Rees. (2013). A warning from the asteroid hunters. *Wall Street Journal*, 14 February 2013, p. A19. Also published online 13 February 2013, http://online.wsj.com/article/SB10001424127887324196204578297823983416036.html, accessed June 25, 2013.

Marshall, E. (2005). Flawed statistics in murder trial may cost expert his medical license. *Science*, 309 (22 July 2005), p. 543.

Phillips, D. P., C. A. Van Voorhies, and T. E. Ruth. (1992). The birthday: Lifeline or deadline? *Psychosomatic Medicine* 54, pp. 532–542.

CHAPTER **15**

Understanding Uncertainty through Simulation

Thought Questions

1. A computer is asked to "randomly generate a two-digit number from the integer choices 00 to 99." Explain what this means.

2. Weather forecasters sometimes use computer simulation to predict the path of a storm. How do you think they do this? What information might they give the computer to help with this prediction?

3. Suppose you have a coin that says "boy" on one side and "girl" on the other side. The coin is weighted so that when you toss it, the probability that "boy" lands face up is .512, the historical probability that a birth results in a boy. If you toss the coin four times and record which side lands face up each time, that's equivalent to simulating the births in a family of four children. What are the options for the number of boys that could result from this simulation of four births? Which of those options do you think is the most likely to occur?

4. Suppose you use the coin described in Question 3 to simulate births for 100 families of four children each. And suppose that out of those 100 simulated families, six of them have all girls and no boys. How would you use this information to estimate the probability that a family of four children has no boys? What is the estimated probability? Is your answer a personal probability or a relative-frequency probability?

5. For the simulation described in Question 4, do you think the estimated probability of having no boys in a family of four children would be more accurate with 100 simulated families or with 10,000 simulated families? Explain. (Of course, to generate 10,000 simulated families you would definitely want a computer to take over the task instead of flipping your boy/girl coin!)

15.1   Mimicking Reality through Simulation

Many random events are so complicated that it is difficult to use the probability rules from Chapter 14 to predict their outcomes. **Simulation** uses computer models to mimic what might happen in the real world. For example, weather forecasters use simulation to predict the path of a storm, by combining knowledge of meteorology, past behavior under similar conditions, and a random, unexplainable component. Economists use simulation to predict the outcome of a change in monetary policy, by combining knowledge of economics, past performance under similar conditions, some understanding of human behavior, and a random unexplainable component.

In general, the results of simulations are only as good as the instructions and information provided to the computer to carry them out. Fortunately, for situations we encounter in probability and statistics, we often can provide accurate models of the real-world situation. Simulation can then be used to describe how likely we are to encounter various possible outcomes by including a random, unexplainable component. For instance, suppose you are in a meeting with 10 people and discover that five of you were born on the 13th of the month, all in different months. (This happened to the author of this book.) By making certain assumptions about the likelihood of being born on various days of the year and then adding in a random component, you could simulate the birth day of the month (1 to 31) for a collection of 10 people and see how often you should expect to find that five of them were all born on the same day of the month, in different months. (The probability would be quite low.) We could compute such probabilities by hand, but it would be a very tedious task because there are lots of combinations of possibilities. Asking the computer to simulate this situation is a relatively simple task.

15.2   Simulating Probabilities

As illustrated by the coincidence of five out of 10 people all being born on the 13th of the month, some probabilities are very difficult to calculate using the rules in Chapter 14, but the information needed to calculate them is entirely available. In those cases, it may be possible to **simulate** the situation repeatedly using a computer or calculator and observe the relative frequency with which the desired outcome occurs. If you simulate the situation n times and the outcome of interest occurs in k of those times, then the probability of that outcome is approximately k/n. To get an accurate answer, n needs to be quite large. In general, determining the accuracy of simulation for estimating a probability is similar to determining the margin of error for a survey. If you have n repetitions of the situation, your answer will be within $\frac{1}{\sqrt{n}}$ of the true probability most of the time. If the true probability is close to either extreme (0 or 1), then the results will be even more accurate.

To simulate a situation, you need a computer or calculator that generates **random numbers**. To generate random numbers, you first specify the numbers you want to have as possibilities. For instance, if you wanted to simulate birthdays, you could use the numbers from 1 to 31 for the day and 1 to 12 for the month, or you could use the numbers from 1 to 366 to represent the unique choices.

It is most common to set things up so that all choices of numbers are equally likely, but you could also weight the choices differently if that's what a particular application requires. For instance, in simulating birthdays, we know that they are not equally likely to occur on all days from 1 to 31, so you would need to tell the computer to weight the choices 1 to 28 differently than 29, 30, and 31, which would also each have different weights. Each birthday chosen would have a specified probability of being each of the 31 possibilities.

Once you specify the numbers you want to have as possibilities, and their relative weights, you tell the computer how many of them you want generated. Then the computer creates a set of numbers such that each value in the set is randomly chosen using the probabilities you specified. For example, here are two sets of results from asking the computer software Minitab to generate five random numbers from the integer choices 00 to 99 with equal probability for each of them:

Set 1: 90 12 48 67 90

Set 2: 38 31 50 49 00

One feature you may notice in the first set is that the number 90 appears twice. That might lead you to wonder about the probability of the same number appearing twice in this situation. Let's investigate that question with an example.

| EXAMPLE 15.1 | License Plate Lottery |

As a promotion, a radio station holds a contest on Monday to Friday of one week by randomly choosing a two-digit number from 00 to 99 each day. If the last two digits on your car's license plate match the number chosen that day, you have 24 hours to send a copy of your registration to the radio station, and they will send you a free coffee mug. They probably should have specified that the same number can't win twice, but they didn't do so. What is the probability that the same number gets chosen in two or more of the 5 days?

The probability of the same number being chosen more than once can be calculated, but it's a somewhat tedious calculation, so let's simulate the event instead. Minitab was asked to generate 100 sets of five numbers chosen from 00 to 99, allowing the same number to be chosen more than once. Each set of five numbers represents what *could* have happened during the five-day contest. It would take too much space to show the results of all 100 sets of numbers, so Table 15.1 displays the first 10 of the 100 sets, followed by all of the sets that had duplicates. (No set had a triplicate or more.) The numbers within each set have been put in order to make it easier to identify duplicate numbers. Duplicates are shown in bold italics.

The results show that there are five sets with duplicate numbers. Therefore, based on this simulation, we would estimate that the probability of the same number being chosen two or more times during the week of the contest is about 5/100, or .05. (We generated 100 sets, and five of them contained duplicates.) In fact, the actual probability is .04, so our probability based on simulating the situation is not far off from the exact

TABLE 15.1	Simulating possible sets of winning numbers for the license plate lottery

Winning numbers for 5 days (numerical order)

First ten sets (no duplicate numbers appeared)

Set 1	8	13	46	82	97
Set 2	12	15	26	51	57
Set 3	3	24	46	69	96
Set 4	9	19	27	64	79
Set 5	7	12	69	82	89
Set 6	23	32	53	69	71
Set 7	24	42	52	56	66
Set 8	20	34	38	45	93
Set 9	16	35	59	64	73
Set 10	2	35	76	93	99

All sets with duplicate numbers

Set 15	14	*27*	*27*	40	47
Set 23	19	69	85	*96*	*96*
Set 38	30	43	50	*73*	*73*
Set 64	8	11	*64*	*64*	92
Set 86	3	9	*18*	*18*	37

answer. Remember that only one set of numbers would actually be chosen during the contest. The simulation generated 100 of the possibilities. The total number of possibilities is quite large, at $(100)^5$, if the order is taken into account. ■

Simulation Using www.randomizer.org

The website www.randomizer.org (Urbaniak and Plous, 2013) can be used for simulating situations such as the one in Example 15.1. Here are the steps to use for that example:

How many sets of numbers do you want to generate? 100

How many numbers per set? 5

Number range (e.g., 1-50): From: 0
To: 99

Do you wish each number in a set to remain unique? No

Do you wish to sort the numbers that are generated?   Yes: Least to Greatest

15.3   Simulating the Chi-Square Test

When we discussed the chi-square test in Chapter 13, we left the connection between the chi-square statistic and the *p*-value as somewhat of a mystery. We gave specific cut-off points for *p*-values of 0.10, 0.05 and 0.01 but did not explain how they were found.

Simulation can be used to explain a different method of finding *p*-values that will give almost the same result as the mysterious method used in Chapter 13 but that is easier to understand. Let's illustrate the method with an example.

EXAMPLE 15.2   Using Simulation to Find the *p*-value for Drinking and Driving

In Example 13.2, we looked at data provided to the Supreme Court to determine whether differential laws for beer sales should be applied to young men and women. The hypotheses were:

Null hypothesis: Males and females in the population are equally likely to drive within 2 hours of drinking alcohol.
Alternative hypothesis: Males and females in the population are not equally likely to drive within 2 hours of drinking alcohol.

The data are shown again in Table 15.2. The results of a chi-square test are shown in Figure 13.2 on page 290, where the chi-square statistic is given as 1.637 and the *p*-value as 0.201.

TABLE 15.2   Drinking and Driving for Young Men and Women

	Drove within 2 hours of drinking?		
	Yes	**No**	**Total**
Male	77	404	481
Female	16	122	138
Total	93	526	619

How can we simulate this situation and the *p*-value? Remember that the *p*-value is the probability of observing a chi-square statistic as large as the one observed or larger, *if* the null hypothesis is true. In this situation, we need to answer this question: *What is the probability of observing a chi-square statistic of 1.637 or larger, if in fact males and females in the population are equally likely to drive within 2 hours of drinking alcohol?*

To answer this question, let's simulate samples for which the following facts are true:

- The numbers of male and female drivers who would be stopped at the roadside survey are fixed at 481 and 138, respectively, as they were for the actual data.
- The probability that any driver would have been drinking in the past two hours is 93/619 = 0.15, which is the combined proportion for males and females in the actual data.

- The null hypothesis is true, so the probability that a driver would have been drinking is the same, 0.15, for both males and females.

This last assumption is necessary because the *p*-value is computed by *assuming* the null hypothesis is true.

Fixing all of those facts, we simulated 10,000 samples, each with 481 males and 138 females, and set the probability that each person had been drinking to be 0.15. We computed the chi-square statistic for each of the 10,000 samples. (We used the R statistical software and adapted computer code provided by Howell, 2013.) In the simulation, 2,021 of the 10,000 samples had a chi-square statistic of 1.637 or higher. Figure 15.1 shows a histogram of all 10,000 chi-square statistics, with the chi-square statistic of 1.637 (observed in the actual study) marked. Therefore, the probability of obtaining a chi-square value that large or larger, when the null hypothesis is true, is estimated to be 2021/10,000 = 0.2021. This value is very close to the *p*-value of 0.201 provided in Chapter 13. ∎

Example 15.2 illustrated how to simulate a *p*-value when the totals in the rows are considered to be fixed. Similar methods can be used when only the overall total is assumed to be fixed or when both the row and column totals are assumed to be fixed. In each case, thousands of possible samples are generated by assuming the null hypothesis is true. The chi-square statistic is calculated for each one of these thousands of samples, and the distribution of them, called the **randomization distribution**, is created as illustrated by Figure 15.1. The randomization distribution is used to find the *p*-value for the chi-square statistic from the real sample. This is done by determining where the chi-square statistic for that sample falls within the randomization distribution. The *p*-value is the proportion of the distribution that falls at or above that chi-square statistic.

Figure 15.1
Simulation of 10,000 samples illustrating p-value for Example 15.2

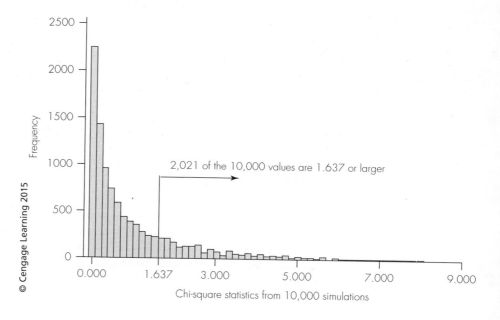

© Cengage Learning 2015

15.4 Randomization Tests

The method used to simulate the *p*-value for a chi-square test in the previous section is a special case of a more general category of **randomization tests.** These tests are particularly useful in situations for which the conditions necessary for carrying out a more traditional test are not met. For instance, one of the conditions required to use the chi-square test in Chapter 13 is that all of the expected counts be at least 5. If that condition is not met, a randomization test can still be used to find a *p*-value.

As another example, let's look at a correlation that looks real, and see what happens when subjected to a randomization test.

EXAMPLE 15.3 ### Are Age and Body Temperature Correlated?

The ages and body temperatures were recorded for eight people who donated blood at a blood bank. The values, shown in Table 15.3, have a correlation of −0.372, indicating that body temperature decreases somewhat with age for this sample of individuals. A scatterplot of the data with the least squares regression line is shown in Figure 15.2. How likely would we be to observe a correlation of that magnitude or higher just by chance if these two variables are not correlated in the population? We can set this up as a hypothesis test with the following hypotheses:

> *Null hypothesis:* Age and body temperature are not correlated in the population.
> *Alternative hypothesis:* Age and body temperature are correlated in the population.

Notice that the alternative hypothesis does not specify a direction for the correlation. That's because it isn't fair to look at the data first and then decide on a direction for the hypothesis.

Figure 15.2
Age versus body temperature for 8 blood donors; correlation is −0.372

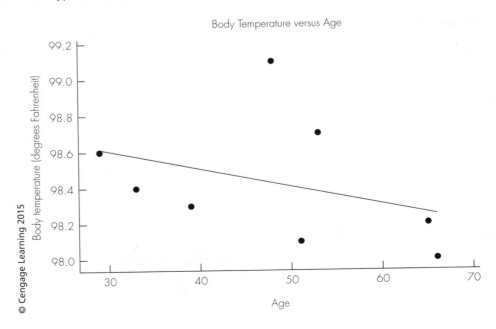

Body Temperature versus Age

© Cengage Learning 2015

To find the *p*-value for this test, we need to answer this question: *If there is no correlation between age and body temperature in the population, what proportion of samples of eight people would show a correlation with absolute value of at least 0.372?* Again, because the alternative hypothesis includes both directions, we are interested in correlations higher than 0.372 in addition to those lower than −0.372.

One way to examine the *p*-value question is to mix up the ages and temperatures. In other words, suppose that the eight temperatures in the data set could equally have been distributed across the ages in any order. Let's see what happens if we randomly scramble the temperatures across the eight ages. If the null hypothesis is true, then the correlation for the actual order should not stand out as being unusual compared to the correlations when the order is scrambled.

The outcomes of two such random scrambles are shown in Table 15.4. For the first one, the correlation is even more extreme in the negative direction than the original data, at −0.389. But for the second one, the correlation is smaller and the direction has reversed, for a correlation of +0.164. Yet in both of these cases, the eight temperatures were just randomly assigned across the eight ages, so any observed pattern is meaningless.

TABLE 15.3	Age and body temperature for eight blood donors; correlation is −0.372							
Age	29	33	39	48	51	53	65	66
Temperature	98.6	98.4	98.3	99.1	98.1	98.7	98.2	98.0

TABLE 15.4	Two random permutations of the temperatures for the eight blood donors							
AGE	29	33	39	48	51	51	65	65	Correlation
Temperature	98.4	98.7	98.6	98.1	98.3	99.1	98.0	98.2	−0.389
Temperature	98.7	98.1	98.4	98.6	98.3	98.0	98.2	99.1	+0.164

We can continue to randomly scramble the temperature values and find correlations to create a distribution of possible correlations. These represent possible correlations for randomly matching the eight temperatures to the eight ages. In other words, suppose that those eight ages and those eight temperatures are fixed values, but that the matching of them is just random. What correlation values should we expect? Simulating this process 5000 times resulted in correlations ranging from −0.915 to +0.924. There were 1804 of the correlations with an absolute magnitude equal to or greater than the observed correlation of −0.372. That result provides a *p*-value of 1804/5000 = 0.361 (36.1%) Therefore, even if there were no correlation between age and temperature in the population, we would be likely to observe a sample correlation for eight people with absolute value of 0.372 or larger about 36% of the time, given these ages and temperatures. The observed correlation based on the sample of eight people is not convincing evidence of a correlation between age and temperature in the population.

As a side note to this example, the eight individuals in this small sample actually were part of a larger sample of 100 individuals. The correlation between age and body temperature for the larger sample was -0.313. Because that sample was so much larger, the corresponding p-value for testing the hypotheses above was 0.002, indicating that there is a correlation between age and body temperature in the population. The fact that the non-zero correlation for the sample of eight people was not convincing illustrates one of the points made in Chapter 13. Small samples generally cannot provide convincing evidence of true relationships. That's why we don't accept the null hypothesis, especially with a small sample, even when the data do not provide convincing evidence to refute it.

For the situation in Example 15.3, it is actually possible to find correlations for all possible permutations of the temperatures across the ages. A **permutation** is a scrambling of the temperature values. Two of the possible permutations are shown in Table 15.4. There are a total of $8 \times 7 \times 6 \times 5 \times 4 \times 3 \times 2 \times 1 = 40{,}320$ possibilities. That's because there are eight possible temperatures to assign to the first age, then a choice of seven remaining ones to assign to the next age, and so on. So, instead of simulating the situation 5000 times, we could have asked the computer to calculate the correlation for all 40,320 possibilities. Because of this option, randomization tests are sometimes called **permutation tests**. In this example, we would find that the greatest possible positive correlation is $+0.940$, the correlation if the eight temperatures are placed across the ages in increasing order. The greatest possible negative correlation is -0.957, the correlation if the eight temperatures are placed across the ages in decreasing order. In our simulation, the correlations ranged from -0.915 to $+0.924$, spanning almost the entire possible range.

CASE STUDY 15.1   **Is it Just Chance, or Do Major Earthquakes Occur in Clusters?**

When the magnitude 9.0 earthquake hit off the coast of Japan in March, 2011, it came just a little more than a year after the historic 8.8 magnitude quake in Chile, on February 27, 2010. And just slightly more than 5 years earlier, on December 26, 2004, one of the largest earthquakes in history, with magnitude 9.1, struck off the coast of northern Sumatra in Indonesia. Perhaps none of this would have surprised scientists if it hadn't been for the fact that there had been no earthquakes of magnitude 8.5 or higher since February 4, 1965, a period of almost 40 years. Were earthquakes occurring in clusters? Or was it just chance that there were three mega-quakes in a 6 ½-year period, after a 40-year period without any?

Scientists looking at the historical record starting in 1900 (when accurate data recording began) noticed that there had been one other cluster of large quakes, from 1950 to 1965, when there were seven earthquakes with magnitudes of 8.5 or greater (Kerr, 2011). Given that there were only 10 earthquakes that large in the entire 20th century, seven large quakes in a 15-year period would certainly seem to qualify as a cluster.

But let's suppose that large earthquakes really do occur at random across time. How unlikely would these types of clusters be, just by chance? That's the question that generated debate in the scientific community after the 2011 earthquake in Japan. A simulation study published in 2005 by Charles Bufe and David Perkins had assessed the likelihood of the 1950 to 1965 cluster occurring just by chance to be very small (Bufe and Perkins, 2005). An updated simulation they presented at the Seismological Society of America conference in 2011, after the Japanese quake, found the probability of both observed clusters occurring just by chance to be just 2% (Kerr, 2011). They based that figure on 100,000 simulations of the timing of earthquakes that have occurred, allowing them to be distributed across time just by chance. They found clusters of mega-quakes as extreme as the ones observed or more so in only 2% of those random sequences. Does that mean chance can be ruled out as the explanation for the two observed clusters?

The simulation study by Bufe and Perkins was criticized by seismologist Andrew Michael, who claimed that Bufe and Perkins had made "a serious statistical mistake… We can't run experiments, so we're stuck testing our hypotheses on the same data we developed them on" (Kerr, 2011). The basic point Michael was trying to make was that *any* very specific pattern, such as exactly two clusters of sizes seven and three spaced 40 years apart, would have a small probability. The more appropriate statistical question would examine the probability that the timing of large earthquakes exhibits *any* pattern that distinguishes it from a random sequence of timings, not just the specific pattern that happened to be observed. As an analogy, consider the cluster of birthdays found by the author in a meeting of 10 people, where it was discovered that five of them were born on the 13th of the month, all in different months. That specific outcome has a very low probability. But a more appropriate question would address the probability of any unusual combination of birthdays, such as five people being born in the same month, or on the 15th of the month, or in a sequence of adjoining days, and so on. Almost any random sequence includes patterns of *some* kind, just by chance.

To address the more general question of unusual patterns, Shearer and Stark (2011) simulated 100,000 random sequences of earthquake timings based on all earthquakes of magnitude 7.0 or greater occurring between 1900 and August, 2011. They removed foreshocks and aftershocks from the data, since those create clustering by their very nature. For each of the 100,000 simulated sequences, they computed a (somewhat complicated) test statistic that measures how much the simulated sequence differs from what's expected by chance. They found that about 17% of the 100,000 random sequences had a stronger deviation from chance than the actual, observed sequence of real earthquakes. Therefore, they concluded that the observed clusters were not statistically significantly different from what would be expected by chance. They ran other tests as well, and the results of those were even more consistent with the chance hypothesis. ∎

Thinking About Key Concepts

- *Simulation* uses computer models to mimic what might happen in the real world, but the results of a simulation are only as good as the information given to the computer to create them.
- Relative-frequency probabilities can be estimated by simulating a random circumstance thousands of times and noting the proportion of those times the desired outcome occurs. For example, we could estimate the probability that a family of four children would be all girls by simulating thousands of such families and seeing what proportion of them didn't have any boys.
- A *random number generator* on a computer or calculator produces a *random number* from a specified set of choices, where each choice has an assigned probability. Most commonly the assigned probabilities are all equal, so any of the possible numbers is equally likely to be the one produced.
- Simulation can be used in hypothesis testing to find an approximate *p*-value. This is done by simulating samples of the same size as the original sample, under the assumption that the null hypothesis is true. For each simulated sample, the test statistic is calculated, and these values are used to create a *randomization distribution*. The *p*-value is the proportion of the randomization distribution that exceeds the test statistic value from the real sample.
- *Randomization tests* and *permutation tests* use simulation to estimate the *p*-value and are especially useful in complex situations for which the exact *p*-value is not easy to calculate.

Exercises

Exercises with numbers divisible by 3 (3, 6, 9, etc.) are included in the Solutions at the back of the book. They are marked with an asterisk (*).

   For Exercises 1 to 4, assume that birthdays are equally likely to occur on all possible days in any given year, so there are no seasonal variations or day of the week variations.

1. Suppose you wanted to simulate the birthdays (month and day, not year) of the five children in one family. For each child, you tell the computer to choose a number from 1 to 366. This covers all possibilities including February 29.

   a. Would you tell the computer to make all 366 choices equally likely? Explain.

   b. How would you assign the results to birthdays? For example, what number would correspond with a birthday of December 31?

   c. Would it make sense to tell the computer to allow the same number to be chosen twice or not to allow that? Explain.

2. Suppose you wanted to simulate the birthdays (month and day, not year) of three children in one family by first choosing a month and then choosing a day. Assume that none of them were born in a leap year.

   a. What range of numbers would you tell the computer to use to simulate the month? Would you tell it to make all of those choices equally likely? Explain.

b. What range of numbers would you tell the computer to use to simulate the day? Would you tell it to make all of those choices equally likely? Explain.

c. In each case (month and day), would it make sense to tell the computer to allow the same number to be chosen twice, or not to allow that? Explain.

*3. Suppose you wanted to simulate the birthdays (month and day, not year) of three children in one family all of whom were born in years that were not leap years. All you have available is a computer website (like www.randomizer.org) that weights all of the choices in the range you provide equally. How could you use the website to carry out this simulation?

4. Suppose you have two siblings, and all three of you were born in the same month of the year. Explain how you could use simulation to estimate the probability that in a family of three children they are all born in the same month. You do not have to provide exact details about how you would actually simulate each birthday.

5. As a promotion, a cereal brand is offering a prize in each box, and there are four possible prizes. You would like to collect all four prizes, but you only plan to buy six boxes of the cereal before the promotion ends. Assume you have a random number generator that weights all numbers equally in a range you provide.

a. Explain how you could simulate the prizes found in one set of six boxes of cereal.

b. Explain how you could use simulation to estimate the probability of obtaining all four prizes in six boxes of cereal. Assume that all four prizes are equally likely for any given box, and the choice of prize is independent from one box to the next.

*6. Suppose that 55% of the voters in a large city support a particular candidate for mayor and 45% do not support the candidate. A poll of 100 voters will be conducted, and the proportion of them who support the candidate will be found. Assume you have a random number generator that weights all numbers equally in a range you provide. It will not allow you to provide unequal weights.

*a. Explain how you could use the random number generator to simulate the response (support or oppose) for one randomly chosen voter.

*b. Explain how you could simulate the 100 responses (support or oppose) for one poll.

*c. Explain how you could use simulation to estimate the probability that the actual poll will result in less than half (49 or fewer) of those polled supporting the candidate.

7. Suppose that there are 38 students in your class. Each day the teacher randomly selects one student to show the others how to do the previous night's homework. Each day, all students are eligible and equally likely to be chosen, even if they have been chosen that week already.

a. On any given day, what is the probability that you are the student selected?

b. What is the probability that you are selected on two consecutive days? (Hint: Use Rule 3 from Chapter 14.)

c. What is the probability that the same student is chosen on two consecutive days?

d. Number the students from 1 to 38. Assume you have a random number generator that weights all numbers equally in a range you provide. Explain how you could use simulation to identify the students who are chosen on two consecutive days. (It could be the same student on both days.)

e. Explain how you could use simulation to estimate the probability that the same student is chosen on two consecutive days.

Exercises 8 to 12: Refer to the box on page 331, describing the website www.randomizer.org. Parts (a) to (e) in each of these exercises list the questions asked when you use that website. In each case, specify how you would answer those questions. (See instructions on page 339.)

8. In Exercise 5, the following scenario was presented: As a promotion, a cereal brand is offering a prize in each box and there are four possible prizes. You would like to collect all four prizes, but you only plan to buy six boxes of the cereal before the promotion ends. For this exercise, you want to carry out 1000 simulations of the prizes (#1, 2, 3, or 4) you will find in six boxes of cereal. The proportion of simulations for which all four prizes appear is an estimate of the probability of getting all four prizes in the six boxes you plan to purchase. Specify what you would answer for each of the following questions to accomplish this simulation using www.randomizer.org.

   a. How many sets of numbers do you want to generate?

   b. How many numbers per set?

   c. Number range (e.g., 1–50).

   d. Do you wish each number in a set to remain unique?

   e. Do you wish to sort the numbers that are generated?

*9. In Exercise 6, the following scenario was presented. Suppose that 55% of the voters in a large city support a particular candidate for mayor and 45% do not support the candidate. A poll of 100 voters will be conducted, and the proportion of them who support the candidate will be found. For this exercise, you want to simulate 1000 polls of 100 people each, and find the proportion of those polls for which 49 or fewer support the candidate. That will provide an estimate of the probability that the poll shows less than majority support, even though in truth 55% of the population supports the candidate. For parts (a) to (e), specify what you would answer for each of the questions to accomplish this simulation using www.randomizer.org. Remember that the simulator will not allow you to specify different weights for the outcomes, so you need to figure out how to incorporate the 0.55 and 0.45 probabilities into your simulation.

   *a. How many sets of numbers do you want to generate?

   *b. How many numbers per set?

   *c. Number range (e.g., 1–50).

   *d. Do you wish each number in a set to remain unique?

   *e. Do you wish to sort the numbers that are generated?

   *f. Once you had the results of the simulation, how would you use them to estimate the desired probability?

10. The scenario in this exercise is similar to the one in Exercise 7, except the class size is smaller. Suppose that there are 15 students in your class. Each day the teacher randomly selects one student to show the others how to do the previous night's homework. Each day, all students are eligible and equally likely to be chosen, even if they have been chosen that week already. For this exercise, you want to simulate three students being drawn, mimicking the Monday, Wednesday, Friday class schedule for this class. You want to estimate the probability that the same student will be chosen all three times in one week. You plan to simulate results for 10,000 three-day weeks. For parts (a) to (e), specify what you would answer for each of the questions to accomplish this simulation using www.randomizer.org.

   a. How many sets of numbers do you want to generate?

   b. How many numbers per set?

   c. Number range (e.g., 1–50).

   d. Do you wish each number in a set to remain unique?

   e. Do you wish to sort the numbers that are generated?

   f. Suppose that the same student is chosen on all 3 days in 47 of your simulated weeks. What is your estimate of the probability that the same student will be chosen all three times in one week?

g. Use the probability rules from Chapter 14 to find the probability that the same student would be chosen all three times in one week, and compare your answer to the estimated probability in part (f).

11. An airline serves lunch in the first-class cabin. Customers are given a choice of either a sandwich or chicken salad. There are 12 customers in first-class, and the airline loads seven of each meal onboard. You would like to simulate the probability that there will not be enough of one or the other meal to meet the preferences of the customers. Assume the probability is 0.6 that a customer requests the sandwich and 0.4 that he or she requests the salad and is independent from one person to the next. You plan to simulate the choices for 100 sets of 12 customers. For parts (a) to (e), specify what you would answer for each of the questions to accomplish this simulation using www.randomizer.org. Remember that the simulator will not allow you to specify different weights for the outcomes, so you need to figure out how to incorporate the 0.4 and 0.6 probabilities into your simulation.

a. How many sets of numbers do you want to generate?

b. How many numbers per set?

c. Number range (e.g., 1–50).

d. Do you wish each number in a set to remain unique?

e. Do you wish to sort the numbers that are generated?

f. Explain how you would use your results to estimate the desired probability.

*12. You are competing in a swimming race that has eight contestants starting at the same time, one per lane. Two of your friends are in the race as well. The eight lanes are assigned to the eight swimmers at random, but you hope that you and your two friends will be in three adjacent lanes. You plan to simulate 100 races to estimate the probability that you and your friends will be assigned to three adjacent lanes. For parts (a) to (e), specify what you

would answer for each of the questions to accomplish this simulation using www.randomizer.org.

***a.** How many sets of numbers do you want to generate?

***b.** How many numbers per set?

***c.** Number range (e.g., 1–50).

***d.** Do you wish each number in a set to remain unique?

***e.** Do you wish to sort the numbers that are generated?

***f.** Once you had the results of the simulation, how would you use them to estimate the desired probability?

13. An intersection has a four-way stop sign but no traffic light. Currently, about 1200 cars use the intersection a day, and the rate of accidents at the intersection is about one every two weeks. The potential benefit of adding a traffic light was studied using a computer simulation by modeling traffic flow at the intersection if a light were to be installed. The simulation included 100,000 repetitions of 1200 cars using the intersection to mimic 100,000 days of use. Of the 100,000 simulations, an accident occurred in 5230 of them, and no accident occurred in the rest. (There were no simulated days with two or more accidents.)

a. Based on the results of the simulation, what is the estimated number of accidents in a 2-week period if the traffic light were to be installed?

b. Does the simulation show that adding the traffic light would be a good idea? Explain.

14. Suppose that the chi-square statistic for a chi-square test on a table with 2 rows and 2 columns was computed to be 5.3. A simulation was run with 10,000 simulated samples, and 217 of them resulted in chi-square statistics of 5.3 or larger. What is the estimated p-value for the test?

*15. Suppose that the chi-square statistic for a chi-square test on a table with 2 rows and 2 columns was computed to be 2.90. A simulation was run with 1000 simulated samples, and 918 of them resulted in chi-square statistics of *less than* 2.90. What is the estimated p-value for the test?

16. Remember that the *p*-value corresponding to a chi-square statistic of 3.84 for a table with 2 rows and 2 columns is 0.05. If you were to simulate 10,000 samples for a 2×2 table under the assumption that the null hypothesis is true, about how many of them would you expect to result in a chi-square statistics of *less than* 3.84?

17. Suppose that a randomization distribution resulting from the simulation of a chi-square test had 7% of the values at or above the chi-square statistic observed for the real sample.

    a. What is the estimated *p*-value for the test?

    b. What decision would you make for the test, using a level of 0.05?

    c. What decision would you make for the test, using a level of 0.10?

*18. Suppose a teacher observed a correlation of 0.38 between age and number of words children could define on a vocabulary test, for a group of nine children aged 7 to 15. The teacher wanted to confirm that there was a correlation between age and vocabulary test scores for the population of children in this age range. She performed a simulation with 1000 randomized orders, scrambling the test scores across the ages, similar to the simulation used in Example 15.3. She was surprised to find that 29% of the simulated samples resulted in correlations of 0.38 or larger, even though there should not have been any correlation between ages and the scrambled test scores.

    *a. What null and alternative hypotheses was the teacher testing?

    *b. What is the estimated *p*-value for her test?

    *c. Do these results confirm that there is no correlation between age and vocabulary scores for the population of children aged 7 to 15? Explain. (Hint: Remember the legitimate conclusions for testing hypotheses.)

    *d. Would a simulation with 10,000 randomized orders be much more likely, much less likely, or about equally likely to enable the teacher to reject the null hypothesis, compared to the simulation performed with 1000 randomized orders?

*e. The teacher plans to repeat the experiment. What could she do differently to improve the chance of rejecting the null hypothesis?

19. Suppose that a chi-square test is carried out on a table with 2 rows and 2 columns, with a total of 100 individuals in the four cells. The computed chi-square statistic is 2.17, and a simulation of 1000 samples finds that the estimated *p*-value is 0.15. The researcher is almost sure that there really is a relationship between the two variables in the population and is disappointed that the results of the test did not provide evidence to reject the null hypothesis using a level of 0.05. The researcher plans to redo the study choosing a new sample of individuals. Explain whether each of the following actions would increase the likelihood that the null hypothesis would be rejected, decrease the likelihood, or leave the likelihood about the same.

    a. Use 10,000 simulated samples instead of only 1000.

    b. Collect a sample of 500 individuals instead of only 100 individuals.

    c. Use a test with level 0.01 instead of level 0.05.

20. In performing a randomization test, explain why the samples are simulated by using the assumption that the null hypothesis is true.

*21. Would probabilities estimated using simulation be considered to be relative-frequency probabilities or personal probabilities? Explain.

22. Using the information in Case Study 15.1 as an example, explain why it is a statistical mistake to formulate hypotheses based on looking at the data you will use to test them.

23. Three males and three females are given 5 minutes to memorize a list of 25 words, and then asked to recall as many of them as possible. The three males recalled 10, 12, and 14 of the words, for an average of 12 words; the three females recalled 11, 14, and 17 of the words, for an average of 14 words.

    a. In comparing males' and females' ability for memorization, what would be reasonable null and alternative hypotheses to test in this situation?

b. The difference in means for the two groups in this sample was two words. Using that as the "test statistic," explain how you could carry out a permutation test in this situation.

c. Give one example of a permutation from the method you explained in part (b), and what the "test statistic" would be for that permutation.

Mini Projects

1. Use a computer, calculator, or website to simulate one of the situations in Exercises 8 to 12. Use at least 100 repetitions. Use your simulation results to estimate the desired probability.

2. Find a study that was done using simulation to estimate a *p*-value, probability, or random outcome, similar to the earthquake simulation study described in Case Study 15.1. Or, find a simulation website that allows you to conduct your own simulation study based on expert opinion or existing data. (Putting terms such as "simulation economics" or "simulation weather" in an internet search engine may help you find an appropriate source.) Explain what was done in the simulation, and then present and interpret the results.

3. Find out how to do a permutation test for the difference in means for independent samples, and carry out the test using the data in Exercise 23. State the hypotheses you are testing and the results of the test. Explain how the test was done.

4. Simulate the situation described in Thought Questions 3, 4, and 5. Assume that the probability of a boy for each birth is 0.51 and the probability of a girl is 0.49. Simulate at least 100 families with four children. Find and report the proportion of your simulated families for which there are no boys. Using the probability rules from Chapter 14, calculate the exact probability that a family with four children will not have any boys, assuming that the probability of a boy is 0.51 for each birth. Compare the result with the results of your simulation.

References

Bufe, C.G. and D.M. Perkins. (2005). Evidence for a global seismic-moment release sequence. *Bulletin of the Seismological Society of America*, 95, pp. 833–843.

Howell, D. (2013). R Code for simulating chi-square tests, http://www.uvm.edu/~dhowell/StatPages/Chi-Square-Folder/R-Code.html, accessed June 27, 2013.

Kerr, R.A. (2011). More megaquakes on the way? That depends on your statistics. *Science*, 332, 22 April 2011, p. 411.

Shearer, P.M and P.B. Stark. (2011). Global risk of big earthquakes has not recently increased. *Proceedings of the National Academy of Sciences*, 109(3), pp. 717–721.

Urbaniak, G. C. and S. Plous. (2013). Research Randomizer (Version 4.0) [Computer software]. Retrieved on June 28, 2013, from http://www.randomizer.org/.

CHAPTER **16**

Psychological Influences on Personal Probability

Thought Questions

1. During the Cold War, Plous (1993) presented readers with the following test.

   Place a check mark beside the alternative that seems most likely to occur within the next 10 years:

   - An all-out nuclear war between the United States and Russia.
   - An all-out nuclear war between the United States and Russia in which neither country intends to use nuclear weapons, but both sides are drawn into the conflict by the actions of a country such as Iraq, Libya, Israel, or Pakistan.

   Using your intuition, pick the more likely event at that time. Now consider the probability rules discussed in Chapter 14 to try to determine which statement was more likely.

2. Which is a more likely cause of death in the United States, homicide or septicemia? How did you arrive at your answer?

3. Do you think people are more likely to pay to reduce their risk of an undesirable event from 95% to 90% or to reduce it from 5% to zero? Explain whether there should be a preferred choice, based on the material from Chapter 14.

4. A fraternity consists of 30% freshmen and sophomores and 70% juniors and seniors. Bill is a member of the fraternity, he studies hard, he is well liked by his fellow fraternity members, and he will probably be quite successful when he graduates. Is there any way to tell if Bill is more likely to be a lower classman (freshman or sophomore) or an upper classman (junior or senior)?

16.1   Revisiting Personal Probability

In Chapter 14, we assumed that the probabilities of various outcomes were known or could be calculated using the relative-frequency interpretation of probability. But most decisions people make are in situations that require the use of personal probabilities. The situations are not repeatable, nor are there physical assumptions that can be used to calculate potential relative frequencies.

Personal probabilities, remember, are values assigned by individuals based on how likely they think events are to occur. By their very definition, personal probabilities do not have a single correct value. However, they should still follow the rules of probability, which we outlined in Chapter 14; otherwise, decisions based on them can be contradictory. For example, if you believe there is a very high probability that you will be married and have children before age 25, but also believe there is a very high probability that you will travel the world for many years as a single person, then you have not assessed these probabilities in a coherent manner. Your two personal probabilities are not consistent with each other and will lead to contradictory decisions.

In this chapter, we explore research that has shown how personal probabilities can be influenced by psychological factors in ways that lead to incoherent or inconsistent probability assignments. We also examine circumstances in which many people assign personal probabilities that can be shown to be incorrect based on the relative-frequency interpretation. Every day, you are required to make decisions that involve risks and rewards. Understanding the kinds of influences that can affect your decisions adversely should help you make more realistic judgments.

16.2   Equivalent Probabilities; Different Decisions

People like a sure thing. It would be wonderful if we could be guaranteed that cancer would never strike us, for instance, or that we would never be in an automobile accident. For this reason, people are willing to pay a premium to reduce their risk of something to zero, but are not as willing to pay to reduce their risk by the same amount to a nonzero value.

The same effect exists for a positive outcome. People are more willing to pay to increase their probability of a favorable outcome from 95% to 100% than to increase it by that same amount with lesser probabilities, for instance from 40% to 45%. In a classic paper outlining their famous *Prospect Theory*, Kahneman and Tversky (1979) named this the **certainty effect**.

Similarly, in his book *Thinking, Fast and Slow*, Kahneman (2011) explains the **possibility effect**, which describes the impact of increasing a favorable outcome from 0% to a small probability, say 5%. People are willing to pay a higher premium for that 5% increase than they are to increase the probability by that same amount from a moderate probability to a slightly larger probability, say 45% to 50%. Let's consider some examples of these psychological phenomena.

The Certainty Effect and the Possibility Effect

Suppose you are buying a new car. The salesperson explains that you can purchase an optional safety feature for $500 that will reduce your chances of death in a high-speed accident from 50% to 45%. Would you be willing to purchase the device?

Now suppose instead that the salesperson explains that you can purchase an optional safety feature for $500 that will reduce your chances of death in a high-speed accident from 5% to zero. Would you be willing to purchase the device?

In both cases, your chances of death are reduced by 5%, or 1/20. But research has shown that people are more willing to pay to reduce their risk from a fixed amount down to zero than they are to reduce their risk by the same amount when it is not reduced to zero. This is an example of the certainty effect. Let's look at another example that does not involve monetary values.

EXAMPLE 16.1   Double the Chances, One Third the Fun

As an illustration of the certainty effect, Kahneman and Tversky (1979) presented the following choices to 72 college students in the United States:

> *Choice A:* A 50% chance to win a three-week tour of England, France, and Italy
> *Choice B:* A one-week tour of England, with certainty

Which would you choose? If you are like the students in this study, you would be more likely to choose the certainty of the one-week trip, which was chosen by 78% of them. Students were then presented with these two choices:

> *Choice C:* A 5% chance to win a three-week tour of England, France, and Italy
> *Choice D:* A 10% chance to win a one-week tour of England

This time, 2/3 (67%) of the students favored the lower probability three-week tour. In both cases, the one-week tour of England had twice the probability of occurring as the three-week tour of multiple countries. Yet students preferred the longer, presumably more rewarding, trip that had a lower probability of happening when both trips were a gamble. When the shorter, less rewarding trip was a sure thing, they generally opted for that one instead. ∎

In his book *Thinking, Fast and Slow*, Kahneman (2011, p. 317) illustrates the four situations generated by imagining gains and losses combined with the certainty effect and the possibility effect. He labels them "the fourfold pattern." Table 16.1 presents the four scenarios, the likely reaction to them, and an example of how they play out in daily life. Notice that in all four scenarios presented in the table, the expected value of the monetary amount is actually lower for the preferred version than for the other version. For example, in the upper right, most people would prefer to win $9000 with certainty than to take a gamble on winning $10,000 with probability 0.95, even though the expected value in the latter case is higher, at $(.95)($10,000)$ = $9500. In the lower left, most people would prefer to pay $11 than to risk losing $10,000, even with a low probability of 1/1000. But the expected value for the latter choice is only $(1/1000)($10,000)$ = $10, which is lower than the sure loss of $11. That explains why people buy insurance and extended warranties.

TABLE 16.1	Kahneman's Fourfold Pattern with Real Life Situations: Preferences in Italics	
	Gain	**Lose**
Certainty Effect: Increase probability from 95% to 100%	95% chance to gain $10,000 *100% chance to gain $9,000*	*95% chance to lose $10,000* 100% chance to lose $9,000
Explains why people	Accept settlement in court case even with high chance to win	Reject settlement in court case even with high chance to lose
Possibility Effect: Increase probability from 0 to 1/1000	*1/1000 chance to gain $10,000* 100% chance to gain $11	1/1000 chance to lose $10,000 *100% chance to lose $11*
Explains why people	Buy lottery tickets	Buy insurance and warranties

The Pseudocertainty Effect

An idea related to the certainty effect, often, used in marketing, is that of the **pseudocertainty effect** (Slovic, Fischhoff, and Lichtenstein, 1982, p. 480). Rather than being offered a reduced risk on a variety of problems, you are offered a complete reduction of risk on certain problems and no reduction on others.

As an example, consider the extended warranty plans offered on automobiles and appliances. You buy the warranty when you purchase the item and certain problems are covered completely for a number of years. Other problems are not covered at all. If you were offered a plan that covered all problems with 30% probability, you would probably not purchase it. But if you were offered a plan that completely covered 30% of the possible problems, you might consider it. Both plans have the same expected value over the long run, but most people prefer the plan that covers some problems with certainty.

EXAMPLE 16.2 Vaccination Questionnaires

To test the idea that the pseudocertainty effect influences decision making, Slovic and colleagues (1982) administered two different forms of a "vaccination questionnaire." The first form described what the authors called "probabilistic protection," in which a vaccine was available for a disease anticipated to afflict 20% of the population. However, the vaccine would protect people with only 50% probability. Respondents were asked if they would volunteer to receive the vaccine, and only 40% indicated that they would.

The second form described a situation of "pseudocertainty," in which there were two equally likely strains of the disease, each anticipated to afflict 10% of the population. The available vaccine was completely effective against one strain but provided no protection at all against the other strain. This time, 57% of respondents indicated they would volunteer to receive the vaccine.

In both cases, receiving the vaccine would reduce the risk of disease from 20% to 10%. However, the scenario in which there was complete elimination of risk for a subset of problems was perceived much more favorably than the one for which there was the same reduction of risk overall. This is what the pseudocertainty effect predicts. ■

Plous (1993, p. 101) notes that an effect similar to pseudocertainty is found in marketing when some items are given away free rather than having the price reduced on all items. For example, rather than reduce all items by 50%, a merchandiser may instead advertise that you can "buy one, get one free." The overall reduction is the same, but the offer of free merchandise may be perceived as more desirable than the discount.

16.3   How Personal Probabilities Can Be Distorted

In 2002, Daniel Kahneman received the Nobel Prize in Economic Sciences for decades of creative work done with his collaborator Amos Tversky. (Tversky would have shared the honor, but he passed away in 1996.) One of the major focuses of the Kahneman and Tversky work was the exploration of psychological "heuristics" that influence our ability to make good decisions. Many of these heuristics involve our ability to accurately assess probability and risk, and learning about them may help you make better decisions when uncertainty is involved.

Kahneman (2011, p. 98) defines a **heuristic** as "a simple procedure that helps find adequate, though often imperfect, answers to difficult questions." One of the features of the heuristics studied by Kahneman and Tversky is that they operate by replacing the question you are asked with a different, simpler question that is easier to answer. Unfortunately, when the answer to a question involves the formulation of a personal probability, answering the wrong question generally leads to a distorted probability. Let's look at how some of these heuristics can lead you astray.

The Availability Heuristic

Which do you think caused more deaths in the United States in 2010, homicide or septicemia? If you are like respondents to a survey conducted by the author of this book in an introductory statistics class at the University of California, Irvine, you answered that it was homicide. Only 39% (96/248) of the students correctly identified the answer as septicemia, which accounted for more than twice as many deaths that year as did homicides. According to the National Center for Health Statistics, in 2010, there were 34,843 deaths from septicemia and only 16,065 from homicide.

The survey mimics studies reported by Slovic and colleagues (1982), in which they asked students to compare two risks, such as causes of death or types of accidents. Invariably, they found that students misjudged the comparative risks. For instance, at that time (and probably continuing to today), strokes caused almost twice as many deaths as accidents from all causes, yet about 80% of respondents judged accidents to be a more likely cause of death.

The distorted view that homicide or accidents are more common than other causes of death results from the fact that those events receive more attention in the media. Psychologists attribute this incorrect perception to the **availability heuristic**. Kahneman explains that "the availability heuristic, like other heuristics of judgment, substitutes one question for another: you wish to estimate the size of a category or the frequency of an event, but [instead] you report an impression of the ease with which instances come to mind" (Kahneman, 2011, p. 130). Because homicides and accidents get media attention, instances of them readily come to mind. In contrast, deaths from septicemia or stroke are not likely to receive any media attention and thus, unless we personally know someone who recently died of those causes, we are likely to underestimate the probability of their occurrence.

Availability can cloud your judgment in numerous instances. For example, if you are buying a used car, you may be influenced more by the bad luck of a friend or relative who owned a particular model than by statistics provided by consumer groups based on the experience of thousands of owners. The memory of the one bad car in that class is readily available to you.

Similarly, most people know many smokers who don't have lung cancer. Fewer people know someone who has actually contracted lung cancer as a result of smoking. Therefore, it is easier to bring to mind the healthy smokers and, if you smoke, to believe that you too will continue to have good health.

Detailed Imagination

One way to encourage availability, in which the probabilities of risk can be distorted, is by having people vividly imagine an event. Salespeople use this trick when they try to sell you extended warranties or insurance. For example, they may convince you that $500 is a reasonable price to pay for an extended warranty on your new car by having you imagine that if your air conditioner fails, it will cost you more than the price of the policy to get it fixed. They don't mention that it is extremely unlikely that your air conditioner will fail during the period of the extended warranty.

Anchoring

Psychologists have shown that people's perceptions can also be severely distorted when they are provided with a reference point, or an **anchor,** from which they then adjust up or down. Most people tend to stay relatively close to the anchor, or initial value, provided.

EXAMPLE 16.3 ### Anchored Estimates of the Median Household Income of Canada

Two different versions of a survey were given to the author's students in an introductory statistics course at the University of California, Irvine. Each survey asked students to provide a guess about the median household income of Canada in 2008 (the latest year with available information at that time). The two groups of students were provided with a different lead-in to the question. The two versions were as follows:

Version 1: The median household income in Australia in 2008 (in U.S. dollars) was $44,820. What do you think was the median household income in Canada in 2008 (in U.S. dollars)?

Version 2: The median household income in New Zealand in 2008 (in U.S. dollars) was $23,122. What do you think was the median household income in Canada in 2008 (in U.S. dollars)?

Version 1 was administered to 78 students. Notice that in this version, an "anchor" of $44,820 was provided by reporting the median household income in Australia. The median response for Canada was $48,000. (The mean was similar, at $47,181.) Version 2 was administered to 60 students. In this version, an "anchor" of $23,122 was provided by reporting the median household income in New Zealand. The median response for Canada was $29,182. (The mean was $31,859.)

Although both groups of students were asked the same question, about Canada, the information provided about either Australia or New Zealand served as an anchor, from which students adjusted their answers. In fact, the correct answer for the median household income in Canada at that time was $51,951. Both groups correctly moved upward from their respective anchors, but not by much in either case.   ■

Research has shown that anchoring influences real-world decisions as well. For example, jurors who are first told about possible harsh verdicts and then about more lenient ones are more likely to give a harsh verdict than jurors given the choices in reverse order.

Anchoring is commonly used in marketing. For instance, at the grocery store you may notice that the sale price of an item is presented as "10 for $10.00." In fact the price is actually $1.00 per item, but the marketing is designed to have you use 10 as an anchor and decide how many of the item to purchase by moving down (or up) from there. Another version of anchoring in sales at the grocery store is to impose a limit or a ration on an item. For instance, a study at a grocery store in Iowa compared the sales of cans of soup when a sign was posted with the sales price reading "Limit of 12 per person" versus a sign reading "No limit per person." In the first case, customers bought an average of seven cans, but in the second case, they bought only about half as many (Wansink et al., 1998).

The next two examples illustrate two different applications of anchoring in everyday life.

EXAMPLE 16.4   Heuristics and Medical Diagnoses

Can a computer program be a better medical diagnostician than an experienced medical doctor? That was the question addressed by a report in *The New York Times* that noted how both anchoring and availability can affect the accurate diagnosis of health problems (Hafner, 2012). Two examples of diagnostic computer programs were discussed, and in both cases, part of the rationale for developing the programs was the realization that psychological heuristics can impact human diagnostic abilities.

In the first case, the program was created by Jason Maude and named for his daughter, Isabel. When Isabel was 3 years old she had the chicken pox. However, her doctors completely missed the fact that she had a second, more serious flesh-eating condition. By the time they realized the mistake, the disease had progressed so far that Isabel continued to have plastic surgery into her young adult years. As Mr. Maude described it, "[Isabel's] doctors were so stuck in what is called anchoring bias—in this case Isabel's simple chickenpox—they couldn't see beyond it" (Hafner, 2012, p. D6).

In the second example, Dr. Martin Kohn, chief medical scientist for I.B.M. Research, described why that company's "Watson for Healthcare" diagnostic program might be a valuable tool for physicians. Kohn noted, "For physicians, one problem is 'the law of availability.' You aren't going to put anything on a list that you don't think is relevant or didn't know to think of" (Hafner, 2012, p. D6).

In both of these cases, doctors are described as operating under the same influences on personal probabilities as the rest of us. Once they think they know what's wrong with a patient, they may become anchored on that diagnosis and ignore evidence that suggests something different or additional. And in making the original diagnosis, doctors are most likely to think of common health issues because of the availability heuristic. Thus, they are likely to miss rare conditions that might be picked up by a diagnostic computer program with thousands of possibilities cataloged in its memory. ∎

EXAMPLE 16.5   Sales Price of a House

Plous (1993) describes a study conducted by Northcraft and Neale (1987), in which real estate agents were asked to give a recommended selling price for a home. They were given a 10-page packet of information about the property and spent 20 minutes walking through it. Contained in the 10-page packet of information was a listing price. To test the effect of anchoring, four different listing prices, ranging from $119,900 to $149,900, were given to different groups of agents. The house had actually been appraised at $135,000. As the anchoring theory predicts, the agents were heavily influenced by the particular listing price they were given. The four listing prices and the corresponding mean recommended selling prices were

Apparent listed price	$119,900	$129,900	$139,900	$149,900
Mean recommended sales price	$117,745	$127,836	$128,530	$130,981

As you can see, the recommended sales price differed by more than $10,000 just because the agents were anchored at different listing prices. Yet, when asked how they made their judgments, very few of the agents mentioned the listing price as one of their top factors. ∎

Anchoring is most effective when the anchor is extreme in one direction or the other. It does not have to take the form of a numerical assessment either. Be wary when someone describes a worst- or best-case scenario and then asks you to make a decision. For example, an investment counselor may encourage you to invest in a certain commodity by describing how one year it had such incredible growth that you would now be rich if you had only been smart enough to invest in the commodity during that year. If you use that year as your anchor, you'll fail to see that, on average, the price of this commodity has risen no faster than inflation.

The Representativeness Heuristic and the Conjunction Fallacy

In some cases, the **representativeness heuristic** leads people to assign higher probabilities than are warranted to scenarios that are representative of how we imagine things would happen. For example, Tversky and Kahneman (1982a, p. 98) note that "the hypothesis 'the defendant left the scene of the crime' may appear less plausible than the hypothesis 'the defendant left the scene of the crime for fear of being accused of murder,' although the latter account is less probable than the former."

It is the representativeness heuristic that sometimes leads people to fall into the judgment trap called the **conjunction fallacy.** We learned in Chapter 14 (Rule 4) that the probability of two events occurring together, in *conjunction*, cannot be higher than the probability of either event occurring alone. The conjunction fallacy occurs when detailed scenarios involving the conjunction of events are given higher probability assessments than statements of one of the simple events alone.

EXAMPLE 16.6   ### An Active Bank Teller

A classic example, provided by Kahneman and Tversky (1982, p. 496), was a study in which they presented subjects with the following statement:

> *Linda is 31 years old, single, outspoken, and very bright. She majored in philosophy. As a student, she was deeply concerned with issues of discrimination and social justice, and also participated in antinuclear demonstrations.*

Respondents were then asked which of two statements was more probable:

1. Linda is a bank teller.
2. Linda is a bank teller who is active in the feminist movement.

Kahneman and Tversky report that "in a large sample of statistically naive undergraduates, 86% judged the second statement to be more probable" (1982, p. 496).

The problem with that judgment is that the group of people in the world who fit the second statement is a subset of the group who fit the first statement. If Linda falls into the second group (bank tellers who are active in the feminist movement), she must also fall into the first group (bank tellers). Therefore, the *first* statement must have a higher probability of being true.

The misjudgment is based on the fact that the second statement is much more representative of how Linda was described. This example illustrates that intuitive judgments can directly contradict the known laws of probability. In this example, it was easy for respondents to fall into the trap of the conjunction fallacy. ■

The representativeness heuristic can be used to affect your judgment by giving you detailed scenarios about how an event is likely to happen. For example, during the Cold War era between the United States and Russia, Plous (1993, p. 4) asked readers of his book to:

Place a check mark beside the alternative that seems most likely to occur within the next 10 years:

- An all-out nuclear war between the United States and Russia.
- An all-out nuclear war between the United States and Russia in which neither country intends to use nuclear weapons, but both sides are drawn into the conflict by the actions of a country such as Iraq, Libya, Israel, or Pakistan.

Notice that the second alternative describes a subset of the first alternative, and thus the first one must be at least as likely as the second. Yet, according to the representativeness heuristic, most people would see the second alternative as more likely. In fact, in a survey of 138 students at the University of California, Irvine, 68% of the students selected the second option when asked which of these they thought would have been more likely to happen during the Cold War.

Be wary when someone describes a scenario to you in great detail in order to try to convince you of its likelihood. For example, lawyers know that jurors are much more likely to believe a person is guilty if they are provided with a detailed scenario of how the person's guilt could have occurred.

Forgotten Base Rates

The representativeness heuristic can lead people to ignore information they may have about the likelihood of various outcomes. For example, Kahneman and Tversky (1973) conducted a study in which they told subjects that a population consisted of 30 engineers and 70 lawyers. The subjects were first asked to assess the likelihood that a randomly selected individual would be an engineer. The average response was indeed close to the correct 30%.

Subjects were then given the following description, written to give no clues as to whether this individual was an engineer or a lawyer.

> *Dick is a 30-year-old man. He is married with no children. A man of high ability and high motivation, he promises to be quite successful in his field. He is well liked by his colleagues.* (Kahneman and Tversky, 1973, p. 243)

This time, the subjects ignored the base rate. When asked to assess the likelihood that Dick was an engineer, the median response was 50%. Because the individual in question did not appear to represent either group more heavily, the respondents concluded that there must be an equally likely chance that he was either. They ignored the information that only 30% of the population were engineers.

Neglecting base rates can cloud the probability assessments of experts as well. For example, physicians who are confronted with a patient's positive test results for a rare disease routinely overestimate the probability that the patient actually has the disease. They fail to take into account the extremely low base rate in the population. We will examine this phenomenon in more detail in Chapter 17.

16.4 Optimism, Reluctance to Change, and Overconfidence

Psychologists have also found that some people tend to have personal probabilities that are unrealistically optimistic. Further, people are often overconfident about how likely they are to be right and are reluctant to change their views even when presented with contradictory evidence.

Optimism

Slovic and colleagues (1982) cite evidence showing that most people view themselves as personally more immune to risks than other people. They note that "the great majority of individuals believe themselves to be better than average drivers, more likely to live past 80, less likely than average to be harmed by the products they use, and so on" (pp. 469–470).

EXAMPLE 16.7   ### Optimistic College Students

Research on college students confirms that they see themselves as more likely than average to encounter positive life events and less likely to encounter negative ones. Weinstein (1980) asked students at Cook College (part of Rutgers, the state university in New Jersey) to rate how likely certain life events were to happen to them compared to other Cook students of the same sex. Plous (1993) summarizes Weinstein's findings:

> *On the average, students rated themselves as 15 percent more likely than others to experience positive events and 20 percent less likely to experience negative events. To take some extreme examples, they rated themselves as 42 percent more likely to receive a good starting salary after graduation, 44 percent more likely to own their own home, 58 percent less likely to develop a drinking problem, and 38 percent less likely to have a heart attack before the age of 40 (p. 135).*

Notice that if all the respondents were accurate, the median response for each question should have been 0 percent more or less likely because approximately half of the students should be more likely and half less likely than average to experience any event.   ■

The tendency to underestimate one's probability of negative life events can lead to foolish risk taking. Examples are driving while intoxicated and having unprotected sex. Plous (1993, p. 134) calls this phenomenon, "It'll never happen to me," whereas Slovic and colleagues (1982, p. 468) title it, "It won't happen to me." The point is clear: If *everyone* underestimates his or her own personal risk of injury, someone has to be wrong . . . it *will* happen to someone.

Reluctance to Change

In addition to optimism, most people are also guilty of **conservatism.** As Plous (1993) notes, "Conservatism is the tendency to change previous probability estimates more slowly than warranted by new data" (p. 138). This explains the reluctance of the scientific community to accept new paradigms or to examine compelling evidence for phenomena such as extrasensory perception. As noted by Hayward (1984):

> *There seems to be a strong need on the part of conventional science to exclude such phenomena from consideration as legitimate observation. Kuhn and Feyerabend showed that it is always the case with "normal" or conventional science that observations not confirming the current belief system are ignored or dismissed. The colleagues of Galileo who refused to look through his telescope because they "knew" what the moon looked like are an example (pp. 78–79).*

This reluctance to change one's personal-probability assessment or belief based on new evidence is not restricted to scientists. It is notable there only because science is supposed to be "objective."

Overconfidence

Consistent with the reluctance to change personal probabilities in the face of new data is the tendency for people to place too much confidence in their own assessments. In other words, when people venture a guess about something for which they are uncertain, they tend to overestimate the probability that they are correct.

EXAMPLE 16.8   How Accurate Are You?

Fischhoff, Slovic, and Lichtenstein (1977) conducted a study to see how accurate assessments were when people were sure they were correct. They asked people to answer hundreds of questions on general knowledge, such as whether *Time* or *Playboy* had a larger circulation or whether absinthe is a liqueur or a precious stone. They also asked people to rate the odds that they were correct, from 1:1 (50% probability) to 1,000,000:1 (virtually certain).

The researchers found that the more confident the respondents were, the more the true proportion of correct answers deviated from the odds given by the respondents. For example, of those questions for which the respondents gave even odds of being correct (50% probability), 53% of the answers were correct. However, of those questions for which they gave odds of 100:1 (99% probability) of being correct, only 80% of the responses were actually correct.

Researchers have found a way to help eliminate overconfidence. As Plous (1993) notes, "The most effective way to improve calibration seems to be very simple: *Stop to consider reasons why your judgment might be wrong*" (p. 228). In a study by Koriat, Lichtenstein, and Fischhoff (1980), respondents were asked to list reasons to support and to oppose their initial judgments. The authors found that when subjects were asked to list reasons to oppose their initial judgment, their probabilities became extremely well calibrated. In other words, respondents were much better able to judge how much confidence they should have in their answers when they considered reasons why they might be wrong.

16.5   Calibrating Personal Probabilities of Experts

Professionals who need to help others make decisions often use personal probabilities themselves, and their personal probabilities are sometimes subject to the same distortions discussed in this chapter. For example, your doctor may observe your symptoms and give you an assessment of the likelihood that you have a certain disease—but fail to take into account the baseline rate for the disease.

Weather forecasters routinely use personal probabilities to deliver their predictions of tomorrow's weather. They attach a number to the likelihood that it will rain in a certain area, for example. Those numbers are a composite of information about what has happened in similar circumstances in the past and the forecaster's knowledge of meteorology.

Using Relative Frequency to Check Personal Probabilities

As consumers, we would like to know how accurate the probabilities delivered by physicians, weather forecasters, and similar professionals are likely to be. To discuss what we mean by accuracy, we need to revert to the relative-frequency interpretation of probability. For example, if we routinely listen to the same professional weather forecaster, we could check his or her predictions using the relative-frequency measure. Each evening, we could record the forecaster's probability of rain for the next day, and then the next day we could record whether it actually rained.

For a *perfectly calibrated* forecaster, of the many times he or she gave a 30% chance of rain, it would actually rain 30% of the time. Of the many times the forecaster gave a 90% chance of rain, it would rain 90% of the time, and so on. We can assert that personal probabilities are *well calibrated* if they come close to meeting this standard.

Notice that we can assess whether probabilities are well calibrated only if we have enough repetitions of the event to apply the relative-frequency definition. For instance, we will never be able to ascertain whether the late Carl Sagan was well calibrated when he made the assessment we saw in Section 14.3 that "the probability that the Earth will be hit by a civilization-threatening small world in the next century is a little less than one in a thousand." This event is obviously not one that will be repeated numerous times.

| CASE STUDY 16.1 | **Calibrating Weather Forecasters and Physicians** |

Studies have been conducted of how well calibrated various professionals are. Figure 16.1 displays the results of two such studies, one for weather forecasters and one for physicians. The open circles indicate actual relative frequencies of rain, plotted against various forecast probabilities. The dark circles indicate the relative frequency with which a patient actually had pneumonia versus his or her physician's personal probability that the patient had it.

The plot indicates that the weather forecasters were generally quite accurate but that, at least for the data presented here, the physicians were not. The weather forecasters were slightly off at the very high end, when they predicted rain with almost certainty. For example, of the times they were sure it was going to rain, and gave a probability of 1 (or 100%), it rained only about 91% of the time. Still, the weather forecasters were well calibrated enough that you could use their assessments to make reliable decisions about how to plan tomorrow's events.

The physicians were not at all well calibrated. The actual probability of pneumonia rose only slightly and remained under 15% even when physicians placed it almost as high as 90%. As we will see in an example in Section 17.4, physicians tend to overestimate the probability of disease, especially when the baseline risk is low. When your physician quotes a probability to you, you should ask if it is a personal probability or one based on data from many individuals in your circumstances. ■

Figure 16.1

Calibrating weather forecasters and physicians

Source: Plous, 1993, p. 223, using data from Murphy and Winkler (1984) for the weather forecasters and Christensen-Szalanski and Bushyhead (1981) for the physicians.

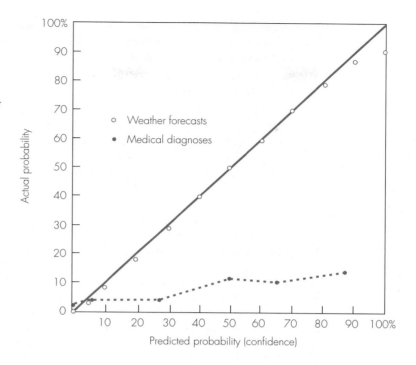

16.6   Tips for Improving Your Personal Probabilities and Judgments

The research summarized in this chapter suggests methods for improving your own decision making when uncertainty and risks are involved. Here are some tips to consider when making judgments:

1. Think of the big picture, including risks and rewards that are not presented to you. For example, when comparing insurance policies, be sure to compare coverage as well as cost.

2. When considering how a decision changes your risk, try to find out what the baseline risk is to begin with. Try to determine risks on an equal scale, such as the drop in *number* of deaths per 100,000 people rather than the *percent* drop in death rate.

3. Don't be fooled by highly detailed scenarios. Remember that excess detail actually decreases the probability that something is true, yet the representativeness heuristic leads people to increase their personal probability that it is true.

4. Remember to list reasons why your judgment might be wrong, to provide a more realistic confidence assessment.

5. Do not fall into the trap of thinking that bad things only happen to other people. Try to be realistic in assessing your own individual risks, and make decisions accordingly.

6. Be aware that the techniques discussed in this chapter are often used in marketing. For example, watch out for the anchoring effect when someone tries to anchor your personal assessment to an unrealistically high or low value.

7. If possible, break events into pieces and try to assess probabilities using the information in Chapter 14 and from publicly available sources. For example, Slovic and colleagues (1982, p. 480) note that because the risk of a disabling injury on any particular auto trip is only about 1 in 100,000, the need to wear a seat belt on a specific trip would seem to be small. However, using the techniques described in Chapter 14, they calculated that over a lifetime of riding in an automobile the risk is about .33. It thus becomes much more reasonable to wear a seat belt at all times.

Thinking About Key Concepts

- Psychologists have identified a number of ways in which our judgment about probability can be distorted. Becoming aware of them can help you make better decisions in uncertain situations.
- The *certainty effect* refers to the tendency to give more value to a fixed amount of change in probability if that change results in 100% assurance of a good thing happening or 100% assurance of a bad thing not happening.
- The *possibility effect* refers to the tendency to give more value to a small change in probability when it increases the probability of a good outcome from 0 to a small non-zero amount.
- The *pseudocertainty effect* says that people will pay more to reduce some of possible risks to zero and not reduce others at all, rather than reducing all risks by some amount that results in the same overall reduction.
- The *availability heuristic* distorts probability estimates by tying them to how readily situations can be brought to mind.
- *Anchoring* refers to the fact that judgments can be distorted when a reference point or *anchor* is provided. Subsequent judgments tend to stay close to the anchor.
- The *representativeness heuristic* distorts probability estimates by tying them to how well a scenario matches what we imagine could happen. It leads to the *conjunction fallacy* in which people assign higher probabilities to a combination of events than to one of the events alone.
- In assessing probabilities for situations that do not start out with equal base rates, *the base rate is often ignored*, such as when physicians assign a high likelihood of disease to a positive test result, even when the disease is rare.
- People tend to be overconfident and optimistic, assigning higher personal probabilities than warranted to positive events and lower than warranted to negative events.
- For a *well calibrated* probability assessor, the relative frequency with which an event happens over the long run should match the probability given for that event. Many experts are not well-calibrated.

Exercises

Exercises with numbers divisible by 3 (3, 6, 9, etc.) are included in the Solutions at the back of the book. They are marked with an asterisk (*).

1. Explain how the pseudocertainty effect differs from the certainty effect.

2. Refer to the scenarios in Table 16.1 on page 347. Suppose someone stands to gain $100,000 in a lawsuit, but there is a 5% chance that something will go wrong and they will get nothing. They are offered a settlement of $90,000 if they are willing to drop the lawsuit.

   a. Using the definition of "expected value" from Chapter 14, what is the expected value for the amount gained under each decision?

   b. Using the material in this chapter, explain whether they are likely to accept the settlement.

   c. Now suppose that instead of a single individual, a company encounters lawsuits of this sort quite frequently. Explain why the decision of the company might be different from the decision of the single individual. Use the concept of expected value in your explanation.

*3. Suppose a television advertisement were to show viewers a product and then say, "You might expect to pay $25, $30, or even more for this product. But we are offering it for only $16.99." Explain which of the ideas in this chapter is being used to try to exploit viewers.

4. There are many more termites in the world than there are mosquitoes, but most of the termites live in tropical forests. Using the ideas in this chapter, explain why most people would think there were more mosquitoes in the world than termites.

5. Suppose a defense attorney is trying to convince the jury that his client's wallet, found at the scene of the crime, was actually planted there by his client's gardener. Here are two possible ways he might present this to the jury:

   Statement A: The gardener dropped the wallet when no one was looking.

   Statement B: The gardener hid the wallet in his sock and when no one was looking he quickly reached down and lowered his sock, allowing the wallet to drop to the ground.

   a. Explain why the second statement cannot have a higher probability of being true than the first statement.

   b. Based on the material in this chapter, to which statement are members of the jury likely to assign higher personal probabilities? Explain.

*6. Explain why you should be cautious when someone tries to convince you of something by presenting a detailed scenario. Give an example.

7. A telephone solicitor recently contacted the author to ask for money for a charity in which typical contributions are in the range of $25 to $50. The solicitor said, "We are asking for as much as you can give, up to $300.00." Do you think the amount people give would be different if the solicitor said, "We typically get $25 to $50, but give as much as you can." Explain, using the relevant material from this chapter.

8. Research by Slovic and colleagues (1982) found that people judged that accidents and diseases cause about the same number of deaths in the United States, whereas, in truth, diseases cause about 16 times as many deaths as accidents. Using the material from this chapter, explain why the researchers found this misperception.

*9. Determine which statement (A or B) has a higher probability of being true and explain your answer. Using the material in this chapter, also explain which statement you think a statistically naive person would think had a higher probability.

   A. A car traveling 120 miles per hour on a two-lane highway will have an accident.

   B. A car traveling 120 miles per hour on a two-lane highway will have a major accident in which all occupants are instantly killed.

10. Explain how an insurance salesperson might try to use each of the following concepts to sell you insurance:

    a. Anchoring

    b. Pseudocertainty

    c. Availability

11. In the early 1990s, there were approximately 5 billion people in the world. Plous (1993, p. 5) asked readers to estimate how wide a cube-shaped tank would have to be to hold all of the human blood in the world. The correct answer is about 870 feet, but most people give much higher answers. Explain which of the concepts covered in this chapter leads people to give higher answers. (*Hint*: Compare 870 to 870^3.)

*12. Kahneman and Tversky (1979) asked students if they would be likely to buy "probabilistic insurance." This insurance would cost half as much as regular insurance but would only cover losses with 50% probability. The majority of respondents (80%) indicated that they would not be interested in such insurance. Notice that the expected value for the return on this insurance is the same as on the regular policy. Explain which of the concepts in this chapter is illustrated by this example.

13. Barnett (1990) examined front page stories in the *New York Times* for 1 year, beginning with October 1, 1988, and found four stories related to automobile deaths but 51 related to deaths from flying on a commercial jet. These correspond to 0.08 story per thousand U.S. deaths by automobile and 138.2 stories per thousand U.S. deaths for commercial jets. He also reported a mid-August 1989 Gallup Poll finding that 63% of Americans had lost confidence in recent years in the safety of airlines. Discuss this finding in the context of the material in this chapter.

14. Explain how the concepts in this chapter account for each of the following scenarios:

    a. Most people rate death by shark attacks to be much more likely than death by falling airplane parts, yet the chances of dying from

the latter are actually 30 times greater (Plous, 1993, p. 121).

    b. You are a juror on a case at your school involving an accusation that a student cheated on an exam. The jury is asked to assess the likelihood of the statement, "Even though he knew it was wrong, the student copied from the person sitting next to him because he desperately wants to get into medical school." The other jurors give the statement a high probability assessment although they know nothing about the accused student.

    c. Research by Tversky and Kahneman (1982b) has shown that people think that words beginning with the letter *k* are more likely to appear in a typical segment of text in English than words with *k* as the third letter. In fact, there are about twice as many words with *k* as the third letter than words that begin with *k*.

    d. A 45-year-old man dies of a heart attack and does not leave a will.

*15. Suppose you go to your doctor for a routine examination, without any complaints of problems. A blood test reveals that you have tested positive for a certain disease. Based on the ideas in this chapter, what should you ask your doctor in order to assess how worried you should be?

16. Give one example of how each of the following concepts has had or might have an unwanted effect on a decision or action in your daily life:

    a. Conservatism

    b. Optimism

    c. Forgotten base rates

    d. Availability

17. In this chapter, we learned that one way to lower personal-probability assessments that are too high is to list reasons why you might be wrong. Explain how the availability heuristic might account for this phenomenon.

*18. Which of the concepts in this chapter might contribute to the decision to buy a lottery ticket? Explain.

19. Suppose you have a friend who is willing to ask her friends a few questions and then, based on their answers, is willing to assess the probability that those friends will get an A in each of their classes. She always assesses the probability to be either .10 or .90. She has made hundreds of these assessments and has kept track of whether her friends actually received A's. How would you determine if she is well calibrated?

20. Guess at the probability that if you ask five people when their birthdays are, you will find someone born in the same month as you. For simplicity, assume that the probability that a randomly selected person will have the same birth month you have is 1/12. Now use the material from Chapter 14 to make a table listing the numbers from 1 to 5, and then fill in the probabilities that you will *first* encounter someone with your birth month by asking that many people. Determine the *accumulated* probability that you will have found someone with your birth month by the time you ask the fifth person. How well calibrated was your initial guess?

Exercises 21 to 24: The scenarios in Exercises 21 to 24 are all taken from the book "Thinking, Fast and Slow" by Daniel Kahneman (2011). Page numbers from the book are given in each exercise.

*21. People were asked to give a personal probability assessment of the possible occurrence of each of the following two events (Kahneman, 2011, p. 159):

- "A massive flood somewhere in North America next year, in which more than 1000 people drown."
- "An earthquake in California sometime next year, causing a flood in which more than 1000 people drown."

Using the material from this chapter, explain which of the two possible events you think people found more likely and whether that event is, in fact, more likely.

22. "In a famous study, spouses were asked, 'How large was your personal contribution to keeping the place tidy, in percentages?' . . . As expected, the self-assessed contributions [from the two spouses] added up to more than 100%.... The bias is not necessarily self-serving: spouses also overestimated their contributions to causing quarrels" (Kahneman, 2011, p. 131). Use the material in this chapter to explain why the percentages given by the two spouses would add up to more than 100%.

23. "You see a person reading *The New York Times* on the New York subway. Which of the following is a better bet about the reading stranger?

- She has a PhD.
- She does not have a college degree" (Kahneman, 2011, p. 151).

a. Which of the two statements do you think people would state as more likely, and why?

b. Which statement do you think actually is most likely?

c. Use the material in this chapter to explain why the answers to parts (a) and (b) might differ.

*24. "In 2002, a survey of American homeowners who had remodeled their kitchens found that, on average, they had expected the job to cost $18,658; in fact, they ended up paying an average of $38,769" (Kahneman, 2011, p. 250). The quote is an example of what Kahneman calls the *planning fallacy*, which describes "plans and forecasts that are unrealistically close to best case scenarios, [and that] could be improved by consulting the statistics of similar cases" (p. 250). Explain how each of the following concepts from this chapter could have contributed to the planning fallacy for the kitchen remodeling example.

*a. Optimism

*b. Availability

*c. Anchoring

Mini-Projects

1. Design and conduct an experiment to try to elicit misjudgments based on one of the phenomena described in this chapter. Explain exactly what you did and your results.

2. Find and explain an example of a marketing strategy that uses one of the techniques in this chapter to try to increase the chances that someone will purchase something. Do not use an exact example from the chapter, such as "buy one, get one free."

3. Find a journal article that describes an experiment designed to test the kinds of biases described in this chapter. Summarize the article, and discuss what conclusions can be made from the research. You can find such articles by searching the web using key words from this chapter.

4. Estimate the probability of some event in your life using a personal probability, such as the probability that a person who passes you on the street will be wearing a hat, or the probability that traffic lights will be red when you get to them. Use an event for which you can keep a record of the relative frequency of occurrence over the next week. How well calibrated were you?

References

Barnett, A. (1990). Air safety: End of the golden age? *Chance* 3, no. 2, pp. 8–12.

Christensen-Szalanski, J. J. J., and J. B. Bushyhead. (1981). Physicians' use of probabilistic information in a real clinical setting. *Journal of Experimental Psychology: Human Perception and Performance* 7, pp. 928–935.

Fischhoff, B., P. Slovic, and S. Lichtenstein. (1977). Knowing with certainty: The appropriateness of extreme confidence. *Journal of Experimental Psychology: Human Perception and Performance* 3, pp. 552–564.

Hafner, Katie. (2012). Could a computer outthink this doctor? *The New York Times*, 4 December 2012, p. D1 and D6.

Hayward, J. W. (1984). *Perceiving ordinary magic. Science and intuitive wisdom*. Boulder, CO: New Science Library.

Kahneman, D. (2011). *Thinking, fast and slow*. New York: Farrar, Straus and Giroux.

Kahneman, D., and A. Tversky. (1973). On the psychology of prediction. *Psychological Review* 80, pp. 237–251.

Kahneman, D., and A. Tversky. (1979). Prospect theory: An analysis of decision under risk. *Econometrica* 47, pp. 263–291.

Kahneman, D., and A. Tversky. (1982). On the study of statistical intuitions. In D. Kahneman, P. Slovic, and A. Tversky (eds.), *Judgment under uncertainty: Heuristics and biases* (Chapter 34). Cambridge, England: Cambridge University Press.

Koriat, A., S. Lichtenstein, and B. Fischhoff. (1980). Reasons for confidence. *Journal of Experimental Psychology: Human Learning and Memory* 6, pp. 107–118.

Murphy, A. H., and R. L. Winkler. (1984). Probability forecasting in meteorology. *Journal of the American Statistical Association* 79, pp. 489–500.

Northcraft, G. B., and M. A. Neale. (1987). Experts, amateurs, and real estate: An anchoring and adjustment perspective on property pricing decisions. *Organizational Behavior and Human Decision Processes* 39, pp. 84–97.

Plous, S. (1993). *The psychology of judgment and decision making.* New York: McGraw-Hill.

Slovic, P., B. Fischhoff, and S. Lichtenstein. (1982). Facts versus fears: Understanding perceived risk. In D. Kahneman, P. Slovic, and A. Tversky (eds.), *Judgment under uncertainty: Heuristics and biases* (Chapter 33). Cambridge, England: Cambridge University Press.

Tversky, A. and D. Kahneman. (1982a). Judgment under uncertainty: Heuristics and biases. In D. Kahneman, P. Slovic, and A. Tversky (eds.), *Judgment under uncertainty: Heuristics and biases* (Chapter 1). Cambridge, England: Cambridge University Press.

Tversky, A., and D. Kahneman. (1982b). Availability: A heuristic for judging frequency and probability. In D. Kahneman, P. Slovic, and A. Tversky (eds.), *Judgment under uncertainty. Heuristics and biases* (Chapter 11). Cambridge, England: Cambridge University Press.

Wansink, B., R.J. Kent and S.J. Hoch. (1998). An anchoring and adjustment model of purchase quantity decisions. *Journal of Marketing Research*, 35, pp. 71–81.

Weinstein, N. D. (1980). Unrealistic optimism about future life events. *Journal of Personality and Social Psychology* 39, pp. 806–820.

CHAPTER 17

When Intuition Differs from Relative Frequency

Thought Questions

1. Do you think it likely that anyone will ever win a state lottery twice in a lifetime?

2. How many people do you think would need to be in a group in order to be at least 50% certain that two of them will have the same birthday?

3. Suppose you test positive for a rare disease, and your original chances of having the disease are no higher than anyone else's—say, close to 1 in 1000. You are told that the test has a 10% false positive rate and a 10% false negative rate. In other words, whether you have the disease or not, the test is 90% likely to give a correct answer. Given that you tested positive, what do you think is the probability that you actually have the disease? Do you think the chances are higher or lower than 50%?

4. If you were to flip a fair coin six times, which sequence do you think would be most likely: HHHHHH or HHTHTH or HHHTTT?

5. If you were faced with the following sets of alternatives, which one would you choose in each set? Choose either A or B and either C or D. Explain your answer.

   A. A gift of $240, guaranteed

   B. A 25% chance to win $1000 and a 75% chance of getting nothing

   C. A sure loss of $740

   D. A 75% chance to lose $1000 and a 25% chance to lose nothing

17.1   Revisiting Relative Frequency

Recall that the relative-frequency interpretation of probability provides a precise answer to certain probability questions. As long as we agree on the physical assumptions underlying an uncertain process, we should also agree on the probabilities of various outcomes. For example, if we agree that lottery numbers are drawn fairly, then we should agree on the probability that a particular ticket will be a winner.

In many instances, the physical situation lends itself to computing a relative-frequency probability, but people ignore that information. In this chapter, we examine how probability assessments that should be objective are instead confused by incorrect thinking.

17.2   Coincidences

When I was in college in upstate New York, I visited Disney World in Florida during summer break. While there, I ran into three people I knew from college, none of whom were there together. A few years later, I visited the top of the Empire State Building in New York City and ran into two friends (who were there together) and two additional unrelated acquaintances. Years later, I was traveling from London to Stockholm and ran into a friend at the airport in London while waiting for the flight. Not only did the friend turn out to be taking the same flight but we had been assigned adjacent seats.

Are Coincidences Improbable?

These events are all examples of what would commonly be called *coincidences.* They are certainly surprising, but are they improbable? Most people think that coincidences have low probabilities of occurring, but we shall see that our intuition can be quite misleading regarding such phenomena.

We will adopt the definition of **coincidence** proposed by Diaconis and Mosteller:

A coincidence is a surprising concurrence of events, perceived as meaningfully related, with no apparent causal connection (1989, p. 853).

The mathematically sophisticated reader may wish to consult the article by Diaconis and Mosteller, in which they provide some instructions on how to compute probabilities for coincidences. For our purposes, we need nothing more sophisticated than the simple probability rules we encountered in Chapter 14.

Here are some examples of coincidences that at first glance seem highly improbable:

EXAMPLE 17.1   Two George D. Brysons

"My next-door neighbor, Mr. George D. Bryson, was making a business trip some years ago from St. Louis to New York. Since this involved weekend travel and he was in no hurry, . . . and since his train went through Louisville, he asked the conductor, after he had boarded the train, whether he might have a stopover in Louisville.

"This was possible, and on arrival at Louisville, he inquired at the station for the leading hotel. He accordingly went to the Brown Hotel and registered. And then, just as a lark, he stepped up to the mail desk and asked if there was any mail for him. The girl calmly handed him a letter addressed to Mr. George D. Bryson, Room 307, that being the number of the room to which he had just been assigned. It turned out that the preceding resident of Room 307 was another George D. Bryson" (Weaver, 1963, pp. 282–283). ◼

EXAMPLE 17.2 Identical Cars and Matching Keys

Plous (1993, p. 154) reprinted an Associated Press news story describing a coincidence in which a man named Richard Baker and his wife were shopping on April Fool's Day at a Wisconsin shopping center. Mr. Baker went out to get their car, a 1978 maroon Concord, and drove it around to pick up his wife. After driving for a short while, they noticed items in the car that did not look familiar. They checked the license plate, and sure enough, they had someone else's car. When they drove back to the shopping center (to find the police waiting for them), they discovered that the owner of the car they were driving was a Mr. Thomas Baker, no relation to Richard Baker. Thus, both Mr. Bakers were at the same shopping center at the same time, with identical cars and with matching keys. The police estimated the odds as "a million to one." ◼

EXAMPLE 17.3 Winning the Lottery Twice

Moore and Notz (2014, p. 407) reported on a *New York Times* story of February 14, 1986, about Evelyn Marie Adams, who won the New Jersey lottery twice in a short time period. Her winnings were $3.9 million the first time and $1.5 million the second time. Then, in May 1988, Robert Humphries won a second Pennsylvania state lottery, bringing his total winnings to $6.8 million. When Ms. Adams won for the second time, the *New York Times* claimed that the odds of one person winning the top prize twice were about 1 in 17 trillion. ◼

Someone, Somewhere, Someday

Most people think that the events just described are exceedingly improbable, and they are. *What is not improbable is that someone, somewhere, someday will experience those events or something similar.*

When we examine the probability of what appears to be a startling coincidence, we ask the wrong question. For example, the figure quoted by the *New York Times* of 1 in 17 trillion is the probability that a *specific* individual who plays the New Jersey state lottery exactly twice will win both times (Diaconis and Mosteller, 1989, p. 859). However, millions of people play the lottery every day, and it is not surprising that someone, somewhere, someday would win twice.

In fact, Purdue professors Stephen Samuels and George McCabe (cited in Diaconis and Mosteller, 1989, p. 859) calculated those odds to be practically a sure thing. They calculated that there was at least a 1 in 30 chance of a double winner in a 4-month period and better than even odds that there would be a double winner in a 7-year period somewhere in the United States. Further, they used conservative assumptions about how many tickets past winners purchase. (Of course there would be no double winners if all past winners quit buying tickets, but they don't.)

When you experience a coincidence, remember that there are over 7 billion people in the world and over 315 million in the United States. If something has only a 1 in 1 million probability of occurring to each individual on a given day, it will occur to an average of over 315 people in the United States *each day* and over 7000 people in the world *each day*. Of course, probabilities of specific events depend on individual circumstances, but you can see that, quite often, it is not unlikely that something surprising will happen.

EXAMPLE 17.4	### Sharing the Same Birthday

Here is a famous example that you can use to test your intuition about surprising events. How many people would need to be gathered together to be at least 50% sure that two of them share the same birthday? Most people provide answers that are much higher than the correct one, which is only 23 people.

There are several reasons why people have trouble with this problem. If your answer was somewhere close to 183, or half the number of birthdays, then you may have confused the question with another one, such as the probability that someone in the group has *your* birthday or that two people have a specific date as their birthday.

It is not difficult to see how to calculate the appropriate probability, using our rules from Chapter 14. Notice that the only way to avoid two people having the same birthday is if all 23 people have different birthdays. To find that probability, we simply use the rule that applies to the word *and* (Rule 3), thus multiplying probabilities.

The probability that the first two people have different birthdays is the probability that the second person does not share a birthday with the first, which is 364/365 (ignoring February 29). The probability that the third person does not share a birthday with either of the first two is 363/365. (Two dates were already taken.) The probability that the first three people all have different birthdays is thus found from Rule 3 as the product of both of these, which is (364/365) × (363/365).

Continuing this line of reasoning, the probability that none of the 23 people share a birthday is

$$(364)(363)(362) \cdots (343)/(365)^{22} = .493$$

Therefore, the probability that at least two people share a birthday is what's left of the probability, or 1 − .493 = .507.

If you find it difficult to follow the arithmetic line of reasoning, simply consider this. Imagine each of the 23 people shaking hands with the remaining 22 people and asking them about their birthday. There would be 253 handshakes and birthday conversations. Surely there is a relatively high probability that at least one of those pairs would discover a common birthday.

By the way, the probability of a shared birthday in a group of 10 people is already better than 1 in 9, at .117. (There would be 45 handshakes.) With only 50 people, it is almost certain, with a probability of .97. (There would be 1225 handshakes.) ∎

EXAMPLE 17.5 **Unusual Hands in Card Games**

As a final example, consider a card game, like bridge, in which a standard 52-card deck is dealt to four players, so they each receive 13 cards. Any specific set of 13 cards is equally likely, each with a probability of about 1 in 635 billion. You would probably not be surprised to get a mixed hand—say, the 4, 7, and 10 of hearts; 3, 8, 9, and jack of spades; 2 and queen of diamonds; and 6, 10, jack, and ace of clubs. Yet, that specific hand is just as unlikely as getting all 13 hearts. The point is that any very specific event, surprising or not, has extremely low probability; however, there are many such events, and their combined probability is quite high.

Magicians sometimes exploit the fact that many small probabilities add up to one large probability by doing a trick in which they don't tell you what to expect in advance. They set it up so that *something* surprising is almost sure to happen. When it does, you are likely to focus on the probability of *that particular outcome*, rather than realizing that a multitude of other outcomes would have also been surprising and that one of them was likely to happen. ∎

Most Coincidences Only Seem Improbable

To summarize, most coincidences seem improbable only if we ask for the probability of that specific event occurring at that time to us. If, instead, we ask the probability of it occurring some time, to someone, the probability can become quite large. Further, because of the multitude of experiences we each have every day, it is not surprising that some of them may appear to be improbable. That specific event is undoubtedly improbable. What is not improbable is that something "unusual" will happen to each of us once in a while.

17.3   The Gambler's Fallacy

Another common misperception about random events is that they should be self-correcting. Another way to state this is that people think the long-run frequency of an event should apply even in the short run. This misperception is called the **gambler's fallacy.**

A classic example of the gambler's fallacy concerns the belief that the number of boys and girls should even out, even in a small number of births. This belief was recognized as long ago as 1796, when the French mathematician Pierre-Simon Laplace wrote (as translated from the original French):

> *"I have seen men, ardently desirous of having a son, who could learn only with anxiety of the births of boys in the month when they expected to become fathers. Imagining that the ratio of these births to those of girls ought to be the same at the end of each month, they judged that the boys already born would render more probable the births next of girls"* (Laplace, 1902 translation, p. 162).

In other words, men who wanted a son were afraid that the birth of other boys in the area would diminish their probability of having a boy, because they thought that the numbers of boys and girls were supposed to even out each month! Laplace hoped to inform his readers of the fallacy by continuing:

> *"Thus the extraction of a white ball from an urn which contains a limited number of white balls and of black balls increases the probability of extracting a black ball at the following drawing. But this ceases to take place when the number of balls in the urn is unlimited, as one must suppose in order to compare this case with that of births"* (Laplace, 1902 translation, pp. 162–163).

The gambler's fallacy can lead to poor decision making, especially when applied to gambling. For example, people tend to believe that a string of good luck must surely follow a string of bad luck, so that things even out. Unfortunately, independent chance events have no memory. Having a string of 10 bad gambles in a row does not change the probability that the next gamble will be bad (or good).

Belief in the Law of Small Numbers

Tversky and Kahneman (1982) identified a more general statistical fallacy called the **belief in the law of small numbers,** "according to which [people believe that] even small samples are highly representative of the populations from which they are drawn" (p. 7). They report that "in considering tosses of a coin for heads and tails, for example, people regard the sequence HTHTTH to be more likely than the sequence HHHTTT, which does not appear random, and also more likely than the sequence HHHHTH, which does not represent the fairness of the coin" (p. 7). Remember that any specific sequence of heads and tails is just as likely as any other if the coin is fair, so the idea that the first sequence is more likely is a misperception.

When the Gambler's Fallacy May Not Apply

Notice that the gambler's fallacy applies to independent events. Recall from Chapter 14 that independent events are those for which occurrence on one occasion gives no information about occurrence on the next occasion, as with successive flips of a coin. The gambler's fallacy may not apply to situations in which knowledge of one outcome affects probabilities of the next. For instance, in card games using a single deck, knowledge of what cards have already been played provides information about what cards are likely to be played next. If you normally receive lots of mail but have received none for two days, you would probably (correctly) assess that you are likely to receive more than usual the next day. If your car has reliably served you without any problems for the last 50,000 miles, you probably would not assume that the next 50,000 miles would be equally likely to be trouble-free.

CASE STUDY 17.1	**Streak Shooting in Basketball: Reality or Illusion?**

SOURCE: Tversky and Gilovich (Winter 1989).

We have learned in this and the previous chapter that people's intuition, when it comes to assessing probabilities, is not very good, particularly when their wishes for certain outcomes are motivated by outside factors. Tversky and Gilovich (Winter 1989) decided to compare basketball fans' impressions of "streak shooting" with the reality evidenced by the records.

First, they generated phony sequences of 21 alleged "hits and misses" in shooting baskets and showed them to 100 knowledgeable basketball fans. Without telling them the sequences were faked, they asked the fans to classify each sequence as "chance shooting," in which the probability of a hit on each shot was unrelated to previous shots; "streak shooting," in which the runs of hits and misses were longer than would be expected by chance; or "alternating shooting," in which runs of hits and misses were shorter than would be expected by chance. They found that people tended to think that streaks had occurred when they had not. In fact, 65% of the respondents thought the sequence that had been generated by "chance shooting" was in fact "streak shooting."

To give you some idea of the task involved, decide which of the following two sequences of 10 successes (S) and 11 failures (F) you think is more likely to be the result of "chance shooting":

Sequence 1: FFSSSFSFFFSSSSFSFFFSF

Sequence 2: FSFFSFSFFFSFSSFSFSSSF

Notice that each sequence represents 21 shots. In "chance shooting," the proportion of throws on which the result is different from the previous throw should be about one-half. If you thought sequence 1 was more likely to be due to chance shooting, you're right. Of the 20 throws that have a preceding throw, exactly 10 are different. In sequence 2, 14 of 20, or 70%, of the shots differ from the previous shot. If you selected sequence 2, you are like the fans tested by Tversky and Gilovich. The sequences with 70% and 80% alternating shots were most likely to be selected (erroneously) as being the result of "chance shooting."

To further test the idea that basketball fans (and players) see patterns in shooting success and failure, Tversky and Gilovich asked questions about the probability of successful hitting after hitting versus after missing. For example, they asked the following question of 100 basketball fans:

When shooting free throws, does a player have a better chance of making his second shot after making his first shot than after missing his first shot? (1989, p. 20)

Sixty-eight percent of the respondents said yes; 32% said no. They asked members of the Philadelphia 76ers basketball team the same question, with similar results. A similar question about ordinary shots elicited even stronger belief in streaks, with 91% responding that the probability of making a shot was higher after having just made the last two or three shots than after having missed them.

What about the data on shooting? The researchers examined data from several NBA teams, including the Philadelphia 76ers, the New Jersey Nets, the New York Knicks, and the Boston Celtics. In this case study, we examine the data they reported for free throws. These are throws in which action stops and the player stands in a fixed position, usually for two successive attempts to put the ball in the basket. Examining free-throw shots removes the possible confounding effect that members of the other team would more heavily guard a player they perceive as being "hot."

Tversky and Gilovich reported free-throw data for nine members of the Boston Celtics basketball team. They examined the long-run frequency of a hit on the second free throw after a hit on the first one, and after a miss on the first one. Of the nine players, five had a higher probability of a hit after a miss, whereas four had a higher probability of a hit after a hit. In other words, the perception of 65% of the fans that the probability of a hit was higher after just receiving a hit was not supported by the actual data.

Tversky and Gilovich looked at other sequences of hits and misses from the NBA teams, in addition to generating their own data in a controlled experiment using players from Cornell University's varsity basketball teams. They analyzed the data in a variety of ways, but they could find no evidence of a "hot hand" or "streak shooting." They conclude:

> *Our research does not tell us anything in general about sports, but it does suggest a generalization about people, namely that they tend to "detect" patterns even where none exist, and to overestimate the degree of clustering in sports events, as in other sequential data. We attribute the discrepancy between the observed basketball statistics and the intuitions of highly interested and informed observers to a general misconception of the laws of chance that induces the expectation that random sequences will be far more balanced than they generally are, and creates the illusion that there are patterns of streaks in independent sequences.* (1989, p. 21).

The research by Tversky and Gilovich has not gone unchallenged. For example, statistician Hal Stern points out that small effects are difficult to detect in general for statistical tests with small sample sizes:

> *"Statisticians working with sports data have generally not found any statistically significant evidence of streaks. But then again, the methods used have also failed to detect some sources of variation in performance that we know exist, [such as] variation in performance due to the varying ability of opponents. The failure to find statistical evidence supporting the hot hand to this point may indicate that there is no such thing. However, the failures are also consistent with a small effect. In situations like this, a person's prior opinion about the issue will make a big difference in how they "see" the data. The hot hand will be difficult to resolve for some time to come"* (Stern, 1997, p. 7).

For additional reading, see the articles by Hooke (1989) and by Larkey, Smith, and Kadane (1989). Similar to Hal Stern's conclusion, they argue that just because Tversky and Gilovich did not find evidence of "streak shooting" in the data they examined doesn't mean that it doesn't exist, sometimes. ∎

17.4   Confusion of the Inverse

Consider the following scenario, discussed by Eddy (1982). You are a physician. One of your patients has a lump in her breast. You are almost certain that it is benign; in fact, you would say there is only a 1% chance that it is malignant. But just to be sure, you have the patient undergo a mammogram, a breast x-ray designed to detect cancer.

You know from the medical literature that mammograms are 80% accurate for malignant lumps and 90% accurate for benign lumps. In other words, if the lump is truly malignant, the test results will say that it is malignant 80% of the time and will falsely say it is benign 20% of the time. If the lump is truly benign, the test results will say so 90% of the time and will falsely declare that it is malignant only 10% of the time.

Sadly, the mammogram for your patient is returned with the news that the lump is malignant. What are the chances that it is truly malignant?

Eddy posed this question to 100 physicians. Most of them thought the probability that the lump was truly malignant was about 75% or .75. In truth, given the probabilities as described, *the probability is only .075, i.e. 7.5%*. The physicians' estimates were 10 times too high!

When he asked them how they arrived at their answers, Eddy realized that the physicians were confusing the actual question with a different question: "When asked about this, the erring physicians usually report that they assumed that the probability of cancer given that the patient has a positive x-ray was approximately equal to the probability of a positive x-ray in a patient with cancer" (1982, p. 254).

Robyn Dawes has called this phenomenon **confusion of the inverse** (Plous, 1993, p. 132). The physicians were confusing the probability of cancer *given a positive x-ray* with its inverse, the probability of a positive x-ray, *given that the patient has cancer.*

Determining the Actual Probability

It is not difficult to see that the correct answer to the question posed to the physicians by Eddy (in the previous section) is indeed .075. Let's construct a hypothetical table of 100,000 women who fit this scenario. (See Table 17.1.) These are women who would present themselves to the physician with a lump for which the probability that it was malignant seemed to be about 1%, which was what Eddy presented to the

TABLE 17.1   Breakdown of Actual Status versus Test Status for a Rare Disease

	Test Shows Malignant	Test Shows Benign	Total
Actually malignant	800	200	1,000
Actually benign	9,900	89,100	99,000
Total	10,700	89,300	100,000

physicians as the initial assessment for this patient. Thus, of the 100,000 women, about 1%, or 1000 of them, would have a malignant lump. The remaining 99%, or 99,000, would have a benign lump.

Further, given that the test was 80% accurate for malignant lumps, it would show a malignancy for 800 of the 1000 women who actually had one. Given that it was 90% accurate for the 99,000 women with benign lumps, it would show benign for 90%, or 89,100 of them and malignant for the remaining 10%, or 9900 of them. Table 17.1 shows how the 100,000 women would fall into these possible categories.

Let's return to the question of interest. Our patient has just received a positive test for malignancy. Given that her test showed malignancy, what is the actual probability that her lump is malignant? Of the 100,000 women, 10,700 of them would have an x-ray showing malignancy. But of those 10,700 women, only 800 of them actually have a malignant lump. Thus, given that the test showed a malignancy, the probability of malignancy is just 800/10,700 = 8/107 = .075.

Forgotten Base Rates

If you look carefully at Table 17.1, you will see why the physicians erred in their logic. The test was indeed relatively accurate for both malignant and benign cases, but the physicians forgot about the vastly different base rates. Because 99% of the women had a benign lump, there were 99 times as many women in the "benign" row of the table than in the "malignant" row. Therefore, even though only 10% of the women in that row had a positive test, the actual *number* of positive tests for them far exceeded the number of positive tests for the women who actually had a malignant lump. This illustrates the problem of "forgotten base rates" discussed in Chapter 16.

The Probability of False Positives

Many physicians are guilty of confusion of the inverse. Remember, in a situation where the *base rate* for a disease is very low and the test for the disease is less than perfect, there will be a relatively high probability that a positive test result is a false positive.

If you ever find yourself in a situation similar to the one just described, you may wish to construct a table like Table 17.1.

To determine the probability of a positive test result being accurate, you need only three pieces of information:

1. The base rate or probability that you are likely to have the disease, without any knowledge of your test results
2. The **sensitivity** of the test, which is the proportion of people who correctly test positive when they actually have the disease
3. The **specificity** of the test, which is the proportion of people who correctly test negative when they don't have the disease

Notice that items 2 and 3 are measures of the accuracy of the test. They do not measure the probability that someone has the disease when they test positive or the probability that they do not have the disease when they test negative. Those probabilities, which are obviously the ones of interest to the patient, can be computed by constructing a table similar to Table 17.1. They can also be computed by using a formula called **Bayes' Rule,** given in the Focus On Formulas section at the end of this chapter.

17.5   Using Expected Values to Make Wise Decisions

In Chapter 14, we learned how to compute the expected value, or long-run average, of numerical outcomes when we know the possible outcomes and their probabilities. Using this information, you would think that people would make decisions that allowed them to maximize their expected monetary return. But people don't behave this way. If they did, they would not buy lottery tickets or insurance.

Businesses like insurance companies and casinos rely on the theory of expected value to stay in business. Insurance companies know that young people are more likely than middle-aged people to have automobile accidents and that older people are more likely to die of nonaccidental causes. They determine the prices of automobile and life insurance policies accordingly.

If individuals were solely interested in maximizing their monetary gains, they would use expected value in a similar manner. For instance, in Example 14.17, we illustrated that for the California Decco lottery game, there was an average loss of 35 cents for each ticket purchased. Most lottery players know that there is an expected loss for every ticket purchased, yet they continue to play. Why? Probably because the excitement of playing and possibly winning has intrinsic, nonmonetary value that compensates for the expected monetary loss. You may recall from Chapter 16 that Kahneman (2011) attributed this behavior to the "possibility effect."

Social scientists have long been intrigued with how people make decisions, and much research has been conducted on the topic. The most popular theory among early researchers, in the 1930s and 1940s, was that people made decisions to maximize their expected *utility*. This may or may not correspond to maximizing their expected dollar amount. The idea was that people would assign a worth or utility to each outcome and choose whatever alternative yielded the highest expected value.

More recent research has shown that decision making is influenced by a number of factors and can be a complicated process. (Plous [1993] presents an excellent summary of much of the research on decision making.) The way in which the decision is presented can make a big difference. For example, Plous (1993, p. 97) discusses experiments in which respondents were presented with scenarios similar to the following:

If you were faced with the following alternatives, which would you choose? Note that you can choose either A or B and either C or D.

A. A gift of $240, guaranteed

B. A 25% chance to win $1000 and a 75% chance of getting nothing

C. A sure loss of $740

D. A 75% chance to lose $1000 and a 25% chance to lose nothing

When asked to choose between A and B, the majority of people chose the sure gain represented by choice A. Notice that the expected value under choice B is $250, which is higher than the sure gain of $240 from choice A, yet people prefer choice A.

When asked to choose between C and D, the majority of people chose the gamble rather than the sure loss. Notice that the expected value under choice D is $750, representing a larger expected loss than the $740 presented in choice C. For dollar amounts and probabilities of this magnitude, people tend to value a sure gain, but are willing to take a risk to prevent a sure loss.

The second set of choices (C and D) is similar to the decision people must make when deciding whether to buy insurance. The cost of the premium is the sure loss. The probabilistic choice represented by alternative D is similar to gambling on whether you will have a fire, burglary, accident, and so on. Why then do people choose to gamble in the scenario just presented, yet tend to buy insurance?

As Plous (1993) explains, one factor seems to be the magnitudes of the probabilities attached to the outcomes. People tend to give small probabilities more weight than they deserve for their face value. Losses connected with most insurance policies have a low probability of actually occurring, yet people worry about them. Plous (1993, p. 99) reports on a study in which people were presented with the following two scenarios. Choose either A or B, and either C or D:

Alternative A: A 1 in 1000 chance of winning $5000

Alternative B: A sure gain of $5

Alternative C: A 1 in 1000 chance of losing $5000

Alternative D: A sure loss of $5

About three-fourths of the respondents presented with scenario A and B chose the risk presented by alternative A. This is similar to the decision to buy a lottery ticket, where the sure gain corresponds to keeping the money rather than using it to buy a ticket.

For scenario C and D, nearly 80% of respondents chose the sure loss (D). This is the situation that results in the success of the insurance industry. Of course, the dollar amounts are also important. A sure loss of $5 may be easy to absorb, while the risk of losing $5000 may be equivalent to the risk of bankruptcy.

CASE STUDY 17.2	**Losing the Least: Sports Betting, Casinos or Lotteries?**

SOURCE: Rudy (2013)

In 2013, New Jersey Governor Chris Christie decided to challenge a law that would not allow sports betting in his state (or most other states in the United States). Presumably, the reason for the law is that betting on sports is a losing proposition, and people need to be protected from it. Minitab blogger and sports statistician Kevin Rudy decided to find out if sports betting was any worse than state lotteries, which are legal in most states, or casino gambling, which is legal in many states including New Jersey (Rudy, 2013). For this case study, we present some of his findings and add a new simulation to mimic the one done by him.

Kevin Rudy presented the following challenge. Suppose you decide to bet $10 a week for a year (which is not a wise use of funds, as you will see). You are trying to decide among these three choices, which are representative of three forms of gambling:

- Sports: Betting the points spread on a National Football League (NFL) game
- Lottery: The Pennsylvania Neon9 instant scratcher lottery game
- Casino: A single number on a roulette wheel

What's your best (or least bad) option? Let's start by finding the expected value for each $10 you spend. Remember, the expected value is the average over the long run, and a negative expected value represents a loss, on average. In Chapter 14, we learned how to compute an expected value:

$$\text{Expected value} = A_1 p_1 + A_2 p_2 + \dots + A_k p_k$$

where A_1 to A_k are the possible amounts won or lost, and the p's are their probabilities.

For the NFL sports bet, according to another blog by Kevin Rudy (2011), you have about a 50% chance of winning and losing, much like a coin toss. If you lose, your $10 is gone, and if you win, you only gain an additional $9.09. So the two possible amounts "won" are −$10 and +$9.09, each with probability of 0.5. For the lottery, you need to find the amounts and probabilities for the specific lottery in question. The Neon9 cost $10 per ticket and had a top prize of $300,000, with smaller prizes ranging from $10 (breakeven) up to $30,000. After consulting the lottery's webpage, Rudy calculated the expected value for each $10 bet to be −$2.78. In other words, for every $10 ticket you buy, you lose an average of $2.78. For roulette, there are 38 numbers, and thus if you bet one of them, your probabilities of winning and losing are 1/38 and 37/38, respectively. If you win, you get your $10 back, plus $350, for a net gain of $350. We can now compare the expected values for each $10 spent on the three bets, remembering that a loss of $10 would be an "amount won" A of −$10.

- Sports: Expected value = (−$10)(.5) + ($9.09)(.5) = −$0.45 (45 cent loss)
- Lottery: Expected value = −$2.78
- Casino: Expected value = (−$10)(37/38) + ($350)(1/38) = −$0.526 (about 53 cent loss)

So, in terms of expected value (average loss per bet), the sports bet seems like the best deal! However, the expected value shows what will happen on average over the very long run. A year isn't a very long run, with only 52 bets, especially for the lottery, where the probability of winning the $300,000 prize is only 1 in 720,000.

Is it possible to come out ahead after a year of betting? To find that out, Mr. Rudy simulated what would happen for the three options if 100 people each bet $10 a week for a year. Because he published a summary of his results and not the full results, we used Minitab to do a new simulation. Figure 17.1 shows boxplots of the results. Each boxplot displays the net total amounts won (positive values) or lost (negative values) for the 100 (simulated) people who played that game for a year. In each box, the mean is also illustrated, with the circled plus sign. You might notice that there is no median line for the roulette box. In fact, the median was equal to the first quartile, at −$160, and thus is hidden at the left end of the box. A line at $0 is shown so you can see who won and who lost.

Although it is not obvious from the boxplot, 43 of the 100 sports betters and 44 of the 100 roulette betters actually came out ahead. Only eight of the 100 lottery players came out ahead. Two of those were only $10 ahead at the end of the year and thus not shown as outliers. The final feature you might notice is that the largest winnings occurred with the roulette game. The people who came out ahead in roulette generally ended the year with much more than the people who came out ahead in the sports betting. But don't forget to look at the left hand side of the plot   those who lost at roulette also lost more. In fact, 22 of the 100 roulette players lost their entire investment of $520. The largest loss for the sports betting was $176.38.

In summary, all three games have a negative expected value, which means that you will eventually lose if you play long enough. However, for just 52 weeks of play, it is possible to come out ahead in any of the three games. The sports betting had the smallest amount of variation, but you are not likely to win (or lose) much. The lottery had the worst performance by far, but none of the 100 players in this simulation won any of the big prizes. In fact, in the entire simulation the largest weekly prize won in the lottery was $1000. That's not surprising because the probability of winning any of the prizes larger than that is slightly less than 1 in 100,000. ∎

Figure 17.1
Amounts won and lost by 100 people betting $10.00 a week for a year on sports, roulette or a lottery

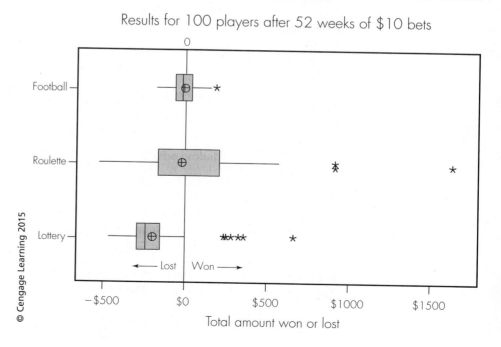

Results for 100 players after 52 weeks of $10 bets

© Cengage Learning 2015

Thinking About Key Concepts

- Specific *coincidences* generally have a low probability of occurrence, but because so many events happen to each of us every day, coincidences are likely to happen to someone, somewhere, someday quite often.
- The *gambler's fallacy* is the mistaken belief that a sequence of independent events should be "self-correcting" so that the long-run frequency of an outcome should happen in the short run.
- The gambler's fallacy is one aspect of the *belief in the law of small numbers,* which is the mistaken belief that the small samples should closely match the population from which they were drawn.
- *Confusion of the inverse* is the mistaken belief that the conditional probability of event A happening given that event B happened is similar to the conditional probability of event B, given event A. The confusion is particularly problematic when the base rate of one event is low. For instance, people confuse the probability of disease given a positive test result with the probability of a positive test result, given that someone has the disease.
- In making decisions involving money, it is sometimes helpful to compute the *expected value* resulting from each decision, which is the average amount gained or lost over the very long run.

Focus On Formulas

Conditional Probability

The *conditional probability* of event A, given knowledge that event B happened, is denoted by $P(A|B)$.

Bayes' Rule (for Reversing the Order of Conditional Probabilities)

Suppose A_1 and A_2 are complementary events with known probabilities. In other words, they are mutually exclusive and their probabilities sum to 1. For example, they might represent presence and absence of a disease in a randomly chosen individual. Suppose B is another event such that the conditional probabilities $P(B|A_1)$ and $P(B|A_2)$ are both known. For example, B might be the probability of testing positive for the disease. We do not need to know $P(B)$.

Then Bayes' Rule determines the conditional probability in the other direction:

$$P(A_1|B) = \frac{P(A_1)P(B|A_1)}{P(A_1)P(B|A_1) + P(A_2)P(B|A_2)}$$

For example, Bayes' Rule can be used to determine the probability of having a disease *given* that the test is positive. The base rate, sensitivity, and specificity would all need to be known. (See definitions on page 373.) Bayes' Rule is easily extended to more than two mutually exclusive events, as long as the probability of each one is known and the probability of B conditional on each one is known.

Exercises

Exercises with numbers divisible by 3 (3, 6, 9, etc.) are included in the Solutions at the back of the book. They are marked with an asterisk (*).

1. Explain why it is not at all unlikely that, in a class of 50 students, two of them will have the same last name.

2. Suppose twin sisters are reunited after not seeing each other since they were 3 years old. They are amazed to find out that they are both married to men named James and that they each have a daughter named Jennifer. Explain why this is not so amazing.

*3. Suppose you are seated next to a stranger on an airplane, and you start discussing various topics such as where you were born (country or state), what your favorite movie of all time is, your spouse's occupation, and so on. For simplicity, assume that the probability that your details match for any given topic is 1/50 and is independent from one topic to the next. If you discuss 10 topics, how surprising would it be to find that you match on at least one of them? Support your answer with a numerical assessment.

4. Why is it not surprising that the night before a major airplane crash several people will have dreams about an airplane disaster? If you were one of those people, would you think that something amazing had occurred?

5. Explain why the story about George D. Bryson, reported in Example 17.1, is not all that surprising.

*6. Find a dollar bill or other item with a serial number. Write down the number. I predict that there is something unusual about it or some pattern to it. Explain what is unusual about it and how I was able to make that prediction.

7. A statistics professor once made a big blunder by announcing to his class of about 50 students that he was fairly certain that someone in the room would share his birthday. We have already learned that there is a 97% chance that there will be two people in a room of 50 with a common birthday. Given that information, why was the professor's announcement a blunder? Do you think he was successful in finding a match? Explain.

8. If you wanted to pretend that you could do psychic readings, you could perform "cold readings" by inviting people you do not know to allow you to tell them about themselves. You would then make a series of statements like

   "I see that there is some distance between you and your mother that bothers you."

   "It seems that you are sometimes less sure of yourself than you indicate."

   "You are thinking of two men in your life [or two women, for a male client], one of whom is sort of light-complexioned and the other of whom is slightly darker. Do you know who I mean?"

   In the context of the material in this chapter, explain why this trick would often work to convince people that you are indeed psychic.

*9. Many people claim that they can often predict who is on the other end of the phone when it rings (obviously without caller ID). Do you think that phenomenon has a normal explanation? Explain.

10. Explain why it would be much more surprising if someone were to flip a coin and get six heads in a row after telling you they were going to do so than it would be to simply watch them flip the coin six times and observe six heads in a row.

11. Comment on the following unusual lottery events, including a probability assessment.

   a. On September 11, 2002, the first anniversary of the 9/11 attack on the World Trade Center, the winning number for the New York State-lottery was 911.

   b. To play the Maryland Pick 4 lottery, players choose four numbers from the digits 0 to 9. The game is played twice every day, at midday and in the evening. In 1999, holiday players who decided to repeat previous winning numbers got lucky. At midday on December 24, the winning numbers were 7535, exactly the same as on the previous evening. And on

New Year's Eve, the evening draw produced the numbers 9521—exactly the same as the previous evening.

*12. Give an example of a sequence of events to which the gambler's fallacy would not apply because the events are not independent.

13. Although it's not quite true, suppose the probability of having a male child (M) is equal to the probability of having a female child (F). A couple has four children.

   a. Are they more likely to have FFFF or to have MFFM? Explain your answer.

   b. Which sequence in part (a) of this exercise would a *belief in the law of small numbers* cause people to say had higher probability? Explain.

   c. Is a couple with four children more likely to have four girls or to have two children of each sex? Explain. (Assume the decision to have four children was independent of the sex of the children.)

14. You are at a casino with a friend, playing a game in which dice are involved. Your friend has just lost six times in a row. She is convinced that she will win on the next bet because she claims that, by the law of averages, it's her turn to win. She explains to you that the probability of winning this game is 40%, and because she has lost six times, she has to win four times to make the odds work out. Is she right? Explain.

*15. Suppose a friend reports that she has just had a string of "bad luck" with her car. She had three major problems in as many months and now has replaced many of the worn parts with new ones. She concludes that it is her turn to be lucky and that she shouldn't have any more problems for a while. Is she using the gambler's fallacy? Explain.

16. A fair coin is flipped six times and the sequence of heads and tails is observed.

   a. Are the sequences HHHHHH and HTHHTT equally likely? Explain.

   b. Are the events "A = 6 heads in the 6 tosses" and "B = 3 heads and 3 tails in the 6 tosses" equally likely? Explain.

   c. Would belief in the law of small numbers lead people to think that the two sequences in part (a) are equally likely, or are not equally likely? Would they be correct? Explain.

   d. Would belief in the law of small numbers lead people to think that the two events in part (b) are equally likely, or are not equally likely? Would they be correct? Explain.

17. The University of California at Berkeley *Wellness Encyclopedia* (1991) contains the following statement in its discussion of HIV testing: "In a high-risk population, virtually all people who test positive will truly be infected, but among people at low risk the false positives will outnumber the true positives. Thus, for every infected person correctly identified in a low-risk population, an estimated 10 noncarriers [of the HIV virus] will test positive" (p. 360).

   a. Suppose you have a friend who is part of this low-risk population but who has just tested positive. Using the numbers in the statement, calculate the probability that the person actually carries the virus.

   b. Your friend is understandably upset and doesn't believe that the probability of being infected with HIV isn't really near 1. After all, the test is accurate, and it came out positive. Explain to your friend how the *Wellness Encyclopedia* statement can be true, even though the test is very accurate both for people with HIV and for people who don't carry it. If it's easier, you can make up numbers to put in a table to support your argument. (See Table 17.1.)

*18. Using the data in Table 17.1 about a hypothetical population of 100,000 women tested for breast cancer, find the probability of each of the following events:

   *a. A woman whose test shows a malignant lump actually has a benign lump.

   *b. A woman who actually has a benign lump has a test that shows a malignant lump.

   *c. A woman with unknown status has a test showing a malignant lump.

19. Suppose a rare disease occurs in about 1 out of 1000 people who are like you. A test for the disease has sensitivity of 95% and specificity of 90%. Using the technique described in this chapter, compute the probability that you actually have the disease, given that your test results are positive.

20. Using the data in Table 17.1, give numerical values and explain the meaning of the sensitivity and the specificity of the test.

*21. Which one of the problems with probability assessments discussed in Chapter 16 contributes to confusion of the inverse? Explain.

22. Suppose the sensitivity of a test is .90. Give either the false positive or the false negative rate for the test, and explain which you are providing. Could you provide the other one without additional information? Explain.

23. In financial situations, are businesses or individuals more likely to make use of expected value for making decisions? Explain.

*24. We learned in this chapter that one idea researchers have tested was that when forced to make a decision, people choose the alternative that yields the highest expected value.

 *a. If that were the case, explain which of the following two choices people would make:

 Choice A: Accept a gift of $10.
 Choice B: Take a gamble with probability 1/1000 of winning $9000 and 999/1000 of winning nothing.

 *b. Using the material from Chapter 16, explain which of the two choices in part (a) you think the majority of people would select.

25. Refer to the previous exercise, about choosing between a gift and a gamble. Explain how the situation in part (a) resembles the choices people have when they decide whether to buy lottery tickets.

26. It is time for the end-of-summer sales. One store is offering bathing suits at 50% of their usual cost, and another store is offering to sell you two for the price of one. Assuming the suits originally all cost the same amount, which store is offering a better deal? Explain.

*27. Suppose you believe that the probability that your team will win a game is 1/4. You are willing to bet $1 that your team will win. What amount should you be offered if you win in order to have a break-even expected value?

28. Suppose you are trying to decide whether to park illegally while you attend class. If you get a ticket, the fine is $25. If you assess the probability of getting a ticket to be 1/100, what is the expected value for the fine you will have to pay? Under those circumstances, explain whether you would be willing to take the risk and why. (Note that there is no correct answer to the last part of the question; it is designed to test your reasoning.)

29. You are making a hotel reservation and are offered a choice of two rates. The advanced purchase rate is $100, but your credit card will be charged immediately and there is no refund, even if you don't use the room. The flexible rate is $140 but you don't pay anything if you don't use the room. Suppose p is the probability that you *will* end up using the room.

 a. Suppose $p = 0.70$, so there is a 70% chance you will use the room. What is the expected value of your cost if you reserve the room with the flexible rate? (Hint: What are the two possible amounts you could pay, and what are their probabilities?)

 b. No longer assume a specific value for p. In terms of p, what is the expected value of your cost if you reserve the room with the flexible rate?

 c. What is the expected value of your cost if you choose the advanced purchase rate? (Hint: There is only one possible amount.)

 d. For what value of p are the expected values you found in parts (b) and (c) the same?

 e. For what range of values of p are you better off choosing the advanced purchase rate?

Mini-Projects

1. Find out the sensitivity and specificity of a common medical test. Calculate the probability of a true positive for someone who tests positive with the test, assuming the rate in the population is 1 per 100; then calculate the probability assuming the rate in the population is 1 per 1000.

2. Ask four friends to tell you their most amazing coincidence story. Use the material in this chapter to assess how surprising each of the stories is to you. Pick one of the stories, and try to approximate the probability of that specific event happening to your friend.

3. Conduct a survey in which you ask 20 people the two scenarios presented in Thought Question 5 at the beginning of this chapter and discussed in Section 17.5. Record the percentage who choose alternative A over B and the percentage who choose alternative C over D.

   a. Report your results. Are they consistent with what other researchers have found? (Refer to p. 374.) Explain.

   b. Explain how you conducted your survey. Discuss whether you overcame the potential difficulties with surveys that were discussed in Chapter 4.

References

Diaconis, P., and F. Mosteller. (1989). Methods for studying coincidences. *Journal of the American Statistical Association* 84, pp. 853–861.

Eddy, D. M. (1982). Probabilistic reasoning in clinical medicine: Problems and opportunities. In D. Kahneman, P. Slovic, and A. Tversky (eds.), *Judgment under uncertainty: Heuristics and biases* (Chapter 18). Cambridge, England: Cambridge University Press.

Hooke, R. (1989). Basketball, baseball, and the null hypothesis, *Chance* 2, no. 4, pp. 35–37.

Kahneman, D. (2011). *Thinking, fast and slow*. New York: Farrar, Straus and Giroux.

Laplace, Pierre Simon, (1902). A Philosophical Essay on Probabilities, translated from the 6th French edition by Frederick Wilson Truscott and Frederick Lincoln Emory. New York: John Wiley & Sons.

Larkey, P. D., R. A. Smith, and J. B. Kadane. (1989). It's okay to believe in the "hot hand." *Chance* 2, no. 4, pp. 22–30.

Moore, D. S. and W. I. Notz. (2014). *Statistics: Concepts and controversies*. 8th ed. New York: W. H. Freeman.

Plous, S. (1993). *The psychology of judgment and decision making*. New York: McGraw-Hill.

Rudy, Kevin. (2013). The lottery, the casino or the sportsbook: What's your best bet? The Minitab Blog, 7 June 2013, http://blog.minitab.com/blog/the-statistics-game/the-lottery-the-casino-or-the-sportsbook-whats-your-best-bet, accessed July 4, 2013.

Rudy, Kevin. (2011). Is betting on NFL games the same as betting on a coin flip: The Minitab Blog, 25 November 2011, http://blog.minitab.com/blog/the-statistics-game/is-betting-on-nfl-games-the-same-as-betting-on-a-coin-flip, accessed July 4, 2013.

Stern, H. S. (1997), "Judging who's hot and who's not," in the column "A Statistician Reads the Sports Pages", *Chance*, 10(2), pp. 40–43.

Tversky, A., and T. Gilovich. (Winter 1989). The cold facts about the "hot hand" in basketball. *Chance* 2, no. 1, pp. 16–21.

Tversky, A., and D. Kahneman. (1982). Judgment under uncertainty: Heuristics and biases. In D. Kahneman, P. Slovic, and A. Tversky (eds.), *Judgment under uncertainty: Heuristics and biases* (Chapter 1). Cambridge, England: Cambridge University Press.

University of California, Berkeley. (1991). *The Wellness Encyclopedia.* Boston: Houghton Mifflin.

Weaver, W. (1963). *Lady luck: The theory of probability.* Garden City, NY: Doubleday.

CHAPTER **18**

Understanding the Economic News

Thought Questions

1. The Conference Board, a not-for-profit organization, produces a composite index of leading economic indicators as well as one of coincident and lagging economic indicators. These are supposed to "indicate" the status of the economy. What do you think the terms *leading, coincident,* and *lagging* mean in this context?

2. Suppose you wanted to measure the yearly change in the "cost of living" for a college student living in the dorms for the past 4 years. How would you do it?

3. Suppose you were told that the price of a certain product, measured in 1984 dollars, has not risen. What do you think is meant by "measured in 1984 dollars?"

4. How do you think governments determine the "rate of inflation?"

18.1   Cost of Living: The Consumer Price Index

Everyone is affected by inflation. When the costs of goods and services rise, most workers expect their employers to increase their salaries to compensate. But how do employers know what a fair salary adjustment would be?

The most common measure of change in the cost of living in the United States is the Consumer Price Index (CPI), produced by the Bureau of Labor Statistics (BLS). The CPI was initiated during World War I, a time of rapidly increasing prices, to help determine salary adjustments in the shipbuilding industry. Since the 1950s, the CPI has increased almost every year. It decreased just slightly in 2009 following the major 2008 economic crisis. It was the first time is had decreased since 1955.

As noted by the BLS, the CPI is not a true cost-of-living index:

> *The CPI frequently is called a cost-of-living index, but it differs in important ways from a complete cost-of-living measure. Both the CPI and a cost-of-living index would reflect changes in the prices of goods and services . . . that are directly purchased in the marketplace; but a complete cost-of-living index would go beyond this to also take into account changes in other governmental or environmental factors that affect consumers' well-being* (U.S. Bureau of Labor Statistics, 2013, CPI website).

Nonetheless, the CPI is the best available measure of changes in the cost of living in the United States.

The CPI measures changes in the cost of a "market basket" of goods and services that a typical consumer would be likely to purchase. The cost of that collection of goods and services is measured during a base period, then again at subsequent time periods. The CPI, at any given time period, is simply a comparison of current cost with cost during the base period. It is supposed to measure the changing cost of maintaining the same standard of living that existed during the base period.

There are actually three major versions of the consumer price index, but we will focus on the one that is most widely quoted, the CPI-U, for all urban consumers. It is estimated that this CPI covers about 88% of all U.S. consumers (http://www.bls.gov/news.release/cpi.nr0.htm). The CPI-U was introduced in 1978. The other two major CPIs are the CPI-W (for "wage earners") and the Chained CPI for urban consumers, designated as the C-CPI-U. The CPI-W is a continuation of the original one from the early 1900s. It is based on a subset of households covered by the CPI-U, for which "more than one-half of the household's income must come from clerical or wage occupations, and at least one of the household's earners must have been employed for at least 37 weeks during the previous 12 months" (http://www.bls.gov/cpi/cpifaq.htm). About 29% of the U.S. population is covered by the CPI-W.

The Chained CPI, or C-CPI-U, was introduced in August, 2002, after a report to Congress criticized the CPI for overestimating inflation. The Chained CPI addresses what's called "upper level substitution bias." Substitution bias refers to the fact that consumers are quite happy to substitute one item for another item if one of them is substantially cheaper. For example, if the price of fresh blueberries goes up too much, people might buy frozen blueberries. The CPI does adjust for lower level substitution, in which consumers substitute two foods in the same category, such as one type of cheese for another type. But upper level substitution refers to substitutions across categories, such as buying frozen instead of fresh fruit. The regular CPI-U only adjusts for this type of substitution every 2 years. But the Chained CPI adjusts for substitutions on a monthly basis, creating a continuous "chain" of adjustments, thus the name. Therefore, the Chained CPI is thought to be a more accurate measure of the month-to-month changes in cost-of-living. The use of the Chained CPI as a measure of inflation is controversial, as will be explained later in this chapter.

To understand how the CPI is calculated, let's first introduce the general concept of *price index numbers*. A price index for a given time period allows you to compare costs with another time period for which you also know the price index.

Price Index Numbers

A **price index number** measures prices (such as the cost of a dozen eggs) at one time period relative to another time period, usually as a percentage. For example, if a dozen eggs cost $1.00 in 2003 and $1.80 in 2013, then the egg price index would be ($1.80/$1.00) $\times$ 100 = 180%. In other words, a dozen eggs in 2013 cost 180% of what they cost in 2003. We could also say the price increased by 80%.

Price index numbers are commonly computed on a collection of products instead of just one. For example, we could compute a price index reflecting the increasing cost of attending college. To define a price index number, decisions about the following three components are necessary:

1. The base year or time period
2. The list of goods and services to be included
3. How to weight the particular goods and services

> *The general formula for computing a price index number is*
>
> $$\text{price index number} = (\text{current cost}/\text{base time period cost}) \times 100$$
>
> where "cost" is the weighted cost of the listed goods and services. Weights are usually determined by the relative quantities of each item purchased during either the current period or the base period.

EXAMPLE 18.1 A College Index Number

Suppose a senior graduating from college wanted to determine by how much the cost of attending college had increased for each of the 4 years she was a student. Here is how she might specify the three components:

1. Use her first year as a base.
2. Include yearly tuition and fees, yearly cost of room and board, and yearly average cost of books and related materials.

3. Weight everything equally because the typical student would be required to "buy" one of each category per year.

Table 18.1 illustrates how the calculation would proceed. We use the formula:

college index number = (current year total/first year total) × 100

Notice that the index for her senior year (listed in Table 18.1) is 127. This means that these components of a college education in her senior year cost 127% of what they cost in her first year. Equivalently, they have increased 27% since she started college.

TABLE 18.1 Cost of Attending College

Year	Tuition	Room and Board	Books and Supplies	Total	College Index
First	$9,500	$9,800	$1,400	$20,700	100
Sophomore	$10,260	$10,400	$1,440	$22,100	($22,100/$20,700) × 100 = 107
Junior	$11,700	$11,300	$1,500	$24,500	($24,500/$20,700) × 100 = 118
Senior	$12,500	$12,200	$1,600	$26,300	($26,300/$20,700) × 100 = 127

The Components of the Consumer Price Index

The Base Year (or Years)

The base year (or years) for the CPI changes periodically, partly so that the index does not get ridiculously large. If the original base year of 1913 were still used, the CPI would be well over 2000 and would be difficult to interpret. Since 1988, and continuing as of 2013, the base period in use for the CPI-U and CPI-W is the years 1982–1984. The previous base was the year 1967. Prior to that time, the base period was changed about once a decade. In December, 1996, the Bureau of Labor Statistics announced that, beginning with the January, 1999 CPI, the base period would change to 1993–1995. After it made that announcement, however, the BLS decided that it would retain the 1982–1984 base for the CPI-U and CPI-W (U.S. Dept. of Labor, 15 June 1998). The base year for the Chained CPI is 1999, because it did not exist in 1984.

The Goods and Services Included

As in the case with the base year(s), the market basket of goods and services is updated about once every 10 years. Items are added, deleted, and reorganized to represent current buying patterns. In 1998, a major revision occurred that included the addition of a new category called "Education and communication." The market basket as of 2013 consists of over 200 types of goods and services. It was established primarily based on the 2007–2008 Consumer Expenditure Survey, in which a multi-stage sampling plan was used to select families who reported their expenditures. That expenditure information, from about 30,000 individuals and families, was then used to determine the items included in the index.

The market basket includes most things that would be routinely purchased. These are divided into eight major categories, each of which is subdivided into varying numbers of smaller categories. The eight major categories are shown in Table 18.2. As noted, these categories are broken down into smaller ones. For example, here is the breakdown leading to the item "Ice cream and related products":

Food and beverages → Food at home → Dairy and related products → Ice cream and related products

Relative Quantities of Particular Goods and Services

Because consumers spend more on some items than on others, it makes sense to weight those items more heavily in the CPI. The weights assigned to each item are the relative quantities spent, as determined by the Consumer Expenditure Survey. The weights for the CPI-U are updated in January of even numbered years and then used for 2 years, whereas the Chained CPI weights are updated monthly. The weights are supposed to reflect the relative amount spent on each category, averaged across all consumers. Table 18.2 shows the CPI categories and the weights that went into effect for the CPI-U in 2002 and in 2012.

TABLE 18.2   Categories and Category Weights for the CPI-U in 2002 and 2012

Category	Weight in 2002	Weight in 2012
1. Food and beverages	15.6%	15.3%
2. Housing	40.9%	41.0%
3. Apparel	4.2%	3.6%
4. Transportation	17.3%	16.8%
5. Medical care	6.0%	7.2%
6. Recreation	5.9%	6.0%
7. Education and communication	5.8%	6.8%
8. Other goods and services	4.3%	3.4%
Total	100%	100.1% (due to rounding)

Source: http://www.bls.gov/cpi

You can see that housing is by far the most heavily weighted category. This makes sense, especially because costs associated with diverse items such as utilities and furnishings are included under the general heading of housing. You can also see that weights remained relatively constant for these broad categories from 2002 to 2012. They increased slightly for Medical care and for Education and communication and decreased slightly for Apparel and Transportation. Weights within subcategories would be more variable. For example, the weight for "wireless telephone services" rose from 0.65% in 2002 to 1.45% in 2012.

Obtaining the Data for the CPI

It is, of course, not possible to actually measure the average price of items paid by all families. The CPI is composed of samples taken from around the United States.

Each month the sampling occurs at about 26,000 retail and service establishments in 87 urban areas, and prices are measured on about 80,000 items. Rents are measured from about 50,000 landlords and tenants.

Obviously, determining the Consumer Price Index and trying to keep it current represent a major investment of government time and money. We now examine ways in which the index is used, as well as some of its problems.

18.2   Uses of the Consumer Price Index

Most Americans, whether they realize it or not, are affected by the Consumer Price Index. It is the most widely used measure of inflation in the United States.

Major Uses of the Consumer Price Index

> *There are four major uses of the Consumer Price Index:*
>
> **1.** The CPI is used to evaluate and determine economic policy.
> **2.** The CPI is used to compare prices in different years.
> **3.** The CPI is used to adjust other economic data for inflation.
> **4.** The CPI is used to determine salary and price adjustments.

The CPI is used to evaluate and determine economic policy

As a measure of inflation, the CPI is of interest to the president, Congress, the Federal Reserve Board, private companies, and individuals. Government officials use it to evaluate how well current economic policies are working. Private companies and individuals also use it to make economic decisions.

The CPI is used to compare prices in different years

If your parents bought a new car in 1983 for $10,000, what would you have paid in 2012 for a similar quality car, using the CPI to adjust for inflation? The CPI in 1983 was very close to 100 (depending on the month), and the average CPI in 2012 was about 230. Therefore, an equivalent price in 2012 would be about $10,000 × (230/100) = $23,000. The general formula for determining the comparable price for two different time periods is

price in time 2 dollars = (price in time 1 dollars) × [(CPI at time 2)/(CPI at time 1)]

For this formula to work, all CPIs must be adjusted to the same base period. When the base period is updated, past CPIs are all adjusted to the new period. Thus, the CPIs of years that precede the current base period are generally less than 100; those of years that follow the current base period are generally over 100. Table 18.3 (next page) shows the CPI-U at the start of each decade, using the 1982–1984 base period. (Numbers represent the average CPI for the year.)

TABLE 18.3   Consumer Price Index at Start of Each Decade, 1982–84 = 100

Year	1920	1930	1940	1950	1960	1970	1980	1990	2000	2010
CPI	20.0	16.7	14.0	24.1	29.6	38.8	82.4	130.7	172.2	218.1

Source: ftp://ftp.bls.gov/pub/special.requests/cpi/cpiai.txt

It is striking how much the CPI increased from 1970 to 2010, after relatively modest increases and decreases from 1920 to 1970. Here are two examples that illustrate how prices have changed.

EXAMPLE 18.2   How to Become a 1970s Millionaire

When the baby boomers were growing up in the 1960s and 1970s, becoming a millionaire seemed completely out of reach to the average citizen. There were about 450,000 millionaires in the United States in 1978 (Kangas, 2013). That figure represented about 1 out of every 500 people. But by 2013, there were reportedly about 3.7 *million* millionaires in the United States, according to rt.com (2013), representing more than 1 out of every 100 people. Of course, that's still a very small proportion, but it should be clear that a million dollars isn't what it used to be! Let's see what a million dollars in 1978 would be worth in 2012, and vice versa. (Any guesses, before you look?) The average annual CPI-U for 1978 and 2012 were about 65.2 and 229.6, respectively, using the 1982–84 base. Here's how to compute the 2012 value of a million dollars in 1978, and the 1978 value of a million dollars in 2012.

Value in 2012 dollars = (value in 1978 dollars) × (CPI in 2012)/(CPI in 1978) = $1,000,000 × (229.6/65.2) = $3,521,472.

Value in 1978 dollars = (value in 2012 dollars) × (CPI in 1978)/(CPI in 2012) = $1,000,000 × (65.2/229.6) = $283,972.

The first result tells us that if you wanted to be as wealthy in 2012 as a millionaire was in 1978, you would need to have over $3.5 million dollars! Another way to express this is to say that if you wanted to be a millionaire in 1978, the requirement in 2012 dollars would be about $3.5 million.

The second result tells us that if you wanted to be as wealthy in 1978 as a millionaire was in 2012, you would need only $283,972. Or we could say that to be a millionaire in 2012 would require $1,000,000 in 2012 dollars, but only $283,972 in 1978 dollars. ▪

EXAMPLE 18.3   The Cost of Room and Board at Big Ten Universities

In 2001–02, the average cost of room and board for a double room in the dorms at the 12 universities in the "Big Ten" Conference was $5506 (University of Illinois, 2012). What would the equivalent cost be in 2013–14 dollars, using the CPI-U to adjust for inflation? Let's use the CPI-U for May, 2001 (177.7) and May, 2013 (232.9) to find the answer:

Cost in 2013 dollars = (cost in 2001 dollars) × (CPI in May 2013)/(CPI in May 2001) = $5506 × (232.9/177.7) = $5506 × 1.31 = $7213.

One way to say this is that the 2001–02 cost of room and board in 2013 dollars was $7213. We could also say that in 2013, the 2001–02 cost of room and board adjusted for inflation would be $7213. Data for the 2013–14 year are not available as of this writing, but the average cost in 2010–11 was about $8895, already exceeding what the inflation-adjusted cost should have been in 2013–14. Clearly, room and board costs at these universities have increased faster than the inflation rate as defined by the CPI. ▪

The CPI is used to adjust other economic data for inflation

If you were to plot just about any economic measure over time, you would see an increase simply because—at least historically—the value of a dollar decreases every year. To provide a true picture of changes in conditions over time, most economic data are presented in values adjusted for inflation. You should always check reports and plots of economic data over time to see if they have been adjusted for inflation.

EXAMPLE 18.4

Why the Dow Jones Industrial Average Keeps Reaching All-time Highs

The Dow Jones Industrial Average (DJIA) is a weighted average of the price of 30 major stocks on the New York Stock Exchange. It used to be a straight dollar average, but as stocks split it got more complicated. Nonetheless, it still represents the monetary value of the stocks. On May 7, 2013, the DJIA reached an all-time high by closing over 15,000 for the first time, at 15,056. On May 3, 1999, it had reached an all-time high by closing over 11,000 for the first time, at just below 11,015. In fact, it reaches an all-time high almost every year! Why? Because the DJIA is not adjusted for inflation; it is simply reported in current dollars. Thus, to compare the high in one year with that in another, we need to adjust it using the CPI.

Let's compare the two all-time highs already quoted with the DJIA high when it crossed 1000 for the first time, to close just over 1003 on November 14, 1972. Using the 1982–84 base, the CPI in November, 1972, May, 1999, and May, 2013 were, respectively, 42.4, 166.2, and 232.9. Did the DJIA rise faster than inflation? To determine if it did, let's calculate what the November, 1972 high of 1003 would have been in 1999 and 2013 dollars:

Value in 1999 = (value in 1972) × (CPI in 1999)/(CPI in 1972) = (1003) × (166.2/42.4) = 3932

Value in 2013 = (value in 1972) × (CPI in 2013)/(CPI in 1972) = (1003) × (232.9/42.4) = 5509

Therefore, the Dow Jones Industrial Average clearly did rise faster than inflation between 1972 and each of the two target years. In fact, comparing the actual 1999 value of 11,015 to the 1972 value in 1999 dollars of gives a ratio of 11015/3932 = 2.8. The DJIA increased at a rate of almost three times the rate of inflation! Did it continue to get that much better between 1999 and 2013? Let's look:

Value in 2013 = (value in 1999) × (CPI in 2013)/(CPI in 1999) = (11015) × (232.9/166.2) = 15,436.

Oops! The high of 15,056 does not exceed the inflation-adjusted value of 15,436. Now the ratio is 15,056/15,436 = 0.975. Therefore, it may sound impressive that the DJIA increased from about 11,000 to 15,000 between 1999 and 2013, but in fact that increase is actually less than the rate of inflation. It's important to notice whether monetary values have been inflation-adjusted. ▪

The CPI is Used to Determine Salary and Price Adjustments

The CPI is commonly used to determine cost of living increases. For instance, the annual cost of living adjustment for Social Security recipients is based on the annual change in the CPI-W. According to the Bureau of Labor Statistics:

> *As a result of statutory action, the CPI affects the income of millions of Americans. Over 50 million Social Security beneficiaries, and military and Federal Civil Service retirees, have cost-of-living adjustments tied to the CPI. In addition, eligibility criteria for millions of food stamp recipients and children who eat lunch at school are affected by changes in the CPI. Many collective bargaining agreements also tie wage increases to the CPI* (http://www.bls.gov/cpi/cpiadd.htm).

In 1996, an Advisory Commission to the U.S. Senate Committee on Finance, commonly known as the Boskin Commission, made numerous recommendations for changes in the CPI (U.S. Senate, 1996). The recommendations resulted in some major changes, the most notable of which was the introduction of the Chained CPI described on page 386. The major changes and their consequences 10 years after the Boskin Commission are discussed in a report by statistician David Johnson and colleagues (Johnson, Reed, and Stewart, 2006). The report provides details of some of the controversies surrounding the CPI. We turn to those next.

18.3   Criticisms of the Consumer Price Index

Although the Consumer Price Index may be the best measure of inflation available, it does have problems. The Boskin Commission estimated that the CPI overestimates inflation by about 1.1% annually, mostly because of substitutions of one product for another, and changes in quality, such as the increasing sophistication of electronics (Johnson et al., 2006). Further, the CPI may overstate the effect of inflated prices for the items it covers on the average worker's standard of living. The following criticisms of the CPI should help you understand these and other problems with its use.

Some criticisms of the CPI:

1. The market basket used in the CPI may not reflect current spending priorities.

2. If the price of one item rises, consumers are likely to substitute another.

3. The CPI does not adjust for changes in quality.

4. The CPI does not take advantage of sale prices and online shopping.

5. The CPI does not measure prices for rural Americans.

---

1. *The market basket used in the CPI may not reflect current spending priorities.* Remember that the market basket of goods and the weights assigned to them are changed infrequently. This does not reflect rapid changes in lifestyle and technology.

2. *If the price of one item rises, consumers are likely to substitute another.* If the price of beef rises substantially, consumers will buy chicken instead. If the price of fresh vegetables goes up due to poor weather conditions, consumers will use canned or frozen vegetables until prices go back down. When the price of lettuce tripled a few years ago, consumers were likely to buy fresh spinach for salads instead. In the past, the CPI has not taken these substitutions into account. Starting with the January, 1999 CPI, substitutions *within* a subcategory such as "Ice cream and related products" have been taken into account through the use of a new statistical method for combining data. It is estimated that this change reduces the annual rate of increase in the CPI by approximately 0.2 percentage point (U.S. Dept. of Labor, 16 April 1998). But the regular CPI-U and CPI-W do not take into account substitutions *across* subcategories. As previously discussed, that's the reason for the creation of the Chained CPI.

3. *The CPI does not account for changes in quality.* The CPI assumes that if you purchase the same items in the current year as you did in the base year, your standard of living will be the same. That may apply to food and clothing but does not apply to many other goods and services. For example, computers were not only more expensive in 1982–1984, they were also much less powerful. Owning a new computer now would add more to your standard of living than owning one in 1982 would have done.

4. *The CPI does not take advantage of sale prices and online shopping.* The outlets (stores, restaurants, etc.) used to measure prices for the CPI are chosen by random sampling methods. The outlets consumers choose are more likely to be based on the best price that week, or on comparing prices online. Further, if a supermarket is having a sale on an item you use often, you will probably stock up on the item at the sale price and then not need to purchase it for a while. The CPI does not take this kind of money-saving behavior into account.

5. *The CPI does not measure prices for rural Americans.* As mentioned earlier, the CPI-U is relevant for about 88% of the population: those who live in and around urban areas. It does not measure prices for the rural population, and we don't know whether it can be extended to that group. The costs of certain goods and services are likely to be relatively consistent. However, if the rise in the CPI in a given time period is mostly due to rising costs particular to urban dwellers, such as the cost of public transportation and apartment rents, then it may not be applicable to rural consumers.

The introduction of the Chained CPI has created what may be the biggest controversy in the history of the Consumer Price Index, which is whether it should replace the CPI-W for determining cost of living increases for Social Security recipients. By the time you read this, the controversy may be resolved, but the issues will remain. At the heart of the problem is whether one or the other—or neither—is truly a good measure of inflation for Social Security recipients, who are mostly senior citizens. The Chained CPI has consistently measured inflation at a lower rate than the CPI-W. Therefore, advocates for senior citizens argue against making a change, while advocates for reduced government spending argue in favor of making the change. Even neutral observers in the controversy know that neither index uses category weights that accurately reflect the spending patterns of senior citizens. For instance, they are likely to spend more on medical care than younger consumers, and if they no longer have a mortgage, may spend less on housing.

The Bureau of Labor Statistics notes that the CPI is not a cost-of-living index and should not be interpreted as such. It is most useful for comparing prices of similar products in the same geographic area across time. The BLS routinely studies and implements changes in methods that lead to improvements in the CPI. Current information about the CPI can be found on the CPI pages of the BLS website; that address, as of this writing, is www.bls.gov/cpi/.

18.4   Seasonal Adjustments: Reporting the Consumer Price Index

You may recall from Chapter 9 that most time series involving economic data or data related to people's behavior have **seasonal components.** In other words, they tend to be high in certain months or seasons and low in others every year. For example, new housing starts are much higher in warmer months. Sales of toys and other standard gifts are much higher just before Christmas. U.S. unemployment rates tend to rise in January, when outdoor jobs are minimal and the Christmas season is over, and again in June, when a new graduating class enters the job market.

The CPI and most of the other economic indicators discussed in the next section are subject to seasonal fluctuations. They are usually reported after they have been *seasonally adjusted.*

Economists have sophisticated methods for seasonally adjusting time series. They use data from the same month or season in prior years to construct a *seasonal factor,* which is a number either greater than one or less than one by which the current figure is multiplied. According to the U.S. Department of Labor (1992, p. 243), "the standard practice at BLS for current seasonal adjustment of data, as it is initially released, is to use projected seasonal factors which are published ahead of time." In other words, when figures such as the Consumer Price Index become available for a given month, the BLS already knows the amount by which the figures should be adjusted up or down to account for the seasonal component.

It is unusual to see the Consumer Price Index itself reported in the news. More commonly, what you see reported is the *change* from the previous month, which is generally reported in the middle of each month. Following is an example of how it was reported by Bloomberg News for the May, 2013 CPI:

Consumer Prices in U.S. Increased Less Than Forecast in May   *The cost of living in the U.S. rose less than forecast in May, restrained by the first drop in food prices in almost four years and signaling inflation remains under control. The consumer price index was up 0.1 percent after falling 0.4 percent in April, the Labor Department reported today in Washington* (Woellert, 2013).

Most news reports never tell you the actual value of the CPI. The article just quoted did not reveal that the CPI for that month (May, 2013) was 232.9. This is not surprising, because the Bureau of Labor Statistics press release, on which all such articles are based, does not give the actual CPI value until near the end. What makes news is the change, not the actual CPI value.

A more important omission in the article quoted, and others like it, is the failure to mention that the reported change in CPI has been *seasonally adjusted*. In other words, the change of 0.1% does not represent an absolute change in the Consumer Price Index; rather it represents a change *after* seasonal adjustments have been made. Adjustments have already been made for the fact that certain items are expected to cost more during certain months of the year. The press release from the Bureau of Labor Statistics explained the reason for using seasonal adjustments:

> *For analyzing general price trends in the economy, seasonally adjusted changes are usually preferred since they eliminate the effect of changes that normally occur at the same time and in about the same magnitude every year--such as price movements resulting from changing climatic conditions, production cycles, model changeovers, holidays, and sales. The unadjusted data are of primary interest to consumers concerned about the prices they actually pay* (http://www.bls.gov/news.release/cpi.nr0.htm, accessed July 5, 2013).

The Bureau of Labor Statistics press release is thus recognizing what you should acknowledge as common sense. It is important that economic indicators be reported with seasonal adjustments; otherwise, it would be impossible to determine the direction of the real trend. For example, it would probably always appear as if new housing starts dipped in February and jumped in May. Therefore, it is prudent reporting to include a seasonal adjustment. Always check to see if economic statistics include the adjustment.

Why Are Changes in the CPI Big News?

You may wonder why it's the changes in the CPI that are reported as the big news. The reason is that financial markets are extremely sensitive to changes in the rate of inflation. For example, on June 17, 2009, a CNN Money News headline read "Consumer price index: Largest drop in 59 years." The article went on to report:

> *A key index of prices paid by consumers showed the largest year-over-year decline since April 1950, primarily due to sinking energy prices, the government said Wednesday. The Consumer Price Index, the Labor Department's key measure of inflation, has fallen 1.3% over the past year. That's the largest decline in nearly 60 years, and is due mainly to a 27.3% decline in the energy index* (Pepitone, 2009).

Probably no one would have noticed the article if the headline had read "Consumer Price Index was 213.8 in May." Like anything else in the world, it is the changes that attract concern and attention, not the continuation of the status quo.

18.5   Economic Indicators

The Consumer Price Index is only one of many economic indicators produced or used by the U.S. government. Historically, the Bureau of Economic Analysis (BEA), part of the Department of Commerce, classified and monitored a whole host of such indicators. In 1995, the Department of Commerce turned over the job of producing and monitoring some of these indicators to the not-for-profit, private organization called The Conference Board.

Most economic indicators are series of data collected across time, like the CPI. Some of them measure financial data, others measure production, and yet others measure consumer confidence and behavior. Here is a list of 10 series, randomly selected by the author from a table of 103 economic indicators accepted by the BEA (provided by Stratford and Stratford, 1992), to give you an idea of the variety. The letters in parentheses will be explained in the following section.

09   Construction contracts awarded for commercial and industrial buildings, floor space (L,C,U)

10   Contracts and orders for plant and equipment in current dollars (L,L,L)

14   Current liabilities of business failure (L,L,L)

25   Changes in manufacturers' unfilled orders, durable goods industries (L,L,L)

27   Manufacturers' new orders in 1982 dollars, nondefense capital goods industries (L,L,L)

39   Percent of consumer installment loans delinquent 30 days or over (L,L,L)

51   Personal income less transfer payments in 1982 dollars (C,C,C)

84   Capacity utilization rate, manufacturing (L,C,U)

110   Funds raised by private nonfinancial borrowers in credit markets (L,L,L)

114   Discount rate on new issues of 91-day Treasury bills (C,LG,LG)

You can see that even this randomly selected subset covers a wide range of information about government, business and consumer behavior, and economic status.

Leading, Coincident, and Lagging Indicators

Most indicators move with the general health of the economy. The Conference Board classifies economic indicators according to whether their changes precede, coincide with, or lag behind changes in the economy.

A **leading economic indicator** is one in which the highs, lows, and changes tend to *precede* or *lead* similar changes in the economy. (Contrary to what you may have thought, the term does not convey that it is one of the most *important* economic indicators.) A **coincident economic indicator** is one with changes that *coincide* with those in the economy. A **lagging economic indicator** is one whose changes *lag behind* or follow changes in the economy.

To further complicate the situation, some economic indicators have highs that precede or lead the highs in the economy but have lows that are coincident with or lag behind the lows in the economy. Therefore, the indicators are further classified according to how their highs, lows, and changes correspond to similar behavior in the economy. The sample of 10 indicators shown in the previous section are classified this way. The letters following each indicator show how the highs, lows, and changes, respectively, are classified for that series.

The code letters are L = Leading, LG = Lagging, C = Coincident, and U = Unclassified. For example, the code letters in indicator 10, "Contracts and orders for plant and equipment in current dollars (L,L,L)," show that this indicator leads the economy in all respects. When this series remains high, remains low, or changes, the economy tends to follow. In contrast, the code letters in indicator 114, "Discount rate on new issues of 91-day Treasury bills (C,LG,LG)," show that this indicator has highs that are coincident with the economy but has lows and changes that tend to lag behind the economy.

Composite Indexes

Rather than require decision makers to follow all of these series separately, the Conference Board produces composite indexes. The *Index of Leading Economic Indicators* for the United States is comprised of 10 series, listed in Table 18.4. Most, but not all, of the individual component series are collected by the U.S. government. For instance, the S&P 500 (Item #7) is provided by Standard and Poor's Corporation, whereas Item #10, Average consumer expectations for business conditions, is part of a broader consumer confidence index provided by the University of Michigan's Survey Research Center.

The *Index of Coincident Economic Indicators* is comprised of four series; the *Index of Lagging Economic Indicators,* seven series. These indexes are produced monthly, quarterly, and annually. The Conference Board now publishes global economic indices and separate indices for a number of other countries as well.

Behavior of the Index of Leading Economic Indicators is thought to precede that of the general economy by about 6 to 9 months. This is based on observing past

---

TABLE 18.4   Components of the Index of Leading Economic Indicators

1. Average weekly hours, manufacturing
2. Average weekly initial claims for unemployment insurance
3. Manufacturers' new orders, consumer goods, and materials
4. Institute for Supply Management Index of New Orders
5. Manufacturers' new orders, nondefense capital goods excluding aircraft orders
6. Building permits, new private housing units
7. Stock prices, 500 common stocks (the S&P 500)
8. Leading Credit Index™
9. Interest rate spread, 10-year Treasury bonds less federal funds
10. Average consumer expectations for business conditions

Source: www.conference-board.org

performance and not on a causal explanation—that is, it may not hold in the future because there is no obvious cause and effect relationship. In addition, monthly changes can be influenced by external events that may not predict later changes in the economy. There are other reasons why the Index maybe limited as a predictor of the economic future. For instance, the Index focuses on manufacturing, yet it is no longer the case that the majority of jobs in the United States are related to manufacturing. Nevertheless, although the Index may not be ideal, it is still the most commonly quoted source of predictions about future economic behavior.

CASE STUDY 18.1   **Did Wages Really Go Up in the Reagan–Bush Years?**

It was the fall of 1992, and the United States presidential election was imminent. The Republican incumbent, George Bush (Senior), had been president for the past 4 years, and vice president to Ronald Reagan for 8 years before that. One of the major themes of the campaign was the economy. Despite the fact that the federal budget deficit had grown astronomically during those 12 Reagan–Bush years, the Republicans argued that Americans were better off in 1992 than they had been 12 years earlier. One of the measures they used to illustrate their point of view was the average earnings of workers. The average wages of workers in private, nonagricultural production had risen from $235.10 per week in 1980 to $345.35 in 1991 (*World Almanac and Book of Facts,* 1995, p. 150).

Were those workers really better off because they were earning almost 50% more in 1991 than they had been in 1980? Supporters of Democratic challenger Bill Clinton didn't think so. They began to counter the argument with some facts of their own.

Based on the material in this chapter, you can decide for yourself. The Consumer Price Index in 1980, measured with the 1982–1984 baseline, was 82.4. For 1991, it was 136.2. Let's see what average weekly earnings in 1991 should have been, adjusting for inflation, to have remained constant with the 1980 average:

$$\text{salary at time 2} = (\text{salary at time 1}) \times [(\text{CPI at time 2})/(\text{CPI at time 1})]$$
$$\text{salary in 1991} = (\$235.10) \times [(136.2)/(82.4)] = \$388.60$$

Therefore, the average weekly salary actually dropped during those 11 years, adjusted for inflation, from the equivalent of $388.60 to $345.35. The actual average was only 89% of what it should have been to have kept up with inflation.

There is another reason why the argument made by the Republicans would sound convincing to individual voters. Those voters who had been working in 1980 may very well have been better off in 1991, even adjusting for inflation, than they had been in 1980. That's because those workers would have had an additional 11 years of seniority in the workforce, during which their relative positions should have improved. A meaningful comparison, which uses average wages adjusted for inflation, would not apply to an individual worker. ∎

Thinking About Key Concepts

- A *price index number* is the ratio of the current value of something to its value at another time, usually a *base period*, multiplied by 100. The index number for the base period is 100.
- To find a price index number for a collection of items, the items receive *weights* that reflect their relative portion of the total.
- The United States *Consumer Price Index (CPI)* is based on a collection of over 200 items divided into eight categories, with weights reflecting actual consumer spending patterns. The *base period* as of 2013 is 1982 to 1984.
- The CPI is used to determine economic policy, to compare prices across time, to measure inflation, and to provide cost of living increases.
- The CPI has been criticized as a measure of inflation because it is not adjusted as quickly as consumers change their spending behavior.
- The *Chained CPI (C-CPI-U)* was created to address the criticism that the standard CPI overestimates inflation by not responding quickly enough to substitutions made by consumers. The C-CPI-U "chains" consumer behavior from one month to the next to make adjustments in weights applied to items.
- The CPI should be reported as *seasonally adjusted* if it is to be used as a measure of economic health.
- The *index of leading economic indicators* is a list of 10 financial measures that tend to predict changes that might happen in the economy in the next 6 to 9 months.

Exercises

Exercises with numbers divisible by 3 (3, 6, 9, etc.) are included in the Solutions at the back of the book. They are marked with an asterisk (*).

1. The price of a first-class stamp in the United States in 1970 was 8 cents, whereas in 2012, it was 44 cents. The Consumer Price Index for 1970 was 38.8, whereas for 2012, it was 229.6. If the true cost of a first-class stamp did not increase between 1970 and 2012, what should it have cost in 2012? In other words, what would an 8-cent stamp in 1970 cost in 2012, when adjusted for inflation?

2. As shown in Table 18.2, when the CPI was computed for 2012, the relative weight for the food and beverages category was 15.3%, whereas for the recreation category, it was only 6.0%. Explain why food and beverages received higher weight than recreation.

*3. A paperback novel cost $3.49 in 1981, $6.99 in 1995, and $14.00 in 2013. Compute a "paperback novel price index" for 1995 and 2013 using 1981 as the base year. In words that can be understood by someone with no training in statistics, explain what the resulting numbers mean.

4. If you wanted to present a time series of the yearly cost of tuition at your local college for the past 30 years, adjusted for inflation, how would you do the adjustment?

5. The Dow Jones Industrial Average reached a high of $7801.63 on December 29, 1997. Recall from Example 18.4 that it reached a high of $1003 on November 14, 1972. The Consumer Price Index for November 1972 was 42.4; for December 1997, it was 161.3. By what percentage did the high in the DJIA increase from November 14, 1972, to December 29, 1997, after adjusting for inflation?

*6. The CPI in July, 1977, was 60.9; in July, 1994, it was 148.4.

  *a. The salary of the governor of California in July, 1977, was $49,100; in July 1994, it was $120,000. Compute what the July, 1977, salary would be in July, 1994, adjusted for inflation, and compare it with the actual salary in July, 1994.

  *b. The salary of the president of the United States in July, 1977, was $200,000. In July, 1994, it was still $200,000. Compute what the July, 1977, salary would be in July, 1994, adjusted for inflation, and compare it with the actual salary.

7. Find out what the tuition and fees were for your school for the previous 4 years and the current year. Using the cost 5 years ago as the base, create a "tuition index" for each year since then. Write a short summary of your results that would be understood by someone who does not know what an index number is.

8. As mentioned in this chapter, both the base year and the relative weights used for the Consumer Price Index are periodically updated.

  a. Why is it important to update the relative weights used for the CPI?

  b. Explain why the base year is periodically updated.

*9. The CPIs at the start of each decade from 1940 to 2010 were are shown in Table 18.5.

  *a. Determine the percentage increase in the CPI for each decade.

  *b. During which decade was inflation the highest, as measured by the percentage change in the CPI?

  *c. During which decade was inflation the lowest, as measured by the percentage change in the CPI?

10. In addition to the overall CPI, the BLS reports the index for the subcategories. The overall CPI in May, 2013, was 232.9. Following are the values for some of the subcategories, taken from the CPI website:

Dairy products	216.3
Fruits, vegetables	289.2
Alcoholic beverages	234.4
Rent (primary residence)	266.6
House furnishings	125.4
Footwear	136.4
Tobacco products	869.0

All of these values are based on using 1982–1984 as the base period, the same period used by the overall CPI.

  a. Find the subcategory in the list that has the least amount of inflation, compared to the base period. Write a few sentences comparing it to the overall CPI that could be understood by someone who does not know what index numbers are.

  b. Find the subcategory in the list that has the most amount of inflation, compared to the base period. Write a few sentences comparing

TABLE 18.5

Year	1940	1950	1960	1970	1980	1990	2000	2010
CPI	14.0	24.1	29.6	38.8	82.4	130.7	172.2	218.1

TABLE 18.6

Year	1987	1988	1989	1990	1991	1992	1993
Amount spent	$1649	$1991	$2173	$2396	$2585	$2763	$2950

it to the overall CPI that could be understood by someone who does not know what index numbers are.

c. Write a brief report discussing all of the information that could be understood by someone who does not know what index numbers are.

11. Table 18.6 shows the amounts Americans spent for medical care, per capita, between 1987 and 1993 (*World Almanac and Book of Facts*, 1995, p. 128 and Census Bureau).

   a. Create a "medical care index" for each of these years, using 1987 as a base.

   b. Comment on how the cost of medical care changed between 1987 and 1993, relative to the change in the Consumer Price Index, which was 113.6 in 1987 and 144.5 in 1993.

*12. Suppose that you gave your niece a check for $50 on her 13th birthday in 2010, when the CPI was 218.06. Your nephew is now about to turn 13. You discover that the CPI is now 244.53. How much should you give your nephew if you want to give him the same amount you gave your niece, adjusted for inflation?

13. An article in the *Sacramento Bee* (Stafford, 2003) on July 7, 2003, reported that the current minimum wage at that time was only $5.15 an hour and that it had not kept pace with inflation. The Consumer Price Index at the time (the end of June, 2003) was 183.7.

   a. One of the quotes in the article was "to keep pace with inflation since 1968, the minimum wage should be $8.45 an hour today." In 1968, the minimum wage was $1.60 an hour, and the Consumer Price Index was 34.8.

Explain how the author of the article determined that the minimum wage should be $8.45 an hour.

   b. The minimum wage was initiated in October, 1938, at $0.25 an hour. The Consumer Price Index in 1938 was 14.1, using the 1982–1984 base years. If the minimum wage had kept pace with inflation from its origination, what should it have been at the end of June, 2003? Compare your answer to the actual minimum wage of $5.15 an hour in June, 2003.

   c. The minimum wage of $5.15 an hour quoted in the article was set in September, 1997, when the Consumer Price Index was 161.2. If it had kept pace with inflation, what should it have been at the end of June, 2003?

   d. Based on your answers to parts (a) to (c), has the minimum wage always kept pace with inflation, never kept pace with inflation, or some combination?

14. Refer to the previous exercise. Find out the current minimum wage and the current Consumer Price Index. (These were available as of July, 2013, at the websites http://www.dol.gov/dol/topic/wages/ and http://www.bls.gov/cpi/, respectively.) Determine what the minimum wage should be at the current time if it had kept pace with inflation from

   a. 1950, when the CPI was 24.1 and the minimum wage was $0.75 an hour.

   b. 1960, when the CPI was 29.6 and the minimum wage was $1.00 an hour.

   c. 1990, when the CPI was 130.7 and the minimum wage was $3.80 an hour.

   d. 2013, when the CPI was 183.7 and the minimum wage was $5.15 an hour.

*15. In 1950, being a millionaire was touted as a goal that would be achievable by very few people. The CPI in 1950 was 24.1, and in 2002, it was 179.9. How much money would one need to have in 2002 to be the equivalent of a millionaire in 1950, adjusted for inflation? Would it still have seemed like a goal achievable by very few people in 2002?

16. The United States Census Bureau, *Statistical Abstract of the United States 1999* (p. 877) contains a table listing median family income for each year from 1947 to 1997. The incomes are presented "in current dollars" and "in constant (1997) dollars." As an example, the median income in 1985 in "current dollars" was $27,735 and in "constant (1997) dollars" it was $41,371. The CPI in 1985 was 107.6 and in 1997, it was 160.5.

    a. Using these figures for 1985 as an illustration, explain what is meant by "in constant (1997) dollars."

    b. The median family income in 1997 was $44,568. After adjusting for inflation, compare the 1985 and 1997 median incomes. Report the percent increase or decrease from 1985 to 1997.

    c. Name one advantage to reporting the incomes in "current dollars" and one advantage to reporting the incomes in "constant dollars."

17. In explaining why it is a costly mistake to have the CPI overestimate inflation, the Associated Press (20 October 1994) reported, "Every 1 percentage point increase in the CPI raises the federal budget deficit by an estimated $6.5 billion." Explain why that would happen.

*18. Explain which of the criticisms of the CPI given in Section 18.3 is supposed to be addressed by the use of the Chained CPI.

19. Many U.S. government payments, such as Social Security benefits, are increased each year by the percentage change in the CPI. In 1995, the government started discussions about lowering these increases or changing the way the CPI is calculated.

As discussed in this chapter, one result was the Chained CPI. According to an article in the *New York Times*, "most economists who have studied the issue closely say the current system is too generous to Federal beneficiaries… the pain of lower COLAs [cost of living adjustments] would be unavoidable but nonetheless appropriate" (Gilpin, 1995, p. D19). Explain in what sense some economists believe the current system (in place in 1995 and continuing in 2013) is too generous.

20. Remember that the CPI is supposed to measure the change in what it costs to maintain the same standard of living that was in effect during the base year(s). Using the material in Section 18.3, explain why it may not do so accurately.

*21. Most news accounts of the Consumer Price Index report the percentage change in the CPI from the previous month rather than the value of the CPI itself. Why do you think that is the case?

22. Recall from Chapter 9 that time series (such as the CPI) have three nonrandom components. Which of the three nonrandom components (trend, seasonal, or cycles) is likely to contribute the most to the unadjusted Consumer Price Index? Explain.

23. The Bureau of Labor Statistics reports that one use of the Consumer Price Index is to periodically adjust the federal income tax structure, which sets higher tax rates for higher income brackets. According to the BLS, "these adjustments prevent inflation-induced increases in tax rates, an effect called 'bracket creep'" (U.S. Dept. of Labor, 2003, CPI website). Explain what is meant by "bracket creep" and how you think the CPI is used to prevent it.

*24. Two of the economic indicators measured by the U.S. government are "Number of employees on nonagricultural payrolls" and "Average duration of unemployment, in weeks." One of these is designated as a "lagging economic indicator," and the other is a "coincident economic indicator." Explain which you think is which and why.

25. One of the components of the Index of Leading Economic Indicators is the "Average consumer expectations for business conditions." Why do you think this index would be a leading economic indicator?

26. In February of 1994, the Index of Leading Economic Indicators dropped slightly, and economists blamed it on unusually severe winter weather that month. Examine the 10 series that make up the Index of Leading Economic Indicators, listed in Table 18.4. Choose at least two of the series to support the explanation that the drop in these indicators in February was partially due to unusually severe winter weather.

Mini-Projects

1. Numerous economic indicators are compiled and reported by the U.S. government and by private companies. Various sources are available at the library and on the Internet to explain these indicators. Write a report on one of the following. Explain how it is calculated, what its uses are, and what limitations it might have.

  a. The Producer Price Index

  b. The Gross Domestic Product

  c. The Dow Jones Industrial Average

2. Find a news story that reports on current values for one of the indexes discussed in this chapter. Discuss the news report in the context of what you have learned in this chapter. For example, does the report contain any information that might be misleading to an uneducated reader? Does it omit any information that you would find useful? Does it provide an accurate picture of the current situation?

3. Refer to Example 18.4. In addition to the Dow Jones Industrial Average, there are other indicators of fluctuation in stock prices. Two examples are the New York Stock Exchange Composite Index and the Standard and Poor's 500. Choose a stock index (other than the Dow Jones), and write a report about it. Include whether it is adjusted for inflation, seasonally adjusted, or both. Give information about its recent performance, and compare it with performance a few decades ago. Make a conclusion about whether the stock market has gone up or down in that time period, based on the index you are using, adjusted for inflation.

References

Associated Press. (20 October 1994). U.S. ready to overhaul measure of inflation. *The Press of Atlantic City*, p. A-8.

Cage, R., J. Greenlees and P. Jackman. (2003). Introducing the chained consumer price index, http://www.bls.gov/cpi/super_paris.pdf, accessed July 5, 2013.

Gilpin, Kenneth N. (22 February 1995). Changing an inflation gauge is tougher than it sounds. *New York Times,* pp. D1, D19.

Johnson, D. S., S. B. Reed, and K. J. Stewart. (2006). Price measurement in the United States: A decade after the Boskin Report, *Monthly Labor Review,* May 2006, pp. 10–19.

Kangas, Steve. (2013). Income and wealth inequality. http://www.huppi.com/kangaroo/4Inequality .htm, accessed July 5, 2013.

Pepitone, J. (2009). Consumer price index: Largest drop in 59 years. http://money.cnn .com/2009/06/17/news/economy/cpi_consumer_price_index/, 17 June 2009, accessed July 5, 2013.

rt.com (2013). World millionaires' wealth totals $46.2 trillion, over 3 times US GDP. http:// rt.com/business/millionaires-wealth-us-gdp-941/, 19 June 2013, accessed July 5, 2013.

Stafford, Diane. (7 July 2003). Minimum wage job—it's a struggle. *Sacramento Bee,* p. D3.

Stratford, J. S., and J. Stratford. (1992). *Major U.S. statistical series: Definitions, publications, limitations.* Chicago: American Library Association.

University of Illinois (2012). Background information concerning tuition and financial aid: An update for FY2012. http://www.pb.uillinois.edu/Documents/tuitionenrollment/FY-2012-Tuition-Book.pdf, accessed July 5, 2013.

U.S. Bureau of Labor Statistics. (16 April 1998). Planned changes in the Consumer Price Index formula. News release.

U.S. Department of Labor. Bureau of Labor Statistics. (15 June 1998). Consumer Price Index summary. News release.

U.S. Bureau of Labor Statistics. (2013). CPI Web site: http://www.bls.gov/dolfaq/bls_ques2.htm.

U.S. Senate. Committee on Finance. (1996). Final report of the Advisory Commission to Study the Consumer Price Index. Print 104-72, 104 Congress, 2 session. Washington, D.C.: Government Printing Office. (Available from http://www.finance.senate.gov)

Woellert, L. (2013). Consumer prices in U.S. increased less than forecast in May. http://www .bloomberg.com/news/2013-06-18/consumer-prices-in-u-s-increased-less-than-forecast-in-may.html, accessed July 5, 2013.

World almanac and book of facts. (1995). Edited by Robert Famighetti. Mahwah, NJ: Funk and Wagnalls.

PART 4

Making Judgments from Surveys and Experiments

In Part 1, you learned how data should be collected in order to be meaningful. In Part 2, you learned some simple things you could do with data, and in Part 3, you learned that uncertainty can be quantified and can lead to worthwhile information about the aggregate.

In Part 4, you will learn about the final steps that allow us to turn data into useful information. You will learn how to use samples collected in surveys and experiments to say something intelligent about what is probably happening in an entire population.

Chapters 19 to 24 are somewhat more technical than previous chapters. Try not to get bogged down in the details. Remember that the purpose of this material is to enable you to say something about a whole population after examining just a small piece of it in the form of a sample. The book concludes with Chapter 27, which provides 11 case studies that will reinforce your awareness that you have indeed become an educated consumer of statistical information.

CHAPTER 19

The Diversity of Samples from the Same Population

Thought Questions

1. Suppose that 40% of a large population disagree with a proposed new law. In parts (a) and (b), think about the role of the sample size when you answer the question.

    a. If you randomly sample 10 people, will exactly four (40%) disagree with the law? Would you be surprised if only two of the people in the sample disagreed with the law? How about if none of the sample disagreed with it?

    b. Now suppose you randomly sample 1000 people. Will exactly 400 (40%) disagree with the law? Would you be surprised if only 200 of the people in the sample disagreed with the law? How about if none of the sample disagreed with it?

    c. Explain how the long-run relative-frequency interpretation of probability and the lack of belief in the law of small numbers helped you answer parts (a) and (b).

2. Suppose the weights of women at a large university come from a bell-shaped curve with a mean of 135 pounds and a standard deviation of 10 pounds.

    a. Recalling the Empirical Rule from Chapter 8, about bell-shaped curves, in what range would you expect 95% of the women's weights to fall?

    b. If you were to randomly sample 10 women at the university, how close do you think their *average* weight would be to 135 pounds? If you sample 1000 women, would you expect the average weight to be closer to 135 pounds than it would be for the sample of only 10 women?

3. Recall from Chapter 4 that a survey of 1000 randomly selected individuals has a margin of error of about 3%, so that the results are accurate to within plus or minus 3% most of the time. Suppose 25% of adults believe in reincarnation. If 10 polls are taken independently, each asking a different random sample of 1000 adults about belief in reincarnation, would you expect each poll to find exactly 25% of respondents expressing belief in reincarnation? If not, into what range would you expect the 10 sample proportions to reasonably fall?

19.1 Setting the Stage

This chapter serves as an introduction to the reasoning that allows pollsters and researchers to make conclusions about entire populations on the basis of a relatively small sample of individuals. The reward for understanding the material presented in this chapter will come in the remaining chapters of this book, as you begin to realize the power of the statistical tools in use today.

Working Backward from Samples to Populations

The first step in this process is to work backward: from a sample to a population. We start with a question about a population, such as: How many teenagers are infected with HIV? At what average age do left-handed people die? What is the average income of all students at a large university? We collect a sample from the population about which we have the question, and we measure the variable of interest. We can then answer the question of interest for the sample. Finally, based on what statisticians have worked out, we will be able to determine how close the answer from our sample is to what we really want to know: the actual answer for the population.

Understanding Dissimilarity among Samples

The secret to understanding how things work is to understand what kind of dissimilarity we should expect to see in various samples from the same population. For example, suppose we knew that most samples were likely to provide an answer that is within 10% of the population answer. Then we would also know the reverse —the population answer should be within 10% of whatever our specific sample gave. Armed only with our sample value, we could make a good guess about the population value. You have already seen this idea at work in Chapter 4, when we used the margin of error for a sample survey to estimate results for the entire population. Statisticians have worked out similar techniques for a variety of sample measurements. In this and the next two chapters, we will cover some of these techniques in detail.

Understanding Proportions, Percentages and Probabilities

In this and subsequent chapters, we will learn about the tools that researchers use to estimate the proportion of a population that has a certain trait, opinion, disease, and so on. Therefore, it's important to understand that when talking about population proportions, it sometimes makes sense to express them as percentages or probabilities instead of proportions. A *percentage* is simply a proportion multiplied by 100%. *Probabilities* are simply proportions expressed as the long run relative frequency of an outcome. For example, if the proportion of births that result in a male is about 0.512, then without any additional information, the probability that the next birth in a hospital will be a male is about 0.512. If a coin is fair, then the probability that a coin toss will result in heads is 0.5.

Probability also makes sense when discussing random selection from a population. For instance, if a person is randomly selected from a population in which 65% supports same-sex marriage, then the probability that a person who supports it will be selected is 0.65. We could also say that the proportion of the population that supports same-sex marriage is 0.65.

EXAMPLE 19.1 Bachelor's Degrees by Sex

According to the National Center for Education Statistics, there were 1,602,480 bachelor's degrees earned by U.S. residents in 2009–10, and 57.4% of them were earned by women (http://nces.ed.gov/fastfacts). Here are three ways we could write the information about the proportion of degrees earned by women.

- Women earned 57.4% of the bachelor's degrees awarded to U.S. residents in 2009–10.
- The proportion of bachelor's degrees in the United States in 2009–10 that were awarded to women was 0.574.
- The probability that a randomly selected 2009–10 U.S. bachelor's degree recipient would be a woman is 0.574. ∎

When we discuss proportions in this and subsequent chapters, we will express them as proportions, percentages, or probabilities, as appropriate in context. Just remember that proportions and probabilities are always between 0 and 1, whereas percentages are those same numbers multiplied by 100%, and thus are always between 0% and 100%. To find a proportion from a percentage, divide by 100. For instance, 40% is equivalent to a proportion of 0.40.

19.2   What to Expect of Sample Proportions

Suppose we want to know what proportion of a population carries the gene for a certain disease. We sample 25 people, and from that sample we make an estimate of the true answer.

Suppose that 40% of the population actually carries the gene. We can think of the population as consisting of two types of people: those who do not carry the gene, represented as ☺, and those who do carry the gene, represented as ⊠. Figure 19.1 (next page) is a conceptual illustration of part of such a population.

Possible Samples and Sample Proportions

What would we find if we randomly sampled 25 people from this population? Would we always find 10 people (40%) with the gene and 15 people (60%) without? You should know from our discussion of the gambler's fallacy and erroneous belief in the law of small numbers in Chapter 17 that we would not.

Each person we chose for our sample would have a 40% chance (i.e., a 0.4 probability) of carrying the gene. But remember that the relative-frequency interpretation

Figure 19.1

A slice of a population where 40% are ⊠

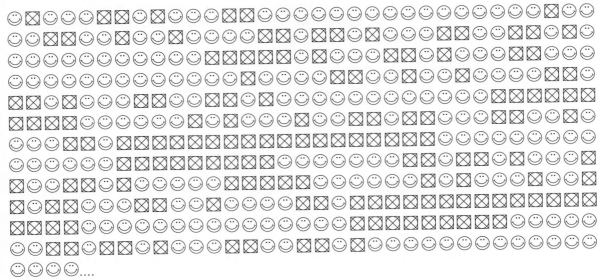

of probability only ensures that we would see 40% of our sample with the gene *in the very long run.* A sample of only 25 people does not qualify as "the very long run."

What should we expect to see? Figure 19.2 shows four different random samples of 25 people taken from the population shown in Figure 19.1. Here is what we would have concluded about the proportion of people who carry the gene, given each of those samples:

Sample 1: Proportion with gene = 12/25 = .48 = or 48%

Sample 2: Proportion with gene = 9/25 = .36 = or 36%

Sample 3: Proportion with gene = 10/25 = .40 = or 40%

Sample 4: Proportion with gene = 7/25 = .28 = or 28%

Figure 19.2

Four possible random samples from Figure 19.1

Notice that each sample gives a different answer, and the sample answer may or may not actually match the truth about the population.

In practice, when a researcher conducts a study similar to this one or a pollster randomly samples a group of people to measure public opinion, only one sample is collected. There is no way to determine whether the sample is an accurate reflection of the population. However, statisticians have calculated what to expect for possible samples. We call the applicable rule the **Rule for Sample Proportions.**

Conditions for Which the Rule for Sample Proportions Applies

The following three conditions must all be met for the Rule for Sample Proportions to apply:

1. There exists an actual population with a fixed proportion who have a certain trait, opinion, disease, and so on.

   or

   There exists a repeatable situation for which an outcome of interest is likely to occur with a fixed relative-frequency probability.

2. A random sample is selected from the population, thus ensuring that the probability of observing the characteristic is the same for each sample unit.

   or

   The situation is repeated numerous times, with the outcome each time independent of all other times.

3. The size of the sample or the number of repetitions is relatively large. The necessary size depends on the proportion or probability under investigation. It must be large enough so that we are likely to see at least ten with and ten without the specified trait in the sample.

Examples of Situations for Which the Rule for Sample Proportions Applies

Here are some examples of situations that meet these conditions.

EXAMPLE 19.2   Election Polls

A pollster wants to estimate the proportion of voters who favor a certain candidate. The voters are the population units, and favoring the candidate is the opinion of interest. ■

EXAMPLE 19.3 Television Ratings

A television rating firm wants to estimate the proportion of households with television sets that are tuned to a certain television program. The collection of all households with television sets makes up the population, and being tuned to that particular program is the trait of interest. ∎

EXAMPLE 19.4 Consumer Preferences

A manufacturer of soft drinks wants to know what proportion of consumers prefers a new mixture of ingredients compared with the old recipe. The population consists of all consumers, and the response of interest is preference for the new formula over the old one. ∎

EXAMPLE 19.5 Testing ESP

A researcher studying extrasensory perception (ESP) wants to know the probability that people can successfully guess which of five symbols is on a hidden card. Each card is equally likely to contain each of the five symbols. There is no physical population. The repeatable situation of interest is a guess, and the response of interest is a successful guess. The researcher wants to see if the probability of a correct guess is higher than 1/5 (0.20 or 20%), which is what it would be if there were no such thing as extrasensory perception. ∎

Defining the Rule for Sample Proportions

The following is what statisticians have determined to be approximately true for the situations that have just been described in Examples 19.2 to 19.5 and for similar ones.

> If numerous samples or repetitions of the same size are taken, the frequency curve made from proportions from the various samples will be approximately bell-shaped. The mean of those sample proportions will be the true proportion from the population. The standard deviation will be
>
> the square root of: (true proportion) $\times$ (1 − true proportion)/(sample size)

EXAMPLE 19.6 Using the Rule for Sample Proportions for Election Polls

Suppose of all voters in the United States, 40% (or .40) are in favor of Candidate X for president. Pollsters take a sample of 2400 people. What proportion of the sample would be expected to favor Candidate X? The rule tells us that the proportion of the sample who favor Candidate X could be anything from a bell-shaped curve with mean of .40 (40%) and standard deviation of

$$\text{the square root of } (.40) \times (1 - .40)/2400 = \sqrt{\frac{(.4)(.6)}{2400}} = .01$$

Figure 19.3

Possible sample proportions when $n = 2400$ and truth $= .4$

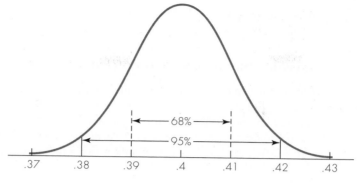

Possible sample proportions

Thus, the mean is .40, and the standard deviation is .01 or 1/100 or 1%.

Figure 19.3 shows what we can expect of the sample proportion in this situation. Recalling the rule we learned in Chapter 8 about bell-shaped distributions (the Empirical Rule), we can also specify that for our sample of 2400 people:

There is a *68%* chance that the sample proportion is between *.39 and .41 (or 39% and 41%)*.

There is a *95%* chance that the sample proportion is between *.38 and .42 (or 38% and 42%)*.

It is almost certain that the sample proportion is between *.37 and .43 (or 37% and 43%)*. ∎

In practice, we have only one sample proportion, and we don't know the true population proportion. However, we do know how far apart the sample proportion and the true proportion are likely to be. That information is contained in the standard deviation, which can be estimated using the sample proportion combined with the known sample size. Therefore, when all we have is a sample proportion, we can indeed say something about the true proportion.

19.3   What to Expect of Sample Means

In the previous section, the question of interest was the proportion falling into one category of a categorical variable. We saw that we could determine an interval of values that was likely to cover the sample proportion if we knew the size of the sample and the magnitude of the true proportion.

We now turn to the case where the information of interest involves the mean or means of measurement variables. For example, researchers might want to compare the mean age at death for left- and right-handed people. A company that sells oat products might want to know the mean cholesterol level people would have if everyone had a certain amount of oat bran in their diet. To help determine financial aid levels, a large university might want to know the mean income of all students on campus who work.

Possible Samples and Sample Means

Suppose a population consists of thousands or millions of individuals, and we are interested in estimating the mean of a measurement variable. If we sample 25 people and compute the mean of the variable, how close will that sample mean be to the population mean we are trying to estimate? Each time we take a sample we will get a different sample mean. Can we say anything about what we expect those means to be?

For example, suppose we are interested in estimating the average weight loss for everyone who attends a national weight-loss clinic for 10 weeks. Suppose, unknown to us, the weight losses for everyone have a mean of 8 pounds, with a standard deviation of 5 pounds. If the weight losses are approximately bell-shaped, we know from Chapter 8 that 95% of the individuals will fall within 2 standard deviations, or 10 pounds, of the mean of 8 pounds. In other words, 95% of the individual weight losses will fall between −2 (a gain of 2 pounds) and 18 pounds lost.

Figure 19.4 lists some possible samples that could result from randomly sampling 25 people from this population; these were indeed the first four samples produced by a computer that is capable of simulating such things. The weight losses have been put into increasing order for ease of reading. A negative value indicates a weight gain.

Following are the sample means and standard deviations, computed for each of the four samples. You can see that the sample means, although all different, are relatively close to the population mean of 8. You can also see that the sample standard deviations are relatively close to the population standard deviation of 5.

Sample 1: Mean = 8.32 pounds, standard deviation = 4.74 pounds

Sample 2: Mean = 6.76 pounds, standard deviation = 4.73 pounds

Sample 3: Mean = 8.48 pounds, standard deviation = 5.27 pounds

Sample 4: Mean = 7.16 pounds, standard deviation = 5.93 pounds

Figure 19.4
Four potential samples from a population with mean = 8, standard deviation = 5

Sample 1: 1,1,2,3,4,4,4,5,6,7,7,7,8,8,9,9,11,11,13,13,14,14,15,16,16

Sample 2: −2,−2,0,0,3,4,4,4,5,5,6,6,8,8,9,9,9,9,9,10,11,12,13,13,16

Sample 3: −4,−4,2,3,4,5,7,8,8,9,9,9,9,9,10,10,11,11,11,12,12,13,14,16,18

Sample 4: −3,−3,−2,0,1,2,2,4,4,5,7,7,9,9,10,10,10,11,11,12,12,14,14,14,19

Conditions to Which the Rule for Sample Means Applies

As with sample proportions, statisticians have developed a rule to tell us what to expect of sample means.

The Rule for Sample Means *applies in either of the following situations:*

1. The population of the measurements of interest is bell-shaped, and a random sample of any size is measured.

2. The population of measurements of interest is not bell-shaped, but a *large* random sample is measured. A sample of size 30 is usually considered "large," but if there are extreme outliers, it is better to have a larger sample.

There are only a limited number of situations for which the Rule for Sample Means does *not* apply. It does not apply if the sample is not random (or at least representative of the population), and it does not apply for small random samples unless the original population is bell-shaped. In practice, it is often difficult to get a random sample. Researchers are usually willing to use the Rule for Sample Means as long as they can get a representative sample with no obvious sources of confounding or bias.

Examples of Situations to Which the Rule for Sample Means Applies

Following are some examples of situations that meet the conditions for applying the Rule for Sample Means.

EXAMPLE 19.7 Average Weight Loss

A weight-loss clinic is interested in measuring the average weight loss for participants in its program. The clinic makes the assumption that the weight losses will be bell-shaped, so the Rule for Sample Means will apply for any sample size. The population of interest is all current and potential clients, and the measurement of interest is weight loss.

EXAMPLE 19.8 Average Age at Death

A researcher is interested in estimating the average age at which left-handed adults die, assuming they have lived to be at least 50. Because ages at death are not bell-shaped, the researcher should measure at least 30 such ages at death. The population of interest is all left-handed people who live to be at least 50 years old. The measurement of interest is age at death.

EXAMPLE 19.9 Average Student Income

A large university wants to know the mean monthly income of students who work. The population consists of all students at the university who work. The measurement of interest is monthly income. Because incomes are not bell-shaped and there are likely to be outliers (a few people with high incomes), the university should use a large random sample of students. The researchers should take particular care to reach the people who are actually selected to be in the sample. A large bias could be created if, for example, they sent an e-mail survey and only used the sample of people who responded immediately without a reminder. The students working the longest hours, and thus making the most money, would probably be hardest to get to respond. Follow-up e-mails and phone calls should be used to get as many people as possible to respond. ∎

Defining the Rule for Sample Means

The Rule for Sample Means is simple:

If numerous samples of the same size are taken, the frequency curve of means from the various samples will be approximately bell-shaped. The mean of this collection of sample means will be the same as the mean of the population. The standard deviation will be

population standard deviation/square root of sample size

EXAMPLE 19.10 **Using the Rule for Sample Means for Weight Loss Clinic Customers**

For our hypothetical weight-loss example, the population mean and standard deviation were 8 pounds and 5 pounds, respectively, and we were taking random samples of size 25. The rule tells us that potential sample means are represented by a bell-shaped curve with a mean of 8 pounds and standard deviation of 5/5 = 1.0. (We divide the population standard deviation of 5 by the square root of 25, which also happens to be 5.)

Therefore, we know the following facts about possible sample means in this situation, based on intervals extending 1, 2, and 3 standard deviations from the mean of 8:

There is a *68%* chance that the sample mean will be between *7 and 9*.

There is a *95%* chance that the sample mean will be between *6 and 10*.

It is *almost certain* that the sample mean will be between *5 and 11*.

Figure 19.5 illustrates this situation. If you look at the four hypothetical samples we chose (see Figure 19.4), you will see that the sample means range from 6.76 to 8.48, well within the range we expect to see using these criteria. ∎

Figure 19.5
Possible sample means for samples of 25 from a bell-shaped population with mean = 8 and standard deviation = 5

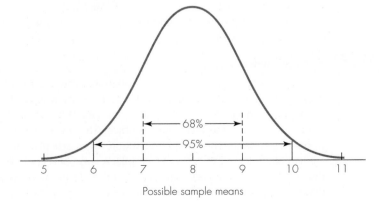

Possible sample means

Increasing the Size of the Sample

Suppose we had taken a sample of 100 people instead of 25. Notice that the mean of the possible sample means would not change; it would still be 8 pounds, but the standard deviation would decrease. It would now be $5/10 = .5$, instead of 1.0. Therefore, for samples of size 100, here is what we would expect of sample means for the weight-loss situation:

There is a *68%* chance that the sample mean will be between *7.5 and 8.5.*

There is a *95%* chance that the sample mean will be between *7 and 9.*

It is *almost certain* that the sample mean will be between *6.5 and 9.5.*

It's obvious that the Rule for Sample Means tells us the same thing our common sense tells us: Larger samples tend to result in more accurate estimates of population values than do smaller samples.

This discussion presumed that we know the population mean and the population standard deviation. Obviously, that's not much use to us in real situations when the population mean is what we are trying to determine. In Chapter 21, we will see how to use the Rule for Sample Means to accurately estimate the population mean when all we have available is a single sample for which we can compute the sample mean and standard deviation.

19.4 What to Expect in Other Situations

We have discussed two common situations that arise in assessing public opinion, conducting medical research, and so on. The first situation arises when we want to know what proportion of a population falls into one category of a categorical variable. The second situation occurs when we want to know the mean of a population for a measurement variable.

There are numerous other situations for which researchers would like to use results from a sample to say something about a population or to compare two or more populations. Statisticians have determined rules similar to those in this chapter for most of the situations researchers are likely to encounter. Those rules are too complicated for a book of this nature. However, once you understand the basic ideas for the two common scenarios covered here (one proportion or one mean), you will be able to understand the results researchers present in more complicated situations. The basic ideas we explore apply equally to most other situations. You may not understand exactly how researchers determined their results, but you will understand the terminology and some of the potential misinterpretations.

In the next several chapters, we explore the two basic techniques researchers use to summarize their statistical results: confidence intervals and hypothesis testing.

Confidence Intervals

One basic technique researchers use is to create a **confidence interval,** which is an interval of values that the researcher is fairly sure covers the true value for the population.

We encountered confidence intervals in Chapter 4, when we learned about the margin of error. Adding and subtracting the margin of error to the reported sample proportion creates an interval that we are 95% "confident" covers the truth. That interval is the confidence interval. We will explore confidence intervals further in Chapters 20 and 21.

Hypothesis Testing

The second statistical technique researchers use is called **hypothesis testing** or **significance testing.** Hypothesis testing uses sample data to attempt to reject the hypothesis that nothing interesting is happening—that is, to reject the notion that chance alone can explain the sample results.

We encountered this idea in Chapter 13, when we learned how to determine whether the relationship between two categorical variables is "statistically significant." The hypothesis that researchers set about to reject in that setting was that two categorical variables are unrelated to each other. In most research settings, the *desired* conclusion is that the variables under scrutiny are related. Achieving statistical significance is equivalent to rejecting the idea that chance alone can explain the observed results. We will explore hypothesis testing further in Chapters 22, 23, and 24.

19.5 Simulated Proportions and Means

In Chapter 15, we learned how to use computers to simulate probabilities and samples. Simulation is an excellent technique for visualizing what would happen if you could take thousands of samples of the same size and find the proportion or mean of interest for each sample. Actually taking thousands of real samples, each one with a few thousand people, would take a very long time! But a computer can simulate doing so in a few seconds.

EXAMPLE 19.11 Simulating the Proportion of Breadwinner Moms

In May, 2013, data from the Census Bureau revealed that 40% of all households in the United States with children under 18 had a woman as the sole or primary source of income (Wang, Parker, and Taylor, 2013). Suppose we take a random sample of 2400 households and ask each household if they fit in this category. For what percentage of our 2400 household sample would we expect to have a woman as the primary earner? Would it always be 40%? This is equivalent to the situation we encountered in Example 19.6, because we are taking samples of size $n = 2400$ from a population where the proportion with the trait is 0.40 (or 40%). The possible sample proportions we should expect to see are shown in Figure 19.3 on page 413, accompanying that example.

Remember that we expected the frequency curve of possible sample proportions to be approximately bell-shaped and that of all the samples of size 2400, we expect about:

68% to have a sample proportion between .39 and .41
95% to have a sample proportion between .38 and .42
Almost all to have a sample proportion between .37 and .43

Let's simulate taking repeated polls of 2400 households and asking each household in the poll whether the primary earner is a woman. Asking the Minitab statistical software to do this is easy. The first sample of 2400 households we asked it to simulate had 958 households with a woman as primary earner, for a proportion of 958/2400 = .3992, or 39.92%. The next sample of 2400 households we asked it to simulate had 1006 such households, or .4192 (41.92%). The next one had 41.67%, and so on. We asked Minitab to do this simulation 1000 times, asking 2400 households each time. The results for all 1000 simulated polls are shown in a histogram in Figure 19.6. A bell-shaped curve is superimposed over the histogram. (Note that the scale on the vertical axis is for the histogram, not the bell-shaped curve, which would have a different scale.) The bell-shaped curve uses the mean and standard deviation of the entire collection of 1000 simulated proportions, which are .3998 and .01037. The simulated versions are very close to the mean of .40 and standard deviation of .01 that the Rule for Sample Proportions told us to expect. The histogram very closely mimics what we expected as well. Almost all of the sample proportions are between .37 and .43, with just a few of them below .37, and none of them above .43. Although it's not obvious from the histogram, 673 of the proportions, or 67.3%, are between .39 and .41, close to the 68% we expected. And 953 of them, or 95.3%, are between .38 and .42, close to the 95% we expected.   ∎

Figure 19.6
Simulated sample proportions for a poll of 2400 when population proportion is .40.

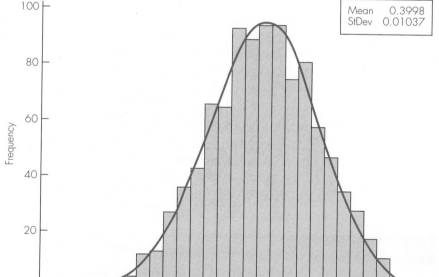

Results of 1000 polls with 2400 people each

| Mean | 0.3998 |
| StDev | 0.01037 |

Sample proportions with woman as primary earner (out of 2400)

EXAMPLE 19.12   Simulating the Average Weight of Carry-On Luggage

A study conducted for the European Aviation Safety Agency reported on the weights of carry-on luggage for different categories of passengers in summer and in winter (Berdowski et al., 2009). For this example, let's use the results for men traveling in the summer. Based on numbers from the study, we will assume that carry-on luggage of the population of men who travel by air in the summer has a mean of 13 pounds and a standard deviation of 10 pounds. Obviously the values are not bell-shaped, because if they were, two standard deviations below the mean would result in weights down to −7 pounds, and carry-on luggage cannot have negative weight! Let's assume that the distribution of carry-on weights starts at zero pounds and has some outliers at the high end.

Because the original population is not bell-shaped, the Rule for Sample Means will only work for large samples. Let's simulate the average weight of carry-on luggage for 100 of these passengers. (An airline would probably be more interested in the *total* weight for these passengers, but that can be found by multiplying the mean by the number of people—100 in this case.) First, let's see what the Rule for Sample Means tells us to expect for the mean of 100 weights.

> *If numerous samples of the same size are taken (100 in this case), the frequency curve of means from the various samples will be approximately bell-shaped. The mean of this collection of sample means will be the same as the mean of the population (13 pounds, in this case). The standard deviation will be (population standard deviation/ square root of sample size), which in this case is $10/\sqrt{100} = 10/10 = 1$ pound.*

Therefore, we expect the frequency curve for the possible sample means to be approximately bell-shaped with a mean of 13 pounds and a standard deviation of 1 pound.

Asking the Minitab software to simulate the carry-on luggage for a single sample of 100 men resulted in mean weight of 11.18 pounds. Asking it to simulate two more samples of 100 men resulted in mean weights of 13.67 pounds and 12.90 pounds. Finally, asking Minitab to simulate 1000 samples, each with 100 men, resulted in the 1000 sample means shown in the histogram in Figure 19.7. A bell-shaped curve has been superimposed on top of the histogram, using the mean (12.96) and standard deviation (1.000) for the 1000 simulated means. (The scale on the vertical axis is for the histogram, not the bell-shaped curve, which would have a different scale.) Notice that the simulated means fit what the Rule for Sample Means told us to expect. Almost all of them are within 3 standard deviations of 13 (i.e., between 10 and 16). The airlines might be interested in knowing that the mean for the carry-on luggage for 100 men could actually be as high as 16 pounds, so the total weight could be as high as about 1600 pounds. They could find similar bounds for the total weight of passengers and all of their carry-on and checked luggage, which would be useful information for calculating fuel requirements. ∎

CASE STUDY 19.1   **Do Americans Really Vote When They Say They Do?**

On November 8, 1994, a historic election took place, in which the Republican Party won control of both houses of Congress for the first time since 1952. But how many people actually voted? On November 28, 1994, *Time* magazine (p. 20) reported that in a telephone poll of 800 adults taken during the two days following the election,

Figure 19.7
Sample means for
carry-on luggage, 1000
samples with n = 100,
mean = 13 pounds, s.d.
= 10 pounds

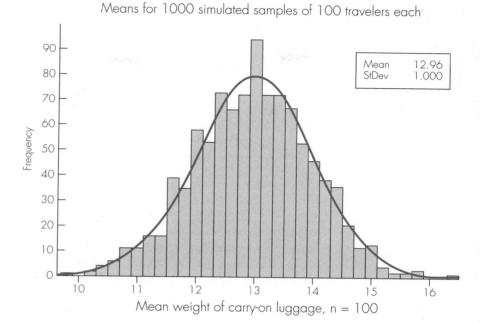

Means for 1000 simulated samples of 100 travelers each

| Mean | 12.96 |
| StDev | 1.000 |

Mean weight of carry-on luggage, n = 100

56% reported that they had voted. Considering that only about 68% of adults are registered to vote, that isn't a bad turnout.

But, along with these numbers, *Time* reported another disturbing fact. They reported that, in fact, only 39% of American adults had voted, based on information from the Committee for the Study of the American Electorate.

Could it be the case that the results of the poll simply reflected a sample that, by chance, voted with greater frequency than the general population? The Rule for Sample Proportions can answer that question. Let's suppose that the truth about the population is, as reported by *Time,* that only 39% of all American adults voted. Then the Rule for Sample Proportions tells us what kind of sample proportions we can expect in samples of 800 adults, the size used by the *Time* magazine poll. The mean of the possibilities is .39, or 39%. The standard deviation is the square root of (.39)(.61)/800, which is .017, or 1.7%.

Therefore, we are almost certain that the sample proportion based on a sample of 800 adults should fall within $3 \times 1.7\% = 5.1\%$ of the truth of 39%. In other words, if respondents were telling the truth, the sample proportion should be no higher than .441 (44.1%), nowhere near the reported percentage of 56%. Figure 19.8 (next page) illustrates the situation.

In fact, if we combine the Rule for Sample Proportions with what we learned about bell-shaped curves in Chapter 8, we can say even more about how unlikely this sample result would be. If, in truth, only 39% of the population voted, the standardized score for the sample proportion of 56% is $(.56 - .39)/.017 = 10$. We know from Chapter 8 that it is virtually impossible to obtain a standardized score

Figure 19.8
Likely sample proportions who voted, if polls of 800 are taken from a population in which .39 (39%) voted

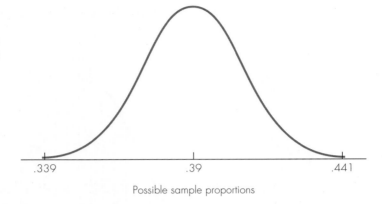

Possible sample proportions

of 10 by chance alone. So we can be virtually certain that many of the respondents in the poll who said they voted did not actually vote.

Another example of the fact that *reported* voting tends to exceed *actual* voting occurred in the 1992 U.S. presidential election. According to the *World Almanac and Book of Facts* (1995, p. 631), 61.3% of American adults reported voting in the 1992 election. In a footnote, the Almanac explains:

> *Total reporting voting compares with 55.9 percent of population actually voting for president, as reported by Voter News Service. Differences between data may be the result of a variety of factors, including sample size, differences in the respondents' interpretation of the questions, and the respondents' inability or unwillingness to provide correct information or recall correct information.*

Unfortunately, because figures are not provided for the size of the sample, we cannot assess whether the difference between the actual percentage of 55.9 and the reported percentage of 61.3 can be explained by the natural variability among possible sample proportions. ∎

Thinking About Key Concepts

- The *proportion* of a sample that has a particular opinion or trait is called a *sample proportion*. It is an estimate of the *population proportion,* but will not generally be exactly equal to the population proportion.
- The *Rule for Sample Proportions* describes the pattern (frequency curve) of sample proportions that would result from taking repeated samples of the same size. The rule says that these sample proportions would have a bell-shaped frequency curve with the mean equal to the actual population proportion. The standard deviation depends on the sample size and decreases as the sample size increases.
- For measurement variables, the *sample mean* can be used to estimate the *population mean* for the population the sample represents.

- The *Rule for Sample Means* describes the pattern (frequency curve) of sample means that would result from taking repeated samples of the same size. The rule says that these sample means would have a bell-shaped frequency curve with mean equal to the actual population mean. The standard deviation depends on the population standard deviation and on the size of the sample. It decreases as the sample size increases.

Focus On Formulas

Notation for Population and Sample Proportions

Sample size $= n$

Population proportion $= p$

Sample proportion $= \hat{p}$, which is read "p-hat" because the p appears to have a little hat on it.

The Rule for Sample Proportions

If numerous samples or repetitions of size n are taken, the frequency curve of the $\hat{p}$'s from the various samples will be approximately bell-shaped. The mean of those $\hat{p}$'s will be p. The standard deviation will be

$$\sqrt{\frac{p(1 - p)}{n}}$$

Notation for Population and Sample Means and Standard Deviations

Population mean $= \mu$ (read "mu"), population standard deviation $= \sigma$ (read "sigma")

Sample mean $= \bar{X}$ (read "X-bar"), sample standard deviation $= s$

The Rule for Sample Means

If numerous samples of size n are taken, the frequency curve of the $\bar{X}$'s from the various samples is approximately bell-shaped with mean μ and standard deviation $\sigma/\sqrt{n}$

Another way to write these rules is using the notation for normal distributions from Chapter 8:

$$\hat{p} \sim N\left(p, \frac{p(1 - p)}{n}\right) \quad \text{and} \quad \bar{X} \sim N\left(\mu, \frac{\sigma^2}{n}\right)$$

Exercises

Exercises with numbers divisible by 3 (3, 6, 9, etc.) are included in the Solutions at the back of the book. They are marked with an asterisk (*).

1. Suppose you want to estimate the proportion of students at your college who are left-handed. You decide to collect a random sample of 200 students and ask them which hand is dominant. Go through the conditions for which the rule for sample proportions applies (p. 411) and explain why the rule would apply to this situation.

2. Refer to Exercise 1. Suppose the truth is that .12, or 12%, of the students are left-handed, and you take a random sample of 200 students. Use the Rule for Sample Proportions to draw a picture similar to Figure 19.3, showing the possible sample proportions for this situation.

*3. A Gallup Poll found that of 800 randomly selected drivers surveyed, 70% thought they were better-than-average drivers. In truth, in the population, no more than 50% of all drivers can be "better than average," assuming "average" is equivalent to the median level of skill.

   a. Draw a picture of the possible sample proportions that would result from samples of 800 people from a population with a true proportion of .50.

   *b. Would we be unlikely to see a sample proportion of .70, based on a sample of 800 people, from a population with a proportion of .50? Explain, using your picture from part (a).

   *c. Which one of the psychological issues discussed in Chapter 16 would help explain the results of this survey?

4. Suppose you are interested in estimating the average number of miles per gallon of gasoline your car can get. You calculate the miles per gallon for each of the next nine times you fill the tank. Suppose, in truth, the values for your car are bell-shaped, with a mean of 25 miles per gallon and a standard deviation of 1. Draw a picture of the possible sample means you are likely to get based on your sample of nine observations. Include the intervals into which 68%, 95%, and almost all of the potential sample means will fall.

5. Refer to Exercise 4. Redraw the picture under the assumption that you will collect 100 measurements instead of only nine. Discuss how the picture differs from the one in Exercise 4.

*6. Give an example of a situation of interest to you for which the Rule for Sample Proportions would apply. Explain how the conditions allowing the rule to be applied are satisfied for your example.

7. Suppose that 35% of the students at a university favor the semester system, 60% favor the quarter system, and 5% have no preference. Is a random sample of 100 students large enough to provide convincing evidence that the quarter system is favored? Explain.

8. According to *USA Today* (20 April 1998, Snapshot), a poll of 8709 adults taken in 1976 found that 9% believed in reincarnation, whereas a poll of 1000 adults taken in 1997 found that 25% held that belief.

   a. Assuming a proper random sample was used, verify that the sample proportion for the poll taken in 1976 almost certainly represents the population proportion to within about 1%.

   b. Based on these results, would you conclude that the proportion of all adults who believe in reincarnation was higher in 1997 than it was in 1976? Explain.

*9. Suppose 20% of all television viewers in the country watch a particular program.

   *a. For a random sample of 2500 households measured by a rating agency, describe the frequency curve for the possible sample proportions who watch the program.

   *b. The program will be canceled if the ratings show less than 17% watching in a random sample of households. Given that 2500 households are used for the ratings, is the program in danger of getting canceled? Explain.

c. Draw a picture of the possible sample proportions, similar to Figure 19.3. Illustrate where the sample proportion of .17 falls on the picture. Use this to confirm your answer in part (b).

10. Use the Rule for Sample Means to explain why it is desirable to take as large a sample as possible when trying to estimate a population value.

11. According to the *Sacramento Bee* (2 April 1998, p. F5), "A 1997–98 survey of 1027 Americans conducted by the National Sleep Foundation found that 23% of adults say they have fallen asleep at the wheel in the last year."

   a. Conditions 2 and 3 needed to apply the Rule for Sample Proportions are met because this result is based on a large random sample of adults. Explain how condition 1 is also met.

   b. The article also said that (based on the same survey) "37 percent of adults report being so sleepy during the day that it interferes with their daytime activities." If, in truth, 40% of all adults have this problem, find the interval in which about 95% of all sample proportions should fall, based on samples of size 1027. Does the result of this survey fall into that interval?

   c. Suppose a survey based on a random sample of 1027 college students was conducted and 25% reported being so sleepy during the day that it interferes with their daytime activities. Would it be reasonable to conclude that the population proportion of college students who have this problem differs from the proportion of all adults who have the problem? Explain.

*12. According to the *Sacramento Bee* (2 April 1998, p. F5), Americans get an average of 6 hours and 57 minutes of sleep per night. A survey of a class of 190 statistics students at a large university found that they averaged 7.1 hours of sleep the previous night, with a standard deviation of 1.95 hours.

   a. Assume that the population average for adults is 6 hours and 57 minutes, or 6.95 hours of sleep per night, with a standard deviation of 2 hours. Draw a picture similar to Figure 19.5, illustrating how the Rule for Sample Means would apply to sample means for random samples of 190 adults.

   *b. Would the mean of 7.1 hours of sleep obtained from the statistics students be a reasonable value to expect for the sample mean of a random sample of 190 adults? Explain.

   *c. Can the sample taken in the statistics class be considered to be a representative sample of all adults? Explain.

13. Suppose the population of IQ scores in the town or city where you live is bell-shaped, with a mean of 105 and a standard deviation of 15. Describe the frequency curve for possible sample means that would result from random samples of 100 IQ scores.

14. Explain whether each of the following situations meets the conditions for which the Rule for Sample Proportions applies. If not, explain which condition is violated.

   a. You are interested in knowing what proportion of days in typical years have rain or snow in the area where you live. For the months of January and February, you record whether there is rain or snow each day, and then you calculate the proportion of those days that had rain or snow.

   b. A large company wants to determine what proportion of its employees are interested in on-site day care. The company asks a random sample of 100 employees and calculates the sample proportion who are interested.

*15. Explain whether each of the following situations meets the conditions for which the Rule for Sample Proportions applies. If not, explain which condition is violated.

   *a. Unknown to the government, 10% of all cars in a certain city do not meet appropriate emissions standards. The government wants to estimate that percentage, so they take a random sample of 30 cars and compute the sample proportion that do not meet the standards.

   *b. The Census Bureau would like to estimate what proportion of households have someone at home between 7 P.M. and 7:30 P.M.

on weeknights, to determine whether that would be an efficient time to collect census data. The Bureau surveys a random sample of 2000 households and visits them during that time to see whether someone is at home.

16. Explain whether you think the Rule for Sample Means applies to each of the following situations. If it does apply, specify the population of interest and the measurement of interest. If it does not apply, explain why not.

   a. A researcher is interested in what the average cholesterol level would be if people restricted their fat intake to 30% of calories. He gets a group of patients who have had heart attacks to volunteer to participate, puts them on a restricted diet for a few months, and then measures their cholesterol.

   b. A large corporation would like to know the average income of the spouses of its workers. Rather than go to the trouble to collect a random sample, they post someone at the exit of the building at 5 P.M. Everyone who leaves between 5 P.M. and 5:30 P.M. is asked to complete a short questionnaire on the issue; there are 70 responses.

17. Explain whether you think the Rule for Sample Means applies to each of the following situations. If it does apply, specify the population of interest and the measurement of interest. If it does not apply, explain why not.

   a. A university wants to know the average income of its alumni. Staff members select a random sample of 200 alumni and mail them a questionnaire. They follow up with a phone call to those who do not respond within 30 days.

   b. An automobile manufacturer wants to know the average price for which used cars of a particular model and year are selling in a certain state. They are able to obtain a list of buyers from the state motor vehicle division, from which they select a random sample of 20 buyers. They make every effort to find out what those people paid for the cars and are successful in doing so.

*18. Suppose the population of grade-point averages (GPAs) for students at the end of their first year at a large university has a mean of 3.1 and a standard deviation of .5. Draw a picture of the frequency curve for the mean GPA of a random sample of 100 students, similar to Figure 19.5.

19. In Case Study 19.1, we learned that about 56% of American adults actually voted in the presidential election of 1992, whereas about 61% of a random sample claimed that they had voted. The size of the sample was not specified, but suppose it were based on 1600 American adults, a common size for such studies.

   a. Into what interval of values should the sample proportion fall 68%, 95%, and almost all of the time?

   b. Is the observed value of 61% reasonable, based on your answer to part (a)?

   c. Now suppose the sample had been of only 400 people. Compute a standardized score to correspond to the reported percentage of 61%. Comment on whether you believe people in the sample could all have been telling the truth, based on your result.

20. The administration of a large university wants to use a random sample of students to measure student opinion of a new food service on campus. Administrators plan to use a continuous scale from 1 to 100, where 1 is complete dissatisfaction and 100 is complete satisfaction. They know from past experience with such questions that the standard deviation for the responses is going to be about 5, but they do not know what to expect for the mean. They want to be almost sure that the sample mean is within plus or minus 1 point of the true population mean value. How large will their random sample have to be?

Mini-Projects

1. The goal of this mini-project is to help you verify the Rule for Sample Proportions firsthand, using a physical simulation. You will use the population represented in Figure 19.1 to do so. It contains 400 individuals, of whom 160 (40%) are ⊠—that is, carry the gene for a disease—and the remaining 240 (60%) are ☺—that is, do not carry the gene. You are going to draw 20 samples of size 15 from this population. Here are the steps you should follow:

   Step 1: Develop a method for drawing simple random samples from this population. One way to do this is to cut up the symbols and put them all into a paper bag, shake well, and draw from the bag. There are less tedious methods, but make sure you actually get random samples. *Explain your method.*

   Step 2: Draw a random sample of size 15 and record the number and percentage who carry the gene.

   Step 3: Repeat step 2 a total of 20 times, thus accumulating 20 samples, each of size 15. Make sure to start over each time; for example, if you used the method of drawing symbols from a paper bag, then put the symbols back into the bag after each sample of size 15 is drawn so they are available for the next sample as well.

   Step 4: Create a stemplot or histogram of your 20 sample proportions. Compute the mean.

   Step 5: Explain what the Rule for Sample Proportions tells you to expect for this situation.

   Step 6: Compare your results with what the Rule for Sample Proportions tells you to expect. Be sure to mention mean, standard deviation, shape, and the intervals into which you expect 68%, 95%, and almost all of the sample proportions to fall.

2. The purpose of this mini-project is to help you verify the Rule for Sample Means, using a physical simulation. Suppose you are interested in measuring the average amount of blood contained in the bodies of adult women, in ounces. Suppose, in truth, the population consists of the following listed values. (Each value would be repeated millions of times, but in the same proportions as they exist in this list.) The actual mean and standard deviation for these numbers are 110 ounces and 5 ounces, respectively. The values are bell-shaped.

Population Values for Ounces of Blood in Adult Women

97	100	101	102	103	103	104	104	104	105	106
106	106	107	107	108	108	109	109	109	110	110
110	110	110	111	112	112	112	112	113	113	113
113	113	113	114	114	114	114	114	114	115	115
116	116	116	117	118	118					

Step 1: Develop a method for drawing simple random samples from this population. One way to do this is to write each number on a slip of paper, put all the slips into a paper bag, shake well, and draw from the bag. If a number occurs multiple times, make sure you include it that many times. Make sure you actually get random samples. *Explain your method.*

Step 2: Draw a random sample of size 9. Calculate and record the mean for your sample.

Step 3: Repeat step 2 a total of 20 times, thus accumulating 20 samples, each of size 9. Make sure to start over each time; for example, if you drew numbers from a paper bag, put the numbers back after each sample of size 9 so they are available for the next sample as well.

Step 4: Create a stemplot or histogram of your 20 sample means. Compute the mean of those sample means.

Step 5: Explain what the Rule for Sample Means tells you to expect for this situation.

Step 6: Compare your results with what the Rule for Sample Means tells you to expect. Be sure to mention mean, standard deviation, shape, and the intervals into which you expect 68%, 95%, and almost all of the sample means to fall.

3. Carry out the simulation in Mini-Project #1 using a computer or website instead of physically doing so. Go through the same steps and answer the same questions.

4. Carry out the simulation in Mini-Project #2 using a computer or website instead of physically doing so. Go through the same steps and answer the same questions.

References

Berdowski, Z., F.N. van den Broek-Serle, J.T. Jetten, Y. Kawabata, J.T. Schoemaker, R. Bersteegh. (2009). Survey on standard weights of passengers and baggage, final report. European Aviation Safety Agency technical report, http://www.easa.europa.eu/.

Wang, W., K. Parker and P. Taylor. (2013). Breadwinner moms. http://www.pewsocialtrends.org/2013/05/29/breadwinner-moms/, accessed July 7, 2013.

World Almanac and Book of Facts. (1995). Edited by Robert Famighetti. Mahwah, NJ: Funk and Wagnalls.

CHAPTER 20

Estimating Proportions with Confidence

Thought Questions

1. One example we see in this chapter is a 95% confidence interval for the proportion of British couples in which the wife is taller than the husband. The interval extends from .02 to .08, or 2% to 8%. What do you think it means to say that the interval from .02 to .08 represents a 95% confidence interval for the proportion of couples in which the wife is taller than the husband?

2. Do you think a 99% confidence interval for the proportion described in Question 1 would be wider or narrower than the 95% interval given? Explain.

3. In a Gallup poll of 503 adults taken in March, 2012, 57% reported that they think nuclear power plants are safe (http://www.pollingreport.com/energy.htm). Based on this survey, a "95% confidence interval" for the percentage in the population who believe they are safe is about 52.5% to 61.5%. If this poll had been based on 5000 adults instead, do you think the "95% confidence interval" would be wider or narrower than the interval given? Explain.

4. How do you think the concept of *margin of error*, explained in Chapter 4, relates to confidence intervals for proportions (or percentages)? As a concrete example, can you determine the margin of error for the situation in Question 1 from the information given? In Question 3?

20.1 Confidence Intervals

In the previous chapter, we saw that we get different summary values (such as means and proportions) each time we take a sample from a population. We also learned that statisticians have been able to quantify the amount by which those sample values are likely to differ from each other and from the population value.

In practice, statistical methods are used in situations where only one sample is taken, and that sample is used to make a conclusion or an inference about numbers (such as means and proportions) for the population from which it was taken. One of the most common types of inferences is to construct what is called a **confidence interval,** which is defined as

an interval of values computed from sample data that is almost sure to cover the true population number.

The most common level of confidence used is 95%. In other words, researchers define "almost sure" to mean that they are 95% certain. They are willing to take a 5% risk that the interval does not actually cover the true value.

It would be impossible to construct an interval in which we could be 100% confident unless we actually measured the entire population. Sometimes, as we shall see in one of the examples in the next section, researchers employ only 90% confidence. In other words, they are willing to take a 10% chance that their interval will not cover the truth. Other **confidence levels** such as 99% are used as well, but the standard is 95%.

Methods for actually constructing confidence intervals differ, depending on the type of question asked and the type of sample used. In this chapter, we learn to construct confidence intervals for proportions and percentages, and in the next chapter, we learn to construct confidence intervals for means. If you understand the kinds of confidence intervals we study in this chapter and the next, you will understand any other type of confidence interval as well.

In most applications, we never know whether the confidence interval covers the truth; we can only apply the long-run frequency interpretation of probability. All we can know is that, in the long run, 95% of all confidence intervals tagged with 95% confidence will be correct (cover the truth) and 5% of them will be wrong. There is no way to know for sure which kind we have in any given situation. A humorous phrase among statisticians is: "Being a statistician means never having to say you're certain," as a parody of the famous line from the movie *Love Story*, "Love means never having to say you're sorry."

20.2 Three Examples of Confidence Intervals from the Media

When the media report the results of a statistical study, they often supply the information necessary to construct a confidence interval. Sometimes they even provide a confidence interval directly. The most commonly reported information that can be used to construct a confidence interval is the *margin of error*. It is sometimes called the *margin of sampling error* to distinguish it from the sources of error caused by bias, described in Chapter 4. Most public opinion polls report a margin of error along with the percentage of the sample that had each opinion. To use that information, you need to know this fact:

> *To construct a 95% confidence interval for a population percentage, simply add and subtract the margin of error to the sample percentage.*

The margin of error is often reported using the symbol "±," which is read "plus or minus." The formula for a 95% confidence interval can thus be expressed as

$$\text{sample percentage} \pm \text{margin of error}$$

Let's examine three examples from the media, in which confidence intervals are either reported directly or can easily be derived.

EXAMPLE 20.1   A Public Opinion Poll on Marijuana

In a Field Poll reported in the *Sacramento Bee* (27 February 2013, p. 3A), 54% of respondents said that they think California should legalize marijuana beyond it's use for medical purposes. In an accompanying graphic on the *Bee's* website, the following information was provided about the poll:

> *The poll was conducted Feb. 5-17, in English and Spanish, via landline and cell-phone, of 834 registered California voters. The results carry a margin of error of plus or minus 3.5 percentage points.* (http://www.sacbee.com/2013/02/27/5220454_a5220468/field-poll-california-voters-favor.html)

What percentage of the entire population of California registered voters at that time would agree that marijuana should be legalized? A 95% confidence interval for that proportion can be found by taking

$$\text{sample percentage} \pm \text{margin of error}$$
$$54\% \pm 3.5\%$$
$$50.5\% \text{ to } 57.5\%$$

Notice that this interval does not cover 50%; it (just barely!) resides completely above 50%. Therefore, it would be fair to conclude, with high confidence, that a majority of California voters in 2013 believed that marijuana should be legalized. ∎

EXAMPLE 20.2   Uncertain Relative Risk of HIV Infection

A news story from HealthDay News (2012) reported the following information:

> *Researchers studied nearly 3300 HIV-discordant couples (one partner has HIV and the other is HIV-free) in sub-Saharan Africa and found that the average rate of HIV-1 transmission was about one per 900 acts of sexual intercourse…The investigators also found that older age was associated with a reduced rate of HIV transmission and that male circumcision reduced female-to-male transmission by about 47 percent.*

What does a reduction of "about 47 percent" for male circumcision mean? To find out, let's consult the original journal article on which the news story was based. It was published in the January 12, 2012, issue of *The Journal of Infectious Diseases*. Here is how the journal reported the same result:

> *Circumcision in male HIV-1–uninfected partners was associated with significantly lower infectivity (RR, 0.53 [95% CI, .29-.96])* (Hughes et al., 2012, p. 361).

You may recall from Chapter 12 that when comparing risks, an appropriate statistic is the relative risk. That's what "RR" means in this report. A relative risk of 0.53 indicates that the risk of HIV infection for the participants in this study whose partners were circumcised was only 0.53 times the risk of infection for those whose partners were not circumcised. The reduction of 47% quoted in the original news story is found by subtracting the relative risk of 0.53 from a relative risk of 1.0, for a difference of 0.47 or 47%. A relative risk of 1.0 would imply equal risk for both groups.

But the study was based on only 3300 couples, so the relative risk of 0.53 may not apply to the larger population. That's why it's important to find a confidence interval. In this case, the journal article reported a 95% confidence interval for the relative risk as 0.29 to 0.96. Now the picture is not as clear. This confidence interval says that the actual risk of infection with a circumcised partner could be as low as 0.29 times the risk with an uncircumcised partner, or it could be as high as 0.96 times the risk. In other words, the risks could be almost the same.

You may notice that the quote from the journal article says that the risk is "significantly lower." How can that be, if the risk could be almost as high for either group? The answer is that the word "significantly" is being used here to mean "statistically significantly lower." In a hypothesis test, the null hypothesis of interest would be that the risks are the same for both groups, so the relative risk is 1.0. But the confidence interval tells us that we can be 95% confident that the highest the relative risk could be is 0.96. The null hypothesis would be rejected. It is only in that sense that the risk is "significantly" lower.   ∎

EXAMPLE 20.3   The Debate over Passive Smoking

On July 28, 1993, the *Wall Street Journal* featured an article by Jerry E. Bishop with the headline, "Statisticians occupy front lines in battle over passive smoking" (pp. B-1, B-4). The interesting feature of this article was that it not only reported a confidence interval but it highlighted a debate between the U.S. Environmental Protection Agency (EPA) and the tobacco industry over what level of confidence should prevail in a confidence interval.

Here is the first side of the story, as reported in the article:

> *The U.S. Environmental Protection Agency says there is a 90% probability that the risk of lung cancer for passive smokers is somewhere between 4% and 35% higher than for those who aren't exposed to environmental smoke. To statisticians, this calculation is called the "90% confidence interval."*

Now, for the other side of the story:

> *And that, say tobacco-company statisticians, is the rub. "Ninety-nine percent of all epidemiological studies use a 95% confidence interval," says Gio B. Gori, director of*

the Health Policy Center in Bethesda, Md., who has frequently served as a consultant and an expert witness for the tobacco industry.

The problem underlying this controversy is that the amount of data available at the time did not allow an extremely accurate estimate of the true change in risk of lung cancer for passive smokers. The EPA statisticians were afraid that the public would not understand the interpretation of a confidence interval. If a 95% confidence interval actually went below zero percent, which it might do, the tobacco industry could argue that passive smoke might *reduce* the risk of lung cancer. As noted by one of the EPA's statisticians:

> *Dr. Gori is correct in saying that using a 95% confidence interval would hint that passive smoking might reduce the risk of cancer. But, he says, this is exactly why it wasn't used. The EPA believes it is inconceivable that breathing in smoke containing cancer-causing substances could be healthy and any hint in the report that it might be would be meaningless and confusing.* (p. B-4)
>
> SOURCE: Bishop, Jerry E., "Statisticians occupy front lines in battle over passive smoking," July 28, 1993, *Wall Street Journal*. Copyright 1993 Dow Jones and Co. All rights reserved. Reprinted with permission.

In Chapter 24, we study in more detail the issue of making erroneous conclusions based on too little data. Often the amount of data available does not allow us to conclusively detect a problem, but that is not the same as concluding that no problem exists.   ■

20.3   Constructing a Confidence Interval for a Proportion

You can easily learn to construct your own confidence intervals for some simple situations. One of those situations is the one we encountered in the previous chapter, in which a simple random sample is taken for a categorical variable. It is easy to construct a confidence interval for the proportion or percentage of the population who fall into one of the categories. Following are some examples of situations where this would apply. After presenting the examples, we develop the method, and then return to the examples to compute confidence intervals.

EXAMPLE 20.4 **How Often Is the Wife Taller than the Husband?**

In Chapter 10, we displayed data representing the heights of husbands and wives for a random sample of 200 British couples. From that set of data, we can count the number of couples for whom the wife is taller than the husband. We can then construct a confidence interval for the true proportion of British couples for whom that would be the case.   ■

EXAMPLE 20.5 **An Experiment in Extrasensory Perception**

In Chapter 22, we will describe in detail an experiment that was conducted to test for extrasensory perception (ESP). For one part of the experiment, participants were asked to describe a video being watched by a "sender" in another room. The participants were then shown four videos and asked to pick the one they thought the "sender" had been watching. Without ESP, the probability of a correct guess should be .25, or one-fourth, because there were four equally likely choices. We will use the data from the experiment to construct a confidence interval for the true probability of a correct guess and see if the interval includes the .25 value that would be expected if there were no ESP. ∎

The Proportion Who Would Quit Smoking with a Nicotine Patch

In Case Study 5.1, we examined an experiment in which 120 volunteers were given nicotine patches. After 8 weeks, 55 of them had quit smoking. Although the volunteers were not a random sample from a population, we can estimate the proportion of people who would quit if they were recruited and treated exactly as these individuals were treated. ∎

Converting between Proportions and Percentages

You may have noticed that throughout this chapter we have alternated between talking about proportions and percentages. Remember that the proportion of a population with a certain trait or opinion is a number between 0 and 1, while a percentage is between 0% and 100%. If you know a proportion, you can convert it to a percentage by multiplying by 100%. If you know a percentage, you can convert it to a proportion by dividing by 100. The same conversion can be made to the endpoints of a confidence interval. For instance, if a confidence interval for the population percentage with a certain opinion is 50.5% to 57.5% (as it was in Example 20.1) the a confidence interval for the corresponding population proportion is .505 to .575.

When we derive the formula for a confidence interval in this situation, it is easier to do so for the population proportion rather than the population percentage. But remember that you can easily find the corresponding confidence interval for the percentage by multiplying the endpoints by 100%. When the margin of error is reported with a survey or poll in the media, it is almost always reported for the percentage rather than for the proportion. It is important that you do not confuse the two situations, and that you know how to convert from one to the other.

Developing the Formula for a 95% Confidence Interval

We develop the formula for a 95% confidence interval only and discuss what we would do differently if we wanted higher or lower confidence.

The formula will follow directly from the Rule for Sample Proportions:

If numerous samples or repetitions of the same size are taken, the frequency curve made from proportions from the various samples will be approximately bell-shaped. The mean will be the true proportion from the population. The standard deviation will be

the square root of: (true proportion) × (1 − true proportion)/(sample size)

Because the possible sample proportions are bell-shaped, we can make the following statement:

In 95% of all samples, the sample *proportion will fall within* 2 standard deviations *of the mean, which is the* true proportion *for the population.*

This statement allows us to easily construct a 95% confidence interval for the true population proportion. Notice that we can rewrite the statement slightly, as follows:

In 95% of all samples, the true proportion *will fall* within 2 standard deviations of the sample proportion.

In other words, if we simply add and subtract 2 standard deviations to the sample proportion, in 95% of all cases we will have captured the true population proportion.

There is just one hurdle left. If you examine the Rule for Sample Proportions to find the standard deviation, you will notice that it uses the "true proportion." But we don't know the true proportion; in fact, that's what we are trying to estimate. There is a simple solution to this dilemma. We can get a fairly accurate answer if we substitute the sample proportion for the true proportion in the formula for the standard deviation.

Standard Error of the Sample Proportions

To avoid confusion, when we substitute the sample proportion for the population proportion in the standard deviation formula, we also give in a new name. The **standard error of the sample proportions**, abbreviated as just **standard error** or **SEP** is the name used when the sample proportion is substituted for the population proportion. The formula is shown in the box below.

Putting all of this together, here is the formula for a 95% confidence interval for a population proportion:

sample proportion ± 2(*SEP*)

where

SEP = the square root of:
(sample proportion) × (1 − sample proportion)/(sample size)

To find a confidence interval for the population percentage, multiply the endpoints by 100%.

A technical note: To be exact, we would actually add and subtract 1.96(*SEP*) instead of 2(*SEP*) because 95% of the values for a bell-shaped curve fall within 1.96 standard deviations of the mean. However, in most practical applications, rounding 1.96 off to 2.0 will not make much difference and this is common practice. We have been rounding off when we use 2.0 for the Empirical Rule. To be exact, 95.44% of the normal curve falls within 2.0 standard deviations of the mean.

Continuing the Examples

Let us now apply this formula to the examples presented at the beginning of this section.

EXAMPLE 20.4 CONTINUED

How Often Is the Wife Taller than the Husband?

The data presented in Chapter 10, on the heights of 200 British couples, showed that in only 10 couples was the wife taller than the husband. Therefore, we find the following numbers:

sample proportion = 10/200 = .05, or 5%

standard error = square root of (.05)(.95)/200 = .015

confidence interval = .05 ± 2(.015) = .05 ± .03 = .02 to .08

In other words, we are 95% confident that of all British couples, between .02 (2%) and .08 (8%) are such that the wife is taller than her husband. ∎

EXAMPLE 20.5 CONTINUED

An Experiment in Extrasensory Perception

The data we will examine in detail in Chapter 22 include 165 cases of experiments in which a participant tried to guess which of four videos the "sender" was watching in another room. Of the 165 cases, 61 resulted in successful guesses. Therefore, we find the following numbers:

sample proportion = 61/165 = .37, or 37%

standard error = square root of (.37)(.63)/165 = .038

confidence interval = .37 ± 2(.038) = .37 ± .08 = .29 to .45

In other words, we are 95% confident that the probability of a successful guess in this situation is between .29 (29%) and .45 (45%). Notice that this interval lies entirely above the 25% value expected by chance. ∎

EXAMPLE 20.6 CONTINUED

The Proportion Who Would Quit Smoking with a Nicotine Patch

In Case Study 5.1, we learned that of 120 volunteers randomly assigned to use a nicotine patch, 55 of them had quit smoking after 8 weeks. We use this information to estimate the probability that a smoker recruited and treated in an identical fashion would quit smoking after 8 weeks:

sample proportion = 55/120 = .46, or 46%

standard error = square root of (.46)(.54)/120 = .045

confidence interval = .46 ± 2(.045) = .46 ± .09 = .37 to .55

In other words, we are 95% confident that between 37% and 55% of smokers treated in this way would quit smoking after 8 weeks. Remember that a placebo group was included for this experiment, in which 24 people, or 20%, quit smoking after 8 weeks. A confidence interval surrounding that value runs from 13% to 27% and thus does not overlap with the confidence interval for those using the nicotine patch. ∎

Other Levels of Confidence

If you wanted to present a narrower interval, you would have to settle for less confidence. Applying the reasoning we used to construct the formula for a 95% confidence interval and using the information about bell-shaped curves from Chapter 8, we could have constructed a 68% confidence interval, for example. We would simply add and subtract 1 standard deviation to the sample proportion instead of 2. Similarly, if we added and subtracted 3 standard deviations, we would have a 99.7% confidence interval.

Although 95% confidence intervals are by far the most common, you will sometimes see 90% or 99% intervals as well. To construct those, you simply replace the value 2 in the formula with 1.645 for a 90% confidence interval or with the value 2.576 for a 99% confidence interval.

The general formula for a **confidence interval for a proportion** is as follows:

Sample proportion ± multiplier × standard error

The multiplier depends on the desired *confidence level*. Common levels and multipliers are:

Confidence level	Multiplier
90%	1.645
95%	1.96 or 2.0
99%	2.576

How the Margin of Error Was Derived in Chapter 4

We have already noted that you can construct a 95% confidence interval for a proportion or percentage if you know the margin of error. You simply add and subtract the margin of error to the sample proportion or percentage.

Some polls report a margin of error with their results that might not match the margin of error presented in this chapter. That's because many polls use a multistage sample (see Chapter 4). Therefore, the simple formulas for confidence intervals given in this chapter, which are based on simple random samples, do not give exactly the same answers as those using the margin of error stated. For polls based on multistage samples, it is more appropriate to use the stated margin of error than to use the formula given in this chapter.

In Chapter 4, we presented an approximate method for computing the margin of error. Using the letter n to represent the sample size, we then said that a conservative way to compute the margin of error was to use $1/\sqrt{n}$. Thus, we now have two apparently different formulas for finding a 95% confidence interval:

$$\text{sample proportion} \pm \text{margin of error} = \text{sample proportion} \pm 1/\sqrt{n}$$

or

$$\text{sample proportion} \pm 2(SEP)$$

How do we reconcile the two different formulas? In order to reconcile them, it should follow that

$$\text{margin of error} = 1/\sqrt{n} = 2(SEP)$$

It turns out that these two formulas are equivalent when the proportion used in the formula for the standard error is .5. In other words, when

$$\text{standard error} = SEP = \text{square root of } (.5)(.5)/n = (.5)/\sqrt{n}$$

In that case, $2(SEP)$ is simply $1/\sqrt{n}$, which is our conservative formula for margin of error. This is called a *conservative* formula because the true margin of error is actually likely to be smaller. If you use any value other than .5 as the proportion in the formula for standard error, you will get a smaller answer than you get using .5. You will be asked to confirm this fact in Exercise 13 at the end of this chapter.

Simulated Confidence Intervals for a Proportion

Remember that not all confidence intervals will cover the true population proportion, and thus we could be misled by a confidence interval. The confidence level represents the long-run relative frequency for which the intervals are correct. For a confidence level of 95%, we expect that in the long run, 5% will not cover the proportion they are designed to estimate. For example, in the few months before a major election, many polling agencies independently conduct polls asking voters who they support. Results generally provide a margin of error that allows readers to compute a 95% confidence interval. But even if all of the polls are done correctly, over the long run, about 5% of those intervals will get it wrong.

Let's use the Minitab statistical software to simulate 20 polls, each asking 400 voters if they plan to vote for Candidate X. Suppose that exactly half of voters (0.5 or 50%) in the population support Candidate X. The confidence intervals resulting from simulating the 20 polls are displayed in Figure 20.1. The first poll of 400 voters found 215 of them supporting the candidate, so the sample proportion was 215/400 = 0.5375, or 53.75%. The margin of error is about 0.05, so the 95% confidence interval is 0.5375 ± .05, which is 0.4875 to 0.5875. The interval is shown as the bottom line in Figure 20.1. Notice that the interval does cover the true population proportion of 0.5.

The true population proportion remains fixed at 0.5, but the sample proportions and resulting confidence intervals differ for each poll. Most of the intervals do cover the true value of 0.5. Of the 20 polls, only Poll #11 (identified in Figure 20.1) resulted in a confidence interval that did not cover the actual population proportion of 0.5. For that interval, just by chance, the sample obtained did not contain enough supporters. Only 171 of the voters polled supported the candidate, for a sample proportion of 171/400 = 0.4275. Adding the margin of error of 0.05 resulted in an upper endpoint of only 0.4775, somewhat short of the true value of 0.5.

In the simulation resulting in Figure 20.1, we happened to obtain exactly 95% correct intervals (19 out of 20) and one incorrect. In general, with 20 polls, it also is possible that all of them would be correct or that more than one of them would not be correct. Remember that it is only over the long run that we expect 95% of them to work.

Figure 20.1
Twenty simulated confidence intervals for
n = 400 and p = 0.5

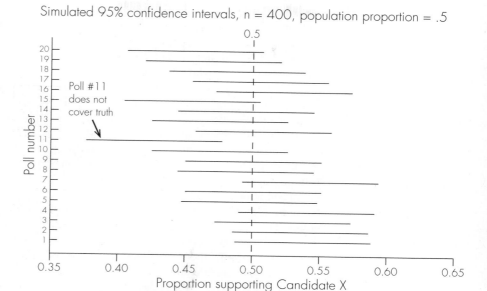

CASE STUDY 20.1   **A Winning Confidence Interval Loses in Court**

Gastwirth (1988, p. 495) describes a court case in which Sears, Roebuck and Company, a large department store chain, tried to use a confidence interval to determine the amount by which it had overpaid city taxes at stores in Inglewood, California. Unfortunately, the judge did not think the confidence interval was appropriate and required Sears to examine all the sales records for the period in question. This case study provides an example of a situation where the answer became known, so we can compare the results from the sample with the true answer.

The problem arose because Sears had erroneously collected and paid city sales taxes for sales made to individuals outside the city limits. The company discovered the mistake during a routine audit, and asked the city for a refund of $27,000, the amount by which it estimated it had overpaid.

Realizing that it needed data to substantiate this amount, Sears decided to take a random sample of sales slips for the period in question and then, on the basis of the sample proportion, try to estimate the proportion of all sales that had been made to people outside of city limits. It used a multistage sampling plan, in which the 33-month period was divided into eleven 3-month periods to ensure that seasonal effects were considered. It then took a random sample of 3 days in each period, for a total of 33 days, and examined all sales slips for those days.

Based on the data, Sears derived a 95% confidence interval for the true proportion of all sales that were made to out-of-city customers. The confidence interval was .367 ± .03, or .337 to .397. To determine the amount of tax Sears believed it was owed, the proportion of out-of-city sales was multiplied by the total tax paid, which was $76,975. The result was $28,250, with a 95% confidence interval extending from $25,940 to $30,559.

The judge did not accept the use of sampling despite testimony from accounting experts who noted that it was common practice in auditing. The judge required Sears to examine all of the sales records. In doing so, Sears discovered that about one month's worth of slips were missing; however, based on the available slips, it had overpaid $26,750.22. This figure is slightly under the true amount due to the miss-ing month, but you can see that the sampling method Sears had used provided a fairly accurate estimate of the amount it was owed. If we assume that the dollar amount from the missing month was similar to those for the months counted, we find that the total Sears was owed would have been about $27,586.

Sampling methods and confidence intervals are routinely used for financial audits. These techniques have two main advantages over studying all of the records. First, they are much cheaper. It took Sears about 300 person-hours to conduct the sample and 3384 hours to do the full audit. Second, a sample can be done more carefully than a complete audit. In the case of Sears, it could have two well-trained people conduct the sample in less than a month. The full audit would require either having those same two people work for 10 months or training 10 times as many people. As Gastwirth (1988, p. 496) concludes in his discussion of the Sears case, "A well designed sampling audit may yield a more accurate estimate than a less carefully carried out complete audit or census." In fairness, the judge in this case was simply following the law; the sales tax return required a sale-by-sale computation.   ■

Thinking About Key Concepts

- A *confidence interval* is an interval of values used to estimate a population value. It is computed from sample data and is fairly certain to cover the true population value it is designed to estimate.
- A *confidence level* accompanies a confidence interval. It provides the long-run relative frequency for which the confidence interval procedure works.
- A 95% *confidence interval for a proportion* is found by adding and subtracting a *margin of error* equal to two standard errors to the sample proportion. Other confidence levels can be used as well, in which case "two" is replaced by a different multiplier, appropriate for the desired confidence level.
- A *conservative margin of error* for a 95% confidence interval is $1/\sqrt{n}$ where n is the number of individuals used to find the sample proportion.
- The higher the desired confidence level, the larger the multiplier and thus the wider the confidence interval.
- Larger sample sizes result in smaller standard errors and thus more narrow confidence intervals.

Focus On Formulas

Notation for Population and Sample Proportions (from Chapter 19)

Sample size $= n$

Population proportion $= p$

Sample proportion $= \hat{p}$

Notation for the Multiplier for a Confidence Interval

For reasons that will become clear in later chapters, we specify the *level* of confidence for a confidence interval as $(1 - \alpha$ (read "alpha")) $\times$ 100%. For example, for a 95% confidence interval, $\alpha = .05$ so the confidence level is $(1 - .05) \times 100\% = 95\%$. Let $z_{\alpha/2} =$ standardized normal score with area $\alpha/2$ above it. Then the area between $z_{\alpha/2}$ and $-z_{\alpha/2}$ is $1 - \alpha$. For example, when $\alpha = .05$, as for a 95% confidence interval, $z_{\alpha/2} = 1.96$, or about 2.

Formula for a $(1 - \alpha) \times$ 100% Confidence Interval for a Proportion

$$\hat{p} \pm z_{\alpha/2}\sqrt{\frac{\hat{p}(1 - \hat{p})}{n}}$$

Common Values of $z_{\alpha/2}$

1.0   for a 68% confidence interval

1.96 or 2.0 for a 95% confidence interval

1.645 for a 90% confidence interval

2.576 for a 99% confidence interval

3.0 for a 99.7% confidence interval

Exercises

Exercises with numbers divisible by 3 (3, 6, 9, etc.) are included in the Solutions at the back of the book. They are marked with an asterisk (*).

1. One of the questions asked in a Gallup poll taken in May, 2012, in the United States was: "All in all, if you had your choice, would you want to be rich, or not?" (Newport, 2012). (A previous question had asked respondents if they thought they were rich, and the 2% who said yes were not asked this question.) Of the 1008 people asked this question, 635 said yes, they would want to be rich.
   a. What is the sample proportion of respondents who said yes, they would want to be rich?
   b. Find the standard error to accompany the proportion you found in part (a).

c. Use your answers from parts (a) and (b) to construct a 95% confidence interval for the population proportion who want to be rich.

d. Interpret the confidence interval found in part (c) by writing a few sentences explaining what it means.

2. Another question asked in the Gallup poll in Exercise 1 was: "Do you think that rich people in America today are happier than you, less happy, or about the same?" (Newport, 2012). Of the 1008 respondents, only 11% said "happier," 27% said "less happy," 57% said "about the same," and the rest were "unsure."

a. Find a 95% confidence interval for the *proportion* of the population who would have answered "less happy" if asked.

b. Convert the interval in part (a) into a 95% confidence interval for the *percentage* of the population who would have responded "less happy" if asked.

c. Write a few sentences interpreting the interval you found in part (b).

*__*3.__ Does a confidence interval for a proportion estimate a sample proportion, or a population proportion?

4. On September 10, 1998, the "Starr Report," alleging impeachable offenses by President Bill Clinton, was released to Congress. That evening, the Gallup Organization conducted a poll of 645 adults nationwide to assess initial reaction (reported at www.gallup.com). One of the questions asked was: "Based on what you know at this point, do you think that Bill Clinton should or should not be impeached and removed from office?" The response "Yes, should" was selected by 31% of the respondents.

a. The Gallup webpage said, "For results based on the total sample of adults nationwide, one can say with 95% confidence that the margin of sampling error is no greater than ± 4 percentage points." Explain what this means and verify that the statement is accurate.

b. Give a 95% confidence interval for the percentage of all adults who would have said President Clinton should be impeached had they been asked that evening.

c. A similar Gallup Poll taken a few months earlier, in June 1998, found that 19% responded that President Clinton should be impeached. Do you think the difference between the results of the two polls can be attributed to chance variation in the samples taken, or does it represent a real difference of opinion in the population in June versus mid-September? Explain.

5. A telephone poll reported in *Time* magazine (6 February 1995, p. 24) asked 359 adult Americans the question: "Do you think Congress should maintain or repeal last year's ban on several types of assault weapons?" Seventy-five percent responded "maintain."

a. Compute the standard error that accompanies the sample proportion of .75.

b. Time reported that the "sampling error is ± 4.5%." Verify that 4.5% is approximately what would be added and subtracted to the sample percentage to create a 95% confidence interval. (Note that it is equivalent to a margin of error of .045 for a confidence interval for the population *proportion*.)

c. Use the information reported by *Time* to create a 95% confidence interval for the population proportion. Interpret the interval in words that would be understood by someone with no training in statistics. Be sure to specify the population to which it applies.

d. The ban was in place for 10 years, but repealed in 2004. In 2013, after some high-profile cases of gun violence, the topic was in the news again. In March, 2013, a Quinnipiac University poll asked 1711 adults: "Do you support or oppose a nationwide ban on the sale of assault weapons?" Fifty-nine percent said "Support." (*Source:* http://www.pollingreport.com/guns.htm). Find a 95% confidence interval for the population proportion that supported this ban in March, 2013.

e. Compare the confidence intervals found in parts (c) and (d). Write a few sentences explaining whether support for banning these weapons seemed to increase or decrease between 1994 and 2013.

*6. What level of confidence would accompany each of the following intervals?

*a. Sample proportion ± 1.0(*SEP*)

*b. Sample proportion ± 1.645(*SEP*)

*c. Sample proportion ± 1.96(*SEP*)

*d. Sample proportion ± 2.576(*SEP*)

7. Use the Empirical Rule to specify what level of confidence would accompany each of the following confidence intervals.

a. Sample proportion ± 1(*SEP*)

b. Sample proportion ± 2(*SEP*)

c. Sample proportion ± 3(*SEP*)

8. Explain whether the width of a confidence interval would increase, decrease, or remain the same as a result of each of the following changes:

a. The sample size is doubled, from 400 to 800.

b. The population size is doubled, from 25 million to 50 million.

c. The level of confidence is lowered from 95% to 90%.

*9. *Parade Magazine* reported that "nearly 3200 readers dialed a 900 number to respond to a survey in our Jan. 8 cover story on America's young people and violence" (19 February 1995, p. 20). Of those responding, "63.3% say they have been victims or personally know a victim of violent crime." Can the results quoted and methods in this chapter legitimately be used to compute a 95% confidence interval for the proportion of Americans who fit that description? If so, compute the interval. If not, explain why not. (Note that calling a 900 number was not a free call at the time; the caller was charged a fee.)

10. Refer to Example 20.6. It is claimed that a 95% confidence interval for the percentage of placebo-patch users who quit smoking by the eighth week covers 13% to 27%. There were 120 placebo-patch users, and 24 quit smoking by the eighth week. Verify that the confidence interval given is correct.

11. Find the results of a poll reported in a magazine, newspaper, or on the Internet in which a margin of error is also reported. (A good source of polls conducted by various organizations is www.pollingreport.com.) Explain what question was asked and what margin of error was reported; then present a 95% confidence interval for the results. Explain in words what the interval means for your example.

*12. Suppose that a survey is planned to estimate the proportion of a population that is left-handed. The sample data will be used to form a confidence interval. Explain which one of the following combinations of sample size and confidence level will give the widest interval.

(i) $n = 400$, confidence level = 90%

(ii) $n = 400$, confidence level = 95%

(iii) $n = 1000$, confidence level = 90%

(iv) $n = 1000$, confidence level = 95%

13. Confirm that the standard deviation for sample proportions is largest when the proportion used to calculate it is .50. Do this by using other values above and below .50 and comparing the answers to what you would get using .50. Try three values above and three values below .50.

14. A university is contemplating switching from the quarter system to the semester system. The administration conducts a survey of a random sample of 400 students and finds that 240 of them prefer to remain on the quarter system.

a. Construct a 95% confidence interval for the true proportion of all students who would prefer to remain on the quarter system.

b. Does the interval you computed in part (a) provide convincing evidence that the majority of students prefer to remain on the quarter system? Explain.

c. Now suppose that only 50 students had been surveyed and that 30 said they preferred the quarter system. Compute a 95% confidence interval for the true proportion that prefers to remain on the quarter system. Does the interval provide convincing evidence that the majority of students prefer to remain on the quarter system?

d. Compare the sample proportions and the confidence intervals found in parts (a) and (c). Use these results to discuss the role sample size plays in helping make decisions from sample data.

*15. A study first reported in the *Journal of the American Medical Association* (7 December 1994) received widespread attention as the first wide-scale study of the use of alcohol on American college campuses and was the subject of an article in *Time* magazine (19 December 1994, p. 16). The researchers surveyed 17,592 students at 140 four-year colleges in 40 states. One of the results they found was that about 8.8%, or about 1550 respondents, were frequent binge drinkers. They defined frequent binge drinking as having had at least four (for women) or five (for men) drinks at a single sitting at least three times during the previous 2 weeks.

*a. *Time* magazine (19 December 1994, p. 66) reported that of the approximately 1550 frequent binge drinkers in this study, 22% reported having had unprotected sex. Find a 95% confidence interval for the population proportion of all frequent binge drinkers who had unprotected sex, and interpret the interval for someone who has no knowledge of statistics.

*b. Notice that the results quoted in part (a) indicate that about 341 students out of the 17,592 interviewed said they were frequent binge drinkers and had unprotected sex. Compute a 95% confidence interval for the proportion of college students who are

frequent binge drinkers and who also had unprotected sex.

c. Using the results from parts (a) and (b), write two short news articles on the problem of binge drinking and unprotected sex. In one, make the situation sound as disastrous as you can. In the other, try to minimize the problem.

16. In a special double issue of *Time* magazine, the cover story featured Pope John Paul II as "Man of the Year" (26 December 1994–2 January 1995, pp. 74–76). As part of the story, *Time* reported on the results of a survey of 507 adult American Catholics, taken by telephone on December 7–8. It was also reported that "sampling error is ± 4.4%."

a. One question asked was, "Do you favor allowing women to be priests?" to which 59% of the respondents answered yes. Using the reported margin of error of 4.4%, calculate a 95% confidence interval for the response to this question. Write a sentence interpreting the interval that could be understood by someone who knows nothing about statistics. Be careful about specifying the correct population.

b. Calculate a 95% confidence interval for the question in part (a), using the formula in this chapter rather than the reported margin of error. Compare your answer to the answer in part (a).

c. Another question in the survey was, "Is it possible to disagree with the Pope and still be a good Catholic?" to which 89% of respondents said yes. Using the formula in this chapter, compute a 95% confidence interval for the true percentage who would answer yes to the question. Now compute a 95% confidence interval using the reported margin of error of 4.4%. Compare your two intervals.

d. If you computed your intervals correctly, you would have found that the two intervals in parts (a) and (b) were quite similar to each other, whereas the two intervals in part (c) were not. In part (c), the interval computed using the reported margin of error was wider than the one computed using the formula. Explain why the two methods for computing the intervals agreed more closely for the survey question in parts (a) and (b) than for the survey question in part (c).

17. In a 2008 Scripps Howard News/Ohio University survey of 1003 randomly selected adults, 56% said that it is either very or somewhat likely that there is intelligent life on other planets (Source: www.scrippsnews.com/node/34758). For each part, use the desired confidence level to find a confidence interval for the proportion of all adults in 2008 who thought it was either very or somewhat likely that there is intelligent life on other planets.

   a. 90%

   b. 95%

   c. 99%

   d. Using the 99% confidence interval for part (c), would you conclude with 99% confidence that a majority of adults (more than 50%) in 2008 thought it was either very or somewhat likely that there is intelligent life on other planets? Explain.

*18. In Example 20.5, we found a 95% confidence interval for the proportion of successes likely in a certain kind of ESP test. Construct a 99.7% confidence interval for that example. Explain why a skeptic of ESP would prefer to report the 99.7% confidence interval rather than the 95% confidence interval.

19. Refer to the formula for a confidence interval in the Focus on Formulas section.

   a. Write the formula for a 90% confidence interval for a proportion.

   b. Refer to Example 20.6. Construct a 90% confidence interval for the proportion of smokers who would quit after 8 weeks using a nicotine patch.

   c. Compare the 90% confidence interval you found in part (b) to the 95% confidence interval used in the example. Explain which

one you would present if your company were advertising the effectiveness of nicotine patches.

20. One of the questions asked in a Pew Research Center/USA Today poll taken in June, 2013, was: "Overall, do you approve or disapprove of the government's collection of telephone and internet data as part of anti-terrorism efforts?" (http://www.pollingreport.com/terror.htm). Of the 1512 respondents, 48% said "approve," 47% said "disapprove," and the rest were unsure.

   a. What is the conservative margin of error for this poll?

   b. Use the conservative margin of error to find a 95% confidence interval for the *proportion* of all adults who would have answered "approve" if asked.

   c. Repeat part (b) for the answer "disapprove."

   d. Compare the confidence intervals in parts (b) and (c). From those intervals, can you conclude that one of the two opinions (approve or disapprove) was held by a majority of the population at that time? Explain.

*21. In each situation, explain why you think that the sample proportion should or should not be used to estimate the population proportion.

   *a. An Internet news organization asks visitors to its website to respond to the question: "Are you satisfied with the president's job performance?" Of 3500 respondents, 61% say that they are not satisfied with the president's performance. On the basis of this survey, the organization writes an article saying that a majority of Americans are not satisfied with the president.

   *b. A convenience sample of 400 college students in two classes at the same university is used to estimate the proportion that is left-handed in the nationwide population of college students.

22. An advertisement for Seldane-D, a (now discontinued) drug prescribed for seasonal allergic rhinitis, reported results of a double-blind

study in which 374 patients took Seldane-D and 193 took a placebo (*Time*, 27 March 1995, p. 18). Headaches were reported as a side effect by 65 of those taking Seldane-D.

   a. What is the sample proportion of Seldane-D takers who reported headaches?

   b. What is the standard error accompanying the proportion computed in part (a)?

   c. Construct a 95% confidence interval for the population proportion based on the information from parts (a) and (b).

   d. Interpret the confidence interval from part (c) by writing a few sentences explaining what it means.

23. Refer to Exercise 22. Of the 193 placebo takers, 43 reported headaches.

   a. Compute a 95% confidence interval for the true population proportion that would get headaches after taking a placebo.

   b. Notice that a higher proportion of placebo takers than Seldane-D takers reported headaches. Use that information to explain why it is important to have a group taking placebos when studying the potential side effects of medications.

*24. Suppose 200 different researchers all randomly select samples of 400 individuals from a population. Each researcher uses his or her sample to compute a 95% confidence interval for the proportion that has blue eyes in the population. About how many of the confidence intervals will cover the population proportion? About how many of the intervals will not cover the population proportion? Briefly explain how you determined your answers.

25. Refer to News Story 2 in the Appendix and on the book's website, "Research shows women harder hit by hangovers," and the accompanying Original Source 2 on the website, "Development and initial validation of the Hangover Symptoms Scale: Prevalence and correlates of hangover symptoms in college students." Table 3 of the journal article reports that 13% of the 1216 college students in the study said that they had not experienced any hangover symptoms in the past year.

   a. Assuming that the participants in this study are a representative sample of college students, find a 95% confidence interval for the proportion of college students who have not experienced any hangover symptoms in the past year. Use the formula in this chapter.

   b. Write a sentence or two interpreting the interval you found in part (a) that would be understood by someone without training in statistics.

   c. The journal article also reported that the study originally had 1474 participants, but only 1234 reported drinking any alcohol in the past year. (Only those who reported drinking in the past year were retained for the hangover symptom questions.) Use this information to find a 95% confidence interval for the proportion of all students who would report drinking any alcohol in the past year.

   d. Refer to the journal article to determine how students were selected for this study. Based on that information, to what population of students do you think the intervals in this exercise apply? Explain.

26. Refer to News Story 13 and the accompanying report on the book's website, "2003 CASA National Survey of American Attitudes on Substance Abuse VIII: Teens and Parents."

   a. The margin of error for the teens and for the parents are reported in the news story. What are they reported to be?

   b. Refer to page 30 of the Original Source 13 report. The margins of error are reported there as well. What are they reported to be? Are they the same as those reported in the news story? Explain.

   c. The 1987 teens in the survey were asked, "How harmful to the health of someone your age is the regular use of alcohol—very harmful, fairly harmful, not too harmful, or not harmful at all?" Forty-nine percent responded that it was very harmful. Find a 95% confidence interval for the proportion of all teens who would respond that way.

d. The 504 parents in the survey were asked, "How harmful to the health of a teenager is the regular use of alcohol—very harmful, fairly harmful, not too harmful, or not harmful at all?" Seventy-seven percent responded that it was very harmful. Find a 95% confidence interval for the proportion of all parents (similar to those in this study) who would respond that way.

e. Compare the confidence intervals in parts (c) and (d). In particular, do they indicate that there is a difference in the population proportions of teens and parents who think alcohol is very harmful to teens?

f. Write a short news story reporting the results you found in parts (c) to (e).

Mini-Projects

1. You are going to use the methods discussed in this chapter to estimate the proportion of all cars in your area that are red. Stand on a busy street and count cars as they pass by. Count 100 cars and keep track of how many are red.

a. Using your data, compute a 95% confidence interval for the proportion of cars in your area that are red.

b. Based on how you conducted the survey, are any biases likely to influence your results? Explain.

2. Collect data and construct a confidence interval for a proportion for which you already know the answer. Use a sample of at least 100. You can select the situation for which you would like to do this. For example, you could flip a coin 100 times and construct a confidence interval for the proportion of heads, knowing that the true proportion is .5. Report how you collected the data and the results you found. Explain the meaning of your confidence interval, and compare it to what you know to be the truth about the proportion of interest.

3. Choose a categorical variable for which you would like to estimate the true proportion that fall into a certain category. Conduct an experiment or a survey that allows you to find a 95% confidence interval for the proportion of interest. Explain exactly what you did, how you computed your results, and how you would interpret the results.

References

Gastwirth, Joseph L. (1988). *Statistical reasoning in law and public policy*. Vol. 2. *Tort law, evidence and health*. Boston: Academic Press.

HealthDay News. (2012). Partner's 'Viral Load' a Major Factor in HIV Transmission: Study. http://health.usnews.com/health-news/family-health/womens-health/articles/2012/01/12/partners-viral-load-a-major-factor-in-hiv-transmission-study, accessed July 8, 2013.

Hughes, J. P., J.M. Baeten, J.R. Lingappa, A.S. Margaret, A. Wald, G. de Bruyn, J. Kiarie, M. Inambao, W. Kilembe, C. Farguhar, C. Celum, and the Partners in Prevention HSV/HIV Transmission Study Team. (2012). Determinants of per-coital-act HIV-1 infectivity among African HIV-1-serodiscordant couples. *Journal of Infectious Diseases,* 205(3), pp. 358–365.

Newport, Frank. (2012). Americans Like Having a Rich Class, as They Did 22 Years Ago. http.://www.gallup.com/poll/154619/Americans-Having-Rich-Class-Years-Ago.aspx, posted May 11, 2012, accessed July 8, 2013.

CHAPTER **21**

The Role of Confidence Intervals in Research

Thought Questions

1. In this chapter, Example 21.1 compares weight loss (over 1 year) for men who diet but do not exercise, and vice versa. The results show that a 95% confidence interval for the mean weight loss for men who diet but do not exercise extends from 13.4 to 18.3 pounds. A 95% confidence interval for the mean weight loss for men who exercise but do not diet extends from 6.4 to 11.2 pounds.

   a. Do you think this means that 95% of all men who diet will lose between 13.4 and 18.3 pounds? Explain.

   b. On the basis of these results, do you think you can conclude that men who diet without exercising lose more weight, on average, than men who exercise but do not diet?

2. The first confidence interval in Question 1 was based on results from 42 men. The confidence interval spans a range of almost 5 pounds. If the results had been based on a much larger sample, do you think the confidence interval for the mean weight loss would have been wider, narrower, or about the same? Explain your reasoning.

3. In Question 1, we compared average weight loss for dieting and for exercising by computing separate confidence intervals for the two means and comparing the intervals. What would be a more direct value to examine to make the comparison between the mean weight loss for the two methods?

4. In Case Study 5.4, we examined the relationship between baldness and heart attacks, and provided details for one study. Many of the results reported in the original journal article for that study were expressed in terms of relative risk of a heart attack for men with severe vertex baldness compared to men with no hair loss. One result reported was a 95% confidence interval for the relative risk for men under 45 years old. It extended from 1.1 to 8.2.

   a. Recalling the material from Chapter 12, explain what it means to have a relative risk of 1.1 in this example.

   b. Interpret the result given by the confidence interval.

449

21.1   Confidence Intervals for Population Means

In Chapter 19, we learned what to expect of sample means, assuming we knew the mean and the standard deviation of the population from which the sample was drawn. In this section, we try to estimate a population mean when all we have available is a sample of measurements from the population. All we need from the sample are its mean, standard deviation, and number of observations.

EXAMPLE 21.1   **Do Men Lose More Weight by Diet or by Exercise?**
Wood and colleagues (1988), also reported by Iman (1994, p. 258), studied a group of 89 sedentary men for a year. Forty-two men were placed on a diet; the remaining 47 were put on an exercise routine.

The group on a diet lost an average of 7.2 kg, with a standard deviation of 3.7 kg. The men who exercised lost an average of 4.0 kg, with a standard deviation of 3.9 kg (Wood et al., 1988, Table 2). Before we discuss how to compare the groups, let's determine how to extend the sample results to what would happen if the entire population of men of this type were to diet or exercise exclusively. We will return to this example after we learn the general method. ∎

The Rule for Sample Means, Revisited

In Chapter 19, we learned how sample means behave.

> *The Rule for Sample Means is*
>
> If numerous samples of the same size are taken, the frequency curve of means from the various samples will be approximately bell-shaped. The mean of this collection of sample means will be the same as the mean of the population. The standard deviation will be
>
> > population standard deviation/square root of sample size

Standard Error of the Mean

Before proceeding, we need to distinguish between the *population* standard deviation and the standard deviation for the sample means, which is the population standard deviation/$\sqrt{n}$. (Recall that n is the number of observations in the sample.) Consistent with the distinction made by most researchers, we use terminology as follows for these two different kinds of standard deviations.

> The standard deviation for the possible sample means is called the **standard error of the mean.** It is sometimes abbreviated as SEM or just "standard error." In other words,
>
> > SEM = standard error = population standard deviation/$\sqrt{n}$

In practice, the population standard deviation is usually unknown and is replaced by the sample standard deviation, computed from the data. The term *standard error of the mean* or *standard error* is still used, although some authors call it the *estimated standard error* when the sample version is used.

Population versus Sample Standard Deviation and Error

An example will help clarify the distinctions among these terms. In Chapter 19, we considered a hypothetical population of people who visited a weight-loss clinic. We said that the weight losses for the thousands of people in the *population* were bell-shaped, with a mean of 8 pounds and a standard deviation of 5 pounds. Further, we considered *samples* of $n = 25$ people. For one sample, we found the mean and standard deviation for the 25 people to be mean $= 8.32$ pounds, standard deviation $= 4.74$ pounds.

Thus, we have the following numbers:

population standard deviation $= 5$ pounds

sample standard deviation $= 4.74$ pounds

standard error of the mean (using population *SD*) $= 5/\sqrt{25} = 1$

standard error of the mean (using sample *SD*) $= 4.74/\sqrt{25} = 0.95$

Let's now return to our discussion of the Rule for Sample Means. It is important to remember the conditions under which this rule applies:

1. The population of measurements of interest is bell-shaped, and a random sample of any size is measured.

or

2. The population of measurements of interest is not bell-shaped, but a *large* random sample is measured. A sample of size 30 is usually considered "large," but if there are extreme outliers, it is better to have a larger sample.

Constructing an Approximate 95% Confidence Interval for a Mean

We can use the same reasoning we used in Chapter 20, where we constructed a 95% confidence interval for a proportion, to construct an approximate 95% confidence interval for a mean. The Rule for Sample Means and the Empirical Rule from Chapter 8 allow us to make the following statement:

> *In about 95% of all samples, the* sample mean *will fall within* 2 standard errors *of the* true population mean.

Now let's rewrite the statement in a more useful form:

> *In about 95% of all samples, the* true population mean *will be within* 2 standard errors *of the* sample mean.

In other words, if we simply add and subtract 2 standard errors to the sample mean, in about 95% of all cases we will have captured the true population mean.

> Putting this all together, here is the formula for an approximate **95% confidence interval for a population mean:**
>
> $$\text{sample mean} \pm 2 \text{ standard errors}$$
>
> where
>
> $$\text{standard error} = \text{standard deviation}/\sqrt{n}$$

Important note: This formula should be used only if there are at least 30 observations in the sample. To compute a 95% confidence interval for the population mean based on smaller samples, a multiplier larger than 2 is used, which is found from a "*t*-distribution." The *t*-distribution can be used to construct confidence intervals for confidence levels other than 95% as well. We turn to this idea following Example 21.1.

EXAMPLE 21.1
CONTINUED

Comparing Diet and Exercise for Weight Loss

Let's construct an approximate 95% confidence interval for the mean weight losses for all men who might diet or who might exercise, based on the sample information given in Example 21.1. We will be constructing two separate confidence intervals, one for each condition. Notice the switch from kilograms to pounds at the end of the computation; 1 kg = 2.2 lb. The results could be expressed in either unit, but pounds are more familiar to many readers.

Diet Only

sample mean = 7.2 kg
sample standard deviation = 3.7 kg
number of participants = n = 42
standard error = $3.7/\sqrt{42}$ = 0.571
2 × standard error = 2(0.571) = 1.1

Exercise Only

sample mean = 4.0 kg
sample standard deviation = 3.9 kg
number of participants = n = 47
standard error = $3.9/\sqrt{47}$ = 0.569
2 × standard error = 2(0.569) = 1.1

Approximate 95% confidence interval for the population mean:
sample mean ± 2 × standard error

Diet Only

7.2 ± 1.1
6.1 kg to 8.3 kg
13.4 lb to 18.3 lb

Exercise Only

4.0 ± 1.1
2.9 kg to 5.1 kg
6.4 lb to 11.2 lb

These results indicate that men similar to those in this study would lose an average of somewhere between 13.4 and 18.3 pounds on a diet but would lose only an average of 6.4 to 11.2 pounds with exercise alone. Notice that these intervals are trying to capture the true mean or average value for the population. They do not encompass the full range of weight loss that would be experienced by most individuals. Also, remember that these intervals could be wrong. Ninety-five percent of intervals constructed this way will contain the correct population mean value, but 5% will not. We will never know which are which.

Based on these results, it appears that dieting probably results in a larger weight loss than exercise because there is no overlap in the two intervals. Comparing the endpoints of these intervals, we are fairly certain that the average weight loss from dieting is no lower than 13.4 pounds and the average weight loss from exercising no higher than 11.2 pounds. In the next section, we learn a more efficient method for making the comparison, one that will enable us to estimate with 95% confidence the actual difference in the two averages for the population. ■

Constructing a General Confidence Interval for a Mean

The formula (in the box on page 452) for an approximate 95% confidence interval for a mean is a special case of a more general formula. Similar to the construction of a confidence interval for a proportion in Chapter 20, the general form of a **confidence interval for a population mean** is:

$$\text{sample mean} \pm \text{multiplier} \times \text{standard error}$$

As with proportions, the multiplier depends on the desired *confidence level*. But when working with means instead of proportions, there is an additional complication. Substituting the sample standard deviation for the population standard deviation in the "standard error" creates a new twist. The appropriate multiplier comes from a slightly different bell-shaped curve than the one we've been using. It's called **Student's *t* distribution**, so we will call the multiplier a "***t*-multiplier**."

The confidence interval formula using the *t*-multiplier is only valid under one (or both) of the following two conditions:

- The original population of measurements is approximately bell-shaped.
- The sample size is at least 30, and even larger if there are extreme outliers in the data.

The technical details for finding the *t*-multiplier are summarized below, but you can skip over them unless you actually need to compute a confidence interval other than the approximate 95% version given previously. In the remainder of this chapter, we will use the formula for the approximate 95% confidence interval, with a multiplier of 2.

Technical Details for a t-Multiplier: There are two pieces of information you need to find the appropriate *t*-multiplier. One is a number called "degrees of freedom," abbreviated "df." For computing a confidence interval for a mean, this number is $df = n - 1$, where n is the number of observations in the data set. You also need to specify the confidence level. Sometime computer software requires you to provide $(1 - \text{confidence level})$ instead. For instance, to find the appropriate *t*-multiplier for a 95% confidence interval in Excel, use "T.INV.2T(.05,df)", where "df" is the number $n - 1$, and .05 is found as $1 - .95$. As an example, for $n = 100$ and a 95% confidence level, type into a cell in Excel "=T.INV.2T(.05,99)" and the result will be 1.984217. Thus, the *t*-multiplier for a 95% confidence interval for the population mean when the sample has 100 observations is 1.984217. You can see why a multiplier of 2 provides a good approximation.

> The general formula for a **confidence interval for a population mean** is:
>
> $$\text{sample mean} \pm t\text{-multiplier} \times \text{standard error}$$
>
> The t-multiplier depends on the confidence level and the size of the sample.
>
> The standard error = sample standard deviation$/\sqrt{n}$.

21.2 Confidence Intervals for the Difference Between Two Means

In many instances, such as in the preceding example, we are interested in comparing the population means under two conditions or for two groups. One way to do that is to construct separate confidence intervals for the two conditions and then compare them. That's what we did in Section 21.1 for the weight-loss example. A more direct and efficient approach is to construct a single confidence interval for the *difference* in the population means for the two groups or conditions. In this section, we learn how to do that.

You may have noticed that the formats are similar for the two types of approximate 95% confidence intervals we have discussed so far. That is, they were both used to estimate a population value, either a proportion or a mean. They were both built around the corresponding sample value, the sample proportion or the sample mean. They both had the form:

$$\text{sample value} \pm 2 \times \text{standard error}$$

This format was based on the fact that the "sample value" over repeated samples was predicted to follow a bell-shaped curve centered on the "population value." All we needed to know in addition was the "standard deviation" for that specific bell-shaped curve, which we called the "standard error" of either a proportion or a mean. The Empirical Rule from Chapter 8 tells us that an interval spanning 2 standard deviations on either side of the center will cover 95% of the possible values.

The same is true for calculating a 95% confidence interval for the difference in two means. Here is the recipe you follow:

> **Constructing an Approximate 95% Confidence Interval for the Difference in Two Population Means**
>
> 1. Collect a large sample of observations (at least 30), independently, under each condition or from each group. Compute the mean and the standard deviation for each sample.
> 2. Compute the standard error of the mean (SEM) for each sample by dividing the sample standard deviation by the square root of the sample size.

3. Square the two SEMs and add them together. Then take the square root. This will give you the necessary "standard error," which is called the **standard error of the difference in two means.** We abbreviate it as "standard error of difference," or simply "SED." In other words,

SED = standard error of difference = square root of $[(SEM_1)^2 + (SEM_2)^2]$

4. An approximate 95% confidence interval for the difference in the two population means is

difference in sample means $\pm$ 2 × standard error of difference

or

difference in sample means $\pm$ 2 × square root of $[(SEM_1)^2 + (SEM_2)^2]$

EXAMPLE 21.2   A Direct Comparison of Diet and Exercise

We are now in a position to compute an approximate 95% confidence interval for the *difference* in population means for weight loss from dieting only and weight loss from exercising only. Let's follow the steps outlined, using the data from the previous section.

Steps 1 and 2. Compute sample means, standard deviations, and SEMs:

Diet Only

sample mean = 7.2 kg
sample standard deviation = 3.7 kg
number of participants = n = 42
standard error = SEM_1 =
  $3.7/\sqrt{47} = 0.571$

Exercise Only

sample mean = 4.0 kg
sample standard deviation = 3.9 kg
number of participants = n = 47
standard error = SEM_2 =
  $3.9/\sqrt{47} = 0.569$

Step 3. Square the two standard errors, and add them together. Take the square root:

SED = standard error of difference = square root of $[(0.571)^2 + (0.569)^2] = 0.81$

Step 4. Compute the interval. An approximate 95% confidence interval for the difference in the two population means is

difference in sample means $\pm$ 2 × standard error of difference

$(7.2 - 4.0) \pm 2(0.81)$

3.2 ± 1.6

1.6 kg to 4.8 kg

3.5 lb to 10.6 lb

Notice that this interval is entirely above zero. Therefore, we can be highly confident that there really is a difference in average weight loss for the populations, with higher weight loss for dieting alone than for exercise alone. The "populations" in this example represent the amount of weight that would be lost by all men like the ones in this study, *if* they were to diet versus *if* they were to exercise. In other words, we are 95% confident that the interval captures the true difference in the means of these hypothetical populations and that the true difference is at least 3.5 pounds. Remember that this interval estimates the difference in *averages,* not in weight losses for individuals.   ■

A Caution about Using This Method

The method described in this section is valid only when *independent* measurements are taken from the two groups. For instance, if matched pairs are used and one treatment is randomly assigned to each half of the pair, the measurements would not be independent. In that case, differences should be taken for each pair of measurements, and then a confidence interval computed for the mean of those differences. Assuming the matched observations were positively correlated, using the method in this section would result in a standard error that was too large.

21.3   Revisiting Case Studies and Examples: How Journals Present Confidence Intervals

Many of the case studies and examples we examined in the early part of this book involved making conclusions about differences in means. The original journal articles from which those examples were drawn each presented results in a slightly different way. Some provided confidence intervals directly; others gave the information necessary for you to construct your own interval. In this section, we discuss how some of the journal articles presented results, as examples of the kinds of information researchers provide for readers.

Direct Reporting of Confidence Intervals: Case Study 6.4

Case Study 6.4 examined the relationship between smoking during pregnancy and subsequent IQ of the child. The journal article in which that study was reported (Olds, Henderson, and Tatelbaum, 1994) provided 95% confidence intervals to accompany all the results it reported. Most of the confidence intervals were based on a comparison of the means for mothers who didn't smoke and mothers who smoked 10 or more cigarettes per day, hereafter called "smokers." Some of the results were presented in tables and others were presented in the text. Table 21.1 gives some of the results from the tables contained in the paper.

Let's interpret these results. The first confidence interval compares the mean educational levels for the smokers and nonsmokers. The result tells us that, in this sample, the average educational level for nonsmokers was 0.67 year higher than for smokers. The confidence interval extends this value to what the difference might be

TABLE 21.1	Some 95% Confidence Intervals from Case Study 6.4		
	Sample Means		
	0 Cigarettes	**10+ Cigarettes**	**Difference (95% CI)**
Maternal education, grades	11.57	10.89	0.67 (0.15,1.19)
Stanford-Binet (IQ), 48 mo	113.28	103.12	10.16 (5.04,15.30)
Birthweight, g	3416	3035	381.0 (167.1,594.9)

Source: "Study: Smoking May Lower Kids' IQs." Associated Press, February 11, 1994. Reprinted with permission.

for the populations from which these samples were drawn. The interval tells us that the difference in the population is probably between 0.15 and 1.19 years of education. In other words, mothers who did not smoke were also likely to have had more education. Maternal education was a *confounding variable* in this study and was part of what the researchers used to try to explain the differences observed in the children's IQs for the two groups.

The second row of Table 21.1 compares the mean IQs for the children of the nonsmokers and smokers at 48 months of age. The difference in means for the sample was 10.16 points. Based on the data from the two groups, the researchers computed a 95% confidence interval and concluded that there is probably a difference of somewhere between 5.04 and 15.30 points for the entire population. In other words, the children of nonsmokers in the population probably have IQs that are between 5.04 and 15.30 points higher than the children of mothers who smoke 10 or more cigarettes per day.

The third row of Table 21.1 (birthweight) represents an example of the kind of explanatory confounding variables that may have been present. Smoking may have caused lower birthweights, which in turn may have caused lower IQs. The result shown here is that the average difference in birthweight for babies of nonsmokers and smokers in the sample was 381 grams. Further, with 95% confidence, we can state that there could be a difference as low as 167.1 grams or as high as 594.9 grams for the population from which this sample was drawn. Thus, we are fairly certain that mothers in the population who smoke are likely to have babies with lower birthweight, on average, than those who don't.

Olds and colleagues (1994) also included numerous confidence intervals in the text. For example, they realized that they needed to control for confounding variables to make a more realistic comparison between the IQs of children of smokers and nonsmokers. They wrote: "After control for confounding background variables (Table 3), the average difference observed at 12 and 24 months was 2.59 points (95% CI: −3.03, 8.20); the difference observed at 36 and 48 months was reduced to 4.35 points (95% CI: 0.02, 8.68)" (pp. 223–224).

The result, as reported in the news story quoted in Chapter 6, was that the gap in average IQ at 3 and 4 years of age narrowed to 4 points when "a wide range of interrelated factors were controlled." You are now in an excellent position to understand a much broader picture than that reported to you in the news story. For example, as you can see from the reported confidence intervals, we can't rule out the possibility that the differences in IQ at 1 and 2 years of age were in the other direction because the interval covers some negative values. Further, even at 3 and 4 years of age, the confidence interval tells us that the gap *could* have been just slightly above zero in the population.

Reporting Standard Errors of the Mean: Example 12.4

In Example 12.4 and other examples in Chapter 12, we discussed an Australian research report that compared separation rates for couples whose parents had or had not been divorced. The primary question the report sought to answer was whether couples whose parents had divorced had a higher likelihood of divorcing as well. Age information for the couples was provided in the report too, so it makes sense to ask whether the couples who separated were, on average, younger, older, or about the same age as couples who did not separate.

Table 21.2 provides the age information for the males as it was presented by Butterworth and colleagues (2008, Table 2, p. 22). The report included the mean age for the males in the intact couples as 39.18 years, and the mean age for the males in the couples who separated as 37.65. (Corresponding mean ages for the females in the two groups were 36.62 and 35.32.) The size of the sample and the standard error of the mean (SEM) are provided for each group as well. You might wonder why the SEM is so much higher for the separating couples (0.75 compared with 0.19). Is it because their ages are more variable? Remember that $SEM = s/\sqrt{n}$, and in this instance, n is 1384 for the intact couples but only 114 for the separating couples. The fact that n is so much larger for the intact couples explains the difference in magnitude of the standard errors. In both cases, s is around 7 or 8 years.

TABLE 21.2	Mean and Standard Errors for Male Ages in Australian Couples	
	Intact couples	**Separating Couple**
Number of couples	1384	114
Mean age of males	39.18	37.65
Standard error	0.19	0.75

We can use the information in Table 21.2 to create an approximate 95% confidence interval for the difference in the *population* mean ages for Australian couples who remained intact and who separated during that time period, assuming that the couples in this study were representative of all such couples. The approximate 95% confidence interval is:

$$\text{difference in sample means} \pm 2 \times \text{square root of } [(SEM_1)^2 + (SEM_2)^2]$$
$$(39.18 - 37.65) \pm 2 \times \text{square root of } [(0.19)^2 + (0.75)^2]$$
$$1.53 \pm 2 \times (0.774)$$
$$1.53 \pm 1.55$$
$$-0.02 \text{ to } 3.08$$

The lower end point of this interval is essentially zero. So with approximate 95% confidence, we can say that the mean age of the population of males in intact couples could be as low as the mean age of the population of males in separating couples, or as high as three years older than the mean for the separating couples. These numbers indicate that the males in the intact couples may have been slightly older, on average, than the males in the couples who separated. Certainly, they do not appear to have been younger.

Reporting Standard Deviations: Case Study 5.1

In Case Study 5.1, we looked at the comparison of smoking cessation rates for patients using nicotine patches versus those using placebo patches. The main variables of interest were categorical and thus should be studied using methods for proportions. However, the authors also presented numerical summaries of a variety of other characteristics of the subjects. That information is useful to make sure that the randomization procedure distributed those variables fairly across the two treatment conditions.

The authors reported means, standard deviations (*SD*), and ranges (low to high) for a variety of characteristics. As an example, Figure 21.1 shows part of a table given by Hurt and colleagues (23 February 1994). In the table, the ages for the group wearing placebo patches had a mean of 43.6 years and a standard deviation of 10.6 years, and ranged from 21 to 65 years.

Figure 21.1

Part of a table from Case Study 5.1

Table 1 Baseline Characteristics		
Mean ± *SD* (Range)	**Active**	**Placebo**
Age, y	42.8 ± 11.1(20–65)	43.6 ± 10.6(21–65)
Cigarettes /d(n = 119/119)*	28.8 ± 9.4(20–60)	30.6 ± 9.4(20–60)

*The notation (n = 119/119) means that there were 119 people in each group for these calculations.
Source: Hurt et al., 23 February 1994, p. 596.

Notice that the intervals given in the table are *not* 95% confidence intervals, despite the fact that they are presented in the standard format used for such intervals. Be sure to read carefully when you examine results presented in journals so that you are not misled into thinking results presented in this format always represent 95% confidence intervals.

From the information presented, we notice a slight difference in the mean ages for each group and in the mean number of cigarettes each group smoked per day at the start of the study. Let's use the information given in the table to compute a 95% confidence interval for the difference in number of cigarettes smoked per day, to find out if it represents a substantial difference in the populations represented by the two groups. (It should be obvious to you that if the placebo group started out smoking significantly more, the results of the study would be questionable.) To compute the confidence interval, we need the means, standard deviations, and the sample sizes (*n*) presented in the table. Here is how we would proceed with the computation.

Steps 1 and 2. Compute sample means, standard deviations, and SEMs:

Active Group

sample mean = 28.8 cigarettes per day

sample standard deviation =
9.4 cigarettes

number of participants = n = 119

standard error = SEM_1 =
$9.4/\sqrt{119} = 0.86$

Placebo Group

sample mean = 30.6 cigarettes per day

sample standard deviation =
9.4 cigarettes

number of participants = n = 119

standard error = SEM_2 =
$9.4/\sqrt{119} = 0.86$

Step 3. Square the two standard errors, and add them together. Take the square root:

$$SED = \text{standard error of difference} = \text{square root of } [(0.86)^2 + (0.86)^2] = 1.2$$

Step 4. Compute the interval. An approximate 95% confidence interval for the difference in the two population means is

$$\text{difference in sample means} \pm 2 \times \text{standard error of difference}$$
$$(28.8 - 30.6) \pm 2(1.2)$$
$$-1.8 \pm 2.4$$
$$-4.2 \text{ to } +0.60$$

It appears that there could have been slightly fewer cigarettes smoked per day by the group that received the nicotine patches, but the interval covers zero, allowing for the possibility that the difference observed in the sample means was *opposite* in direction from the difference in the population means. In other words, we simply can't tell if the difference observed in the sample means represents a real difference in the populations.

Summary of the Variety of Information Given in Journals

There is no standard for how journal articles present results. However, you can determine confidence intervals for individual means or for the difference in two means (for large, independent samples) as long as you are given one of the following sets of information:

1. Direct confidence intervals

2. Means and standard errors of the means

3. Means, standard deviations, and sample sizes

21.4   Understanding Any Confidence Interval

Not all results are reported as proportions, means, or differences in means. Numerous other statistics can be computed to make comparisons, almost all of which have corresponding formulas for computing confidence intervals. Some of those formulas are quite complex, however.

In cases where a complicated procedure is needed to compute a confidence interval, authors of journal articles usually provide the completed intervals. Your job is to be able to interpret those intervals. The principles you have learned for understanding confidence intervals for means and proportions are directly applicable to understanding any confidence interval. As an example, let's consider the confidence intervals reported in another of our earlier case studies.

Confidence Interval for Relative Risk: Case Study 5.4

In Case Study 5.4, we investigated a study relating baldness and heart disease. The measure of interest in that study was the relative risk of heart disease based on degree of baldness. The investigators focused on the relative risk (RR) of myocardial infarction (MI)—that is, a heart attack—for men with baldness compared to men without any baldness. Here is how they reported some of the results:

> *For mild or moderate vertex baldness, the age-adjusted RR estimates were approximately 1.3, while for extreme baldness the estimate was 3.4 (95% CI, 1.7 to 7.0). . . . For any vertex baldness (i.e., mild, moderate, and severe combined), the age-adjusted RR was 1.4 (95% CI, 1.2 to 1.9)* (Lesko et al., 1993, p. 1000).

The confidence intervals for age-adjusted relative risk are not simple to compute. You may notice that they are not of the form "sample value $\pm$ 2 $\times$ (standard error)," evidenced by the fact that they are not symmetric about the sample values given. However, these intervals can be interpreted in the same way as any other confidence interval. For instance, with 95% certainty we can say that men with extreme baldness are at higher risk of heart attack than men with no baldness and that the ratio of risks is probably between 1.7 and 7.0. In other words, men with extreme baldness are probably anywhere from 1.7 to 7 times more likely to experience a heart attack than men of the same age without any baldness. Of course, these results assume that the men in the study are representative of the larger population.

Understanding the Confidence Level

For a 95% confidence interval, the value of 95% is called the **confidence level.** In general, the confidence level is a measure of how much confidence we have that the *procedure* used to generate the interval worked. For a confidence level of 95%, we expect that about 95% of all such intervals will actually cover the true population value. The remaining 5% will not. Any particular numerical interval computed this way either covers the truth or not. The trouble is, when we have an interval in hand, we don't know if it's one of the 95% "good" ones or the 5% "bad" ones. Thus, we say we are 95% confident that the interval captures the true population value.

There is nothing sacred about 95%, although it is the value most commonly used. As noted in Chapter 20, if we changed the multiplier to 1.645 (instead of 2), we would construct an approximate 90% confidence interval. To construct an approximate 99% confidence interval, we would use a multiplier of 2.576. As you should be aware from the Empirical Rule in Chapter 8, a multiplier of 3 would produce a

99.7% confidence interval. In general, the larger the multiplier, the higher the level of confidence that we have correctly captured the truth. Of course, the trade-off is that a larger multiplier produces a wider interval.

| CASE STUDY 21.1 | **Premenstrual Syndrome? Try Calcium** |

It was front page news in the *Sacramento Bee*. The headline read "Study says calcium can help ease PMS," and the article continued: "Daily doses of calcium can reduce both the physical and psychological symptoms of premenstrual syndrome by at least half, according to new research that points toward a low-cost, simple remedy for a condition that affects millions of women" (Maugh, 26 August 1998). The article described a randomized, double-blind experiment in which women who suffered from premenstrual syndrome (PMS) were randomly assigned to take either a placebo or 1200 mg of calcium per day in the form of four Tums E-X tablets (Thys-Jacobs et al., 1998). Participants included 466 women with a history of PMS: 231 in the calcium treatment group and 235 in the placebo group.

The primary measure of interest was a composite score based on 17 PMS symptoms, including six that were mood-related, five involving water retention, two involving food cravings, three related to pain, and insomnia. Participants were asked to rate each of the 17 symptoms daily on a scale from 0 (absent) to 3 (severe). A composite "symptom complex score" was created by using the mean rating for the 17 symptoms. Thus, a score of 0 would imply that all symptoms were absent, and a score of 3 would indicate that all symptoms were severe. The original article (Thys-Jacobs et al., 1998) presents results individually for each of the 17 symptoms plus the composite score.

One interesting outcome of this study was that the severity of symptoms was substantially reduced for both the placebo and the calcium-treated groups. Therefore, comparisons should be made between those two groups rather than examining the reduction in scores before and after taking calcium for the treatment group alone. In other words, part of the total reduction in symptoms for the calcium-treated group could be the result of a "placebo effect." We are interested in knowing the additional influence of taking calcium.

Let's compare the severity of symptoms as measured by the composite score for the placebo and calcium-treated groups. The treatments were continued for three menstrual cycles; we report the symptom scores for the premenstrual period (7 days) before treatments began (baseline) and before the third cycle.

Table 21.3 presents results as given in the journal article, including sample sizes and the mean symptom complex scores ± 1 standard deviation. Notice that sample sizes were slightly reduced by the third cycle due to patients dropping out of the study. Let's use the results in the table to compute a confidence interval for what the mean differences would be for the entire population of PMS sufferers.

The purpose of the experiment is to see if taking calcium diminishes symptom severity. Because we know that placebos alone can be responsible for reducing symptoms, the appropriate comparison is between the placebo and calcium-treated

TABLE 21.3   Results for Case Study 21.1

	Symptom Complex Score: Mean ± *SD*	
	Placebo Group	**Calcium-Treated Group**
Baseline	0.92 ± 0.55 (*n* = 235)	0.90 ± 0.52 (*n* = 231)
Third cycle	0.60 ± 0.52 (*n* = 228)	0.43 ± 0.40 (*n* = 212)

groups rather than between the baseline and third cycle symptoms for the calcium-treated group alone.

The difference in means (placebo − calcium) for the third cycle is $(0.60 - 0.43) = 0.17$. The "standard error" is about 0.039, so a 95% confidence interval for the difference is $0.17 \pm 2(0.039)$, or about 0.09 to 0.25.

To put this in perspective, remember that the scores are averages over the 17 symptoms. Therefore, a reduction from a mean of 0.60 to a mean of 0.43 would, for instance, correspond to a reduction from $(0.6)(17) = 10$ mild symptoms (rating of 1) to $(0.43)(17) = 7.31$, or just over seven mild symptoms. In fact, examination of the full results shows that all 17 symptoms had reduced severity in the calcium-treated group compared with the placebo group. And, because this is a randomized experiment and not an observational study, we can conclude that the calcium actually *caused* the reduction in symptoms.

As a final note, Table 21.3 also indicates a striking drop in the mean symptom score from baseline to the third cycle for both groups. For the placebo group, the symptom scores dropped by about a third; for the calcium-treated group, they were more than cut in half. Thus, it appears that placebos can help reduce the severity of PMS symptoms. ∎

Thinking About Key Concepts

- Confidence intervals provide an interval of plausible values for an unknown population value, such as a mean, proportion, or difference in two means.
- The *general format for a confidence interval* for a population proportion, mean, or difference in two means is (sample estimate) ± multiplier × (standard error), where the sample estimate and standard error are different for each situation, and the multiplier is determined by the desired confidence level.
- The multiplier for an approximate 95% confidence interval is 2.
- Confidence intervals can be found for other population characteristics, such as relative risk, but the format is different than the one used for proportions and means. The interpretation is the same.
- A *confidence level* accompanies any confidence interval. It provides the long-run relative frequency for which the confidence interval procedure works, meaning that the interval actually covers the population number is was designed to estimate.

- The higher the desired confidence level, the larger the multiplier, and thus the wider the confidence interval.
- Larger sample sizes result in smaller standard errors and thus more narrow confidence intervals.

Focus on Formulas

Review of Notation from Previous Chapters

Population mean $= \mu$, sample mean $= \overline{X}$, sample standard deviation $= s$

$z_{\alpha/2} = $ standardized normal score with area $\alpha/2$ above it

Standard Error of the Mean

Standard error of the mean $=$ SEM $= s/\sqrt{n}$

Approximate 95% Confidence Interval for a Single Mean μ

$$\overline{X} \pm 2 \times SEM$$

General Confidence Interval for a Single Mean μ

$$\overline{X} \pm t\left(\frac{\alpha}{2}, df\right) \times SEM$$

where $t(\alpha/2, df)$ is the t-multiplier for a $(1 - \alpha)$ confidence level and degrees of freedom df.

Notation for Two Populations and Samples

Population mean $= \mu_i$, $i = 1$ or 2

Sample mean $= \overline{X}_i$, $i = 1$ or 2

Sample standard deviation $= s_i$, $i = 1$ or 2

Sample size $= n_i$, $i = 1$ or 2

Standard error of the mean $=$ SEM$_i = s_i/\sqrt{n_i}$, $i = 1$ or 2

Standard error of the difference $=$ SED $= \sqrt{SEM_1^2 + SEM_2^2}$

Approximate 95% Confidence Interval for the Difference in Two Population Means, Independent Samples

$$(\overline{X}_1 - \overline{X}_2) \pm 2 \times SED$$

$$(\overline{X}_1 - \overline{X}_2) \pm 2 \times \sqrt{\frac{s_1^2}{n_1} + \frac{s_2^2}{n_2}}$$

Exercises

Exercises with numbers divisible by 3 (3, 6, 9, etc.) are included in the Solutions at the back of the book. They are marked with an asterisk (*).

1. In Chapter 20, we saw that to construct a confidence interval for a population proportion it was enough to know the sample proportion and the sample size. Is the same true for constructing a confidence interval for a population mean? That is, is it enough to know the sample mean and sample size? Explain.

2. Explain the difference between a population mean and a sample mean using one of the studies discussed in the chapter as an example.

*3. The *Baltimore Sun* (Haney, 21 February 1995) reported on a study by Dr. Sara Harkness in which she compared the sleep patterns of 6-month-old infants in the United States and the Netherlands. She found that the 36 U.S. infants slept an average of just under 13 hours out of every 24, whereas the 66 Dutch infants slept an average of almost 15 hours.

   *a. The article did not report a standard deviation, but suppose it was 0.5 hour for each group. Compute the standard error of the mean (SEM) for the U.S. babies.

   *b. Continuing to assume that the standard deviation is 0.5 hour, compute an approximate 95% confidence interval for the mean sleep time for 6-month-old babies in the United States.

   *c. Continuing to assume that the standard deviation for each group is 0.5 hour, compute an approximate 95% confidence interval for the *difference* in average sleep time for 6-month-old Dutch and U.S. infants.

4. What is the probability that a 95% confidence interval will not cover the true population value?

5. Suppose a university wants to know the average income of its students who work, and all students supply that information when they register. Would the university need to use the methods in this chapter to compute a confidence interval for the population mean income? Explain. (*Hint:* What is the sample mean and what is the population mean?)

*6. Suppose you were given a 95% confidence interval for the difference in two population means. What could you conclude about the population means if

   *a. The confidence interval *did not* cover zero.

   *b. The confidence interval *did* cover zero.

7. Suppose you were given a 95% confidence interval for the relative risk of disease under two different conditions. What could you conclude about the risk of disease under the two conditions if

   a. The confidence interval *did not* cover 1.0.

   b. The confidence interval *did* cover 1.0.

8. Ever since President John F. Kennedy was assassinated in 1963, there has been speculation about whether there was a single assassin or more than one. Periodically, public opinion polls ask this question: "Turning now to the assassination of John F. Kennedy in 1963—Do you think that one man was responsible for the assassination of President Kennedy, or do you think that others were involved in a conspiracy?" (http://www.pollingreport.com/news3.htm#Kennedy). In a poll of 533 adults in the United States in 2003, 400 (i.e., 75%) of them responded "Others involved." In a poll of 1004 U.S. adults in 2013, 592 (i.e., 59%) of them responded "Others involved." An approximate 95% confidence interval for the *difference* in population percentages who felt that way in 2003 versus 2013 is 16% ± 5%, with the higher percentage in 2003.

   a. Compute the endpoints of the interval.

   b. Write a sentence or two interpreting the interval in words that would be understood by someone with no training in statistics.

   c. Does the interval indicate that opinion on this issue changed between 2003 and 2013 for the population of U.S. adults? Explain.

*9. On page 453 in this chapter technical details were presented for finding a *t*-multiplier for constructing a confidence interval for a population mean. Use appropriate software, a calculator, or a website to find the *t*-multiplier for the following situations.

  *a. 95% confidence interval, *n* = 200

  *b. 95% confidence interval, *n* = 50

  *c. 90% confidence interval, *n* = 200

  *d. 90% confidence interval, *n* = 50

10. On page 453 in this chapter technical details were presented for finding a *t*-multiplier for constructing a confidence interval for a population mean. Use appropriate software, a calculator, or a website to find a confidence interval for the population mean for the following situations. Use the *t*-multiplier in each case.

  a. Refer to Table 21.2. Find a 95% confidence interval for the population mean age of Australian males in couples who separated during the time period of the study by Butterworth et al (2008). Relevant numbers from the table are the sample mean of 37.65 years, SEM of 0.75, and sample size of 114.

  b. Repeat part (a) for the population of males in couples who remained intact. Relevant numbers from the table are the sample mean of 39.18 years, SEM of 0.19, and sample size of 1384.

  c. Repeat part (a) using a confidence level of 90%.

11. In Case Study 6.4, which examined maternal smoking and child's IQ, one of the results reported in the journal article was the average number of days the infant spent in the neonatal intensive care unit. The results showed an average of 0.35 day for infants of nonsmokers and an average of 0.58 day for the infants of women who smoked 10 or more cigarettes per day. In other words, the infants of smokers spent an average of 0.23 day more in neonatal intensive care. A 95% confidence interval for the difference in the two means extended from −3.02 days to +2.57 days. Explain why it would have been misleading to report, "these results show that the infants of smokers spend more time in neonatal intensive care than do the infants of nonsmokers," even though it was true for the infants in this study.

*12. In a study comparing age of death for left- and right-handed baseball players, Coren and Halpern (1991, p. 93) provided the following information: "Mean age of death for strong right-handers was 64.64 years (SD = 15.5, *n* = 1472); mean age of death for strong left-handers [was] 63.97 years (SD = 15.4, *n* = 236)." The term "strong handers" applies to baseball players who both threw and batted with the same hand. The data were actually taken from entries in *The Baseball Encyclopedia* (6th ed., New York: Macmillan, 1985), but, for the purposes of this exercise, pretend that the data were from a sample drawn from a larger population.

  *a. Compute an approximate 95% confidence interval for the mean age of death for the population of strong right-handers from which this sample was drawn.

  *b. Repeat part (a) for the strong left-handers.

  *c. Compare the results from parts (a) and (b) in two ways. First, explain why one confidence interval is substantially wider than the other. Second, explain whether you would conclude that there is a difference in the mean ages of death for left- and right-handers on the basis of these results.

  *d. Compute an approximate 95% confidence interval for the difference in mean ages of death for the strong right- and left-handers. Interpret the result.

13. In revisiting Case Study 5.4, we quoted the original journal article as reporting that "for any vertex baldness (i.e., mild, moderate, and severe combined), the age-adjusted RR was 1.4 (95% CI, 1.2 to 1.9)" (Lesko et al., 1993, p. 1000). Interpret this result. Explain in words that someone with no training in statistics would understand.

14. In a report titled, "Secondhand Smoke: Is It a Hazard?" (*Consumer Reports,* January 1995, pp. 27–33), 26 studies linking secondhand smoke and lung cancer were summarized by

noting, "those studies estimated that people breathing secondhand smoke were 8 to 150 percent more likely to get lung cancer sometime later" (p. 28). Although it is not explicit, assume that the statement refers to a 95% confidence interval and interpret what this means.

*15. Refer to Case Study 21.1, illustrating the role of calcium in reducing the symptoms of PMS. Using the caution given at the end of the case study, explain why we cannot use the method presented in Section 21.2 to compare baseline symptom scores with third-cycle symptom scores for the calcium-treated group alone.

16. Parts (a) through (d) below provide additional results for Case Study 21.1. For each of the parts, compute an approximate 95% confidence interval for the difference in mean symptom scores between the placebo and calcium-treated conditions for the symptom listed. In each case, the results given are mean ± standard deviation. There were 228 participants in the placebo group and 212 in the calcium-treated group.

   a. Mood swings: placebo = 0.70 ± 0.75; calcium = 0.50 ± 0.58

   b. Crying spells: placebo = 0.37 ± 0.57; calcium = 0.23 ± 0.40

   c. Aches and pains: placebo = 0.49 ± 0.60; calcium = 0.31 ± 0.49

   d. Craving sweets or salts: placebo = 0.60 ± 0.78; calcium = 0.43 ± 0.64

17. In Chapter 20, we learned that to compute an approximate 90% confidence interval, the appropriate multiplier is 1.645 instead of 2.0. This works for confidence intervals for one mean or the difference in two means as well. Consult Table 21.2 on page 458 to find the means and standard errors for the ages of males in couples that remained intact and that separated in the study of Australian couples.

   a. Compute an approximate 90% confidence interval for the difference in mean ages for the populations of men in couples that remained intact and that separated.

   b. Based on your result in part (a) could you conclude with 90% confidence that the difference in mean ages observed in the samples represents a real difference in mean ages in the population? Explain.

*18. Using the data presented by Hand and colleagues (1994) and discussed in previous chapters, we would like to estimate the average age difference between husbands and wives in Britain. Recall that the data consisted of a random sample of 200 couples. Following are two methods that were used to construct a confidence interval for the difference in ages. Your job is to figure out which method is correct:

   Method 1: Take the difference between the husband's age and the wife's age for each couple, and use the differences to construct an approximate 95% confidence interval for a single mean. The result was an interval from 1.6 to 2.9 years.

   Method 2: Use the method presented in this chapter for constructing an approximate confidence interval for the difference in two means for two independent samples. The result was an interval from −0.4 to 4.3 years.

   Explain which method is correct, and why. Then interpret the confidence interval that resulted from the correct method.

19. Refer to Exercise 18. Suppose that from that same data set, we want to compute the average difference between the heights of adult British men and adult British women—not the average difference within married couples. Which of the two methods in Exercise 18 would be appropriate for this situation? Explain.

20. Refer to Exercises 18 and 19. The 200 men in the sample had a mean height of 68.2 inches, with a standard deviation of 2.7 inches. The 200 women had a mean height of 63.1 inches, with a standard deviation of 2.5 inches. Assuming these were independent samples, compute an approximate 95% confidence interval for the mean difference in heights between British males and females. Interpret the resulting

interval in words that a statistically naive reader would understand.

*21. Refer to Case Study 21.1 and the material in Part 1 of this book.

    *a. In their original report, Thys-Jacobs and colleagues (1998) noted that the study was "double-blind." Explain what that means in the context of this example.

    *b. Explain why it is possible to conclude that, based on this study, calcium actually *causes* a reduction in premenstrual symptoms.

22. Refer to the following statement on page 458: "For example, as you can see from the reported confidence intervals, we can't rule out the possibility that the differences in IQ at 1 and 2 years of age were in the other direction because the interval covers some negative values." The statement refers to a confidence interval given in the previous paragraph, ranging from -3.03 to 8.20. Write a paragraph explaining how to interpret this confidence interval that would be understood by someone with no training in statistics. Make sure you are clear about the population to which the result applies.

The following information is for Exercises 23 to 25: Refer to Original Source 5 on the companion website, "Distractions in Everyday Driving." Table 14 on page 51 provides 95% confidence intervals for the average percent of time drivers in the population would be observed not to have their hands on the wheel during various activities while the vehicle was moving, assuming they were like the drivers in this study. The confidence intervals were computed using a different method than the one presented in this book because of the type of data available, but the interpretation is the same. (See Appendix D of the report if you are interested in the details.)

23. a. The confidence interval accompanying "Reading/writing" is from 4.24% to 34.39%. Write a few sentences interpreting this interval that would be understood by someone with no training in statistics. Make sure you are clear about the population to which the result applies.

    b. Repeat part (a) for the confidence interval for "Conversing."

*24. Refer to the interval in part (a) of Exercise 23. Write it as a 95% confidence interval for the average *number of minutes* out of an hour that drivers who were reading or writing would be observed to have their hands off of the wheel.

25. Notice that in Table 14 (on page 51 of Original Source 5) some of the intervals are much wider than others. For instance, the 95% confidence interval for the percent of time with hands off the wheel while *not* reading or writing is from 0.97 to 1.93, compared with the interval from 4.24 to 34.39 while reading or writing. What two features of the data do you think are responsible for some intervals being wider than others? (*Hint:* What two features of the data determine the width of a 95% confidence interval for a mean? The answer is the same in this situation.)

26. Refer to Original Source 9 on the companion website, "Suicide Rates in Clinical Trials of SSRIs, Other Antidepressants, and Placebo: Analysis of FDA Reports." Table 1 in that paper (page 791) provides confidence intervals for suicide rates for patients taking three different kinds of medications. For example, the 95% confidence interval for "Placebo" is 0.01 to 0.19. This interval means that we can be 95% confident that between 0.01% and 0.19% of the population of people similar to the ones in these studies would commit suicide while taking a placebo. Notice that 0.01% is about 1 out of 10,000 people and 0.19% is about 1 out of 526 people, so this is a wide interval in terms of the actual suicide rate. Also note that the population from which the samples were drawn consisted of depressed patients who sought medical help.

    a. Report (from Table 1 in the article) the 95% confidence interval for the suicide rate for patients taking "selective serotonin reuptake inhibitors" (SSRIs). Write a sentence interpreting this interval.

    b. Compare the interval for patients taking placebos with the one of those taking SSRIs.

What can you conclude about the effectiveness of SSRIs in preventing suicide, based on a comparison between the two intervals?

c. Read the article to determine whether the results are based on observational studies or randomized experiments. Based on that determination, can you conclude that differences that might have been observed in suicide rates for the different groups were *caused* by the difference in type of drug treatment? Explain.

The following information is for Exercises 27 to 30: Original Source 10 (not on the companion website), titled "Religious attendance and cause of death over 31 years," provides 95% confidence intervals for the relative risk of death by various causes for those who attend religious services less than weekly versus weekly. (The relative risk is called "relative hazard" in the article and adjustments already have been made for some confounding factors such as age.)

*27. The 95% confidence interval for "All Causes" is 1.06 to 1.37. Explain what this interval means in a few sentences that would be understood by someone with no training in statistics. Make sure your explanation applies to the correct population.

28. What value of relative risk would indicate equal risk for those who attend and do not attend religious services at least weekly?

29. Confidence intervals are given for the relative risk of death for five specific causes of death, as follows: Circulatory (1.02 to 1.45), Cancer (0.83 to 1.37), Digestive (0.98 to 4.03), Respiratory (0.92, 3.02) and External (0.60 to 2.12). For which of these causes can it be concluded that the risk of death in the population is lower for those who attend religious services at least weekly? Explain what criterion you need to determine your answer. (*Hint:* Refer to Exercise 28.)

*30. Read Additional News Story 10 in the Appendix to determine whether the results are based on an observational study or a randomized experiment. Using that information, explain whether it can be concluded that attending religious services at least weekly *causes* the risk of death to change, for causes of death for which the confidence interval indicates there is such a change.

31. Refer to the link to Original Source 11 on the companion website, "Driving impairment due to sleepiness is exacerbated by low alcohol intake." (If the link no longer works, do a search on the title of the article to locate it.) Table 1 on the top of page 691 of the article presents mean blood alcohol concentrations (BAC) and standard errors of the mean (SE in Table 1) for the participants before driving. The values are given for the "With sleep restriction" and "No sleep restriction" conditions.

a. Compute approximate 95% confidence intervals for the mean BAC before driving for each of the two conditions. (Note that the sample sizes are smaller than may be appropriate for using the method in this chapter, but that will cause the intervals to be just slightly too narrow, or alternatively, the confidence to be slightly less than 95%. You can ignore that detail.)

b. Interpret the interval for the "With sleep restriction" condition. Make sure your explanation refers to the correct population.

c. Ignoring the fact that the sample size is slightly too small, would it be appropriate to compute a confidence interval for the difference in the two means using the method provided in this chapter? If so, compute the interval. If not, explain why not. You may have to read the article to determine whether the appropriate condition is met.

32. Table 1 on page 200 of Original Source 18, "Birth weight and cognitive function in the British 1946 birth cohort: longitudinal population based study" (link available on the companion website), provides 95% confidence intervals for the difference in mean standardized cognitive scores for mid-range birthweight babies (3.01 to 3.50 kg) and babies in each of four other birthweight groups. The comparisons are made at various ages. For

scores at age 8, the intervals are -0.42 to -0.11 for low weight (0 to 2.50 kg at birth); -0.16 to $+0.03$ for low-normal weight (2.51 to 3.00 kg at birth); 0.05 to 0.21 for high-normal weight (3.51 to 4.00 kg at birth); and -0.08 to $+0.14$ for very high-normal weight (4.01 to 5.00 kg at birth).

a. Write a sentence or two interpreting the interval for the low-weight group. Make sure your explanation applies to the correct population.

b. What difference value would indicate that the two means are equal in the population?

c. For which of the birthweight groups is it clear that there is a difference in the population mean standardized cognitive scores for that group and the midrange group? What criterion did you use to decide? In each case, explain whether the difference indicates that the mean standardized cognitive score is higher or lower for that group than for the midrange birthweight group.

d. Do the results of this study imply that low birth weight *causes* lower cognitive scores? Explain.

Mini-Projects

1. Find a journal article that reports at least one 95% confidence interval. Explain what the study was trying to accomplish. Give the results as reported in the article in terms of 95% confidence intervals. Interpret the results. Discuss whether you think the article accomplished its intended purpose. In your discussion, include potential problems with the study, as discussed in Chapters 4 to 6.

2. Collect data on a measurement variable for which the mean is of interest to you. Collect at least 30 observations. Using the data, compute an approximate 95% confidence interval for the mean of the population from which you drew your observations. Explain how you collected your sample, and note whether your method would be likely to result in any biases if you tried to extend your sample results to the population. Interpret the 95% confidence interval to make a conclusion about the mean of the population.

3. Collect data on a measurement variable for which the difference in the means for two conditions or groups is of interest to you. Collect at least 30 observations for each condition or group. Using the data, compute an approximate 95% confidence interval for the difference in the means of the populations from which you drew your observations. Explain how you collected your samples, and note whether your method would be likely to result in any biases if you tried to extend your sample results to the populations. Interpret the 95% confidence interval to make a conclusion about the difference in the means of the populations or conditions.

References

Butterworth, P., T. Oz, B. Rodgers, and H. Berry. (2008). Factors associated with relationship dissolution of Australian families with children. Social Policy Research Paper No. 37, Australian Government, Department of Families, Housing, Community Services and Indigenous Affairs. http://www.fahcsia.gov.au/about-fahcsia/publications-articles/research-publications/

social-policy-research-paper-series/number-37-factors-associated-with-relationship-dissolution-of-australian-families-with-children, accessed June 19, 2013.

Coren, S., and D. Halpern. (1991). Left-handedness: A marker for decreased survival fitness. *Psychological Bulletin* 109, no. 1, pp. 90–106.

Hand, D. J., F. Daly, A. D. Lunn, K. J. McConway, and E. Ostrowski. (1994). *A handbook of small data sets.* London: Chapman and Hall.

Haney, Daniel Q. (21 February 1995). Highly stimulated babies may be sleeping less as a result. *Baltimore Sun,* pp. 1D, 5D.

Hurt, R., L. Dale, P. Fredrickson, C. Caldwell, G. Lee, K. Offord, G. Lauger, Z. Marusic, L. Neese, and T. Lundberg. (23 February 1994). Nicotine patch therapy for smoking cessation combined with physician advice and nurse follow-up. *Journal of the American Medical Association* 271, no. 8, pp. 595–600.

Iman, R. L. (1994). *A data-based approach to statistics.* Belmont, CA: Wadsworth.

Lesko, S. M., L. Rosenberg, and S. Shapiro. (1993). A case-control study of baldness in relation to myocardial infarction in men. *Journal of the American Medical Association* 269, no. 8, pp. 998–1003.

Maugh, Thomas H., II. (26 August 1998). Study says calcium can help ease PMS. *Sacramento Bee,* pp. A1, A9.

Olds, D. L., C. R. Henderson, Jr., and R. Tatelbaum. (1994). Intellectual impairment in children of women who smoke cigarettes during pregnancy. *Pediatrics* 93, no. 2, pp. 221–227.

Thys-Jacobs, S., P. Starkey, D. Bernstein, J. Tian, and the Premenstrual Syndrome Study Group. (1998). Calcium carbonate and the premenstrual syndrome: Effects on premenstrual and menstrual symptoms. *American Journal of Obstetrics and Gynecology* 179, no. 2, pp. 444–452.

Wood, R. D., M. L. Stefanick, D. M. Dreon, B. Frey-Hewitt, S. C. Garay, P. T. Williams, H. R. Superko, S. P. Fortmann, J. J. Albers, K. M. Vranizan, N. M. Ellsworth, R. B. Terry, and W. L. Haskell. (1988). Changes in plasma lipids and lipoproteins in overweight men during weight loss through dieting as compared with exercise. *New England Journal of Medicine* 319, no. 18, pp. 1173–1179.

CHAPTER 22

Rejecting Chance—Testing Hypotheses in Research

Thought Questions

1. In the courtroom, juries must make a decision about the guilt or innocence of a defendant. Suppose you are on the jury in a murder trial. It is obviously a mistake if the jury claims the suspect is guilty when in fact he or she is innocent. What is the other type of mistake the jury could make? Which is more serious?

2. Suppose exactly half, or 0.50, of a certain population would answer yes when asked if they support the death penalty. A random sample of 400 people results in 220, or 0.55, who answer yes. The Rule for Sample Proportions tells us that the potential sample proportions in this situation are approximately bell-shaped, with standard deviation of 0.025. Using the formula on page 176 (Chapter 8), find the standardized score for the observed value of 0.55. Then determine how often you would expect to see a standardized score at least that large or larger.

3. Suppose you are interested in testing a claim you have heard about the proportion of a population who have a certain trait. You collect data and discover that *if* the claim is true, the sample proportion you have observed is so large that it falls at the 99th percentile of possible sample proportions for your sample size. Would you believe the claim and conclude that you just happened to get a weird sample, or would you reject the claim? What if the result was at the 70th percentile? At the 99.99th percentile?

4. Which is generally more serious when getting results of a medical diagnostic test: a false positive, which tells you you have the disease when you don't, or a false negative, which tells you you do not have the disease when you do?

22.1   Using Data to Make Decisions

In Chapters 20 and 21, we computed confidence intervals based on sample data to learn something about the population from which the sample had been taken. We sometimes used those confidence intervals to make a decision about whether there was a difference between two conditions.

Examining Confidence Intervals

When we examined the confidence interval for the relative risk of heart attacks for men with vertex baldness compared with no baldness, we noticed that the interval (1.2 to 1.9) was entirely above a relative risk of 1.0. Remember that a relative risk of 1.0 is equivalent to equal risk for both groups, whereas a relative risk above 1.0 means the risk is higher for the first group. In this example, if the confidence interval had included 1.0, then we could not say whether the risk of heart attack is higher for men with vertex baldness or for men with no hair loss, even with 95% confidence.

Using another example from Chapter 21, we noticed that the confidence interval for the difference in mean weight loss resulting from dieting alone versus exercising alone was entirely above zero. From that, we concluded, with 95% confidence, that the mean weight loss using dieting alone would be higher than it would be with exercise alone. If the interval had covered (included) zero, we would not have been able to say, with high confidence, which method resulted in greater average weight loss in the population.

Hypothesis Tests

As we learned in Chapter 13, researchers interested in answering direct questions often conduct *hypothesis tests*. Remember the basic question researchers ask when they conduct such a test:

> *Is the relationship observed in the sample large enough to be called statistically significant, or could it have been due to chance?*

In this chapter, we learn more about the basic thinking that underlies hypothesis testing. In the next chapter, we learn how to carry out some simple hypothesis tests and examine some in-depth case studies.

EXAMPLE 22.1   Deciding if Students Prefer Quarters or Semesters

To illustrate the idea behind hypothesis testing, let's look at a simple, hypothetical example. There are two basic academic calendar systems in the United States: the quarter system and the semester system. Universities using the quarter system generally have three 10-week terms of classes, whereas those using the semester system have classes for two 15-week terms.

Suppose a university is currently on the quarter system and is trying to decide whether to switch to the semester system. Administrators are leaning toward switching to the semester system, but they have heard that the majority of students may oppose the switch. They decide to conduct a survey to see if there is convincing evidence that a majority of students opposes the plan, in which case they will reconsider their proposed change.

Administrators must choose from two hypotheses:

1. There is no clear preference (or the switch is preferred), so there is no problem.
2. As rumored, a majority of students oppose the switch, so the administrators should stop their plan.

The administrators pick a random sample of 400 students and ask their opinions. Of the 400, 220 of them say they oppose the switch. Thus, a clear majority of the sample, 0.55 (55%), is opposed to the plan. Here is the question that would be answered by a hypothesis test:

If there is really no clear preference, how likely would we be to observe sample results of this magnitude or larger, just by chance?

We already have the tools to answer this question. From the Rule for Sample Proportions, we know what to expect if there is no clear preference—that is, if, in truth, 50% of the students prefer each system:

> If numerous samples of 400 students are taken, the frequency curve for the proportions from the various samples will be approximately bell-shaped. The mean will be the true proportion from the population—in this case, 0.50. The standard deviation will be:
>
> the square root of: (true proportion) $\times$ (1 $-$ true proportion)/(sample size)
>
> In this case, the square root of $[(0.5)(0.5)/400] = 0.025$.

In other words, *if* there is truly no preference, then the observed value of 0.55 must have come from a bell-shaped curve with a mean of 0.50 and a standard deviation of 0.025.

How likely would a value as large as 0.55 be from that particular bell-shaped curve? To answer that, we need to compute the standardized score corresponding to 0.55:

$$\text{standardized score} = z\text{-score} = (0.55 - 0.50)/0.025 = 2.00$$

From Table 8.1 in Chapter 8, we find that a standardized score of 2.00 falls between 1.96 and 2.05, the values for the 97.5th and 98th percentiles. That is, if there is truly no preference, then we would observe a sample proportion as high as this (or higher) between 2% and 2.5% of the time. (Using Excel or other software provides a more precise answer of 2.3%.)

The administration must now make a decision. One of two things has happened:

1. There really is no clear preference, but by the "luck of the draw" this particular sample resulted in an unusually high proportion opposed to the switch. In fact, it is so high that chance would lead to such a high value only slightly more than 2% of the time.

2. There really is a preference against switching to the semester system. The proportion (of all students) against the switch is actually higher than 0.50.

Most researchers agree that, by convention, we can rule out chance if the "luck of the draw" would have produced such extreme results less than 5% of the time. Therefore, in this case, the administrators should probably decide to rule out chance. The proper conclusion is that, indeed, a majority is opposed to switching to the semester system.

When a relationship or value from a sample is so strong that we can effectively rule out chance in this way, we say that the result is *statistically significant*. In this case, we would say that *the percent of students who are opposed to switching to the semester system is statistically significantly higher than 50%.*   ■

22.2   The Basic Steps for Testing Hypotheses

Although the specific computational steps required to test hypotheses depend on the type of variables measured, the basic steps are the same for all hypothesis tests. In this section, we review the basic steps, first presented in Chapter 13.

The basic steps for testing hypotheses are

1. Determine the null hypothesis and the alternative hypothesis.
2. Collect the data and summarize them with a single number called a *test statistic.*
3. Determine how unlikely the test statistic would be if the null hypothesis were true.
4. Make a decision.

Step 1. *Determine the null hypothesis and the alternative hypothesis.* There are always two hypotheses. The first one is called the **null hypothesis**—the hypothesis that says nothing is happening. The exact interpretation varies, but it can generally be thought of as the status quo, no relationship, chance only, or some variation on that theme.

The second hypothesis is called the **alternative hypothesis** or the **research hypothesis.** This hypothesis is usually the reason the data are being collected in the first place. The researcher suspects that the status quo belief is incorrect or that there is indeed a relationship between two variables that has not been established before. The only way to conclude that the alternative hypothesis is the likely one is to have enough evidence to effectively rule out chance, as presented in the null hypothesis.

| EXAMPLE 22.2 | A Jury Trial

If you are on a jury in the American judicial system, you must presume that the defendant is innocent unless there is enough evidence to conclude that he or she is guilty. Therefore, the two hypotheses are

Null hypothesis: The defendant is innocent.

Alternative hypothesis: The defendant is guilty.

The trial is being held because the prosecution believes the status quo assumption of innocence is incorrect. The prosecution collects evidence, much like researchers collect data, in the hope that the jurors will be convinced that such evidence would be extremely unlikely if the assumption of innocence were true. ∎

For Example 22.1, the two hypotheses were

Null hypothesis: There is no clear preference for quarters over semesters.

Alternative hypothesis: The majority opposes the switch to semesters.

The administrators are collecting data because they are concerned that the null hypothesis is incorrect. If, in fact, the results are extreme enough (many more than 50% of the *sample* oppose the switch), then the administrators must conclude that a majority of the *population* of students opposes semesters.

Step 2. *Collect the data and summarize them with a single number called a test statistic.* Recall how we summarized the data in Example 22.1, when we tried to determine whether a clear majority of students opposed the semester system. In the final analysis, we based our decision on only one number, the *standardized score* for our sample proportion.

In general, the decision in a hypothesis test is based on a single summary of the data. This summary is called the **test statistic.** We encountered this idea in Chapter 13, where we used the chi-square statistic to determine whether the relationship between two categorical variables was statistically significant. In that kind of problem, the chi-square statistic is the only summary of the data needed to make the decision. The chi-square statistic is the *test statistic* in that situation. The standardized score was the *test statistic* for Example 22.1.

Step 3. *Determine how unlikely the test statistic would be if the null hypothesis were true.* In order to decide whether the results could be just due to chance, we ask the following question:

If the null hypothesis is really true, how likely would we be to observe sample results of this magnitude or larger (in a direction supporting the alternative hypothesis) just by chance?

Answering that question usually requires special tables, such as Table 8.1 for standardized scores, a computer with Excel or other software, or a calculator with statistical functions. Fortunately, most researchers answer the question for you in their reports, and you must simply learn how to interpret the answer. The numerical value giving the answer to the question is called the *p-value.*

> The *p-value* is computed by *assuming* the null hypothesis is true, and then asking how likely we would be to observe such extreme results (or even more extreme results) under that assumption.

Many statistical novices misinterpret the meaning of a *p*-value. *The p-value does not give the probability that the null hypothesis is true.* There is no way to do that. For example, in Case Study 1.2, we noticed that for the men in that sample, those who took aspirin had fewer heart attacks than those who took a placebo. The *p*-value for that example tells us the probability of observing a relationship that extreme or more so in a sample of that size *if* there really is no difference in heart attack rates for the two conditions in the population. There is *no way* to determine the probability that aspirin actually has no effect on heart attack rates. In other words, there is no way to determine the probability that the null hypothesis is true. Don't fall into the trap of believing that common misinterpretation of the *p*-value.

Step 4. *Make a decision.* Once we know how unlikely the results would have been if the null hypothesis were true, we must make one of two conclusions:

> *Conclusion 1:* The *p*-value is not small enough to convincingly rule out chance. Therefore, we *cannot reject* the null hypothesis as an explanation for the results.

> *Conclusion 2:* The *p*-value is small enough to convincingly rule out chance. We *reject* the null hypothesis and *accept* the alternative hypothesis.

Notice that it is not valid to actually *accept* that the null hypothesis is true. To do so would be to say that we are essentially *convinced* that chance alone produced the observed results. This is another common mistake, which we will explore in Chapter 24.

Making conclusion 2 is equivalent to declaring that the result is *statistically significant.* We can rephrase the two possible conclusions in terms of statistical significance as follows:

> *Conclusion 1:* There is no statistically significant difference or relationship evidenced by the data.

> *Conclusion 2:* There is a statistically significant difference or relationship evidenced by the data.

You may be wondering how small the *p*-value must be in order to be small enough to rule out the null hypothesis. The cut off point is called the **level of significance.** The standard used by most researchers is 5%. However, that is simply a convention that has become accepted over the years, and there are situations for which that value may not be wise. (We explore this issue further in Section 22.4.)

Let's return to the analogy of the jury in a courtroom. In that situation, the information provided is generally not summarized into a single number. However, the two possible conclusions are equivalent to those in hypothesis testing:

> *Conclusion 1:* The evidence is not strong enough to convincingly rule out that the defendant is innocent. Therefore, we *cannot reject* the null hypothesis, or

innocence of the defendant, based on the evidence presented. We say that the defendant is *not guilty.* Notice that we do not conclude that the defendant is *innocent,* which would be akin to accepting the null hypothesis.

Conclusion 2: The evidence was strong enough that we are willing to rule out the possibility that an innocent person (as stated in the null hypothesis) produced the observed data. We *reject* the null hypothesis, that the defendant is innocent, and assert the alternative hypothesis, that he or she is guilty.

Consistent with our thinking in hypothesis testing, for Conclusion 1 we would not *accept* the hypothesis that the defendant is innocent. We would simply conclude that the evidence was not strong enough to rule out the possibility of innocence and conclude that the defendant was not guilty.

22.3   Testing Hypotheses for Proportions

Let's illustrate the four steps of hypothesis testing for the situation in which we are testing whether a population proportion is equal to a specific value, as in Example 22.1. In that example, we outlined the steps informally, and in this section, we will be explicit about them. We will use the following example as we review the four steps of hypothesis testing in the context of one proportion.

EXAMPLE 22.3   Family Structure in the Teen Drug Survey

According to data from the U.S. government (http://www.census.gov/population/www/socdemo/hh-fam/cps2002.html), 67% of children aged 12 to 17 in the United States in 2001 were living with two parents, either biological or step parents. News Story 13 and Original Source 13 on the companion website describe a survey of teens and their parents using a random sample of teens in the United States, aged 12 to 17. In Original Source 13, following question 8 on page 38, a summary is given showing that 84% of the teens in the survey were living with two parents. Does this mean that the population represented by the teens who were willing to participate in the survey had a higher proportion living with both parents than did the general population in that age group at that time? (If that's the case, it could call into question the extension of the drug survey results to the population of all teens.) Or, could the 84% found in the sample for this survey simply represent chance variation based on the fact that there were only 1987 teens in the study? We will test the null hypothesis that the population proportion in this case is 0.67 (67%) versus the alternative hypothesis that it is not, where the population is all teens who would be willing and able to participate in this survey if they had been asked. In other words, we are testing whether the population represented by the survey is the same as the general population in terms of family structure.   ■

Step 1. *Determine the null hypothesis and the alternative hypothesis.* Researchers are often interested in testing whether a population proportion is equal to a specific value, which we will call the **null value.** In all cases, the null hypothesis is that the population proportion is *equal* to the null value. However, the alternative hypothesis depends on whether the researchers have a preconceived idea about which direction the difference will take, if there is one. It is definitely not legitimate to look at the data first and then decide! But, if the original hypothesis of interest is in only one direction, then it is legitimate to include values only in that direction as part of the alternative hypothesis. If values above the null value only or below the null value only are included in the alternative hypothesis, the test is called a **one-sided test** or a **one-tailed test.** If values on either side of the null value are included in the alternative hypothesis, the test is called a **two-sided test** or a **two-tailed test.**

To write the hypothesis, researchers must first specify what population they are measuring and what proportion is of interest. For instance, in Example 22.1, the population of interest was all students at the university, and the proportion of interest was the proportion of them that oppose switching to the semester system. In Example 22.3, the population of interest was all teens who would have been willing and able (parental permission was required) to participate in the survey if they had been asked. The proportion of interest was the proportion of them living with both parents.

Once the population has been specified, the null value is determined by the research question of interest. Then the null hypothesis is written as follows:

Null hypothesis: The population proportion of interest equals the null value.

The alternative hypothesis is one of the following, depending on the research question:

Alternative hypothesis: The population proportion of interest does not equal the null value. (This is a two-sided hypothesis.)

Alternative hypothesis: The population proportion of interest is greater than the null value. (This is a one-sided hypothesis.)

Alternative hypothesis: The population proportion of interest is less than the null value. (This is a one-sided hypothesis.)

EXAMPLE 22.1 CONTINUED

Switching to Semesters

For Example 22.1, the administration was only concerned with discovering the situation in which a majority of students was opposed to switching. So, the test was one-sided with hypotheses:

Null hypothesis: The proportion of students at the university who oppose switching to semesters is 0.50.

Alternative hypothesis: The proportion of students at the university who oppose switching to semesters is greater than 0.50. ∎

EXAMPLE 22.3 CONTINUED

Teen Drug Survey

For Example 22.3, the researchers should be concerned about whether their sample is representative of all teens in terms of family structure. Based on data from the U.S. government, it was known that at the time about 67% of teens lived with both parents. Therefore, the appropriate test would examine whether the population of teens represented by the sample had 67% or something different than 67% of the teens living with both parents. (It would be dishonest to look at the data first, then write the hypothesis to fit it.) Thus, in this case the alternative hypothesis should be two-sided. The appropriate hypotheses are:

> *Null hypothesis:* For the population of teens represented by the survey, the pro-portion living with both parents is 0.67.

> *Alternative hypothesis:* For the population of teens represented by the survey, the proportion living with both parents is not equal to 0.67. ◼

Step 2. *Collect the data and summarize them with a single number called a test sta-tistic.* The key when testing whether a population proportion differs from the null value based on a sample is to compare the null value to the corresponding proportion in the sample, simply called the **sample proportion.** For instance, in Example 22.1, the sample proportion (opposing the switch to semesters) was 220/400 or 0.55, and we want to know how far off that is from the null value of 0.50. In Example 22.3, the sample proportion (living with both parents) was reported as 0.84, and we want to know how far off that is from the null value of 0.67. In general, we want to know how far the sample value is from the null value, if the null hypothesis is true. If we know that, then we can find out how unlikely the sample value would be if the null value is the actual population proportion.

It is, of course, easy to find the actual difference between the sample value and the null value, but that doesn't help us because we don't know how to assess whether the difference is large enough to indicate a real population difference or not. To make that assessment, we need to know how many standard deviations apart the two values are.

For this calculation, we assume that the true population proportion is the null value. For the special case in which we use the null value to compute the standard deviation, we call the result the **null standard error**. (When there is no risk of con-fusion, we will simply call it the standard error.) Therefore, the *test statistic* when testing a single proportion is the *standardized score* measuring the distance between the sample proportion and the null value:

$$\text{test statistic} = \text{standardized score} = z\text{-score} = \frac{\text{sample proportion} - \text{null value}}{\text{null standard error}}$$

where

$$\text{null standard error} = \sqrt{\frac{(\text{null value}) \times (1 - \text{null value})}{\text{sample size}}}.$$

EXAMPLE 22.1 CONTINUED

Switching to Semesters

For Example 22.1, the null value is 0.50, the sample proportion is 0.55, and the sample size is 400. Therefore, the null standard error is the square root of $[(0.5)(0.5)/400] = 0.025$. The test statistic, computed in Example 22.1, is $(0.55 - 0.50)/0.025 = 2.00$. ◼

EXAMPLE 22.3 CONTINUED

Teen Drug Survey

For Example 22.3, the null value is 0.67, the sample proportion is 0.84, and the sample size is 1987. Therefore, the null standard error is the square root of $[(0.67)(1 - 0.67)/1987] = 0.0105$. The test statistic is $(0.84 - 0.67)/0.0105 = 0.17/0.0105 = 16$, which is extremely large for a standardized score. It is virtually impossible that 84% of the sample would be living with both parents if in fact only 67% of the population were doing so.   ▪

Step 3. *Determine how unlikely the test statistic would be if the null hypothesis were true.* In this step, we compute the *p*-value. We find the probability of observing a standardized score as far from the null value or more so, in the direction specified in the alternative hypothesis, if the null hypothesis is true. To find this value, use a table, calculator or computer package. The correct probability depends on whether the test is a one-sided or two-sided test. The method is as follows:

Alternative Hypothesis:	*p*-**value = Proportion of Bell-Shaped Curve:**
Proportion is greater than null value.	Above the *z*-score test statistic value.
Proportion is less than null value.	Below the *z*-score test statistic value.
Proportion is not equal to null value.	[Above absolute value of test statistic] × 2.

EXAMPLE 22.1 CONTINUED

Switching to Semesters

For Example 22.1, the alternative hypothesis was one-sided, that the proportion opposed to switching to semesters is greater than the null value of 0.50. The *z*-score test statistic value was computed to be 2.00. Therefore, the *p*-value is the proportion of the bell-shaped curve above 2.00. From Table 8.1, we see that the proportion *below* 2.00 is between 0.975 and 0.98. Therefore, the proportion *above* 2.00 is between 0.025 (that is, 1 − 0.975) and 0.02. We can find the exact proportion below 2.00 by using the Excel statement NORMSDIST(2.0); the result is 0.9772. Thus, the exact *p*-value is (1 − 0.9772) = 0.0228.   ▪

EXAMPLE 22.3 CONTINUED

Teen Drug Survey

For Example 22.3, the alternative hypothesis was two-sided, that the proportion of the population living with two parents was not equal to 0.67. The *z*-score test statistic was computed to be 16. Therefore, the *p*-value is two times the proportion of the bell-shaped curve above 16. The proportion above 16 is essentially 0. In fact, from Table 8.1, you can see that the proportion above 6.0 is already essentially 0. So, in this case, the *p*-value is essentially 0. It is almost impossible to observe a sample of 1987 teens in which 84% are living with both parents if in fact only 67% of the population are doing so. Clearly, we will reject the null hypothesis that the true population proportion in this case is 0.67.   ▪

Step 4. *Make a decision.* This step is the same for any situation, but the wording may differ slightly based on the context. The researcher specifies the *level of significance* and then determines whether or not the *p*-value is that small or smaller. If so, the result is statistically significant. The most common level of significance is 0.05. There are multiple ways in which the decision can be worded. Here are some generic versions, as well as how it would be worded in the case of testing one proportion.

If the p-value is greater than the level of significance,

- Do not reject the null hypothesis.
- We do not have enough evidence to support the alternative hypothesis.
- The true population proportion is not significantly different from the null value.

If the p-value is less than or equal to the level of significance,

- Reject the null hypothesis.
- Accept the alternative hypothesis.
- The evidence supports the alternative hypothesis.
- The true population proportion is significantly different from the null value.

For the last version, if the test is one-sided, the decision would be worded accordingly.

EXAMPLE 22.1 CONTINUED

Switching to Semesters

In Example 22.1, a one-sided test was used, with the alternative hypothesis specifying that more than 50% opposed switching to semesters. The p-value for the test is 0.023. Therefore, using a level of significance of 0.05, the p-value is less than the level of significance. The conclusion can be worded in these ways:

- Reject the null hypothesis.
- Accept the alternative hypothesis.
- The evidence supports the alternative hypothesis.
- The true population proportion opposing the switch to semesters is significantly *greater* than 0.50.

Therefore, the administration should think twice about making the switch. ■

EXAMPLE 22.3 CONTINUED

Teen Drug Survey

For Example 22.3, the alternative hypothesis was two-sided. The p-value was essentially 0, so no matter what level of significance is used, within reason, the p-value is less. The conclusion can be worded in these ways:

- Reject the null hypothesis.
- Accept the alternative hypothesis.
- The true proportion of teens living with both parents in the population represented by the survey is significantly *different* from 0.67.
- The proportion of teens living with both parents in the population represented by the survey is significantly different than the proportion of teens living with both parents in the United States.

Notice that the conclusion doesn't state the direction of the difference. A confidence interval could be found to estimate the proportion of teens living with both parents for the population represented by the survey, and that interval would provide the direction of the difference. ■

22.4   What Can Go Wrong: The Two Types of Errors

Any time decisions are made in the face of uncertainty, mistakes can be made. In testing hypotheses, there are two potential decisions or choices and each one brings with it the possibility that a mistake, or error, has been made.

The Courtroom Analogy

It is important to consider the seriousness of the two possible errors before making a choice. Let's use the courtroom analogy as an illustration. Here are the possible conclusions and the error that could accompany each:

> *Conclusion 1:* We cannot rule out that the defendant is innocent, so he or she is set free without penalty.
>
> *Potential error:* A criminal has been erroneously freed.
>
> *Conclusion 2:* We believe there is enough evidence to conclude that the defendant is guilty.
>
> *Potential error:* An innocent person is falsely convicted and penalized and the guilty party remains free.

Although the seriousness of the two potential errors depends on the seriousness of the crime and the punishment, conclusion 2 is usually seen as more serious. Not only is an innocent person punished but a guilty one remains completely free and the case is closed.

A Medical Analogy: False Positive versus False Negative

As another example, consider a medical scenario in which you are tested for a disease. Most tests for diseases are not 100% accurate. In reading your results, the lab technician or physician must make a choice between two hypotheses:

> *Null hypothesis:* You do not have the disease.
>
> *Alternative hypothesis:* You have the disease.

Notice the possible errors engendered by each of these decisions:

> *Conclusion 1:* In the opinion of the medical practitioner, you are healthy. The test result was weak enough to be called "negative" for the disease.
>
> *Potential error:* You are actually diseased but have been told you are not. In other words, your test was a **false negative.**
>
> *Conclusion 2:* In the opinion of the medical practitioner, you are diseased. The test results were strong enough to be called "positive" for the disease.
>
> *Potential error:* You are actually healthy but have been told you are diseased. In other words, your test was a **false positive.**

Which error is more serious in medical testing? It depends on the disease and on the consequences of a negative or positive test result. For instance, a false negative in a screening test for cancer could lead to a fatal delay in treatment, whereas a false positive would probably only lead to a retest and short-term concern.

A more troublesome example occurs in testing for HIV or other communicable diseases, in which it is very serious to report a false negative and tell someone they are not infected when in truth they are infected. However, it is quite frightening for the patient to be given a false positive test for something like HIV—that is, the patient is really healthy but is told otherwise. HIV testing tends to err on the side of erroneous false positives. However, before being given the results, people who test positive for HIV with an inexpensive screening test are generally retested using an extremely accurate but more expensive test. Those who test negative are generally not retested, so it is important for the initial test to have a very low false negative rate.

The Two Types of Error in Hypothesis Testing

You can see that there is a trade-off between the two types of errors just discussed. Being too lenient in making conclusions in one direction or the other is not wise. Determining the best direction in which to err depends on the situation and the consequences of each type of potential error.

The courtroom and medical analogies are not much different from the scenario encountered in hypothesis testing in general. Two types of errors can be made. A **type 1 error** can *only* be made if the null hypothesis is actually true. A **type 2 error** can *only* be made if the alternative hypothesis is actually true. Figure 22.1 illustrates the errors that can happen in the courtroom, in medical testing, and in hypothesis testing.

In our example of switching from quarters to semesters, if the university administrators were to make a type 1 error, the decision would correspond to a false positive. The administrators would be creating a false alarm, in which they stopped their planned change for no good reason. A type 2 error would correspond to a false negative, in which the administrators would have decided there isn't a problem when there really is one. ∎

Figure 22.1
Potential errors in the courtroom, in medical testing, and in hypothesis testing

	True State of Nature	
Conclusion Made	**Innocent, Healthy** **Null Hypothesis**	**Guilty, Diseased** **Alternative Hypothesis**

Not guilty
Healthy
Don't reject null
    hypothesis
	Correct
	☺
		{ Undeserved freedom
		  False negative
		  Type 2 error

Guilty
Diseased
Accept alternative
    hypothesis
	{ Undeserved punishment
	  False positive
	  Type 1 error
		Correct
		☺

Probabilities Associated with Type 1 and Type 2 Errors

It would be nice if we could specify the probability that we were indeed making an error with each potential decision. We could then weigh the consequence of the error against its probability. Unfortunately, in most cases, we can only specify the conditional probability of making a type 1 error, given that the null hypothesis is true. That probability is the *level of significance,* usually set at 0.05.

Level of Significance and Type 1 Errors

Remember that two conditions have to hold for a type 1 error to be made. First, it can only happen when the null hypothesis is true. Second, the data must convince us that the alternative hypothesis is true. For this to happen, the p-value must be less than or equal to the level of significance, usually set at 0.05. What is the probability that the p-value will be 0.05 or less by chance? If the null hypothesis is true, that probability is in fact 0.05. No matter what level of significance is chosen, the probability that the p-value will be that small or smaller is equal to the level of significance (when the null hypothesis is true). Let's restate this information about the probability of making a type 1 error:

> *If the null hypothesis is true, the probability of making a type 1 error is equal to the stated level of significance, usually 0.05. If the null hypothesis is not true, a type 1 error cannot be made.*

Type 2 Errors

Notice that making a type 2 error is possible only if the alternative hypothesis is true. A type 2 error is made *if* the alternative hypothesis is true, but you fail to choose it. The probability of doing that depends on exactly which part of the alternative hypothesis is true, so that computing the probability of making a type 2 error is not feasible.

For instance, university administrators in our hypothetical example did not specify what percentage of students would have to oppose switching to semesters in order for the alternative hypothesis to hold. They merely specified that it would be over half. If only 51% of the population of students oppose the switch, then a sample of 400 students could easily result in fewer than half of the sample in opposition, in which case the null hypothesis would not be rejected even though it is false. However, if 80% of the students in the population oppose the switch, the sample proportion would almost surely be large enough to convince the administration that the alternative hypothesis holds. In the first case (51% in opposition), it would be very easy to make a type 2 error, whereas in the second case (80% in opposition), it would be almost impossible. Yet, both are legitimate cases where the alternative hypothesis is true. That's why we can't specify one single value for the probability of making a type 2 error.

The Power of a Test

Researchers should never conclude that the null hypothesis is true just because the data failed to provide enough evidence to reject it. There is no control over the probability that an erroneous decision has been made in that situation. The **power** of a test is the probability of making the correct decision when the alternative hypothesis is true. It should be clear to you that the *power* of the administrators to detect that a majority opposes switching to semesters will be much higher if that majority constitutes 80% of the population than if it constitutes just 51% of the population. In other words, if the population value falls close to the value specified in the null hypothesis, then it may be difficult to get enough evidence from the sample to conclusively choose the alternative hypothesis. There will be a relative high probability of making a type 2 error, and the test will have relatively low power in that case.

Sometimes news reports, especially in science magazines, will insightfully note that a study may have failed to find a relationship between two variables because the test had such low power. As we will see in Chapter 24, this is a common consequence of conducting research with samples that are too small, but it is one that is often overlooked in media reports.

When to Reject the Null Hypothesis

When you read the results of research in a journal, you often are presented with a *p*-value, and the conclusion is left to you. In deciding what level of significance to use to reject the null hypothesis, you should consider the consequences of the two potential types of errors. If you think the consequences of a type 1 error are very serious, then you should choose a very small level of significance. In that case, you would only reject the null hypothesis and accept the alternative hypothesis if the *p*-value is very small. Conversely, if you think a type 2 error is more serious, you should choose a higher level of significance. In other words, you should be willing to reject the null hypothesis with a moderately large *p*-value, typically 0.05 to 0.10.

CASE STUDY 22.1	**Testing for the Existence of Extrasensory Perception**

For centuries, people have reported experiences of knowing or communicating information that cannot have been transmitted through normal sensory channels. There have also been numerous reports of people seeing specific events in visions and dreams that then came to pass. These phenomena are collectively termed *extrasensory perception.*

In Chapter 17, we learned that it is easy to be fooled by underestimating the probability that weird events could have happened just by chance. Therefore, many of these reported episodes of extrasensory perception are probably explainable in terms of chance phenomena and coincidences.

Scientists have been conducting experiments to test extrasensory perception in the laboratory for several decades. As with many experiments, those performed in the laboratory lack the "ecological validity" of the reported anecdotes, but they have the advantage that the results can be quantified and studied using statistics.

Description of the Experiments

In this case study, we focus on one branch of research that has been investigated for a number of years under very carefully controlled conditions. The experiments are reported in detail by Honorton and colleagues (1990) and have also been summarized by Utts (1991, 1996), Bem and Honorton (1994), Bem et al. (2002), and Storm et al. (2010).

The experiments use an experimental setup called the *ganzfeld procedure*. This involves four individuals, two of whom are participants and two, researchers. One of the two participants is designated as the sender and the other as the receiver. One of the two researchers is designated as the experimenter and the other as the assistant.

Each session of this experiment produces a single yes or no data value and takes over an hour to complete. The session begins by sequestering both the receiver and the sender in separate, sound-isolated, electrically shielded rooms. The receiver wears headphones over which "white noise" (which sounds like a continuous hissing sound) is played. He or she is also looking into a red light, with halved Ping-Pong balls taped over the eyes to produce a uniform visual field. The term *ganzfeld* means "total field" and is derived from this visual experience. The reasoning behind this setup is that the senses will be open and expecting meaningful input, but nothing in the room will be providing such input. The mind may, therefore, look elsewhere for input.

Meanwhile, in another room, the sender is looking at either a still picture (a "static target") or a short video (a "dynamic target") on a television screen and attempting to "send" the image (or images) to the receiver. Here is an example of a description of a static target, from Honorton and colleagues (1990, p. 123):

> *Flying Eagle. An eagle with outstretched wings is about to land on a perch; its claws are extended. The eagle's head is white and its wings and body are black.*

The receiver has a microphone into which he or she is supposed to provide a continuous monologue about what images or thoughts are present. The receiver has no idea what kind of picture the sender might be watching. Here is part of what the receiver said during the monologue for the session in which the "Flying Eagle" was the target (from Honorton et al., 1990, p. 123):

> *A black bird. I see a dark shape of a black bird with a very pointed beak with his wings down. . . . Almost needle-like beak. . . . Something that would fly or is flying . . . like a big parrot with long feathers on a perch. Lots of feathers, tail feathers, long, long, long. . . . Flying, a big huge, huge eagle. The wings of an eagle spread out. . . . The head of an eagle. White head and dark feathers. . . . The bottom of a bird.*

The experimenter monitors the whole procedure and listens to the receiver's monologue. The assistant has one task only, to randomly select the target material the sender will view on the television screen. This is done by using a computer, and no one knows the identity of the "target" selected except the sender. For the particular set of experiments we will examine, there were 160 possibilities, of which half were static targets and half were dynamic targets.

(Continued)

Quantifying the Results

Although it may seem as though the receiver provided an excellent description of the Flying Eagle target, remember that the quote given was only a small part of what the receiver said. So far, there is no quantifiable result. Such results are secured only at the end of the session.

To provide results that can be analyzed statistically, the results must be expressed in terms that can be compared with chance. To provide a comparison to chance, a single categorical measure is taken. Before the experiment begins, potential targets are grouped into sets of four of the same type (static or dynamic). For each session, one set of four is chosen. Within the set one is randomly chosen to be the real target, and the other three are "decoys." Note that due to the random selection, any of the four targets (the real one and the three decoys) could equally have been chosen to be the real target.

At the end of the session the receiver is shown the four possible targets. He or she is then asked to decide, on the basis of the monologue provided, which one the sender was watching. If the receiver picks the right one, the session is a success. The example provided, in which the target was a picture of an eagle, was indeed a success.

The Null and Alternative Hypotheses

By chance, one-fourth or 25%, of the sessions should result in a success. (Remember, the correct target is randomly chosen from the four possibilities that are presented to the receiver.) Therefore, the statistical question for this hypothesis test is: Did the sessions result in significantly more than 25% successes?

The hypotheses being tested are thus as follows:

Null hypothesis: There is no extrasensory perception, and the results are due to chance guessing. The probability of a successful session is 0.25.

Alternative hypothesis: The results are not due to chance guessing. The probability of a successful session is higher than 0.25.

The Results

Honorton and his colleagues, who reported their results in the *Journal of Parapsychology* in 1990, ran several experiments, using the setup described, between 1983 and 1989, when the lab closed. There were a total of 355 sessions, of which 122 were successful.

The sample proportion of successful results is $122/355 = 0.344$, or 34.4%. If the null hypothesis were true, the true proportion would be 0.25, and the null standard error would be:

$$\text{null standard error} = \text{square root of } [(0.25)(0.75)/355] = 0.023$$

Therefore, the test statistic is

$$\text{standardized score} = z\text{-score} = (0.344 - 0.25)/0.023 = 4.09$$

It is obvious that such a large standardized score would rarely occur by chance, and indeed the *p*-value is about 0.00005. In other words, if chance alone were operating, we would see results of this magnitude about 5 times in every 100,000 such experiments. Therefore, we would certainly declare this to be a statistically significant result.

Carl Sagan has said that "exceptional claims require exceptional proof," and for some nonbelievers, even a *p*-value this small may not increase their personal probability that extrasensory perception exists. However, as with any area of science, such a striking result should be taken as evidence that something out of the ordinary is definitely happening in these experiments. Further experiments have continued to achieve similar results (Bem et al., 2002; Storm et al., 2010). ■

Thinking About Key Concepts

- *Hypothesis tests* can be used to decide whether sample data provide convincing evidence to reject a particular hypothesized value for a population proportion, mean, etc. The hypothesized value is called the *null value*.
- The *null hypothesis* in this kind of test is that the population value equals the null value. The *alternative hypothesis* is that the population value does not equal the null value and sometimes specifies whether the inequality is in a particular direction (greater than or less than the null value).
- The data is summarized into a *test statistic*, which is a standardized score that represents how far apart the sample data and the null value are from each other.
- The *p-value* for a test is the probability that a standardized score would be as large as the computed test statistic value or even larger, if the null value is actually correct. "Larger" is defined by the direction specified in the alternative hypothesis.
- The *significance level of a test* is the threshold for declaring statistical significance. If the *p*-value is less than or equal to the desired level of the test, then the null hypothesis is rejected, and the alternative hypothesis is accepted. Typically, a level of 0.05 is used.
- Two types of error can be made. A *type 1 error* is equivalent to a false positive and is made when the null hypothesis is true but is rejected. A *type 2 error* is equivalent to a false negative and is made when the alternative hypothesis is true but the data does not provide convincing evidence that it is true.
- When the null hypothesis is true, a type 1 error could be made. The probability of this happening (when the null hypothesis is true) is the same as the significance level of the test and is thus under the researcher or reader's control. If a type 1 error is very serious, a level of significance smaller than the standard of .05 should be used.
- When the alternative hypothesis is true, a type 2 error could be made. The probability of this happening cannot be computed because there are many possible "true" values included in the alternative hypothesis.
- The *power* of a test is the probability of correctly rejecting the null hypothesis when in fact it should be rejected.

Focus on Formulas

The formulas for the material in this chapter are presented along with other formulas for testing hypotheses at the end of Chapter 23.

Exercises

Exercises with numbers divisible by 3 (3, 6, 9, etc.) are included in the Solutions at the back of the book. They are marked with an asterisk (*).

1. When we revisited Case Study 6.4 in Chapter 21, we learned that a 95% confidence interval for the difference in years of education for mothers who did not smoke compared with those who did extended from 0.15 to 1.19 years, with higher education for those who did not smoke. Suppose we had used the data to construct a test instead of a confidence interval, to see if one group in the population was more educated than the other. What would the null and alternative hypotheses have been for the test?

2. Refer to Exercise 1. If we had conducted the hypothesis test, the resulting p-value would be 0.01. Explain what the p-value represents for this example.

*3. Refer to Case Study 21.1, in which women were randomly assigned to receive either a placebo or calcium and severity of premenstrual syndrome (PMS) symptoms was measured.

   *a. What are the null and alternative hypotheses tested in this experiment?

   *b. The researchers concluded that calcium helped reduce the severity of PMS symptoms. Which type of error could they have made?

   *c. What would be the consequences of making a type 1 error in this experiment? What would be the consequences of making a type 2 error?

4. The journal article reporting the experiment described in Case Study 21.1 (see Thys-Jacobs et al., 1998, in Chapter 21) compared the placebo and calcium-treated groups for a number of PMS symptoms, both before the treatment began (baseline) and in the third cycle. A p-value was given for each comparison. For each of the following comparisons, state the null and alternative hypotheses and the appropriate conclusion:

   a. Baseline, mood swings, p-value = 0.484

   b. Third cycle, mood swings, p-value = 0.002

   c. Third cycle, insomnia, p-value = 0.213

5. The news story discussed in Case Study 6.5 (pp. 132-133) was headlined "NIH study finds that coffee drinkers have lower risk of death" (*Source:* http://www.nih.gov). The news story was based on an article published in *The New England Journal of Medicine* (Freedman et. al., 2012) that followed hundreds

of thousands of older adults from 1995 to 2008 and examined the association between drinking coffee and longevity.

a. What are the null and alternative hypotheses for this study?

b. The authors concluded that those who drank coffee had a statistically significantly lower risk of death (during the time period of the study) than those who did not. Which type of error, type 1 or type 2, could have been made in making this conclusion?

c. Explain what a type 1 error and a type 2 error would be in this situation. Which one do you think is more serious? Explain.

*6. An article in *Science News* reported on a study to compare treatments for reducing cocaine use. Part of the results are

> *short-term psychotherapy that offers cocaine abusers practical strategies for maintaining abstinence sparks a marked drop in their overall cocaine use. . . . In contrast, brief treatment with desipramine—a drug thought by some researchers to reduce cocaine cravings—generates much weaker drops in cocaine use* (Bower, 24, 31 December 1994, p. 421).

a. The researchers were obviously interested in comparing the rates of cocaine use following treatment with the two methods. State the null and alternative hypotheses for this situation.

b. Explain what the two types of error could be for this situation and what their consequences would be.

c. Although no *p*-value is given, the researchers presumably concluded that the psychotherapy treatment was superior to the drug treatment. Which type of error could they have made?

7. State the null and alternative hypotheses for each of the following potential research questions:

a. Does working 5 hours a day or more at a computer contribute to deteriorating eyesight?

b. Does placing babies in an incubator during infancy lead to claustrophobia in adult life?

c. Does placing plants in an office lead to fewer sick days?

8. For each of the situations in Exercise 7, explain the two errors that could be made and what the consequences would be.

*9. In Case Study 6.1, a study was described that compared scores on a mock GRE exam for students who had learned and practiced meditation and students who had taken a nutrition class (Mrazek et al., 2013). Although the analysis used by the researchers was somewhat complicated, for simplicity in this exercise, suppose the measure of interest was the difference in mean GRE scores for the two groups. The researchers were interested in showing that practicing meditation boosts mean GRE scores.

a. The null hypothesis can be stated as "Population mean difference = null value." What is the null value?

b. Write the alternative hypothesis for this situation.

c. Describe what a type 1 error would be in this situation.

d. Describe what a type 2 error would be in this situation.

e. Discuss the consequences of the two types of errors in this situation. Which error do you think is more serious? (There is no correct answer; it is your reasoning that counts.)

10. Explain why we can specify the probability of making a type 1 error, given that the null hypothesis is true, but we cannot specify the probability of making a type 2 error, given that the alternative hypothesis is true.

11. Compute a 95% confidence interval for the probability of a successful session in the ganzfeld studies reported in Case Study 22.1.

*12. Specify what a type 1 and a type 2 error would be for the ganzfeld studies reported in Case Study 22.1.

13. Given the convention of declaring that a result is statistically significant if the p-value is 0.05 or less, what decision would be made concerning the null and alternative hypotheses in each of the following cases? Be explicit about the wording of the decision.

a. p-value = 0.35

b. p-value = 0.04

14. For the research described in Case Study 6.2, the goal was to find out if eating breakfast cereal was associated with a reduction in body mass index (BMI) for children (Frantzen et al., 2013). Although the analysis used by the researchers was somewhat complicated, for simplicity in this exercise, suppose the measure of interest was the difference between population mean BMI for children who eat cereal and children who don't.

a. Write the null and alternative hypotheses for this situation.

b. Describe what a type 1 error would be in this situation.

c. Describe what a type 2 error would be in this situation.

d. Which type of error do you think is more serious in this situation? (There is no correct answer; it is your reasoning that counts.)

*15. In previous chapters, we learned that researchers have discovered a link between vertex baldness and heart attacks in men.

*a. State the null hypothesis and the alternative hypothesis used to investigate whether there is such a relationship.

*b. Discuss what would constitute a type 1 error in this situation.

*c. Discuss what would constitute a type 2 error in this situation.

16. A report in the *Davis* (CA) *Enterprise* (6 April 1994, p. A-11) was headlined, "Highly educated people are less likely to develop Alzheimer's disease, a new study suggests."

a. State the null and alternative hypotheses the researchers would have used in this study.

b. What do you think the headline is implying about statistical significance? Restate the headline in terms of statistical significance.

c. Was this a one-sided test or a two-sided test? Explain.

17. Suppose that a study is designed to choose between the hypotheses:

Null hypothesis: Population proportion is 0.25.

Alternative hypothesis: Population proportion is higher than 0.25.

On the basis of a sample of size 500, the sample proportion is 0.29. The null standard error for the potential sample proportions in this case is about 0.02.

a. Compute the standardized score corresponding to the sample proportion of 0.29, assuming the null hypothesis is true.

b. What is the percentile for the standardized score computed in part (a)?

c. What is the p-value for the test?

d. Based on the results of parts (a) to (c), make a conclusion. Be explicit about the wording of your conclusion and justify your answer.

e. To compute the standardized score in part (a), you assumed the null hypothesis was true. Explain why you could not compute a standardized score under the assumption that the *alternative* hypothesis was true.

*18. Consider medical tests in which the null hypothesis is that the patient does not have the disease and the alternative hypothesis is that he or she does.

*a. Give an example of a medical situation in which a type 1 error would be more serious.

*b. Give an example of a medical situation in which a type 2 error would be more serious.

19. An article in the *Los Angeles Times* (24 December 1994, p. A16) announced that a new test for detecting HIV had been approved by the Food and Drug Administration (FDA). The test requires the person to send a saliva sample to a lab. The article described the accuracy of the test as follows:

> The FDA cautioned that the saliva test may miss one or two infected individuals per 100 infected people tested, and may also result in false positives at the same rate in the uninfected. For this reason, the agency recommended that those who test positive by saliva undergo confirmatory blood tests to establish true infection.

a. Do you think it would be wise to use this saliva test to screen blood donated at a blood bank, as long as those who test positive were retested as suggested by the FDA? Explain your reasoning.

b. Suppose that 10,000 students at a university were all tested with this saliva test and that, in truth, 100 of them were infected. Further, suppose the false positive and false negative rates were actually both 1 in 100 for this group. If someone tests positive, what is the probability that he or she is infected?

20. In Case Study 1.2 and in Chapters 12 and 13, we examined a study showing that there appears to be a relationship between taking aspirin and incidence of heart attack. The null hypothesis in that study would be that there is no relationship between the two variables, and the alternative would be that there is a relationship. Explain what a type 1 error and a type 2 error would be for the study and what the consequences of each type would be for the public.

*21. Many researchers decide to reject the null hypothesis as long as the *p*-value is 0.05 or less. In a testing situation for which a type 2 error is much more serious than a type 1 error, should researchers require a higher or a lower *p*-value in order to reject the null hypothesis? Explain your reasoning.

22. In Case Study 1.1, Lee Salk did an experiment to see if hearing the sound of a human heartbeat would help infants gain weight during the first few days of life. By comparing weight gains for two sample groups of infants, he concluded that it did. One group listened to a heartbeat and the other did not.

a. What are the null and alternative hypotheses for this study?

b. Was this a one-sided test or a two-sided test? Explain.

c. What would a type 1 and type 2 error be for this study?

d. Given the conclusion made by Dr. Salk, explain which error he could possibly have committed and which one he could not have committed.

e. Rather than simply knowing whether there was a difference in average weight gains for the two groups, what statistical technique would have provided additional information?

The following information is for Exercises 23 and 24: In Original Source 17 (link provided on the companion website), "Monkeys reject unequal pay," the researchers found that female monkeys were less willing to participate in an exchange of a small token for a piece of cucumber if they noticed that another monkey got a better deal. In one part of the experiment, a condition called "effort control," the monkey witnessed another monkey being given a grape without having to give up a token (or anything else) in return. The first monkey was then cued to perform as usual, giving up a small token in return for a piece of cucumber. The proportion of times the monkeys were willing to do so was recorded, and the results are shown in Figure 1 (page 297) of the report. This "effort control" condition was compared with the "equality condition" in which the first monkey was also required to exchange a token for cucumber. For this part of the study, the researchers reported, "Despite the small number of subjects, the overall exchange tendency varied significantly ... comparing effort controls with equality tests, $P < 0.05$" (p 298). (The symbol $<$ means "less than.")

23. a. The participants in the study were five capuchin monkeys. To what population do you think the results apply? (Note that there is no right or wrong answer.)

b. The researchers were interested in comparing the proportion of times the monkeys would cooperate by trading the token for the cucumber after observing another monkey doing the same, versus observing another monkey receiving a free grape. The null hypothesis in this test is that the proportion of times monkeys in the population would cooperate after observing either of the two conditions is the same. What is the alternative hypothesis?

*24. See the information for Exercises 23 and 24 provided on page 493. Refer to the quote that begins with "Despite the small number..."

*a. Based on the quote, what do you know about the p-value?

*b. Based on the quote, what level of significance are the researchers using?

*c. What conclusion would be made? Write your conclusion in statistical language and in the context of the example.

25. In Original Source 2 on the companion website, "Development and initial validation of the Hangover Symptoms Scale: Prevalence and correlates of hangover symptoms in college students," one of the questions was how many times respondents had experienced at least one hangover symptom in the past year. Table 3 of the paper (page 1445) shows that out of all 1216 respondents, 40% answered that it was two or fewer times. Suppose the researchers are interested in the proportion of the population who would answer that it was two or fewer times. Can they conclude that this population proportion is significantly less than half (50% or 0.5)? Go through the four steps of hypothesis testing for this situation.

26. Refer to the previous exercise. Repeat the test for women only. Refer to Table 3 on page 1445 of the paper for the data.

Mini-Projects

1. Construct a situation for which you can test null and alternative hypotheses for a population proportion. For example, you could see whether you can flip a coin in a manner so as to bias it in favor of heads. Or you could conduct an ESP test in which you ask someone to guess the suits in a deck of cards. (To do the latter experiment properly, you must replace the card and shuffle each time so you don't change the probability of each suit by having only a partial deck, and you must separate the "sender" and "receiver" to rule out normal communication.) Collect data for a sample of at least size 100. Carry out the test. Make sure you follow the four steps given in Section 22.2, and be explicit about your hypotheses, your decision and the reasoning behind your decision.

2. Find two news stories reporting on the results of statistical studies with the following characteristics. First, find one that reports on a study that failed to find a relationship. Next, find one that reports on a study that did find a relationship. For each study, state what hypotheses you think the researchers were trying to test. Then explain what you think the results really imply compared with what is implied in the news reports for each study.

References

Bem, D. J., and C. Honorton. (1994). Does psi exist? Replicable evidence for an anomalous process of information transfer. *Psychological Bulletin* 115, no. 1, pp. 4–18.

Bem, D. J., J. Palmer, R. S. Broughton. (2002). Updating the ganzfeld database: A victim of its own success? *Journal of Parapsychology* 65, no. 3, pp. 207–218.

Bower, B. (24, 31 December 1994). Psychotherapy's road to cocaine control. *Science News* 146, nos. 26 and 27, p. 421.

Frantzen, L. B., R. P. Treviño, R. M. Echon, O. Garcia-Dominic, and N. DiMarco. (2013). Association between frequency of ready-to-eat cereal consumption, nutrient intakes, and body mass index in fourth- to sixth- grade low-income minority children. *Journal of the Academy of Nutrition and Dietetics* 113(4), pp. 511–519.

Freedman, N. D., Y Park, C. C. Abnet, A. R. Hollenbeck, and R. Sinha. (2012). The association of coffee drinking with total and cause-specific mortality. *New England Journal of Medicine 366*, May 17, 2012, pp. 1896–1904.

Honorton, C., R. E. Berger, M. P. Varvoglis, M. Quant, P. Derr, E. I. Schechter, and D. C. Ferrari. (1990). Psi communication in the ganzfeld. *Journal of Parapsychology* 54, no. 2, pp. 99–139.

Storm, L., P. E. Tressoldi, and L. Di Risio. (2010). Meta-analyses of free-response studies 1992-2008: Assessing the noise reduction model in parapsychology. *Psychological Bulletin, 136*(4), 471–485.

Utts, J. (1991). Replication and meta-analysis in parapsychology. *Statistical Science* 6, no. 4, pp. 363–403.

Utts, J. (1996). Exploring psychic functioning: Statistics and other issues. *Stats: The Magazine for Students of Statistics* 16, pp. 3–8.

CHAPTER 23

Hypothesis Testing— Examples and Case Studies

Thought Questions

1. In Chapter 21, we examined a study showing that the difference in sample means for weight loss based on dieting only versus exercising only was 3.2 kg. That same study showed that the difference in average amount of *fat weight* (as opposed to muscle weight) lost was 1.8 kg and that the corresponding standard error was 0.83 kg. Suppose the means are actually equal, so that the mean difference in fat that would be lost for the populations is actually zero. What is the standardized score corresponding to the observed difference of 1.8 kg? Would you expect to see a standardized score that large or larger very often?

2. In the journal article reported on in Case Study 6.4 comparing IQs for children of smokers and nonsmokers, one of the statements made was, "after control for confounding background variables, the average difference [in mean IQs] observed at 12 and 24 months was 2.59 points (95% CI: −3.03, 8.20; $P = 0.37$)" (Olds et al., 1994, p. 223). The reported value of 0.37 is the *p*-value. What do you think are the null and alternative hypotheses being tested?

3. In chi-square tests for two categorical variables, introduced in Chapter 13, we were interested in whether a relationship observed in a sample reflected a real relationship in the population. What are the null and alternative hypotheses?

4. In Chapter 13, we found a statistically significant relationship between smoking (yes or no) and time to pregnancy (one cycle or more than one cycle). Explain what the type 1 and type 2 errors would be for this situation and the consequences of making each type of error.

23.1 How Hypothesis Tests Are Reported in the News

When you read the results of hypothesis tests in the news, you are given very little information about the details of the test. It is therefore important for you to remember the steps occurring behind the scenes, so you can translate what you read into the bigger picture. Remember that the basic steps to hypothesis testing are similar in any setting; they are stated slightly more concisely here than in Chapters 13 and 22:

> **Step 1:** Determine the null and alternative hypotheses.
>
> **Step 2:** Collect and summarize the data into a test statistic.
>
> **Step 3:** Use the test statistic to determine the *p*-value.
>
> **Step 4:** The result is statistically significant if the *p*-value is less than or equal to the level of significance, usually set at .05.

In the presentation of most research results in the media, you are simply told the results of step 4, which isn't usually even presented in statistical language. If the results are statistically significant, you are told that a relationship has been found between two variables or that a difference has been found between two groups. If the results are not statistically significant, you may simply be told that no relationship or difference was found. In Chapter 24, we will revisit the problems that can arise if you are only told whether a statistically significant difference emerged from the research, but not told how large the difference was or how many participants there were in the study.

Let's look at two examples of news stories that present results of statistical studies. In both examples, there is enough information to apply the results to your daily life. But if you consult the original journal article used to write the news stories, you will learn much more about who was involved in the study, what was measured, and how the results were obtained.

EXAMPLE 23.1 Morning People, Age, and Happiness

If you are a night owl you might have scowled at the headline, "Morning people are happier than night owls, study suggests" (Welsh, 2012). The news story was based on a study of 435 young adults (17 to 38) and 297 older adults (59 to 79). The story quoted the lead author of the study, Reneé Biss, as follows:

> *"We found that older adults reported greater positive emotion than younger adults, and older adults were more likely to be morning-type people than younger adults," Biss said. "The 'morningness' was associated with greater happiness emotions in both age groups"* (Welsh, 2012).

The news story did not provide additional details about what was measured, statistical significance, or magnitude of the effects. However, consulting the original journal article revealed many of those details (Biss and Hasher, 2012).

Participants in the study filled out a "Morningness-Eveningness Questionnaire" (MEQ), answering questions about their preferred times to go to sleep, how alert they feel when they wake up, and so on. Scores on the MEQ can range from 16 to 86, with higher scores associated with "morning people." Participants also rated 16 mood adjectives on a scale from 1 to 7, thus creating a "positive affect" or "happiness" score. Although some of the statistical methods used were more complicated than what has been covered in this book, it is not difficult to understand some of the basic results of the study. Here are some examples (both from Biss and Hasher, 2012, p. 3):

> *"Older adults reported better moods, with higher positive affect (M = 20.8, SD = 7.1) compared with younger adults (M = 10.8, SD = 6.4), t(730) = 19.92, p < .001."* The measure considered in this test is the sample mean positive affect score (denoted M), a measure of happiness. Sample standard deviations (SD) are given as well. The implicit null hypothesis is that the population means of the happiness scores would be the same for older and younger adults if everyone were to be tested. A comparison of the sample means of 20.8 and 10.8 reveals that older adults have statistically significantly higher happiness scores. The test statistic is a t standardized score with a value of 19.92, resulting in a p-value less than .001, so the null hypothesis is rejected. (The t-score is similar to a z-score and is discussed later in this chapter.)

> *"Critically, morningness predicted positive affect when age was controlled (β = 0.149, p < .001)."* This statement is not as easy to interpret as the previous one, but represents the results of a regression analysis, using the MEQ score to predict the positive affect score. The regression "controlled for age," which means that the results hold even after adjusting for the fact that older people have higher scores on both the MEQ and the happiness scale. The slope for the *sample* regression line was 0.149. In other words, for the samples, if two people differ by one-unit on the MEQ scale, they are predicted to differ by 0.149 in the happiness score, with the more "morning person" being happier. The implicit null hypothesis is that the *population* slope is 0, which would indicate that the two measures are not related in the population. The null hypothesis is rejected with a p-value less than .001. Additional information about this relationship was given by presenting a correlation between the MEQ and positive affect scores, separately for each age group. For the younger participants, the *sample* correlation was .25 and for the older participants it was .19. In both cases, a test of the null hypothesis that the *population* correlation is 0 resulted in a p-value less than .001. Thus, there is a statistically significant relationship between score on the "morningness" scale and score on the happiness measure, with morning people scoring higher on the happiness scale. It holds for both age groups. ∎

EXAMPLE 23.2   A study, which will be examined in further detail in Chapter 27, found that cranberry juice lived up to its popular reputation of preventing bladder infections, at least in older women. Here is a newspaper article reporting on the study (*Davis* [CA] *Enterprise*, 9 March 1994, p. A9):

> CHICAGO (AP)   *A scientific study has proven what many women have long suspected: Cranberry juice helps protect against bladder infections. Researchers found that elderly*

women who drank 10 ounces of a juice drink containing cranberry juice each day had less than half as many urinary tract infections as those who consumed a look-alike drink without cranberry juice. The study, which appeared today in the Journal of the American Medical Association, was funded by Ocean Spray Cranberries, Inc., but the company had no role in the study's design, analysis or interpretation, JAMA said. "This is the first demonstration that cranberry juice can reduce the presence of bacteria in the urine in humans," said lead researcher Dr. Jerry Avorn, a specialist in medication for the elderly at Harvard Medical School.

Reading the article, you should be able to determine the null and alternative hypotheses and the conclusion. But there is no way for you to determine the value of the test statistic or the *p*-value. The study is attempting to compare the odds of getting an infection for the population of elderly women if they were to follow one of two regimes: 10 ounces of cranberry juice per day or 10 ounces of a placebo drink. The null hypothesis is that the odds ratio is 1; that is, the odds are the same for both groups. The alternative hypothesis is that the odds of infection are higher for the group drinking the placebo. The article indicates that the odds ratio is under 50%. In fact, the original article (Avorn et al., 1994) sets it at 42% and reports that the associated *p*-value is 0.004. The newspaper article captured the most important aspect of the research, but did not make it clear that the *p*-value was extremely low. ◾

23.2   Testing Hypotheses about Proportions and Means

For many situations, performing the computations necessary to find the test statistic —and thus the *p*-value—requires a level of statistical expertise beyond the scope of this text. However, for some simple situations—such as those involving a proportion, a mean, or the difference between two means—you already have all the necessary tools to understand the steps that need to be taken in doing such computations.

Standardized Scores and *p*-Values

If the null and alternative hypotheses can be expressed in terms of a population proportion, mean, or difference between two means, and if the sample sizes are large, then the test statistic is simply the standardized score associated with the sample proportion, mean, or difference between two means. The standardized score is computed assuming the null hypothesis is true, so the null value is used as the population mean. The standardized score is thus written as:

$$\text{Standardized score} = \frac{sample\ value\ -\ null\ value}{standard\ error}$$

The *p*-value is found from a table of percentiles for standardized scores (such as Table 8.1) or with the help of a computer program like Excel. It gives us the percentile at which a particular sample would fall if the null hypothesis represented the truth.

Student's t-Test

In Chapter 22 (p. 480), we presented the details for how to compute the standardized z-score when testing hypotheses about one proportion. We also showed how to find the p-value by using Table 8.1 or computer software that provides areas under the curve for z-scores (p. 481).

When the hypotheses are about one mean or the difference in two means and the sample sizes are large, an approximate p-value can be found using the usual normal curve percentiles such as those presented in Table 8.1. But when the sample sizes are small and the standardized score is calculated using the sample standard deviation, it is more appropriate to use the Student's t-distribution introduced in Chapter 21. This discovery was first made in 1908 and reported in a paper authored by "A Student." The author was actually William Gosset, but his employer, the Guinness Brewing Company, did not want it to be known that they were using statistical methods. Thus, the test based on this standardized score became known as "Student's t-test," and the standardized score is called t instead of z.

Recall from Chapter 21 that the t-distribution is also bell-shaped, but that its values change depending on the accompanying *degrees of freedom*, or *df*. In tests for one mean, df $= n - 1$, where n is the sample size. But in tests comparing two means, the formula is much more complicated, and we will leave its computation to computer software. Results presented in journal articles will always provide this number for you.

Finding p-Values

Because the accuracy of the sample standard deviation improves as the sample size increases, the frequency curve for possible Student's t values gets very close to the standard normal (z-score) curve for large sample sizes. Therefore, in the absence of computer software to do the work, we will use the standard normal curve and Table 8.1 to find approximate p-values as long as the sample sizes are large. If you need

---

Using Excel to Find p-Values when test statistic value is z (for proportions) or t (for means)

Alternative Hypothesis	*p*-value = proportion of curve	z (proportions)	t (means)
Not equal	[Below $-|z|$ or $-|t|$] $\times$ 2	[NORMSDIST $(-|z|)$] $\times$ 2	TDIST $(t, df, 2)$
Less than	Below t or z	NORMSDIST(z)	$t < 0$: TDIST$(|t|, df, 1)$ $t > 0$: $1 -$ TDIST$(t, df, 1)$
Greater Than	Above t or z	$1 -$ NORMSDIST(z)	$t < 0$: $1 -$ TDIST$(|t|, df, 1)$ $t > 0$: TDIST$(t, df, 1)$

to find the *p*-value using the *t*-distribution and know the appropriate degrees of freedom, an appropriate calculator, software, or website will provide it.

The summary box above illustrates the appropriate *p*-value areas for one-sided and two-sided *z*- and *t*-statistics, and the Excel commands for finding them. For example, suppose a two-sided test has a standardized score of $t = 2.17$ and df $= 87$ (the values for an example later in this chapter). The Excel command TDIST(2.17,87,2) returns the *p*-value for a two-tailed test as 0.0327. The general format is TDIST (*t*-value, df, number of tails). One caveat is that if the *t*-value is negative, its absolute value must be used instead, and for a one-tailed test, you need to be careful to find the proportion in the correct tail of the curve.

Let's reexamine two previous examples to illustrate how hypothesis tests work. The examples should help you understand the ideas behind hypothesis testing in general.

A Two-Sided Hypothesis Test for the Difference in Two Means

EXAMPLE 23.3   Weight Loss for Diet versus Exercise

In Chapter 21, we found a confidence interval for the mean difference in weight loss for men who participated in dieting only versus exercising only for the period of a year. However, weight loss occurs in two forms: lost fat and lost muscle. Did the dieters also lose more fat than the exercisers? These were the sample results for amount of fat lost after 1 year:

Diet Only

sample mean $= 5.9$ kg
sample standard deviation $= 4.1$ kg
number of participants $= n = 42$
standard error $=$ SEM$_1 =$
    $4.1/\sqrt{42} = 0.633$

Exercise Only

sample mean $= 4.1$ kg
sample standard deviation $= 3.7$ kg
number of participants $= n = 47$
standard error $=$ SEM$_2 =$
    $3.7/\sqrt{47} = 0.540$

standard error of difference $=$ square root of $[(0.633)^2 + (0.540)^2] = 0.83$

Here are the steps required to determine if there is a difference in average fat that would be lost for the two methods if carried out in the population represented by these men.

Step 1. Determine the null and alternative hypotheses.

Null hypothesis: There is no difference in average fat lost in the population for the two methods. The population mean difference is zero.

Alternative hypothesis: There is a difference in average fat lost in the population for the two methods. The population mean difference is not zero.

Notice that we do not specify which method has higher fat loss in the population. In this study, the researchers were simply trying to ascertain whether there was a difference. They did not specify in advance which method they thought would lead to higher fat loss.

Remember that when the alternative hypothesis includes a possible difference in either direction, the test is called a **two-sided,** or **two-tailed, hypothesis test.** In this example, it made sense for the researchers to construct a two-sided test because they had no preconceived idea of which method would be more effective for fat loss. They were simply interested in knowing whether there was a difference. When the test is two-sided, the *p*-value must account for possible chance differences in both "tails" or both directions. In step 3 for this example, our computation of the *p*-value will take both possible directions into account.

Step 2. Collect and summarize the data into a test statistic. The test statistic is the standardized score for the sample value when the null hypothesis is true. If there is no difference in the two methods, then the mean *population* difference is zero. The *sample* value is the observed difference in the two sample means, namely $5.9 - 4.1 = 1.8$ kg. The standard error for this difference, which we computed above, is 0.83. Thus, the test statistic is

$$\text{standardized score} = (1.8 - 0)/0.83 = 2.17$$

Step 3. Use the test statistic to determine the *p*-value. How extreme is this standardized score? First, we need to define extreme to mean both directions because a standardized score of -2.17 would have been equally informative against the null hypothesis. From Table 8.1, we see that 2.17 is between the 98th and 99th percentiles, at about the 98.5th percentile, so the proportion of the curve above 2.17 is about 0.015. This is also the probability of an extreme result in the other direction—that is, below -2.17. Thus, the test statistic results in a *p*-value of 2(0.015), or 0.03. Using software for either *t* or *z* confirms that this *p*-value is correct (to two decimal places).

Step 4. The result is statistically significant if the *p*-value is less than or equal to the level of significance.

Using the standard 0.05, the *p*-value of 0.03 indicates that the result is statistically significant. If there were really no difference between dieting and exercise as fat-loss methods, we would see such an extreme result only 3% of the time, or 3 times out of 100. Therefore, we conclude that the such a result would be too unlikely, and the null hypothesis is not true. There are various ways to say this. We conclude that there is a statistically significant difference between what the average fat loss for the two methods would be in the population if everyone were to diet or if everyone were to exercise. We reject the null hypothesis. We accept the alternative hypothesis. In the exercises for this chapter, you will be given a chance to test whether the same thing holds true for lean body mass (muscle) weight loss.   ■

A One-Sided Hypothesis Test for a Proportion

EXAMPLE 23.4   Public Opinion about the President

Americans (and probably all nationalities) have always been fascinated by scandals and judgemental of the public figures involved in them. Throughout his presidency, Bill Clinton was plagued with questions about the integrity of his personal life. On May 16, 1994, *Newsweek* reported the results of a public opinion poll that asked: "From everything you know about Bill Clinton, does he have the honesty and integrity you expect in a president?" (p. 23). The poll surveyed 518 adults and 233, or 0.45 of them (clearly less than half), answered yes. Could Clinton's adversaries conclude from this that only a minority (less than half) of the *population* of Americans thought Clinton had the honesty and integrity to be president? Assume the poll took a simple random sample of American adults. (In truth, a multistage sample was used, but the results will be similar.) Here is how we would proceed with testing this question.

Step 1. Determine the null and alternative hypotheses.

> *Null hypothesis:* There is no clear winning opinion on this issue; the proportions who would answer yes or no are each 0.50.

> *Alternative hypothesis:* Fewer than 0.50, or 50%, of the population would answer yes to this question. The majority did not think Clinton had the honesty and integrity to be president.

Notice that, unlike the previous example, this alternative hypothesis includes values on only one side of the null hypothesis. It does not include the possibility that *more* than 50% would answer yes. If the data indicated an uneven split of opinion in that direction, we would not reject the null hypothesis because the data would not be supporting the alternative hypothesis, which is the direction of interest. Some researchers include the other direction as part of the null hypothesis, but others simply don't include that possibility in either hypothesis, as illustrated in this example.

Remember that when the alternative hypothesis includes values in one direction only, the test is called a **one-sided,** or **one-tailed, hypothesis test.** The *p*-value is computed using only the values in the direction specified in the alternative hypothesis, as we shall see for this example.

Step 2. Collect and summarize the data into a test statistic. The test statistic is the standardized score for the sample value when the null hypothesis is true. The sample value is 0.45. If the null hypothesis is true, the population proportion is 0.50. The corresponding null standard error, assuming the null hypothesis is true, is

$$\text{null standard error} = \text{square root of } [(0.5)(0.5)/518] = 0.022$$
$$\text{standardized score} = z\text{-score} = (0.45 - 0.50)/0.022 = -2.27$$

Step 3. Use the test statistic to determine the *p*-value. The *p*-value is the probability of observing a standardized score of -2.27 or less, just by chance. From Table 8.1, we find that -2.27 is between the 1st and 2nd percentiles. In fact, using a more exact method, the statistical function NORMSDIST(-2.27) in Excel, the *p*-value is 0.0116. Notice that, unlike the two-sided hypothesis test, we *do not* double this value. If we were to find an extreme value in the other direction, a standardized score of 2.27 or more, we would *not* reject the null hypothesis. We are performing a one-sided hypothesis test and are only concerned with values in *one tail* of the normal curve when we find the *p*-value. In this example, it is the lower tail because those values would indicate that the true proportion is *less than* the null hypothesis value of 0.50, thus supporting the alternative hypothesis.

Step 4. The result is statistically significant if the *p*-value is less than or equal to the level of significance. Using the 0.05 criterion, we have found a statistically significant result. Therefore, we could conclude that the proportion of American adults in 1994 who believed Bill Clinton had the honesty and integrity they expected in a president was statistically significantly less than a majority. However, we will revisit this example in the next chapter and see that Clinton's supporters could have made a similar claim in the other direction. ∎

23.3   How Journals Present Hypothesis Tests

There are a few common test statistics that you will encounter when you read journal articles, and a brief description of them will help you understand what you are reading. A common method for presenting results of hypothesis tests is to give a numerical value for the test statistic, followed by the p-value. To understand the results, all you need to do is figure out what hypotheses were being tested and then make a conclusion based on the p-value. Let's look at some common test statistics you may encounter and the hypotheses that accompany them.

t-Tests for One or Two Means

When the null hypothesis is that a population mean equals a specific null value or that two population means are equal to each other, the most common test statistic is a t-score, as shown in the previous section. Journals generally will report the comparison and provide the value for t, provide the degrees of freedom, and give the p-value. Often the sample means and standard deviations will be provided as well. For instance, here is a quote from the Original Source 2 (Slutske et al., 2003, p. 1447), comparing the hangover symptom scale for people who had alcohol-related problems and people who did not:

> *Those who reported an alcohol-related problem had significantly higher scores on the HSS than those who did not (no alcohol-related problems: mean = 4.3, SD = 3.3; alcohol-related problems: mean = 7.0, SD = 3.0; t = 13.64, df = 1225, p < 0.001).*

The quoted test statistic is $t = 13.64$. The null hypothesis is that the mean HSS (hangover symptoms scale) scores for the populations of students with and without alcohol-related problems would be the same if everyone in the population were to complete the symptoms scale. Notice that the p-value is quoted as less than 0.001; in fact it is much less than 0.001. Often p-values will be reported in reference to common levels of significance, such as $p < 0.05$, $p < 0.01$ or $p < 0.001$.

F-Tests for Comparing More Than Two Means

When more than two means are to be compared the most common method used is called *analysis of variance*, abbreviated as ANOVA. Test statistics are used to test the null hypothesis that a set of means are all equal versus the alternative hypothesis that at least one mean differs from the others. The letter F is used to denote the test statistic in this situation, and the resulting frequency curve is called an F-distribution. This distribution has two different degrees of freedom values associated with it. Finding degrees of freedom and p-values in this situation is beyond the level of our discussion, but you should now be able to interpret the results when you read them.

For instance, in Original Source 11 (Horne, Reyner and Barrett, 2003, p. 690), one result is

> *For sleep-related driving incidents there was a significant between-conditions effect (F = 11.3, df 3, 33; p < 0.001).*

The means of interest in this case are for the number of lane drifting incidents in a 30-minute simulated driving test. The four conditions compared were normal sleep with no alcohol, normal sleep with lunchtime alcohol, reduced sleep with no alcohol, and reduced sleep with lunchtime alcohol. The null hypothesis is that if people in the population were to drive under each of these four conditions, the mean number of lane drifting incidents would be the same for the four conditions. The alternative hypothesis is that at least one of the means would differ from the others. In this case, you can see that the null hypothesis would be rejected, because the *p*-value is so small (<0.001). Follow-up tests showed that the mean for reduced sleep with alcohol was significantly higher than the rest, and that the means for reduced sleep alone and alcohol alone were both significantly higher than the mean for normal sleep with no alcohol.

Tests for Slope in Regression

Another common test you may encounter is for the slope of a regression equation. If there is no linear relationship between the explanatory and response variables in the population, then the population slope is zero. Therefore, after observing a non-zero slope for the regression line computed from the sample, it may be of interest to test these hypotheses:

Null hypothesis: Population slope is 0 (no linear relationship between the variables)

Alternative hypothesis: Population slope is not 0 (there is a linear relationship)

You may remember the quote from Example 23.1 about whether the degree to which someone is a morning person can help predict how happy they are. Let's repeat the quote here:

> *"Critically, morningness predicted positive affect when age was controlled ($\beta = 0.149$, $p < .001$)."*

The quote says that the sample slope was 0.149, and the *p*-value for testing whether the population slope is 0 is less than .001, so the alternative hypothesis can be accepted. It does appear that the sample results showing the relationship between morningness and happiness can be extended to the population.

EXAMPLE 23.5   ### Cereal and Obesity Revisited

In Case Study 6.2, we discussed a study that concluded that children who eat breakfast cereal had lower body mass index (BMI) than children who did not eat cereal. Here is how one of the results was presented in that study:

> *Frequency of RTEC [ready to eat cereal] consumption significantly ($P = 0.001$) affected a child's BMI with a decrease of 2 percentiles (-1.977 ± 0.209) for every day of RTEC consumption, whereas sex, ethnicity, age, and time had no effect"* (Frantzen et al., 2013, pp. 514–515).

The hypotheses implicitly being tested in this quote are:

Null hypothesis: There is no linear relationship between cereal consumption and BMI in the population; the population slope is 0.

Alternative hypothesis: There is a linear relationship between cereal consumption and BMI in the population; the population slope is not 0.

The results tell us that the slope for the regression line in the *sample* was −1.977. It is not obvious in the quote, but in an accompanying table, we learn that the quoted value of 0.209 is one standard error. To find an approximate 95% confidence interval for the population slope, we could add and subtract two standard errors, giving an interval from −2.39 to −1.56. We are also told that a test of the hypotheses above resulted in a *p*-value of 0.001. (In fact, the *p*-value in this situation is much less than 0.001, but journal results commonly give 0.001 as the upper bound when the *p*-value is really much smaller.) A slope of −1.977 is interpreted in the quote to mean that each increase of one day of eating cereal decreases the BMI percentile by 2. But remember that in Case Study 6.2 we warned about making the conclusion that eating cereal actually *causes* a change in BMI, so it is a misinterpretation to say that consumption "affected" BMI. ∎

Tests for Relative Risk and Odds Ratios

We learned about relative risk and odds ratios in Chapter 12. In most cases, when these statistics are of interest, the researcher wants to know if the population risks of something are equal for two groups or under two conditions. When risks are equal, the relative risk and odds ratio would both be 1.0. Therefore, the hypotheses are:

Null hypothesis: Population relative risk (or odds ratio) = 1.0

Alternative hypothesis: Population relative risk (or odds ratio) ≠ 1.0

Sometimes a one-sided test is used for the alternative hypothesis, such as when testing the relative risk of a heart attack when taking aspirin versus placebo. The original research question was whether aspirin would *decrease* the risk of a heart attack. Therefore, it would not make sense to include the possibility of an increased risk in the alternative hypothesis.

EXAMPLE 23.6 **Baldness and Heart Attacks Revisited**

The meta-analysis presented in Case Study 5.4 examined the relationship between vertex or pattern baldness and the risk of a heart attack. Here is a quote from the original journal article presenting the results:

> *The adjusted RR [relative risk] of men with severe baldness for CHD [coronary heart disease] was 1.32 (95% CI 1.08 to 1.63, p = 0.008) compared to those without baldness. Analysis of younger men (<55 or ≤60 years) showed a similar association of CHD with severe baldness (RR 1.44, 95% CI 1.11 to 1.86, p=0.006)"* (Yamada et al., 2013, p. 1).

This quote provides results for testing the above hypotheses for men with severe baldness in general and for younger men with severe baldness. The *p*-values are given as 0.008 for all men, and 0.006 for younger men, so in both cases, the null hypothesis is rejected. In both cases, we can conclude that the relative risk in the population is not 0. Notice that 95% confidence intervals are provided as well, so you can see what the plausible values are for the population relative risks. When a hypothesis tests rejects a null value, that value will not be contained in the confidence interval of plausible values. We see this fact here because a relative risk of 1.0 is not contained in either of the confidence intervals provided. ∎

Other Tests

We have already discussed chi-square tests in Chapter 13. The symbol for a chi-square test statistic is the Greek letter "chi" with a square, written as χ^2. Other tests you may encounter are a test for whether or not a correlation is significantly different from 0, in which case the test statistic is just the sample correlation r, *nonparametric tests* such as the *sign test, Wilcoxon test* or *Mann–Whitney test,* and tests for an interaction between two variables. In most cases, you will simply need to understand what the hypotheses of interest were and how to interpret the reported p-value. Thus, it's very important that you understand the generic meaning of a p-value.

Revisiting Case Studies

Whereas news stories tend to simply report the decision from hypothesis testing, as noted, journals tend to report p-values as well. This practice allows you to make your own decision, based on the severity of a type 1 error and the magnitude of the p-value. News reports leave you little choice but to accept the convention that the probability of a type 1 error (the level of significance) is 5%.

As a final demonstration of the way in which journals report the results of hypothesis tests, let's return to two more of our earlier case studies.

| CASE STUDY 5.1 REVISITED | **Quitting Smoking with Nicotine Patches** |

This study compared the smoking cessation rates for smokers randomly assigned to use a nicotine patch versus a placebo patch. In the summary at the beginning of the journal article, the results were reported as follows:

> *Higher smoking cessation rates were observed in the active nicotine patch group at 8 weeks (46.7% vs 20%) (P < .001) and at 1 year (27.5% vs 14.2%) (P = .011)* (Hurt et al., 1994, p. 595).

Two sets of hypotheses are being tested, one for the results after 8 weeks and one for the results after 1 year. In both cases, the hypotheses are

Null hypothesis: The proportion of smokers in the population who would quit smoking using a nicotine patch and a placebo patch are the same.

Alternative hypothesis: The proportion of smokers in the population who would quit smoking using a nicotine patch is higher than the proportion who would quit using a placebo patch.

In both cases, the reported *p*-values are quite small: less than 0.001 for the difference after 8 weeks and equal to 0.011 for the difference after a year. Therefore, we would conclude that rates of quitting are significantly higher using a nicotine patch than using a placebo patch after 8 weeks and after 1 year.   ■

CASE STUDY 6.4 REVISITED   **Smoking During Pregnancy and Child's IQ**

In this study, researchers investigated the impact of maternal smoking on subsequent IQ of the child at ages 1, 2, 3, and 4 years of age. Earlier, we reported some of the confidence intervals provided in the journal article reporting the study. Those confidence intervals were actually accompanied by *p*-values. Here is the complete reporting of some of the results:

> *Children born to women who smoked 10+ cigarettes per day during pregnancy had developmental quotients at 12 and 24 months of age that were 6.97 points lower (averaged across these two time points) than children born to women who did not smoke during pregnancy (95% CI: 1.62,12.31, P = .01); at 36 and 48 months they were 9.44 points lower (95% CI: 4.52, 14.35, P = .0002)* (Olds et al., 1994, p. 223).

Notice that we are given more information in this report than in most because we are given both confidence interval and hypothesis testing results. This is excellent reporting because, with this information, we can determine the magnitude of the observed effects instead of just whether they are statistically significant or not.

Again, two sets of null and alternative hypotheses are being tested in the report, one set at 12 and 24 months (1 and 2 years) and another at 36 and 48 months (3 and 4 years of age). In both cases, the hypotheses are

Null hypothesis: The mean IQ scores for children whose mothers smoke 10 or more cigarettes a day during pregnancy are the same as the mean for those whose mothers do not smoke, in populations similar to the one from which this sample was drawn.

Alternative hypothesis: The mean IQ scores for children whose mothers smoke 10 or more cigarettes a day during pregnancy are not the same as the mean for those whose mothers do not smoke, in populations similar to the one from which this sample was drawn.

This is a *two-tailed test* because the researchers included the possibility that the mean IQ score could actually be *higher* for those whose mothers smoke. The confidence interval provides the evidence of the direction in which the difference falls. The *p*-value simply tells us that there is a statistically significant difference.   ■

CASE STUDY 23.1   **An Interpretation of a *p*-Value *Not* Fit to Print**

In an article entitled, "Probability experts may decide Pennsylvania vote," the *New York Times* (Passell, 11 April 1994, p. A15) reported on the use of statistics to try to decide whether there had been fraud in a special election held in Philadelphia. Unfortunately, the newspaper account made a common mistake, misinterpreting a *p*-value to be the probability that the results *could* be explained by chance. The consequence was that readers who did not know how to spot the error would have been led to think that the election probably *was* a fraud.

It all started with the death of a state senator from Pennsylvania's Second Senate District. A special election was held to fill the seat until the end of the unexpired term. The Republican candidate, Bruce Marks, beat the Democratic candidate, William Stinson, in the voting booth but lost the election because Stinson received so many more votes in absentee ballots. The results in the voting booth were very close, with 19,691 votes for Mr. Marks and 19,127 votes for Mr. Stinson. But the absentee ballots were not at all close, with only 366 votes for Mr. Marks and 1391 votes for Mr. Stinson.

The Republicans charged that the election was fraudulent and asked that the courts examine whether the absentee ballot votes could be discounted on the basis of suspicion of fraud. In February, 1994, 3 months after the election, Philadelphia Federal District Court Judge Clarence Newcomer disqualified all absentee ballots and overturned the election. The ruling was appealed, and statisticians were hired to help sort out what might have happened.

One of the statistical experts, Orley Ashenfelter, decided to examine previous senatorial elections in Philadelphia to determine the relationship between votes cast in the voting booth and those cast by absentee ballot. He computed the difference between the Republican and Democratic votes for those who voted in the voting booth, and then for those who voted by absentee ballot. He found there was a positive correlation between the voting booth difference and the absentee ballot difference. Using data from 21 previous elections, he calculated a regression equation to predict one from the other. Using his equation, the difference in votes for the two parties by absentee ballot could be predicted from knowing the difference in votes in the voting booth.

Ashenfelter then used his equation to predict what should have happened in the special election in dispute. There was a difference of 19,691 − 19,127 = 564 votes (in favor of the Republicans) in the voting booth. From that, he predicted a difference of 133 votes in favor of the Republicans in absentee ballots. Instead, a difference of 1025 votes in favor of the *Democrats* was observed in the absentee ballots of the disputed election.

Of course, everyone knows that chance events play a role in determining who votes in any given election. So Ashenfelter decided to set up and test two hypotheses. The null hypothesis was that, given past elections as a guide and given the voting booth difference, the overall difference observed in this election could be explained by chance. The alternative hypothesis was that something other than chance influenced the voting results in this election.

Ashenfelter reported that *if* chance alone was responsible, there was a 6% chance of observing results as extreme as the ones observed in this election, given the voting booth difference. In other words, the *p*-value associated with his test was about 6%. That is not how the result was reported in the *New York Times*. When you read its report, see if you can detect the mistake in interpretation:

> *There is some chance that random variations alone could explain a 1158-vote swing in the 1993 contest—the difference between the predicted 133-vote Republican advantage and the 1025-Democratic edge that was reported. More to the point, there is some larger probability that chance alone would lead to a sufficiently large Democratic edge on the absentee ballots to overcome the Republican margin on the machine balloting. And the probability of such a swing of 697 votes from the expected results, Professor Ashenfelter calculates, was about 6 percent. Putting it another way, if past elections are a reliable guide to current voting behavior,* there is a 94 percent chance that irregularities in the absentee ballots, not chance alone, *swung the election to the Democrat, Professor Ashenfelter concludes* (Passell, 11 April 1994, p. A15; emphasis added).

The author of this article has mistakenly interpreted the *p*-value to be the probability that the null hypothesis is true and has thus reported what he thought to be the probability that the alternative hypothesis was true. We hope you realize that this is not a valid conclusion. The *p*-value can only tell us the probability of observing these results *if* the election was *not* fraudulent. It cannot tell us the probability in the other direction—namely, the probability that the election was fraudulent based on observed results. This is akin to the "confusion of the inverse" discussed in Chapter 17. There we saw that physicians sometimes confuse the (unknown) probability that the patient has a disease, given a positive test, with the (known) probability of a positive test, given that the patient has the disease.

You should also realize that the implication that the results of past elections would hold in this special election may not be correct. This point was raised by another statistician involved with the case. The *New York Times* report notes:

> *Paul Shaman, a professor of statistics at the Wharton School at University of Pennsylvania . . . exploits the limits in Professor Ashenfelter's reasoning. Relationships between machine and absentee voting that held in the past, he argues, need not hold in the present. Could not the difference, he asks, be explained by Mr. Stinson's "engaging in aggressive efforts to obtain absentee votes?"* (Passell, 11 April 1994, p. A15).

The case went to court two more times, but the original decision made by Judge Newcomer was upheld each time. The Republican Bruce Marks held the seat until December 1994. As a footnote, in the regular election in November 1994, Bruce Marks lost by 393 votes to Christina Tartaglione, the daughter of the chair of the board of elections, one of the people allegedly involved in the suspected fraud. This time, both candidates agreed that the election had been conducted fairly (Shaman, 28 November 1994). ■

Thinking About Key Concepts

- When the results of hypothesis tests are reported in the news they rarely are accompanied by hypotheses or a p-value. You should be able to figure out what the implicit hypotheses are and whether the null hypothesis was rejected based on information given in the story.
- Hypothesis tests for means and proportions generally use standardized scores as the test statistic. These scores measure how far the sample mean, proportion, or difference is from a hypothesized population version, standardized by dividing by the standard error. As a general rule, standardized scores greater than about 2.0 (in absolute value) will lead to rejection of the null hypothesis. But the direction of the alternative hypothesis must be considered as well.
- For tests involving means, especially for small samples, the test statistic should be compared to a "Student's t distribution" rather than a normal curve to find the p-value.
- To interpret hypothesis tests presented in journal articles, it is important to understand the general meaning of a p-value. Do not make the common mistake of thinking it represents the probability that the null hypothesis is true. A p-value is the probability of seeing evidence as extreme as that found in the sample or more so, *if* the null hypothesis is true.

Focus on Formulas

Some Notation for Hypothesis Tests

The null hypothesis is denoted by H_0, and the alternative hypothesis is denoted by H_1 or H_a.

"alpha" $= \alpha =$ level of significance $=$ desired probability of making a type 1 error when H_0 is true; we reject H_0 if p-value $\leq \alpha$.

"beta" $= \beta =$ probability of making a type 2 error when H_1 is true; power $= 1 - \beta$

Steps for Testing the Mean of a Single Population

Denote the population mean by μ and the sample mean and standard deviation by $\overline{X}$ and s, respectively.

Step 1. H_0: $\mu = \mu_0$, where μ_0 is the *chance* or *status quo* value.
H_1: $\mu \neq \mu_0$ for a two-sided test; H_1: $\mu < \mu_0$ or H_1: $\mu > \mu_0$ for a one-sided test, with the direction determined by the research hypothesis of interest.

Step 2. The z test statistic can be used as an approximation if the sample is large. The t test statistic is valid if the underlying population is bell-shaped (even with a small sample) and/or the sample is large. The test statistic is

$$z \text{ or } t = \frac{\overline{X} - \mu_0}{s/\sqrt{n}}$$

Step 3. The p-value depends on the form of H_1. In each case, we refer to the proportion of the t-distribution or the standard normal curve above (or below) a value as the "area" above (or below) that value. Then we list the p-values as follows:

Alternative Hypothesis	**p-Value**
$H_1: \mu \neq \mu_0$	$2 \times$ area above $|t|$ or $|z|$
$H_1: \mu > \mu_0$	area above t or z
$H_1: \mu < \mu_0$	area below t or z

Step 4. You must specify the desired α; it is commonly 0.05. Reject H_0 if p-value $\leq \alpha$.

Steps for Testing a Proportion for a Single Population

Steps 1, 3, and 4 are the same, except replace μ with the population proportion p and μ_0 with the hypothesized proportion p_0. The test statistic (step 2) is always z (not t):

$$z = \frac{\hat{p} - p_0}{\sqrt{\dfrac{p_0(1 - p_0)}{n}}}$$

Steps for Testing for Equality of Two Population Means Using Large Independent Samples

Steps 1, 3, and 4 are the same, except replace μ with $(\mu_1 - \mu_2)$ and μ_0 with 0.

Use previous notation for sample sizes, means, and standard deviations; z can be used as an approximation if the sample sizes are both large, otherwise use t, which can be used if it is assumed that both populations are bell-shaped. Test statistic (step 2) is

$$z \text{ or } t = \frac{\overline{X}_1 - \overline{X}_2}{\sqrt{\dfrac{s_1^2}{n_1} + \dfrac{s_2^2}{n_2}}}$$

Exercises

Exercises with numbers divisible by 3 (3, 6, 9, etc.) are included in the Solutions at the back of the book. They are marked with an asterisk (*).

1. Are null and alternative hypotheses statements about populations or samples, or does it depend on the situation?

2. In a February, 2013, CBS News/New York Times poll of 250 American Catholics who attend mass at least once a week, 57% answered yes to the question, "Do you favor allowing women to be priests?" (Source: http://www.nytimes.com/interactive/2013/03/06/us/catholics-america-poll.html?ref=us.)

   a. Set up the null and alternative hypotheses for deciding whether a majority of American

Catholics who attend mass at least once a week favors allowing women to be priests.

b. Using Example 23.4 (p. 502) as a guide, compute the test statistic for this situation.

c. If you have done everything correctly, the *p*-value for the test is about 0.0127. Based on this, make a conclusion for this situation. Write it in both statistical language and in words that someone with no training in statistics would understand.

*3. Refer to Exercise 2. Is the test described there a one-sided or a two-sided test?

4. Explain the difference between statistical significance and significance as used in everyday language.

5. Suppose a one-sided test for a proportion resulted in a *p*-value of 0.03. What would the *p*-value be if the test were two-sided instead?

*6. Suppose a two-sided test for a difference in two means resulted in a *p*-value of 0.08.

*a. Using the usual criterion for hypothesis testing, would we conclude that there was a difference in the population means? Explain.

*b. Suppose the test had been constructed as a one-sided test instead, and the evidence in the sample means was in the direction to support the alternative hypothesis. Using the usual criterion for hypothesis testing, would we be able to conclude that there was a difference in the population means? Explain.

7. Suppose you were given a hypothesized population mean, a sample mean, a sample standard deviation, and a sample size for a study involving a random sample from one population. What formula would you use for the test statistic?

8. In Example 23.4, we showed that the Excel command NORMSDIST(*z*) gives the area below the standardized score *z*. Use Excel or an appropriate calculator, software, website, or table to find the *p*-value for each of the following examples and case studies, taking into account whether the test is one-sided or two-sided:

a. Example 23.3, pages 501–502, $z = 2.17$, two-sided test (using *z*, not *t*)

b. Example 22.1, pages 473–475, $z = 2.00$, one-sided test

c. Case Study 22.1 pages 486–489, $z = 4.09$, one-sided test

*9. Suppose you wanted to see whether a training program helped raise students' scores on a standardized test. You administer the test to a random sample of students, give them the training program, then readminister the test. For each student, you record the increase (or decrease) in the test score from one time to the next.

*a. What would the null and alternative hypotheses be for this situation?

*b. Suppose the mean change for the sample was 10 points and the accompanying standard error was 4 points. What would be the standardized score that corresponded to the sample mean of 10 points?

*c. Based on the information in part (b), what would you conclude about this situation? (Assume the sample size is large and use *z*, not *t*.)

10. Refer to the previous exercise.

a. What explanation might be given for the increased scores, other than the fact that the training program had an impact?

b. What would have been a better way to design the study in order to rule out the explanation you gave in part (a)?

11. On July 1, 1994, *The Press* of Atlantic City, NJ, had a headline reading, "Study: Female hormone makes mind keener" (p. A2). Here is part of the report:

Halbreich said he tested 36 post-menopausal women before and after they started the estrogen therapy. He gave each one a battery of tests that measured such things as memory, hand-eye coordination, reflexes and the ability to learn new information and apply it to a problem. After estrogen therapy started, he said, there was a subtle but statistically significant increase in the mental scores of the patients.

Explain what you learned about the study, in the context of the material in this chapter, by reading this quote. Be sure to specify the hypotheses that were being tested and what you know about the statistical results.

*12. Siegel (1993) reported a study in which she measured the effect of pet ownership on the use of medical care by the elderly. She interviewed 938 elderly adults. One of her results was reported as: "After demographics and health status were controlled for, subjects with pets had fewer total doctor contacts during the one-year period than those without pets (beta $= -.07, p < .05$)" (p. 164).

  *a. State the null and alternative hypotheses Siegel was testing. Be careful to distinguish between a population and a sample.

  *b. State the conclusion you would make. Be explicit about the wording.

13. Refer to Exercise 12. Here is another of the results reported by Siegel: "For subjects without a pet, having an above-average number of stressful life events resulted in about two more doctor contacts during the study year (10.37 vs. 8.38, $p < .005$). In contrast, the number of stressful life events was not significantly related to doctor visits among subjects with a pet" (1993, p. 164).

  a. State the null and alternative hypotheses Siegel is testing in this passage. Notice that two tests are being performed; be sure to cover both.

  b. Pretend you are a news reporter, and write a short story describing the results reported in this exercise. Be sure you do not convey any misleading information. You are writing for a general audience, so do not use statistical jargon that would be unfamiliar to them.

14. Refer to Example 13.2, in which we tested whether there was a relationship between gender and driving after drinking alcohol. Remember that the Supreme Court used the data to determine whether a law was justified. The law differentiated between the ages at which young males and young females could purchase 3.2% beer. Specify what a type 1 and a type 2 error would be for this example. Explain what the consequences of the two types of error would be in that context.

*15. A CNN/ORC poll conducted in January, 2013, asked 814 adults in the United States, "Which of the following do you think is the most pressing issue facing the country today?" and then presented seven choices, one of which was "The Economy." (*Source:* http://www.pollingreport.com/prioriti. htm, accessed July 16, 2013.) "The Economy" was chosen by 46% of the respondents. Suppose an unscrupulous politician wanted to show that the economy was not a pressing issue, and stated "Significantly fewer than half of adults think that the economy is a pressing issue."

  *a. What are the null and alternative hypotheses the politician is implicitly testing in this quote? Make sure you specify the population value being tested and the population to which it applies.

  *b. Using the results of the poll, find the value of the standardized score that would be used as the test statistic.

  *c. If you answered parts (a) and (b) correctly, the *p*-value for the test should be about 0.011. Explain how the politician reached the conclusion stated in the quote.

  *d. Do you think the statement made by the politician is justified? Explain.

16. In Example 23.3, we tested whether the average fat lost from 1 year of dieting versus 1 year of exercise was equivalent. The study also measured lean body weight (muscle) lost or gained. The average for the 47 men who exercised was a *gain* of 0.1 kg, which can be thought of as a loss of -0.1 kg. The standard deviation was 2.2 kg. For the 42 men in the dieting group, there was an average loss of 1.3 kg, with a standard deviation of 2.6 kg. Test to see whether the average lean body mass lost (or gained) would be different for the population of men similar to the ones in this study. Specify all four steps of your hypothesis test. (Use z instead of t.)

17. Professors and other researchers use scholarly journals to publish the results of their research. However, only a small minority of the submitted papers is accepted for publication by the most prestigious journals. In many academic fields, there is a debate as to whether submitted papers written by women are treated as well as those submitted by men. In the January 1994 issue of *European Science Editing* (Maisonneuve, January 1994), there was a report on a study that examined this question. Here is part of that report:

Similarly, no bias was found to exist at JAMA [Journal of the American Medical Association] *in acceptance rates based on the gender of the corresponding author and the assigned editor. In the sample of 1851 articles considered in this study female editors used female reviewers more often than did male editors (P < 0.001).*

That quote actually contains the results of two separate hypothesis tests. Explain what the two sets of hypotheses tested are and what you can conclude about the *p*-value for each set.

*18. Use Excel or an appropriate calculator, software, website, or table to find the *p*-value in each of the following situations.

   *a. Alternative hypothesis is "greater than," $t = +2.5$, df = 40.

   *b. Alternative hypothesis is "less than," $t = -2.5$, df = 40.

   *c. Alternative hypothesis is "not equal," $t = 2.5$, df = 40.

   *d. Alternative hypothesis is "greater than," $z = +1.75$.

   *e. Alternative hypothesis is "less than," $z = -1.75$.

   *f. Alternative hypothesis is "not equal," $z = -1.75$.

19. On January 30, 1995, *Time* magazine reported the results of a poll of adult Americans, in which they were asked, "Have you ever driven a car when you probably had too much alcohol to drive safely?" The exact results were not given, but from the information provided we can guess at what they were. Of the 300 men who answered, 189 (63%) said yes and 108 (36%) said no. The remaining three weren't sure. Of the 300 women, 87 (29%) said yes while 210 (70%) said no, and the remaining three weren't sure.

   a. Ignoring those who said they weren't sure, there were 297 men asked, and 189 said yes, they had driven a car when they probably had too much alcohol. Does this provide statistically significant evidence that a majority of men in the population (that is, more than half) would say that they had driven a car when they probably had too much alcohol, if asked? Go through the four steps to test this hypothesis.

   b. For the test in part (a), you were instructed to perform a one-sided test. Why do you think it would make sense to do so in this situation? If you do not think it made sense, explain why not.

   c. Repeat parts (a) and (b) for the women. (Note that of the 297 women who answered, 87 said yes.)

The following information is for Exercises 20 to 22: In Example 23.3 (p. 501), we tested to see whether dieters and exercisers had significantly different average fat loss. We concluded that they did because the difference for the samples under consideration was 1.8 kg, with a standard error of 0.83 kg and a standardized score of 2.17. Fat loss was higher for the dieters.

20. Construct an approximate 95% confidence interval for the population difference in mean fat loss. Consider the two different methods for presenting results: (1) the *p*-value and conclusion from the hypothesis test or (2) the confidence interval. Which do you think is more informative? Explain.

*21. Suppose the alternative hypothesis had been that men who exercised lost more fat on average than men who dieted. Would the null hypothesis have been rejected? Explain why or why not. If yes, give the *p*-value that would have been used.

22. Suppose the alternative hypothesis had been that men who dieted lost more fat on average than men who exercised. Would the null hypothesis have been rejected? Explain why or why not. If yes, give the *p*-value that would have been used.

Exercises 23 to 29 refer to News Story and Original Source 1, "Alterations in Brain and Immune Function Produced by Mindfulness Meditation," (not available on the companion website). For this study, volunteers were randomly assigned to a meditation group (25 participants) or a control group (16 participants). The meditation group received an 8-week meditation training program. Both groups were given influenza shots and their antibodies were measured. This measurement was taken after the meditation group had been practicing meditation for about 8 weeks. The researchers also measured brain activity, but these exercises will not explore that part of the study.

23. On page 565 of the article we are told that "Participants were right-handed subjects who were employees of a biotechnology corporation in Madison, Wisconsin." To what population do you think the results of this study apply?

*24. Participants were given psychological tests measuring positive and negative affect (mood) as well as anxiety at three time periods. Time 1 was before the meditation training. Time 2 was at the end of the 8 weeks of training, and Time 3 was 4 months later. One of the results reported is:

There was a significant decrease in trait negative affect with the meditators showing less negative affect at Times 2 and 3 compared with their negative affect at Time 1 [t(20) = 2.27 and t(21) = 2.45, respectively, p < .05 for both]. Subjects in the control group showed no change over time in negative affect (t < 1) (Davidson et al., p. 565).

The first sentence of the quote reports the results of two hypothesis tests for the meditators.

Specify in words the null hypothesis for each of the two tests. They are the same except that one is for Time 2 and one is for Time 3. State each one separately, referring to what the time periods were. Make sure you don't confuse the population with the sample and that you state the hypotheses using the correct one.

25. **a.** Refer to Exercise 24. State the alternative hypothesis for each test. Explain whether you decided to use a one-sided or a two-sided test and why.

   b. Write the conclusion for the Time 2 test in statistical language and in plain English.

26. Refer to the quote in the Exercise 24. Explain what is meant by the last sentence, "Subjects in the control group showed no change over time in negative affect (t < 1)." In particular, do you think the sample difference was exactly zero?

*27. Another quote in the article is

In response to the influenza vaccine, the meditators displayed a significantly greater rise in antibody titers from the 4 to 8 week blood draw compared with the controls [t(33) = 2.05, p < .05, Figure 5] (Davidson et al., p. 566).

Specify in words the null hypothesis being tested. Make sure you don't confuse the population with the sample and that you state the hypothesis using the correct one.

28. Refer to the quote in Exercise 27. Specify in words the alternative hypothesis being tested. Explain whether you decided to use a one-sided or a two-sided test and why.

29. Refer to the quote in Exercise 27.

   a. What is the meaning of the word *significantly* in the quote?

   b. Explain in plain English what the results of this test mean.

Exercises 30 to 32 refer to News Story 14 and the journal article Original Source 14, "Sex differences in the neural basis of emotional memories" (not available on the companion website). In this study, 12 men and 12 women were shown 96 pictures each. They rated the pictures for emotional intensity on a scale from 0 (no emotion) to 3 (intense emotion). Without being told in advance that this would happen, 3 weeks later they were shown the same set of pictures, interspersed with an additional 48 pictures, called "foils." They were asked which of the pictures they thought that they had seen previously, and if each one of those was just familiar or was distinctly remembered. Results for each picture were coded as 0 (forgotten), 1 (familiar), or 2 (remembered).

*30. One of the results was "women rated significantly more pictures as highly arousing (rated 3) than did men [$t(22) = 2.41$, $P < 0.025$]" (Canli et al., 2002, p. 10790).

  *a. Specify in words the null hypothesis being tested. Make sure you don't confuse the population with the sample and that you state the hypothesis using the correct one.

  *b. Were the researchers using a one-sided or a two-sided alternative hypothesis? Explain how you know.

  *c. Explain in words what is meant by [$t(22) = 2.41$, $P < 0.025$].

  *d. Is the use of the word *significantly* in the quote the statistical version or the English version or both? Explain how you know.

31. One of the results was "women had better memory for emotional pictures than men; pictures rated as most highly arousing were recognized significantly more often by women than by men as familiar [$t(20) = 2.40$, $P < 0.05$] or

remembered [$t(20) = 2.38$, $P < 0.05$]" (Canli et al., 2002, p. 10790). There are two separate test results reported in this quote.

  a. Specify in words the null hypothesis being tested for each of the two tests. Make sure you don't confuse the population with the sample and that you state the hypothesis using the correct one.

  b. Were the researchers using a one-sided or a two-sided alternative hypothesis? Explain how you know.

  c. Explain in words what is meant by [$t(20) = 2.38$, $P < 0.05$].

  d. Is the use of the word "significantly" in the quote the statistical version or the English version or both? Explain how you know.

32. One of the results was "there were no significant sex differences in memory for pictures rated less intense (0–2) or in false-positive rates (12 and 10% for women and men, respectively)" (Canli et al., 2002, p. 10790).

  a. What is meant by "false positive rates" in this example?

  b. Specify in words the null hypothesis being tested regarding false positives. Make sure you don't confuse the population with the sample and that you state the hypothesis using the correct one.

  c. Is the use of the word *significant* in the quote the statistical version or the English version or both? Explain how you know.

Mini-Projects

1. Find three separate journal articles that report the results of hypothesis tests. For each one, do or answer the following:

  a. State the null and alternative hypotheses.

  b. Based on the *p*-value reported, what conclusion would you make?

  c. What would a type 1 and a type 2 error be for the hypotheses being tested?

2. Conduct a test for extrasensory perception. You can either create target pictures or use a deck of cards and ask people to guess suits or colors. Whatever you

use, be sure to randomize properly. For example, with a deck of cards you should always replace the previous target and shuffle very well.

a. Explain how you conducted the experiment.

b. State the null and alternative hypotheses for your experiment.

c. Report your results.

d. If you do not find a statistically significant result, can you conclude that extra-sensory perception does not exist? Explain.

References

Avorn, J., M. Monane, J. H. Gurwitz, R. J. Glynn, I. Choodnovskiy, and L. A. Lipsitz. (1994). Reduction of bacteriuria and pyuria after ingestion of cranberry juice. *Journal of the American Medical Association* 271, no. 10, pp. 751–754.

Biss, R. K., & Hasher, L. (2012). Happy as a lark: Morning-type younger and older adults are higher in positive affect. *Emotion.* 12(3), Jun 2012, 437–441. Doi: 10.1037/a0027071.

Canli, T., J. E. Desmond, Z. Zhao, and J. D. E. Gabrieli. (2002). Sex differences in the neural basis of emotional memories. *Proceedings of the National Academy of Sciences,* 99, pp. 10789–10794.

Frantzen, L. B., R. P. Treviño, R. M. Echon, O. Garcia-Dominic, and N. DiMarco. (2013). Association between frequency of ready-to-eat cereal consumption, nutrient intakes, and body mass index in fourth- to sixth- grade low-income minority children. *Journal of the Academy of Nutrition and Dietetics* 113(4), pp. 511–519.

Hurt, R., L. Dale, P. Fredrickson, C. Caldwell, G. Lee, K. Offord, G. Lauger, Z. Marusic, L. Neese, and T. Lundberg. (1994). Nicotine patch therapy for smoking cessation combined with physician advice and nurse follow-up. *Journal of the American Medical Association* 271, no. 8, pp. 595–600.

Maisonneuve, H. (January 1994). Peer review congress. *European Science Editing,* no. 51, pp. 7–8.

Olds, D. L., C. R. Henderson, Jr., and R. Tatelbaum. (1994). Intellectual impairment in children of women who smoke cigarettes during pregnancy. *Pediatrics* 93, no. 2, pp. 221–227.

Passell, Peter. (11 April 1994). Probability experts may decide Pennsylvania vote. *New York Times,* p. A15.

Rauscher, F. H., G. L. Shaw, and K. N. Ky. (14 October 1993). Music and spatial task performance. *Nature* 365, p. 611.

Shaman, Paul. (28 November 1994). Personal communication.

Siegel, J. M. (1993). Companion animals: In sickness and in health. *Journal of Social Issues,* 49, no. 1, pp. 157–167.

Slutske, W. S., T. M. Piasecki and E. E. Hunt-Carter. (2003). Development and initial validation of the Hangover Symptoms Scale: Prevalence and correlates of hangover symptoms in college students. *Alcoholism: Clinical and Experimental Research* 27, 1442–1450.

Welsh, Jennifer. (2012). Morning people are happier than night owls, study suggests, http://www.today.com/health/morning-people-are-happier-night-owls-study-suggests-1C9381787?franchiseSlug=todayhealthmain, 11 June, 2012, accessed July 15, 2013.

Yamada, T., K. Hara, H. Umematsu, and T. Kadowaki. (2013). Male pattern baldness and its association with coronary heart disease: a meta-analysis. *British Medical Journal Open,* 3: e002537. Doi:10.1136/bmjopen-2012-002537.

CHAPTER 24

Significance, Importance, and Undetected Differences

Thought Questions

1. Which do you think is more informative when you are given the results of a study: a confidence interval or a *p*-value? Explain.

2. Suppose you were to read that a new study based on 100 men had found that there was *no difference* in heart attack rates for men who exercised regularly and men who did not. What would you suspect was the reason for that finding? Do you think the study found *exactly* the same rate of heart attacks for the two groups of men?

3. An example in Chapter 23 used the results of a public opinion poll to conclude that a majority of Americans did not think Bill Clinton had the honesty and integrity they expected in a president. Would it be fair reporting to claim that "significantly fewer than 50% of American adults in 1994 thought Bill Clinton had the honesty and integrity they expected in a president"? Explain.

4. When reporting the results of a study, explain why a distinction should be made between "statistical significance" and "significance," as the term is used in ordinary language.

5. Remember that a type 2 error is made when a study fails to find a relationship or difference when one actually exists in the population. Is this kind of error more likely to occur in studies with large samples or with small samples? Use your answer to explain why it is important to learn the size of a study that claims to have found no relationship or difference.

24.1   Real Importance versus Statistical Significance

By now, you should realize that a statistically significant relationship or difference does not necessarily mean an important one. Further, a result that is "significant" in the statistical meaning of the word may *not* be "significant" in the common meaning of the word. Whether the results of a test are statistically significant or not, it is helpful to examine a confidence interval so that you can determine the *magnitude* of the effect. From the width of the confidence interval, you will also learn how much uncertainty there was in the sample results. For instance, from a confidence interval for a proportion, we can learn the "margin of error."

EXAMPLE 24.1   Was President Clinton That Bad?

In the previous chapter, we examined the results of a *Newsweek* poll that asked the question: "From everything you know about Bill Clinton, does he have the honesty and integrity you expect in a president?" (16 May 1994, p. 23). The poll surveyed 518 adults, and 233, or 45% of them, said yes. There were 238 no answers (46%), and the rest were not sure. Using a hypothesis test, we determined that the proportion of the population who would answer yes to that question was statistically significantly less than half. From this result, would it be fair to report that "significantly less than 50% of all American adults in 1994 thought Bill Clinton had the honesty and integrity they expected in a president"?

What the Word *Significant* Implies   The word *significant* as used in the quote implies that the proportion who felt that way was *much less* than 50%. Using our computations from Chapter 23 and methods from Chapter 20, let's compute a 95% confidence interval for the true proportion who felt that way:

$$95\% \text{ confidence interval} = \text{sample value} \pm 2 \text{ (standard error)}$$
$$= 0.45 \pm 2(0.022) = 0.45 \pm 0.044 = 0.406 \text{ to } 0.494$$

Therefore, it could be that the true percentage is as high as 49.4%! Although that value is less than 50%, it is certainly not "significantly less" in the usual, nonstatistical meaning of the word.

In addition, if we were to repeat this exercise for the proportion who answered no to the question, we would reach the opposite conclusion. Of the 518 respondents, 238, or 46%, answered no when asked, "From everything you know about Bill Clinton, does he have the honesty and integrity you expect in a president?" If we construct the test as follows:

Null hypothesis: The population proportion who would have answered no was 0.50.

Alternative hypothesis: The population proportion who would have answered no was less than 0.50.

then the test statistic and *p*-value would be -1.82 and 0.034, respectively. Therefore, we would also accept the alternative hypothesis that less than a majority would answer no to the question. In other words, we have now found that less than a majority would have answered yes and less than a majority would have answered no to the question.

The Importance of Learning the Exact Results   The problem is that only 91% of the respondents gave a definitive answer. The rest of them had no opinion. Therefore, it would be misleading to focus only on the yes answers or only on the no answers without also reporting the percentage who were undecided. This example illustrates again the importance of learning *exactly* what was measured and what results were used in a confidence interval or hypothesis test.   ■

EXAMPLE 24.2   **Is Aspirin Worth the Effort?**

In Case Study 1.2, we examined the relationship between taking an aspirin every other day and incidence of heart attack. In testing the null hypothesis (there is no relationship between taking aspirin and incidence of heart attacks) versus the alternative (there is a relationship), the chi-square (test) statistic is over 25. Recall that a chi-square statistic over 3.84 would be enough to reject the null hypothesis at the standard 0.05 level of significance. In fact, the *p*-value for the test is less than 0.00001.

The Magnitude of the Effect   These results leave little doubt that there is a strongly statistically significant relationship between taking aspirin and incidence of heart attack. However, the test statistic and *p*-value do not provide information about the *magnitude* of the effect. And remember, the *p*-value does not indicate the probability that there is a relationship between the two variables. It indicates the probability of observing a sample with such a strong relationship, *if* there is *no* relationship between the two variables in the population. Therefore, it's important to know the *extent* to which aspirin is related to heart attack outcome.

Representing the Size of the Effect   The data show that the rates of heart attack were 9.4 per 1000 for the group who took aspirin and 17.1 per 1000 for those who took the placebo. Thus, there is a difference of slightly less than 8 people per 1000, or about one less heart attack for every 125 individuals who took aspirin. Therefore, if all men who are similar to those in the study were to take an aspirin every other day, the results indicate that out of every 125 men, 1 less would have a heart attack than would otherwise have been the case. Another way to represent the size of the effect is to note that the aspirin group had just over half as many heart attacks as the placebo group, indicating that aspirin could cut someone's risk almost in half. The original report in which these results were presented gave the relative risk as 0.53, with a 95% confidence interval extending from 0.42 to 0.67.

Whether that difference is large enough to convince a given individual to start taking that much aspirin is a personal choice. However, being told only the fact that there is very strong statistical evidence for a relationship between taking aspirin and incidence of heart attack does not, by itself, provide the information needed to make that decision. ■

24.2   The Role of Sample Size in Statistical Significance

If the sample size is large enough, almost any null hypothesis can be rejected. This is because there is almost always a slight relationship between two variables, or a difference between two groups, and if you collect enough data, you will find it.

TABLE 24.1	Age at First Birth and Breast Cancer		

		Developed Breast Cancer?		
		Yes	**No**	**Total**
First child before age 25?	Yes	65	4475	4540
	No	31	1597	1628
	Total	96	6072	6168

Source: Carter et al., 1989, cited in Pagano and Gauvreau, 1993.

EXAMPLE 24.3 How the Same Relative Risk Can Produce Different Conclusions

Consider an example discussed in Chapter 13 determining the relationship between breast cancer and age at which a woman had her first child. The results are shown in Table 24.1. The chi-square statistic for the results (found in Example 13.4) is 1.746 with *p*-value of 0.19. Therefore, we would not reject the null hypothesis. In other words, we have not found a statistically significant relationship between age at which a woman had her first child and the subsequent development of breast cancer. This is despite the fact that the relative risk calculated from the data is 1.33.

Now suppose a larger sample size had been used and the same pattern of results had been found. In fact, suppose three times as many women had been sampled, but the relative risk was still found to be 1.33. In that case, the chi-square statistic would also be increased by a factor of 3. Thus, we would find a chi-square statistic of 5.24 with *p*-value of 0.02! For the same pattern of results, we would now declare that there *is* a relationship between age at which a woman had her first child and the subsequent development of breast cancer. Yet, the reported relative risk would still be 1.33, just as in the earlier result. ∎

24.3 No Difference versus No Statistically Significant Difference

As we have seen, whether the results of a study are statistically significant can depend on the sample size. In the previous section, we found that the results in Example 24.3 were not statistically significant. We showed, however, that a larger sample size with the same pattern of results would have yielded a statistically significant finding.

There is a flip side to that problem. If the sample size is too small, an important relationship or difference can go undetected. In that case, we would say that the *power* of the test is too low. Remember that the power of a test is the probability of making the correct decision when the alternative hypothesis is true. But the null hypothesis is the status quo and is assumed to be true unless the sample values deviate from it

TABLE 24.2   Aspirin, Placebo, and Heart Attacks for a Smaller Study

	Heart Attack	No Heart Attack	Total	Percent Heart Attacks	Rate per 1000
Aspirin	14	1486	1500	0.93	9.3
Placebo	26	1474	1500	1.73	17.3
Total	40	2960	3000		

enough to convince us that chance alone cannot reasonably explain the deviation. If we don't collect enough data (even if the alternative hypothesis is true), we may not have enough evidence to convincingly rule out the null hypothesis. In that case, a relationship that really does exist in a population may go undetected in the sample.

EXAMPLE 24.4   All That Aspirin Paid Off

The relationship between aspirin and incidence of heart attacks was discovered in a long-term project called the Physicians' Health Study (Physicians' Health Study Research Group, 1988). Fortunately, the study included a large number of participants (22,071)—or the relationship may never have been found.

   The chi-square statistic for the study was 25.01, highly statistically significant. But suppose the study had only been based on 3000 participants, still a large number. Further, suppose that approximately the same pattern of results had emerged, with about 9 heart attacks per 1000 in the aspirin condition and about 17 per 1000 in the placebo condition. The result using the smaller sample would be as shown in Table 24.2. For this table, we find the chi-square statistic is 3.65, with a p-value of 0.06. This result is *not* statistically significant using the usual criterion of requiring a p-value of 0.05 or less. Thus, we would have to conclude that there is not a statistically significant relationship between taking aspirin and incidence of heart attacks.   ■

EXAMPLE 24.5   Important, But Not Significant, Differences in Salaries

A number of universities have tried to determine whether male and female faculty members with equivalent seniority earn equivalent salaries. A common method for determining this is to use the salary and seniority data for men to find a regression equation to predict expected salary when seniority is known. The equation is then used to predict what each woman's salary should be, given her seniority; this is then compared with her actual salary. The differences between actual and predicted salaries are next averaged over all women faculty members to see if, on average, they are higher or lower than they would be if the equation based on the men's salaries worked for them.

   Tomlinson-Keasey and colleagues (1994) used this method to study salary differences between male and female faculty members at the University of California at Davis. They divided faculty into 11 separate groups, by subject matter, to make comparisons more useful.

In each of the 11 groups, the researchers found that the women's actual pay was lower than what would be predicted from the regression equation, and they concluded that the situation should be investigated further. However, for some of the subject matter groups, the difference found was not statistically significant. For this reason, the researchers' conclusion generated some criticism.

Let's look at how large a difference would have had to exist for the study to be statistically significant. We use the data from the humanities group as an example. There were 92 men and 51 women included in that analysis. The mean difference between men's and women's salaries, after accounting for seniority and years since Ph.D., was $3612.

If we were to assume that the data came from some larger population of faculty members and test the null hypothesis that men and women were paid equally in the population, then the p-value for the test would be 0.08. Thus, a statistically naive reader might conclude that no problem exists because the study found no statistically significant difference between average salaries for men and for women, adjusted for seniority.

Because of the natural variability in salaries, even after adjusting for seniority, the sample means would have to differ by over $4000 per year for samples of this size to be able to declare the difference to be *statistically significant*. (To put these numbers into better perspective, note that the inflation-adjusted amount in 2013 would be closer to $7000.)

The conclusion that there is not a statistically significant difference between men's and women's salaries does not imply that there is not an important difference. It simply means that the natural variability in salaries is so large that a very large difference in means would be required to achieve statistical significance.

As one student suggested, the male faculty who complained about the study's conclusions because the differences were not statistically significant should donate the "nonsignificant" amount of $3612 to help a student pay fees for the following year. ■

| CASE STUDY 24.1 | Seen a UFO? You May Be Healthier Than Your Friends |

A survey of 5947 adult Americans taken in the summer of 1991 (Roper Organization, 1992) found that 431, or about 7%, reported having seen a UFO. Some authors have suggested that people who make such claims are probably psychologically disturbed and prone to fantasy. Nicholas Spanos and his colleagues (1993) at Carleton University in Ottawa, Canada, decided to test that theory. They recruited 49 volunteers who claimed to have seen or encountered a UFO. They found the volunteers by placing a newspaper ad that read, "Carleton University researcher seeks adults who have seen UFOs. Confidential" (Spanos et al., 1993, p. 625).

Eighteen of these volunteers, who reported seeing only "lights or shapes in the sky that they interpreted as UFOs" (p. 626) were placed in the "UFO nonintense" group. The remaining 31, who reported more complex experiences, were placed in the "UFO intense" group.

For comparison, 74 students and 53 community members were included in the study. The community members were recruited in a manner similar to the UFO groups, except that the ad read "for personality study" in place of "who have seen UFOs." The students received credit in a psychology course for participating.

All participants were given a series of tests and questionnaires. Attributes measured were "UFO beliefs, Esoteric beliefs, Psychological health, Intelligence, Temporal lobe lability [to see if epileptic type episodes could account for the experiences], Imaginal propensities [such as Fantasy proneness], and Hypnotizability" (p. 626).

The *New York Times* reported the results of this work (Sullivan, 29 November 1993) with the headline "Study finds no abnormality in those reporting UFOs." The article described the results as follows:

> *A study of 49 people who have reported encounters with unidentified flying objects, or UFOs, has found no tendency toward abnormality, apart from a previous belief that such visitations from beyond the earth do occur. . . . The tests [given to the participants] included standard psychological tests used to identify subjects with various mental disorders and assess their intelligence. The UFO group proved slightly more intelligent than the others.*
>
> Source: Sullivan, "Study finds no abnormality in those reporting UFOs," November 29, 1993, *New York Times,* p. 37.

Reading the *Times* report would leave the impression that there were no statistically significant differences found in the psychological health of the groups. In fact, that is *not* the case. On many of the psychological measures, the UFO groups scored statistically significantly *better* if a two-tailed test were to be used, meaning they were healthier than the student and community groups. The null hypothesis for this study was that there are no population differences in mean psychological scores for those who have encountered UFOs compared with those who have not. But the alternative hypothesis of interest to Spanos and his colleagues was *one-sided:* the speculation that UFO observers are *less* healthy. The data, indicating that they might be healthier, was not consistent with that alternative hypothesis, so the null hypothesis could not be rejected. Here is how Spanos and colleagues (1993) discussed the results in their report:

> *The most important findings indicate that neither of the UFO groups scored lower on any measures of psychological health than either of the comparison groups. Moreover, both UFO groups attained higher psychological health scores than either one or both of the comparison groups on five of the psychological health variables. In short, these findings provide no support whatsoever for the hypothesis that UFO reporters are psychologically disturbed.* (p. 628)

In case you are curious, the UFO nonintense group scored statistically significantly higher than any of the others on the IQ test, and there were no significant differences in fantasy proneness, other paranormal experiences, or temporal lobe lability. The UFO groups scored better (healthier) on scales measuring self-esteem, schizophrenia, perceptual aberration, stress, well-being and agression. ∎

24.4   Multiple Tests, Multiple Comparisons, and False Positives

There is one more problem with statistical inference that is important to consider when interpreting the results of a research study. Many studies include multiple hypothesis tests and/or confidence intervals. With each test and interval, there is a chance that an erroneous conclusion will be made, so it stands to reason that if multiple tests and intervals are considered, the chance of an erroneous conclusion will increase. This problem of **multiple testing** (conducting many hypothesis tests) or **multiple comparisons** (making many comparisons through either confidence intervals or hypothesis tests) should be acknowledged and accounted for when considering p-values and confidence levels.

As a simple example, suppose that 100 independent 95% confidence intervals are to be computed. How many of them will not cover the truth? Because they are independent and the probability of not covering the truth is 0.05 for each interval, we should expect that about $100 \times (.05) = 5$ will not cover the true population value. The same reasoning holds for hypothesis tests. Remember that when the null hypothesis is true, the level of significance is also the probability of making a Type 1 error. If 100 independent tests are conducted using a significance level of 0.05, and if all of the null hypotheses are true, on average we should expect five false positives in which a type 1 error is made, and the null hypothesis is rejected even though it is true.

The situation is rarely as simple as the illustrations just given because the tests and confidence intervals done in one study generally use the same individuals and thus are not independent. So it is almost impossible to ascertain the actual probability of a false positive. However, statisticians have developed methods for handling multiple comparisons. The simplest (and most conservative) is called the **Bonferroni method**. The method proceeds by dividing up the significance level (or confidence level) and apportioning it across tests (or confidence intervals). For example, if a total type 1 error probability of 0.05 is desired and five tests are to be done, each test is done using a significance level of $0.05/5 = 0.01$. This method controls the overall probability of making at least one type 1 error to be the sum of the individual significance levels used for each test.

Unfortunately, the media rarely mention how many tests and confidence intervals are done in one study, and the focus is often placed on surprising outcomes. It is thus very easy for false positive results to gain wide media attention. This is one reason that replication of studies is important in science. A false positive is unlikely to be replicated in multiple studies. The next case study provides an example of multiple testing, in which much media attention was given to a study purporting to show that eating breakfast cereal before becoming pregnant could increase the chances of having a boy.

CASE STUDY 24.2   **Did Your Mother's Breakfast Determine Your Sex?**

Sources: (http://www.cbsnews.com/stories/2008/04/22/health/webmd/main4036102.shtml)
Mathews et al., 2008 and 2009
Young et al., 2009

You've probably heard that "you are what you eat," but did it ever occur to you that you might be *who* you are because of what your mother ate? A study published in 2008 by the British Royal Society seemed to find just that. The researchers reported that mothers who ate breakfast cereal prior to conception were more likely to have boys than mothers who did not (Mathews et al., 2008). But 9 months later, just enough time for the potential increased cereal sales to have produced a plethora of little baby boys, another study was published that dashed cold milk on the original claim (Young et al., 2009). The dispute was based on the problem of multiple testing.

The authors of the original cereal study had asked 740 women about 133 different foods they might have eaten just before getting pregnant, and then again during the pregnancy. They found that 59% of the women who consumed breakfast cereal daily before getting pregnant gave birth to a boy, compared to only 43% of the women who rarely or never ate cereal. The difference was highly statistically significant. Almost none of the other foods tested showed a statistically significant difference in the ratio of male to female births, and none of them showed a difference that large.

Remember that the more differences that are tested, the more likely it is that one of them will be a false positive, if in fact there are no real differences at all in the population. The criticism by Young et al. was based on this idea. Testing 133 food items at two different time periods resulted in a total of 266 hypothesis tests. With that many tests, it is quite likely that statistically significant differences would show up, even at a significance level of 0.01 or less.

The authors of the original study defended their work (Mathews et al., 2009). They noted that they only tested the individual food items after an initial test based on total preconception calorie consumption showed a difference in male and female births. They found that 56% of the mothers in the top third of calorie consumption had boys, compared with only 45% of the mothers in the bottom third of calorie consumption. That was one of only two initial tests they did; the other had to do with vitamin intake. With only two tests, it is much less likely that either of them would be a false positive. Unfortunately, the media found the cereal connection to be the most interesting result in the study, and that's what received overwhelming publicity. The best way to resolve the debate, as in most areas of science, is to ask the same questions in a new study and see if the results are consistent. The authors of the original study have stated their intention to do that. ∎

24.5 A Summary of Warnings and Key Concepts

From this discussion, you should realize that you can't simply rely on news reports to determine what to conclude from the results of studies. In particular, you should heed the following warnings:

1. If the word *significant* is used to try to convince you that there is an important effect or relationship, determine if the word is being used in the usual sense or in the statistical sense only.

2. If a study is based on a very large sample size, relationships found to be statistically significant may not have much practical importance.

3. If you read that "no difference" or "no relationship" has been found in a study, try to determine the sample size used. Unless the sample size was large, remember that an important relationship may well exist in the population but that not enough data were collected to detect it. In other words, the test could have had very low power.

4. If possible, learn what confidence interval accompanies the hypothesis test, if any. Even then you can be misled into concluding that there is no effect when there really is, but at least you will have more information about the magnitude of the possible difference or relationship.

5. Try to determine whether the test was one-sided or two-sided. If a test is one-sided, as in Case Study 24.1, and details aren't reported, you could be misled into thinking there would be no significant difference in a two-sided test, when in fact there was one in the direction opposite to that hypothesized.

6. Remember that the decision to do a one-sided test must be made *before* looking at the data, based on the research question. Using the same data to both generate and test the hypotheses is cheating. A one-sided test done that way will have a *p*-value smaller than it should, making it easier to reject the null hypothesis.

7. Beware of *multiple testing* and *multiple comparisons*. Sometimes researchers perform a multitude of tests, and the reports focus on those that achieved statistical significance. If all of the null hypotheses tested are true, then over the long run, about 1 in 20 tests should achieve statistical significance just by chance. Beware of reports in which it is evident that many tests were conducted, but in which results of only one or two are presented as "significant."

Exercises

Exercises with numbers divisible by 3 (3, 6, 9, etc.) are included in the Solutions at the back of the book. They are marked with an asterisk (*).

1. Explain why it is important to learn what sample size was used in a study for which "no difference" was found.

2. Suppose that you were to read the following news story: "Researchers compared a new drug to a placebo for treating high blood pressure, and it seemed to work. But the researchers were concerned because they found that significantly more people got headaches when taking the new drug than when taking the placebo. Headaches were the only problem out of the 20 possible side effects the researchers tested." Do you think the researchers are justified in thinking the new drug would cause more headaches in the population than the placebo would? Explain.

*3. An advertisement for Claritin, a drug for seasonal nasal allergies, made this claim: "Clear relief without drowsiness. In studies, the incidence of drowsiness was similar to placebo" (*Time*, 6 February 1995, p. 43). The advertisement also reported that 8% of the 1926 Claritin takers and 6% of the 2545 placebo takers reported drowsiness as a side effect. A one-sided test of whether a higher proportion of Claritin takers than placebo takers would experience drowsiness in the population results in a *p*-value of about 0.005.

*a. Can you conclude that the incidence of drowsiness for the Claritin takers is statistically significantly higher than that for the placebo takers?

*b. Does the answer to part (a) contradict the statement in the advertisement that the "incidence of drowsiness was similar to placebo"? Explain.

4. Use the example in Exercise 3 to discuss the importance of making the distinction between the common use and the statistical use of the word *significant*.

5. An article in *Time* magazine (Gorman, 6 February 1995) reported that an advisory panel recommended that the Food and Drug Administration (FDA) allow an experimental AIDS vaccine to go forward for testing on 5000 volunteers. The vaccine was developed by Jonas Salk, who developed the first effective polio vaccine. The AIDS vaccine was designed to boost the immune system of HIV-infected individuals and had already been tested on a small number of patients, with mixed results but no apparent side effects.

a. In making its recommendation to the FDA, the advisory panel was faced with a choice similar to that in hypothesis testing. The null hypothesis was that the vaccine was not effective and therefore should not be tested further, whereas the alternative hypothesis was that it might have some benefit. Explain the consequences of a type 1 and a type 2 error for the decision the panel was required to make.

b. The chairman of the panel, Dr. Stanley Lemon, was quoted as saying, "I'm not all that excited about the data I've seen . . . [but] the only way the concept is going to be laid to rest . . . is really to try [the vaccine] in a large population" (p. 53). Explain why the vaccine should be tested on a larger group, when it had not proven effective in the initial tests on a small group.

*6. In Example 13.2, we revisited data from Case Study 6.3, regarding testing to see if there was a relationship between gender and driving after drinking. We found that we could not rule out chance as an explanation for the sample data; the chi-square statistic was 1.637, and the *p*-value was 0.20. Now suppose that the sample contained three times as many drivers, but the proportions of males and females who drank before driving were still 16% and 11.6%, respectively.

*a. What would be the value of the chi-square statistic for this hypothetical larger sample? (*Hint:* See Example 24.3.)

*b. The *p*-value for the test based on the larger sample would be about 0.03. Restate the hypotheses being tested and explain which one you would choose on the basis of this larger hypothetical sample.

*c. In both the original test and the hypothetical one based on the larger sample, the probability of making a type 1 error if there really is no relationship is 5%. Assuming there is a relationship in the population, would the power of the test—that is, the probability of correctly finding the relationship—be higher, lower, or the same for the sample three times larger compared with the original sample size?

7. *New Scientist* (Mestel, 12 November 1994) reported a study in which psychiatrist Donald Black used the drug fluvoxamine to treat compulsive shoppers:

In Black's study, patients take the drug for eight weeks, and the effect on their shopping urges is monitored. Then the patients are taken off the drug and watched for another month. In the seven patients examined so far, the results are clear and dramatic, says

Black: the urge to shop and the time spent shopping decrease markedly. When the patient stops taking the drug, however, the symptoms slowly return (p. 7).

a. Explain why it would have been preferable to have a double-blind study, in which shoppers were randomly assigned to take fluvoxamine or a placebo.

b. What are the null and alternative hypotheses for this research?

c. Can you make a conclusion about the hypotheses in part (b) on the basis of this report? Explain.

8. In reporting the results of a study to compare two population means, explain why researchers should report each of the following:

a. A confidence interval for the difference in the means.

b. A *p*-value for the results of the test, as well as whether it was one- or two-sided.

c. The sample sizes used.

d. The number of separate tests they conducted during the course of the research.

*9. The top story in *USA Today* on December 2, 1993, reported that "two research teams, one at Harvard and one in Germany, found that the risk of a heart attack during heavy physical exertion . . . was two to six times greater than during less strenuous activity or inactivity" (Snider, 2 December 1993, p. 1A). It was implicit, but not stated, that these results were for those who do not exercise regularly. There is a confidence interval reported in this statement. Explain what it is and what characteristic of the population it is measuring.

10. The news story quoted in Exercise 9 also stated that "frequent exercisers had slight or no increased risk." (In other words, frequent exercisers had slight or no increased risk of heart attack when they took on a strenuous activity compared with when they didn't.) Do you think that means the relative risk was actually 1.0? If not, discuss what the statement really does mean in the context of what you have learned in this chapter.

11. We have learned that the probability of making a type 1 error when the null hypothesis is true is usually set at 5%. The probability of making a type 2 error when the alternative hypothesis is true is harder to find. Do you think that probability depends on the size of the sample? Explain your answer.

*12. Refer to Case Study 22.1, concerning the ganzfeld procedure for testing ESP. In earlier studies using the ganzfeld procedure, the results were mixed in terms of whether they were statistically significant. In other words, some of the experiments were statistically significant and others were not. Critics used this as evidence that there was really nothing going on, even though more than 5% of the experiments were successful. Typical sample sizes were from 10 to 100 participants. Give a statistical explanation for why the results would have been mixed.

13. When the Steering Committee of the Physicians' Health Study Research Group (1988) reported the results of the effect of aspirin on heart attacks, committee members also reported the results of the same aspirin consumption, for the same sample, on strokes. There were 80 strokes in the aspirin group and only 70 in the placebo group. The relative risk was 1.15, with a 95% confidence interval ranging from 0.84 to 1.58.

a. What value for relative risk would indicate that there is no relationship between taking aspirin and having a stroke? Is that value contained in the confidence interval?

b. Set up the appropriate null and alternative hypotheses for this part of the study. The original report gave a *p*-value of 0.41 for this test. What conclusion would be made on the basis of that value?

c. Compare your results from parts (a) and (b). Explain how they are related.

d. There was actually a higher percentage of strokes in the group that took aspirin than in the group that took a placebo. Why do you think this result did not get much press coverage, whereas the result indicating that aspirin reduces the risk of heart attacks did get substantial coverage?

14. In the study by Lee Salk, reported in Case Study 1.1, he found that infants who listened to the sound of a heartbeat in the first few days of life gained more weight than those who did not. In searching for potential explanations, Salk wrote the following. Discuss Salk's conclusion and the evidence he provided for it. Do you think he effectively ruled out food intake as being important?

> *In terms of actual weight gain the heartbeat group showed a median gain of 40 grams; the control group showed a median loss of 20 grams. There was no significant difference in food intake between the two groups. . . . There was crying 38 percent of the time in the heartbeat group of infants; in the control group one infant or more cried 60 percent of the time. . . . Since there was no difference in food intake between the two groups, it is likely that the weight gain for the heartbeat group was due to a decrease in crying* (Salk, May 1973, p. 29).

*15. The authors of the report in Case Study 24.1, comparing the psychological health of UFO observers and nonobservers, presented a table in which they compared the four groups of volunteers on each of 20 psychological measures. For each of the measures, the null hypothesis was that there were no differences in population means for the various types of people on that measure. If there truly were no population differences, and if all of the measures were independent of each other, for how many of the 20 measures would you expect the null hypothesis to be rejected, using the 0.05 criterion for rejection? Explain how you got your answer.

16. An advertisement for NordicTrack, an exercise machine that simulates cross-country skiing, claimed that "in just 12 weeks, research shows that *people who used a NordicTrack lost an average of 18 pounds.*" Forgetting about the questions surrounding how such a study might have been conducted, what additional statistical information would you want to know about the study and its results before you could come to a reasonable conclusion?

17. Explain why it is not wise to *accept* a null hypothesis.

*18. Refer to the list of warnings on pages 527–528. Explain which ones should be of concern if the sample size(s) for a test are large.

19. Refer to the list of warnings on pages 527–528. Explain which ones should be of concern if the sample size(s) for a test are small.

20. Now that you understand the reasoning behind making inferences about populations based on samples (confidence intervals and hypothesis tests), explain why these methods require the use of random, or at least representative, samples instead of convenience samples.

*21. Would it be easier to reject hypotheses about populations that had a lot of natural variability in the measurements or a little variability in the measurements? Explain.

22. Refer to the sentence in warning 7 on p. 528 that states: "If all of the null hypotheses tested are true, then over the long run about 1 in 20 tests should achieve statistical significance just by chance." Is that statement the same thing as saying that the null hypothesis is likely to be true in about 1 out of 20 tests that have achieved statistical significance? Explain.

 Exercises 23 to 29 refer to News Story and Original Source 1, "Alterations in Brain and Immune Function Produced by Mindfulness Meditation" (not available on the companion website). For this study, volunteers were randomly assigned to a meditation group (25 participants) or a control group (16 participants). The meditation group received an 8-week meditation training program. Both groups were given influenza shots, and their antibodies were measured. This measurement was taken after the meditation group had been practicing meditation for about 8 weeks. The researchers also measured brain activity, but these exercises will not explore that part of the study.

23. One of the quotes in News Story 1 refers to the results of measuring antibodies to the influenza vaccine. It says:

> While everyone who participated in the study had an increased number of antibodies, the meditators had a significantly greater increase than the control group. "The changes were subtle, but statistically it was significant," says [one of the researchers] (Kotansky, 2003, p. 35).

a. Do you think it would have been fair reporting if only the first sentence of the quote had been given, without the second sentence? Explain, including what additional information you learned from the second sentence.

b. Based on this quote, what null hypothesis do you think was being tested? Make sure you use correct terminology regarding populations and samples.

*24. Refer to the previous exercise.

*a. Do you think the alternative hypothesis used by the researchers was one-sided or two-sided? Explain.

*b. Do you think the researchers would be justified in specifying a one-sided alternative hypothesis in this situation? Explain why or why not.

25. In Original Source 1, the researchers addressed some limitations with the study. One of them was:

> First, there was a relatively small number of subjects who participated and this limited our statistical power. A number of our hypothesized effects were in the predicted direction, but failed to reach significance (Davidson et al., p. 569).

Explain what is meant by the first sentence of the quote.

26. Refer to the quote in Exercise 25. Explain what is meant by the second sentence of the quote.

*27. One quote from Original Source 1 compared the ages of the participants who were randomly assigned to the two groups. The researchers

reported: "Average age of subjects was 36 years and did not differ between [the two] groups" (Davidson et al., p. 565). Do you think that the average age of the participants in the two groups was *exactly* the same? Explain.

28. Rewrite the quote in the previous exercise to convey what you think the researchers meant to say.

29. Refer to items 1 and 2 in Section 24.5 "A Summary of Warnings." Explain whether or not they are likely to apply to results from this study.

Exercises 30 and 31 refer to News Story 19 and Additional News Story 19 on the companion website. The following quote is from Additional News Story 19:

> They found that adolescents who were romantically involved during the year experienced a significantly larger increase in symptoms of depression than adolescents who were not romantically involved. They also found that depression levels of romantically involved girls increased more sharply than that of romantically involved boys, especially among younger adolescents (Lang, 2001).

*30. Read News Story 19, "Young romance may lead to depression, study says" and use it to answer these questions.

*a. How many participants were there in this study?

*b. The story actually reports the magnitude of the difference in depression levels for romantically involved and uninvolved teens. What was it?

*c. Given your answers to parts (a) and (b), which two of the seven warnings in Section 24.5 apply to the quote preceding this exercise, from Additional News Story 19?

*d. Do you think the word *significantly* in the quote is being used in the statistical or the English sense of the word?

31. Read News Story 19 and Additional News Story 19. Based on the information contained in them, rewrite and expand the information in the quote preceding Exercise 30 so that it would not be misleading to someone with no training in statistics.

Mini-Projects

1. Find a news story that you think presents the results of a hypothesis test in a misleading way. Explain why you think it is misleading. Rewrite the appropriate part of the article in a way that you consider to be not misleading. If necessary, find the original journal report of the study so you can determine what details are missing from the news account.

2. Find two journal articles, one that reports on a statistically significant result and one that reports on a nonsignificant result. Discuss the role of the sample size in the determination of statistical significance, or lack thereof, in each case. Discuss whether you think the researchers would have reached the same conclusion if they had used a smaller or larger sample size.

References

Carter, C. L., D. Y. Jones, A. Schatzkin, and L. A. Brinton. (1989). A prospective study of reproductive, familial, and socioeconomic risk factors for breast cancer using NHANES I data, *Public Health Reports* 104, January–February, pp. 45–49.

Gorman, Christine. (6 February 1995). Salk vaccine for AIDS. *Time*, p. 53.

Mathews, F., P. J. Johnson, and A. Neil. (2008). You are what your mother eats: evidence for maternal preconception diet influencing foetal sex in humans. *Proceedings of The Royal Society B,* 275, pp. 1661–1668.

Mathews, F., P. Johnson, and A. Neil. (2009). Reply to Comment by Young et al. *Proceedings of The Royal Society B,* 276, pp. 1213–1214.

Mestel, Rosie. (12 November 1994). Drug brings relief to big spenders. *New Scientist,* no. 1951, p. 7.

Pagano, M. and K. Gauvreau. (1993). *Principles of biostatistics.* Belmont, CA: Duxbury Press.

Physicians' Health Study Research Group. Steering Committee. (28 January 1988). Preliminary report: Findings from the aspirin component of the ongoing Physicians' Health Study. *New England Journal of Medicine* 318, no. 4, pp. 262–264.

Roper Organization. (1992). *Unusual personal experiences: An analysis of the data from three national surveys.* Las Vegas: Bigelow Holding Corp.

Salk, Lee. (May 1973). The role of the heartbeat in the relations between mother and infant. *Scientific American,* pp. 26–29.

Snider, M. (2 December 1993). "Weekend warriors" at higher heart risk. *USA Today,* p. 1A.

Spanos, N. P., P. A. Cross, K. Dickson, and S. C. DuBreuil. (1993). Close encounters: An examination of UFO experiences. *Journal of Abnormal Psychology* 102, no. 4, pp. 624–632.

Sullivan, Walter. (29 November 1993). Study finds no abnormality in those reporting U.F.O.s. *New York Times,* p. B7.

Tomlinson-Keasey, C., J. Utts, and J. Strand. (1994). Salary equity study at UC Davis, 1994. Technical Report, Office of the Provost, University of California at Davis.

Young, S. S., H. Bang, and K. Oktay. (2009). Cereal-induced gender selection? Most likely a multiple testing false positive." *Proceedings of The Royal Society B,* 276, pp. 1211–1212.

CHAPTER 25

Meta-Analysis: Resolving Inconsistencies across Studies

Thought Questions

1. Suppose a new study involving 14,000 participants has found a relationship between a particular food and a certain type of cancer. The report on the study notes that "past studies of this relationship have been inconsistent. Some of them have found a relationship, but others have not." What might be the explanation for the inconsistent results from the different studies?

2. Suppose 10 similar studies, all on the same kind of population, have been conducted to determine the relative risk of heart attack for those who take aspirin and those who don't. To get an overall picture of the relative risk, we could compute a separate confidence interval for each study or combine all the data to create one confidence interval. Which method do you think would be preferable? Explain.

3. Suppose two studies have been done to compare surgery versus relaxation for sufferers of chronic back pain. One study was done at a back-care specialty clinic and the other at a suburban medical center. The result of interest in each study was the relative risk of further back problems following surgery versus following relaxation training. To get an overall picture of the relative risk, we could compute a separate confidence interval for each study or combine the data to create one confidence interval. Which method do you think would be preferable? Explain.

4. Refer to Thought Questions 2 and 3. If two or more studies have been done to measure the same relative risk, give one reason why it would be better to combine the results and one reason why it would be better to look at the results separately for each study.

25.1   The Need for Meta-Analysis

The obvious questions were answered long ago. Most research today tends to focus on relationships or differences that are not readily apparent. For example, plants that are obviously poisonous have been absent from our diets for centuries. But we may still consume plants (or animals) that contribute to cancer or other diseases over the long term. The only way to discover these kinds of relationships is through statistical studies.

Because most of the relationships we now study are small or moderate in size, researchers often fail to find a statistically significant result. As we learned in the last chapter, the number of participants in a study is a crucial factor in determining whether the study finds a "significant" relationship or difference, and many studies are simply too small to do so. As a consequence, reports are often published that appear to conflict with earlier results, confusing the public and researchers alike.

The Vote-Counting Method

One way to address this problem is to gather all the studies that have been done on a topic and try to assimilate the results. Until recently, if researchers wanted to examine the accumulated body of evidence for a particular relationship, they would find all studies conducted on the topic and simply count how many of these had found a statistically significant result. They would often discount entirely all studies that had not and then attempt to explain any remaining differences in study results by subjective assessments.

As you should realize from Chapter 24, this **vote-counting method** is seriously flawed unless the number of participants in each study is taken into account. For example, if 10 studies of the effect of aspirin on heart attacks had each been conducted on 2200 men, rather than one study on 22,000 men, none of the 10 studies would be likely to show a relationship. In contrast, as we saw in Chapter 13, the one study on 22,000 men showed a very significant relationship. In other words, the same data, had it been collected by 10 separate researchers, easily could have resulted in exactly the opposite conclusion to what was found in the one large study.

What Is Meta-Analysis?

Since the early 1980s researchers have become increasingly aware of the problems with traditional research synthesis methods. They are now more likely to conduct a *meta-analysis* of the studies. **Meta-analysis** is a collection of statistical techniques for combining studies. These techniques focus on the *magnitude of the effect* in each study, rather than on either a vote count or a subjective evaluation of the available evidence.

Quantitative methods for combining results have been available since the early 1900s, but it wasn't until 1976 that the name "meta-analysis" was coined. The seminal paper was called, "Primary, secondary, and meta-analysis of research." In that paper, Gene Glass (1976) showed how these methods could be used to synthesize the results of studies comparing treatments in psychotherapy. It was an idea whose time had come. Researchers and methodologists set about to create new techniques for combining and comparing studies, and thousands of meta-analyses were undertaken.

What Meta-Analysis Can Do

Meta-analysis has made it possible to find definitive answers to questions about small and moderate relationships by taking into account data from a number of different studies. It is not without its critics, however, and as with most statistical methods, difficulties and disasters can befall the naive user. In the remainder of this chapter, we examine some decisions that can affect the results of a meta-analysis, discuss some benefits and criticisms of meta-analysis, and look at two case studies.

25.2   Two Important Decisions for the Analyst

You have already learned how to do a proper survey, experiment, or observational study. Conducting a meta-analysis is an entirely different enterprise because it does not involve using new participants at all. It is basically a study of the studies that have already been done on a particular topic.

A variety of decisions must be made when conducting a meta-analysis, many of which are beyond the scope of our discussion. However, there are two important decisions of which you should be informed when you read the results of a meta-analysis. The answers will help you determine whether to believe the results.

When reading the results of meta-analysis, you should know

1. Which studies were included
2. Whether the results were compared or combined

Question 1: Which Studies Should Be Included?

Types of Studies

As you learned early in this book, studies are sometimes conducted by special-interest groups. Some studies are conducted by students to satisfy requirements for a Ph.D. degree but then never published in a scholarly journal. Sometimes studies are reported at conferences and published in the proceedings. Often these studies are not carefully reviewed and criticized in the same way they would be for publication in a scholarly journal. Thus, one decision that must be made before conducting a meta-analysis is whether to include all studies on a particular topic or only those that have been published in properly reviewed journals.

Timing of the Studies

Another consideration is the timing of the studies. For example, in Case Study 25.2, we review a meta-analysis that attempted to answer the question of whether mammograms should be used for the detection of breast cancer in women in their 40s. Should the analysis have included studies started many years ago, when neither the equipment nor the technicians may have been as sophisticated as they are today?

Quality Control

Some of the decisions that need to be made when deciding what studies to include focus on the quality of the individual studies. Even when lower quality studies are included, sometimes they are weighted less heavily than higher quality studies.

The Cochrane Collaboration (www.cochrane.org) is a repository of thousands of meta-analyses and systematic reviews on medical topics, with the goal of making the most comprehensive information available to health practitioners. They have a handbook that lists guidelines for conducting a review, and the section on assessing quality includes a long list of features that should be considered when determining the quality of individual studies. Here are some examples (in italics) from Table 8.5.a of the Cochrane Handbook (http://handbook.cochrane.org/) and questions they are designed to address:

- *Random sequence generation*—Was the method used to randomize participants to treatments likely to produce comparable groups?
- *Blinding of participants and personnel*—Was the experiment double blind?
- *Selective reporting*—Did the authors report all of the tests they did, or is the problem of multiple testing likely to be an undetected problem?

As an example of quality assessment used in a meta-analysis, consider a meta-analysis conducted by Eisenberg and colleagues (15 June 1993) on whether behavioral techniques such as biofeedback and relaxation were effective in lowering blood pressure in people whose blood pressure was too high. They identified a total of 857 articles for possible inclusion but used only 26 of them in the meta-analysis. In order to be included, the studies had to meet stringent criteria, including the use of one or more control conditions, random assignment into the experimental and control groups, detailed descriptions of the intervention techniques for treatment and control, and so on. They also excluded studies that used children or that used only pregnant women.

Thus, they decided to exclude studies that had some of the problems you learned about earlier in this book, such as nonrandomized treatments or no control group. Using the remaining 26 studies, they found that both the behavioral techniques and the placebo techniques (such as "sham meditation") produced effects but that there was not a significant difference between the two. Their conclusion was that "cognitive interventions for essential hypertension are superior to no therapy but not superior to credible sham techniques or self-monitoring" (p. 964). If they had included a multitude of studies that did not use placebo techniques, they may not have realized that those techniques could be as effective as the real thing. Therefore, their quality assessment was also a way of limiting the analysis, and thus the conclusions, to studies that had an adequate placebo condition.

Accounting for Differences in Quality

Some researchers believe all possible studies should be included in the meta-analysis. Otherwise, they say, the analyst could be accused of including studies that support the desired outcome and finding excuses to exclude those that don't. One solution to this problem is to include all studies but account for differences in quality in the process of the analysis. A meta-analysis of programs designed to prevent adolescents from smoking (reviewed in Example 25.3) used a compromise approach:

> *Evaluations of 94 separate interventions were included in the meta-analysis. Studies were screened for methodological rigor, and those with weaker methodology were segregated from those with more defensible methodology; major analyses focused on the latter.* (Bruvold, 1993, p. 872).

Assessing Quality

Very early in the evolution of meta-analysis, Chalmers and his co-authors (1981) constructed a generic set of criteria for deciding which papers to include in a meta-analysis and for assessing the quality of studies. Their basic idea is to have two researchers, who are blind as to who conducted the studies and what the results were, independently make decisions about the quality of each study. In other words, there should be two independent assessments about whether to include each study, and those assessments should be made without knowing how the study came out. This technique should eliminate inclusion biases based on whether the results were in a desired direction.

By 1995, Moher and colleagues (1995) had identified 25 scales and nine checklists that had been proposed for assessing quality of studies for inclusion in a meta-analysis. You can see that there are lots of choices for how to decide what studies to include and how to assess their quality. When you read the results of a meta-analysis, you should be told how the authors decided which studies to include. If there was no attempt to discount flawed studies in some way, then you should realize that the combined results may also be flawed.

Question 2: Should Results Be Compared or Combined?

In Chapters 11 and 12, you learned about the perils of combining data from separate studies. Recall that Simpson's Paradox can occur when you combine the results of two separate contingency tables into one. The problem occurs when a relationship is different for one population than it is for another, but the results for samples from the two are combined.

Meta-Analysis and Simpson's Paradox

Meta-analysis is particularly prone to the problem of Simpson's Paradox. For example, consider two studies comparing surgery to relaxation as treatments for chronic back pain. Suppose one is conducted at a back-care specialty clinic, whereas the other is conducted at a suburban medical center. It is likely that the patients with the most severe problems will seek out the specialty clinic. Therefore, relaxation training may be sufficient and preferable for the suburban medical center, but surgery may be preferable for the specialty clinic.

Populations Must Be the Same and Methods Similar

A meta-analysis should never attempt to combine the data from all studies into one hypothesis test or confidence interval unless it is clear that the same populations were sampled and similar methods were used. Otherwise, the analysis should first attempt to see if there are readily explainable differences among the results.

A technical decision that must be made is whether to use a **fixed effects** or a **random effects** model. In a fixed effects model, the assumption is that all of the studies included samples from similar populations, with a fixed but unknown magnitude of the effect being tested. (The effect of interest might be a difference in means, a relative risk, a correlation between an explanatory and response variable, and so on.) The goal is to estimate that effect.

In a random effects model, the assumption is that the population effect is different for each study, but that each one is made up of an overall effect, plus a random component specific to that study. In a random effects model, one of the goals is to estimate how much variability there is from study to study.

For example, in Case Study 25.1, we will investigate a meta-analysis of the effects of cigarette smoking on sperm density. Some of the studies included in the meta-analysis used infertility clinics as the source of participants; others used the general population. The researchers discovered that the relationship between smoking and sperm density was more variable across the studies for which the participants were from infertility clinics. They speculated that "it is possible that smoking has a greater effect on normal men since infertility clinic populations may have other reasons for the lowered sperm density" (Vine et al., January 1994, p. 41). Because of the difference, they reported results separately for the two sources of participants. If they had not, the relationship between smoking and sperm density would have been underestimated for the general population.

When you read the results of a meta-analysis, you need to ascertain whether something like Simpson's Paradox could have been a problem. If dozens of studies have been combined into one result, without any mention of whether the potential for this problem was investigated, you should be suspicious.

CASE STUDY 25.1	**Smoking and Reduced Fertility**

Sources: American Society for Reproductive Medicine, 2012.
Vine et al., 1994.

A 2012 systematic review of research on the relationship between smoking and fertility in both men and women came to the following conclusion:

> *Approximately 30% of women of reproductive age and 35% of men of reproductive age in the United States smoke cigarettes. Substantial harmful effects of cigarette smoke on fecundity and reproduction have become apparent but are not generally appreciated*" (American Society for Reproductive Medicine, 2012, p. 1400).

The article reviewed dozens of other articles showing the relationship between smoking and fertility. Although studies of this relationship must necessarily be observational (smoking cannot be randomly assigned), the article discussed how a cause and effect conclusion may be supported. The reasons were similar to those outlined in Section 11.4 of this book on "Confirming Causation," such as the existence of a dose-response relationship.

Some of the articles discussed in this systematic review consisted of meta-analyses on subgroups or for specific situations, such as in vitro fertilization. One meta-analysis (Augwood et al., 1998) found an odds ratio of 1.6 for infertility for women smokers compared to women nonsmokers. In other words, the odds of infertility to fertility for female smokers are 1.6 times what the odds are for female nonsmokers.

The situation was more complex for men, partly because it is difficult to separate out the effects of smoking from other confounding variables, such as whether the man's partner smokes. However, one meta-analysis in the early 1990s did a comprehensive review of studies that specifically investigated the relationship between smoking and sperm density. The remainder of this case study describes that meta-analysis.

Vine and colleagues (January 1994) identified 20 studies, published between January 1966 and October 1, 1992, that examined whether men who smoked cigarettes had lower sperm density than men who did not smoke. They found that most of the studies reported reduced sperm density for men who smoked, but the difference was statistically significant for only a few of the studies. Thus, in the traditional scientific review, the accumulated results would seem to be inconsistent and inconclusive.

Studies were excluded from the meta-analysis for only two reasons. One was if the study was a subset of a larger study that was already included. The other was if it was obvious that the smokers and nonsmokers differed with respect to fertility status. One study was excluded for that reason. The nonsmokers all had children, whereas the smokers were attending a fertility clinic.

A variety of factors differed among the studies. For example, only ten of the studies reported that the person measuring the sperm density was blind to the smoking status of the participants. In six of the studies, a "smoker" was defined as someone who smoked at least ten cigarettes per day. Thirteen of the studies used infertility clinics as a source of participants; seven did not. All of these factors were checked to see if they resulted in a difference in outcome.

None of these factors resulted in a difference. However, the authors were suspicious of two studies conducted on infertility clinic patients because those two studies showed a much larger relationship between smoking and sperm density than the remaining 11 studies on that same population. When those two studies were omitted, the remaining studies using infertility clinic patients showed a smaller average effect than the studies using men from the general population. The authors conducted an analysis using all of the data, as well as separate analyses on the two types of populations: those attending infertility clinics and those not attending. They omitted the two studies with larger effects when studying only the infertility clinics.

The authors found that there was indeed lower average sperm density for men who smoked. Using all of the data combined, but giving more weight to studies with larger sample sizes, they found that the reduction in sperm density for smokers compared with nonsmokers was 12.6%, with a 95% confidence interval extending from 8.0% to 17.1%. A test of the null hypothesis that the reduction in sperm density for smokers in the population is actually zero resulted in a p-value less than .0001. The estimate of reduction in sperm density for the men from the general population (excluding the fertility clinics) was even higher, at 23.3%; a confidence interval was not provided.

The authors also conducted their own study of this question using 88 volunteers recruited through a newspaper. In summarizing the findings of past reviews, their meta-analysis, and their own study of 88 men, Vine and colleagues (January 1994) illustrate the importance of meta-analysis:

> *The results of this meta-analysis indicate that smokers' sperm density is on average 13% [when studies are weighted by sample size] to 17% [when studies are given equal weight] lower than that of nonsmokers. . . . The reason for the inconsistencies in published findings with regard to the association between smoking and sperm density appears to be the result of random error and small sample sizes in most studies. Consequently, the power is low and the chance of a false negative finding is high. Results of the authors' study support these findings. The authors noted a 22.8% lower sperm density among smokers compared with nonsmokers. However, because of the inherent variability in sperm density among individuals, the study sample size (n = 88) was insufficient to produce statistically significant results.* (p. 40)

As with individual studies, results from a meta-analysis like this one should not be interpreted to imply a causal relationship. As the authors note, there are potential confounding factors. They mention studies that have shown smokers are more likely to consume drugs, alcohol, and caffeine and are more likely to experience sex early and to be divorced. Those are all factors that could contribute to reductions in sperm density, according to the authors.

25.3   Some Benefits of Meta-Analysis

We have just seen one of the major benefits of meta-analysis. When a relationship is small or moderate, it is difficult to detect with small studies. A meta-analysis allows researchers to rule out chance for the combined results, when they could not do so for the individual studies. It also allows researchers to get a much more precise estimate of a relationship—that is, a narrower confidence interval—than they could get with individual small studies.

Some benefits of meta-analysis are

1. Detecting small or moderate relationships
2. Obtaining a more precise estimate of a relationship
3. Determining future research
4. Finding patterns across studies

Determining Future Research

Summarizing what has been learned so far in a research area can lead to insights about what to ask next. In addition, identifying and enumerating flaws from past studies can help illustrate how to better conduct future studies.

EXAMPLE 25.1   Designing Better Experiments

In Case Study 22.1, we examined the ganzfeld experiments used to study extrasensory perception. The experiments in that case study were conducted using an automated procedure that was developed in response to an earlier meta-analysis. The earlier meta-analysis had found highly significant differences when compared with chance (Honorton, 1985). However, a critic who was involved with the analysis insisted that the results were, in fact, due to flaws in the procedure (Hyman, 1985). He identified flaws such as improper randomization of target pictures and potentially sloppy data recording. He also identified more subtle flaws. For example, some of the early studies used photographs that a sender was supposed to transmit mentally to a receiver. If the sender was holding a photograph of the true target and the receiver was later asked to pick which of four pictures had been the true target, the receiver could have seen the sender's fingerprints, and these—and not some psychic information—could have provided the answer. The new experiments, reported in Case Study 22.1, were designed to be free of all the flaws that had been identified in the first meta-analysis.   ■

EXAMPLE 25.2   Changes in Focus of the Research

Another example is provided by a meta-analysis on the impact of sexual abuse on children (Kendall-Tackett et al., 1993). In this case, the authors did not suggest corrections of flaws but rather changes in focus for future research. For example, they noted that many studies of the effects of sexual abuse combine children across all age groups into one search for symptoms. In their analysis across studies, they were able to look at different age groups to see if the consequences of abuse differed. They found:

> For preschoolers, the most common symptoms were anxiety, nightmares, general PTSD [post-traumatic stress disorder], internalizing, externalizing, and inappropriate sexual behavior. For school-age children, the most common symptoms included fear, neurotic and general mental illness, aggression, nightmares, school problems, hyperactivity, and regressive behavior. For adolescents, the most common behaviors were depression; withdrawn, suicidal, or self-injurious behaviors; somatic complaints; illegal acts; running away; and substance abuse (p. 167).

In essence, the authors are warning researchers that they could encounter Simpson's Paradox if they do not recognize the differences among various age groups. They provide advice for anyone conducting future studies in this domain:

> Many researchers have studied subjects from very broad age ranges (e.g., 3–18 years) and grouped boys and girls together . . . this grouping together of all ages can mask particular developmental patterns of the occurrence of some symptoms. At a minimum, future researchers should divide children into preschool, school, and adolescent age ranges when reporting the percentages of victims with symptoms (p. 176).   ■

Finding Patterns across Studies

Sometimes patterns that are not apparent in single studies become apparent in a meta-analysis. This could be due to small sample sizes in the original studies or to the fact that each of the original studies considered only one part of a question.

In Case Study 25.1, in which smokers were found to have lower sperm density, the authors were able to investigate whether the decrease was more pronounced for heavier smokers. In the 12 studies for which relevant information was provided, eight showed evidence that the magnitude of decrease in sperm density was related to increasing numbers of cigarettes smoked. None of these studies alone provided sufficient evidence to detect this pattern.

EXAMPLE 25.3

Grouping Studies According to Orientation

Bruvold (1993) compared adolescent smoking prevention programs. He characterized programs as having one of four orientations. The "rational" orientation focused on lectures and displays of substances; the "developmental" orientation used lectures with discussion, problem solving, and some role-playing; the "social norms" orientation used participation in community and recreational activities; and the "social reinforcement" orientation used discussion, role-playing, and public commitment not to smoke.

Individual studies are likely to focus on one orientation only. But a meta-analysis can group studies according to which orientation they used and do a comparison. Bruvold (1993) found that

> the rational orientation had very little impact on behavior, that the social norms and developmental orientations had approximately the same intermediate impact on behavior, and that the social reinforcement orientation had the greatest impact on behavior (pp. 877–878).

Of course, caution must be applied when comparing studies on one feature only. As with most research, confounding factors are a possibility. For example, if the programs using the social reinforcement orientation had also been the more recent studies, that would have confounded the results because there has been much more societal pressure against smoking in recent years.

25.4   Criticisms of Meta-Analysis

Meta-analysis is a relatively new endeavor, and not all of the surrounding issues have been resolved. Some valid criticisms are still being debated by methodologists. You have already encountered one of them—namely, that Simpson's Paradox or its equivalent could apply.

> *Some possible problems with meta-analysis are*
>
> 1. Simpson's Paradox
> 2. Confounding variables

3. Subtle differences in treatments of the same name
4. The file drawer problem
5. Biased or flawed original studies
6. Statistical significance versus practical importance
7. False findings of "no difference"

The Possibility of Confounding Variables

Because meta-analysis is essentially observational in nature, the various treatments cannot be randomly assigned across studies. Therefore, there may be differences across studies that are confounded with the treatments used. For example, it is common for the studies considered in meta-analysis to have been carried out in a variety of countries. It could be that cultural differences are confounded with treatment differences. For instance, a meta-analysis of studies relating dietary fat to breast cancer may find a strong link, but it may be because the same countries that have high-fat diets are also unhealthful in other ways.

Subtle Differences in Treatments of the Same Name

Another problem encountered in meta-analysis is that different studies may involve subtle differences in treatments but call the different treatments by the same name. For instance, chemotherapy may be applied weekly in one study but biweekly in another. The higher frequency may be too toxic, whereas the lower frequency is beneficial. When researchers are combining hundreds of studies, not uncommon in meta-analysis, they may not take the time to discover these subtle differences, which may result in major differences in the outcomes under study.

The File Drawer Problem

Related to the question of which studies to include in a meta-analysis is the possibility that numerous studies may not be discovered by the meta-analyst. These are likely to be studies that did not achieve statistical significance and thus were never published. This is called the **file drawer problem** because the assumption is that these studies are filed away somewhere but not publicly accessible. If the statistically significant studies of a relationship are the ones that are more likely to be available, then the meta-analysis may overestimate the size of the relationship. This is akin to selecting only those with strong opinions in sample surveys.

One way researchers deal with the file drawer problem is to contact all persons known to be working on a particular topic and ask them if they have done studies that were never published. In smaller fields of research, this is probably an effective method of retrieving studies. There are also sophisticated statistical methods for estimating the extent of the problem. For instance, one method suggested by psychologist Robert Rosenthal (1991, p. 104), is to estimate how many undiscovered studies would need to exist to reduce the relationship to nonsignificance. Rosenthal called this number the "fail-safe N." If the answer is an absurdly large number, then the relationship found in the meta-analysis is probably a real one.

Biased or Flawed Original Studies

If the original studies were flawed or biased, so is the meta-analysis. In *Tainted Truth*, Cynthia Crossen (1994, pp. 43–53) discusses the controversy surrounding whether oat bran lowers cholesterol. She notes that the final word on the controversy came in the form of a meta-analysis published in June 1992 (Ripsin et al., 1992) that was funded by Quaker Oats. The study concluded that "this analysis supports the hypothesis that incorporating oat products into the diet causes a modest reduction in blood cholesterol level" (Ripsin et al., 1992, p. 3317).

Crossen raises questions about the studies that were included in the meta-analysis. First, she comments that "of the entire published literature of scientific research on oat bran in the U.S., the lion's share has been at least partly financed by Quaker Oats" (Crossen, 1994, p. 52). In response to Quaker Oats's defense that no strings were attached to the money and that scientists are not going to risk their reputations for a small research grant, Crossen replies:

> *In most cases, that is true. Nothing is more damaging to scientists' reputations —and their economic survival—than suspicions of fraud, corruption, or dishonesty. But scientists are only human, and in the course of any research project, they make choices. Who were the subjects of the study? Young or old, male or female, high cholesterol or low? How were they chosen? How long did the study go on? . . . Since cholesterol levels vary season to season, when did the study begin and when did it end?* (p. 49).

Statistical Significance versus Practical Importance

We have already learned that a statistically significant result is not necessarily of any practical importance. Meta-analysis is particularly likely to find statistically significant results because typically the combined studies provide very large sample sizes. Thus, it is important to learn the magnitude of an effect found in a meta-analysis and not just that it is statistically significant.

For example, the oat bran meta-analysis included 10 studies, and it did indeed show a statistically significant reduction in cholesterol, a conclusion that led to headlines across the country. However, the magnitude of the reduction was quite small. A headline in the *New York Times* (1992, online) read: "Lots of Oat Bran Found to Cut Cholesterol." But the second sentence of the article conceded that the effect was small. It read, "Cholesterol levels fell an average of 2 to 3 percent in the 1278 adults studied. There were larger drops in people with higher blood cholesterol levels." Much later in the article the effect for people with higher cholesterol levels was given: "People in this group [with cholesterol of at least 229 milligrams per deciliter of blood] had a drop of 6 to 7 percent when they ate 3 grams or more of oat bran." It wasn't until near the end of the article that the story noted, "For people with relatively low cholesterol levels, eating oat bran is not necessary." Consulting the original meta-analysis revealed that a 95% confidence interval for the drop extended from 3.3 mg/dL to 8.4 mg/dL. The average cholesterol level for Americans is about 210 mg/dL. Given the extensive publicity this study received, readers who pay attention to headlines but not to details could easily have been misled into thinking that the effect was much larger than it was.

False Findings of "No Difference"

It is also possible that a meta-analysis will erroneously reach the conclusion that there is no difference or no relationship—when in fact there was simply not enough data to find one that was statistically significant. Because a meta-analysis is supposed to be so comprehensive, there is even greater danger than with single studies that the conclusion will be taken to mean that there really is no difference. We will see an example of this problem in Case Study 25.2.

Most meta-analyses include sufficient data so that important differences will be detected, and that can lead to a false sense of security. The main danger is when not many studies have been done on a particular subset or question. It may be true that there are hundreds of thousands of participants across all studies, but if only a small fraction of them are in a certain age group, received a certain treatment, and so on, then a statistically significant difference may still not be found for that subset. When you read about a meta-analysis, be careful not to confuse the overall sample size with the one used for any particular subgroup.

CASE STUDY 25.2   **Controversy over Breast Cancer Screening for Women Under 50**

The question of whether women in their 40s should have mammography screening for breast cancer continues to be controversial. For example, a *Science News* article in September, 2010, was headlined, "Mammography Reduces Mortality from Breast Cancer in Ages 40–49 Years, Swedish Study Finds" (Umeå University, 2010). Yet just a little over a year later, a Canadian panel recommended against screening for women in that age group. For instance, a *Los Angeles Times* headline in November, 2011, read "Mammograms for women in 40s: Now Canada recommends against them" (Mestel, 2011).

The debate has been going on for decades, and part of the problem is that there are so few women in that age group who get breast cancer that it is difficult to reach a conclusion either way, even with a large meta-analysis. As early as the Fall of 1993 the debate pitted the National Cancer Institute against the American Cancer Society. Up until that time, both organizations had been recommending that women should begin having mammograms at age 40. But in February 1993, the National Cancer Institute convened an international conference to bring together experts from around the world to help conduct a meta-analysis of studies on the effectiveness of mammography as a screening device. Their conclusion about women aged 40–49 years was: "For this age group it is clear that in the first 5–7 years after study entry, there is no reduction in mortality from breast cancer that can be attributed to screening" (Fletcher et al., 20 October 1993, p. 1653).

The results of the meta-analysis amounted to a withdrawal of the National Cancer Institute's support for mammograms for women under 50. The American Cancer Society refuted the study and announced that it would not change its recommendation. As noted in a front-page story in the *San Jose Mercury News* on November 24, 1993:

> *A spokeswoman for the American Cancer Society's national office said Tuesday that the . . . study would not change the group's recommendation because it was not big enough to draw definite conclusions. The study would have to screen 1 million women to get a certain answer because breast cancer is so uncommon in young women.*

In fact, the entire meta-analysis considered eight randomized experiments conducted over 30 years, including nearly 500,000 women. That sounds impressive. But not all of the studies included women under 50. As noted by Sickles and Kopans (20 October 1993):

> *Even pooling the data from all eight randomized controlled trials produces insufficient statistical power to indicate presence or absence of benefit from screening. In the eight trials, there were only 167,000 women (30% of the participants) aged 40–49, a number too small to provide a statistically significant result* (p. 1622).

There were other complaints about the studies as well. Even the participants in the conference recognized that there were methodological flaws in some of the studies. About one of the studies on women aged 40–49, they commented:

> *It is worrisome that more patients in the screening group had advanced tumors, and this fact may be responsible for the results reported to date. . . . The technical quality of mammography early in this trial is of concern, but it is not clear to what extent mammography quality affected the study outcome* (Fletcher et al., 20 October 1993, p. 1653).

There is thus sufficient concern about the studies themselves to make the results inconclusive. But they were obviously statistically inconclusive as well. Here is the full set of results for women aged 40–49:

> *A meta-analysis of six trials found a relative risk of 1.08 (95% confidence interval = 0.85 to 1.39) after 7 years' follow-up. After 10–12 years of follow-up, none of four trials have found a statistically significant benefit in mortality; a combined analysis of Swedish studies showed a statistically insignificant 13% decrease in mortality at 12 years. One trial (Health Insurance Plan) has data beyond 12 years of follow-up, and results show a 25% decrease in mortality at 10–18 years. Statistical significance of this result is disputed, however* (Fletcher et al., 20 October 1993, p. 1644).

A debate on the *MacNeil-Lehrer News Hour* between American Cancer Society spokesman Dr. Harmon Eyre and one of the authors of the meta-analysis, Dr. Barbara Rimer, illustrated the difficulty of interpreting results like these for the public. Dr. Eyre argued that "30% of those who will die could be saved if you look at the data as we do; the 95% confidence limits in the meta-analysis could include a 30% decrease in those who will die."

Dr. Rimer, on the other hand, simply kept reiterating that mammography for women under age 50 simply "hasn't been shown to save lives." In response, Dr. Eyre accused the National Cancer Institute of having a political agenda. If women under 50 do not have regular mammograms, it could save a national health-care program large amounts of money.

By now, you should recognize the real nature of the debate here. The results are inconclusive because of small sample sizes and possible methodological flaws. The question is not a statistical one—it is a public policy one. Given that we do not know for certain that mammograms save lives for women under 50, should health insurers be required to pay for them? ■

Thinking About Key Concepts

- *Meta-analysis* is used to combine results of multiple studies on the same topic.
- *Vote-counting*, the practice of simply counting how many studies on a topic were statistically significant, is seriously flawed because statistical significance is so dependent on the sample sizes used for the studies.
- An important decision in meta-analysis is what studies to include. Quality of studies, where they were published, and when they were conducted are all important considerations.
- Another important decision is how to combine results across studies. If the assumption is that all studies were done on the same population with the same underlying effect, then a *fixed effects* model is used. If the assumption is that there are different underlying population effect sizes in various studies, then a *random effects* model is used. There are more complex models as well, so it is important to find out what assumptions the authors of a meta-analysis used when they chose how to combine the results.
- Benefits of meta-analysis include the ability to detect small effects that individual studies might miss, obtaining more accurate estimates of effects than individual studies could obtain, and finding patterns that may be of interest on their own and that can be used to dictate what questions to ask in future research.
- Problems that can be encountered in a meta-analysis include inappropriate combination of studies, confounding variables, missing studies that were not published (the *file drawer effect*), mixing high and low quality studies, reporting conclusions that are statistically significant but have no practical importance, and reporting that there is no effect when in fact the sample sizes were too small to find a real effect.

Exercises

Exercises with numbers divisible by 3 (3, 6, 9, etc.) are included in the Solutions at the back of the book. They are marked with an asterisk (*).

1. Explain why the vote-counting method is not a good way to summarize the results in a research area.

2. Explain why the person deciding which studies to include in a meta-analysis should be told how each study was conducted but not the results of the study.

*3. When the Cochrane Collaboration (discussed on page 537) was first getting started, a report in *New Scientist* (Vine, 21 January 1995), announced that the UK Cochrane Centre in Oxford would be launching the *Cochrane Database of Systematic Reviews,* which will "focus on a number of diseases, drawing conclusions about which treatments work and which do not from all the available randomized controlled trials" (p. 14).

    *a. Why do you think they only planned to include "randomized controlled trials" (that is, randomized experiments) and not observational studies?

    *b. Do you think they should include studies that did *not* find statistically significant results when they do their meta-analyses? Why or why not?

4. Refer to Exercise 3. Pick one of the benefits of meta-analysis listed in Section 25.3, and explain how that benefit applies to the reviews in the *Cochrane Database.*

5. Refer to Exercise 3. Pick three of the possible problems with meta-analysis listed in Section 25.4, and discuss how they might apply to the reviews in the *Cochrane Database.*

*6. An article in the *Sacramento Bee* (15 April 1998, pp. A1, A12), titled "Drug reactions a top killer, research finds," reported a study estimating that between 76,000 and 137,000 deaths a year in the United States occur due to adverse reactions to medications. Here are two quotes from the article:

The scientists reached their conclusion not in one large study but by combining the

results of 39 smaller studies. This technique, called meta-analysis, can enable researchers to draw statistically significant conclusions from studies that individually are too small (p. A12).

[A critic of the research] said the estimates in Pomeranz's study might be high, because the figures came from large teaching hospitals with the sickest patients, where more drug use and higher rates of drug reactions would be expected than in smaller hospitals (p. A12).

    *a. Pick one of the benefits in Section 25.3 and one of the criticisms in Section 25.4, and apply them to this study.

    *b. The study estimated that between 76,000 and 137,000 deaths a year occur due to adverse reactions to medications, with a mean estimate of 100,000. If this is true, then adverse drug reaction is the fourth-leading cause of death in the United States. Based on these results, which one of the criticisms given in Section 25.4 is definitely not a problem? Explain.

7. In a meta-analysis, researchers can choose either to combine results across studies to produce a single confidence interval or to report separate confidence intervals for each study and compare them. Give one advantage and one disadvantage of combining the results into one confidence interval.

8. The handbook for the Cochrane Collaboration (http://handbook.cochrane.org, Section 2.3.2) includes the following quote:

It is important to let people know when there is no reliable evidence, or no evidence about particular outcomes that are likely to be important to decision makers. No evidence of effect should not be confused with evidence of no effect.

Explain what is meant by the last sentence in the quote.

*9. Give two reasons why researchers might not want to include all possible studies in a meta-analysis of a certain topic.

10. Suppose a meta-analysis on the risks of drinking coffee included 100,000 participants of all ages across 80 studies. One of the conclusions was that a confidence interval for the relative risk of heart attack for women over 70 who drank coffee, compared with those who didn't, was 0.90 to 1.30. The researchers concluded that coffee does not present a problem for women over 70 because a relative risk of 1.0 is included in the interval. Explain why their conclusion is not justified.

11. Eisenberg and colleagues (15 June 1993) used meta-analysis to examine the effect of "cognitive therapies" such as biofeedback, meditation, and relaxation methods on blood pressure. Their abstract states that "cognitive interventions for essential hypertension are superior to no therapy but not superior to credible sham techniques" (p. 964). Here are some of the details from the text of the article:

> Mean blood pressure reductions were smallest when comparing persons experimentally treated with those randomly assigned to a credible placebo or sham intervention and for whom baseline blood pressure assessments were made during a period of more than 1 day. Under these conditions, individuals treated with cognitive behavioral therapy experienced a mean reduction in systolic blood pressure of 2.8 mm Hg (CI, −0.8 to 6.4 mm Hg) . . . relative to controls (p. 967).

The comparison included 12 studies involving 368 subjects. Do you agree with the conclusion in the abstract that cognitive interventions are not superior to credible sham techniques? What would be a better way to word the conclusion?

*12. Would the file drawer problem be more likely to present a substantial difficulty in a research field with 50 researchers or in one with 1000 researchers? Explain.

13. *Science News* (25 January 1995) reported a study on the relationship between levels of radon gas in homes and lung cancer. It was a case-control study, with 538 women with lung cancer and 1183 without lung cancer. The article noted that the average level of radon concentration in the homes of the two groups over a 1-year period was exactly the same. The article also contained the following quote, reporting on an editorial by Jonathan Samet that accompanied the original study:

> For the statistical significance needed to assess accurately whether low residential exposures constitute no risk . . . the study would have required many more participants than the number used in this or any other residential-radon study to date. . . . But those numbers may soon become available, he adds, as researchers complete a spate of new studies whose data can be pooled for reanalysis (p. 26).

Write a short report interpreting the quote for someone with no training in statistics.

14. An article in *Science* (Mann, 11 November 1994) describes two approaches used to try to determine how well programs to improve public schools have worked. The first approach was taken by an economist named Eric Hanushek. Here is part of the description:

> Hanushek reviewed 38 studies and found the "startlingly consistent" result that "there is no strong or systematic relationship between school expenditures and student performance." . . . Hanushek's review used a technique called "vote-counting" (p. 961).

The other approach was a meta-analysis and the results are reported as follows:

> [The researchers] found systematic positive effects. . . . Indeed, decreased class size, increased teacher experience, increased teacher salaries, and increased per-pupil spending were all positively related to academic performance (p. 962).

Explain why the two approaches yielded different answers and which one you think is more credible.

*15. Suppose 10 studies were done to assess the relationship between watching violence on television and subsequent violent behavior in children. Suppose that none of the 10 studies detected a statistically significant relationship. Is it possible for a vote-counting procedure to detect a relationship? Is it possible for a meta-analysis to detect a relationship? Explain.

16. Refer to the summary of News Story 9 in the Appendix. The story appeared in the *Washington Post* on May 7, 2002. You may be able to find it through an internet search of the story's headline, "Against depression, a sugar pill is hard to beat." There is a meta-analysis of "96 antidepressant trials" described in the story and the summary in the Appendix. Use that description to answer these questions.

   a. How did the analyst answer the question "Which studies should be included?"

   b. How did the analyst answer the question "Should results be compared or combined?"

   c. One statement in the story is "in 52 percent of them [the studies], the effect of the antidepressant could not be distinguished from that of the placebo." Do you think this statement is an example of "vote-counting"? What additional information would you need to determine whether or not it is?

   d. Refer to part (c). Assuming the statement is an example of vote-counting, should it be used to conclude that the effects of antidepressants and placebos are the same? Explain.

17. A meta-analysis of 32 studies on the effectiveness of "self-talk" on sports performance was summarized as follows:

   The analysis revealed a positive moderate effect size (ES = .48). The moderator analyses showed that self-talk interventions were more effective for tasks involving relatively fine, compared with relatively gross, motor demands, and for novel, compared with well-learned, tasks (Hatzigeorgiadis et al., 2011, p. 348).

   a. Read the explanation of a fixed effects model on page 539. Based on the quote above, do you think a fixed effects model would be appropriate for this meta-analysis? Explain.

   b. Another quote from the article was: *"The fail-safe statistic indicated that it would be unlikely that a sufficient number of unpublished studies (102) would exist to reduce the effect to a trivial size"* (Hatzigeorgiadis, 2011, p. 351). Explain what the "fail-safe statistic" is, and why it is important to include it in the report.

Mini-Projects

1. Find a journal article that presents a meta-analysis. (They are commonly found in medical and psychology journals.)

   a. Give an overview of the article and its conclusions.

   b. Explain what the researchers used as the criteria for deciding which studies to include.

   c. Explain which of the benefits of meta-analysis were incorporated in the article.

   d. Discuss each of the potential criticisms of meta-analysis and how they were handled in the article.

   e. Summarize your conclusions about the topic based on the article and your answers to parts (b) to (d).

2. Find a news story reporting on a meta-analysis. Discuss whether important information is missing from the report. Critically evaluate the conclusions of the meta-analysis, taking into consideration the criticisms listed in Section 25.4.

References

American Society for Reproductive Medicine. (2012). Smoking and infertility: A committee opinion, *Fertility and Sterility,* 98(6), pp 1400–1406.

Augwood, C., K. Duckitt, A.A. Templeton. (1998). Smoking and female infertility: A systematic review and meta-analysis. *Human Reproduction,* 13, pp. 1532–1539.

Bruvold, W. H. (1993). A meta-analysis of adolescent smoking prevention programs. *American Journal of Public Health* 83, no. 6, pp. 872–880.

Chalmers, T. C., H. Smith, Jr., B. Blackburn, B. Silverman, B. Schroeder, D. Reitman, and A. Ambroz. (1981). A method for assessing the quality of a randomized control trial. *Controlled Clinical Trials* 2, pp. 31–49.

Crossen, C. (1994). *Tainted truth: The manipulation of fact in America.* New York: Simon and Schuster.

Eisenberg, D. M., T. L. Delbance, C. S. Berkey, T. J. Kaptchuk, B. Kupelnick, J. Kuhl, and T. C. Chalmers. (15 June 1993). Cognitive behavioral techniques for hypertension: Are they effective? *Annals of Internal Medicine* 118, no. 12, pp. 964–972.

Fletcher, S. W., B. Black, R. Harris, B. K. Rimer, and S. Shapiro. (20 October 1993). Report of the International Workshop on Screening for Breast Cancer. *Journal of the National Cancer Institute* 85, no. 20, pp. 1644–1656.

Glass, G. V. (1976). Primary, secondary and meta-analysis of research. *Educational Researcher* 5, pp. 3–8.

Hatzigeorgiadis, Antonis, Nikos Zourbanos, Evangelos Galanis and Yiannis Theodorakis. (2011). Self-Talk and Sports Performance: A Meta-Analysis, *Perspectives on Psychological Science* 6(4), pp. 348–356. DOI: 10.1177/1745691611413136

Honorton, C. (1985). Meta-analysis of psi ganzfeld research: A response to Hyman. *Journal of Parapsychology* 49, pp. 51–91.

Hyman, R. (1985). The psi ganzfeld experiment: A critical appraisal. *Journal of Parapsychology* 49, pp. 3–49.

Kendall-Tackett, K. A., L. M. Williams, and D. Finkelhor. (1993). Impact of sexual abuse on children: A review and synthesis of recent empirical studies. *Psychological Bulletin* 113, no. 1, pp. 164–180.

Mann, C. C. (11 November 1994). Can meta-analysis make policy? *Science* 266, pp. 960–962.

Mestel, Rosie. (2011). Mammograms for women in 40s: Now Canada recommends against them, November 21, 2011, http://articles.latimes.com/2011/nov/21/news/la-heb-mammograms-canada-20111121, accessed July 17, 2013.

Moher, D., A. R. Jadad, G. Nichol, M. Penman, P. Tugwell, S. Walsh. (1995). Assessing the quality of randomized controlled trials: An annotated bibliography of scales and checklists. *Controlled Clinical Trials,* 16(1), pp. 62–73.

New York Times (1992, online). Lots of Oat Bran Found to Cut Cholesterol. http://www.nytimes.com/1992/06/24/health/lots-of-oat-bran-found-to-cut-cholesterol.html, 24 June 1992, accessed July 17, 2013.

Ripsin, C. M., J. M. Keenan, D. R. Jacobs, Jr., P. J. Elmer, R. R. Welch, L. Van Horn, K. Liu, W. H. Turnbull, F. W. Thye, and M. Kestin et al. (1992). Oat products and lipid lowering: A meta-analysis. *Journal of the American Medical Association* 267, no. 24, pp. 3317–3325.

Rosenthal, R. (1991). *Meta-analytic procedures for social research.* Rev. ed. Newbury Park, CA: Sage Publications.

Sickles, E. A., and D. B. Kopans. (20 October 1993). Deficiencies in the analysis of breast cancer screening data. *Journal of the National Cancer Institute* 85, no. 20, pp. 1621–1624.

Umeå University. (2010, September 29). Mammography reduces mortality from breast cancer in ages 40–49 years, Swedish study finds. *ScienceDaily.* Retrieved July 17, 2013, from http://www.sciencedaily.com/releases/2010/09/100929142007.htm

Vine, Gail (21 January 1995). Is there a database in the house? *New Scientist,* no. 1961, pp. 14–15.

Vine, M. F., B. H. Margolin, H. I. Morrison, and B. S. Hulka. (January 1994). Cigarette smoking and sperm density: A meta-analysis. *Fertility and Sterility* 61, no. 1, pp. 35–43.

CHAPTER **26**

Ethics in Statistical Studies

As with any human endeavor, in statistical studies there are ethical considerations that must be taken into account. These considerations fall into multiple categories, and we discuss the following issues here:

1. Ethical treatment of human and animal participants
2. Assurance of data quality
3. Appropriate statistical analyses
4. Fair reporting of results

Most professional societies have a code of ethics that their members are asked to follow. Guidelines about the conduct of research studies are often included. For instance, the American Psychological Association first published a code of ethics in 1953 and has updated it at least every 10 years since then. In 1999, the American Statistical Association, one of the largest organizations of professional statisticians, published *Ethical Guidelines for Statistical Practice* listing 67 recommendations, divided into categories such as "Professionalism" and "Responsibilities to Research Subjects."

26.1 Ethical Treatment of Human and Animal Participants

There has been increasing attention paid to the ethics of experimental work, partly based on some examples of deceptive experiments that led to unintended harm to participants. Here is a classic example of an experiment conducted in the early 1960s that would almost surely be considered unethical today.

EXAMPLE 26.1 Stanley Milgram's "Obedience and Individual Responsibility" Experiment

Social psychologist Stanley Milgram was interested in the extent to which ordinary citizens would obey an authority figure, even if it meant injuring another human being. Through newspaper ads offering people money to participate in an experiment on learning, he recruited people in the area surrounding Yale University, where he was a member of the faculty. When participants arrived they were greeted by an authoritative researcher in a white lab coat, and introduced to another person who they were told was a participant like them but who was actually an actor. Lots were drawn to see who would be the "teacher" and who would be the "student," but in fact it had been predetermined that the actor would be the "student" and the local citizen would be the "teacher."

The student/actor was placed in a chair with restricted movement and hooked up to what was alleged to be an electrode that administered an electric shock. The teacher conducted a memory task with the student and was instructed to administer a shock when the student missed an answer. The shocking mechanism was shown to start at 15 volts and increase in intensity with each wrong answer, up to 450 volts. When the alleged voltage reached 375, it was labeled as "Danger/Severe" and when it reached 435 it was labeled "XXX." The experimenter sat at a nearby table, encouraging the teacher to continue to administer the shocks. The student/actor would respond with visible and increasing distress. The teacher was told that the experimenter would take responsibility for any harm that came to the student.

The disturbing outcome of the experiment was that 65% of the participants continued to administer the alleged shocks up to the full intensity, even though many of them were quite distressed and nervous about doing so. Even at "very strong" intensity, 80% of the participants were still administering the electric shocks. (*Sources:* http://doi.apa.org/psycinfo/1964-03472-001 and Milgram (1983).) ∎

This experiment would be considered unethical today because of the stress it caused the participants. Based on this and similar experiments, the American Psychological Association continues to update its *Ethical Principles of Psychologists and Code of Conduct.* The latest version can be found at http://www.apa.org/ethics/code. Section 8.07 of the code is called "Deception in Research" and includes three instructions (http://www.apa.org/ethics/code, Section 8.07); Milgram's experiment would most likely fail the criterion in part (b):

(a) Psychologists do not conduct a study involving deception unless they have determined that the use of deceptive techniques is justified by the study's significant prospective scientific, educational, or applied value and that effective nondeceptive alternative procedures are not feasible.

(b) Psychologists do not deceive prospective participants about research that is reasonably expected to cause physical pain or severe emotional distress.

(c) Psychologists explain any deception that is an integral feature of the design and conduct of an experiment to participants as early as is feasible, preferably at the conclusion of their participation, but no later than at the conclusion of the data collection, and permit participants to withdraw their data.

Informed Consent

Virtually all experiments with human participants require that the researchers obtain the **informed consent** of the participants. In other words, participants are to be told what the research is about and given an opportunity to make an informed choice about whether to participate. If you were a potential participant in a research study, what would you want to know in advance to make an informed choice about participation? Because of such issues as the need for a control group and the use of double-blinding, it is often the case that participants cannot be told everything in advance. For instance, it would be antithetical to good experimental procedure to tell participants in advance if they were taking a drug or a placebo, or to tell them if they were in the treatment group or control group. Instead, the use of multiple groups is explained and participants are told that they will be randomly assigned to a group but will not know what it is until the conclusion of the experiment.

The information provided in this process is slightly different in research such as psychology experiments than it is in medical research. In both cases, participants are supposed to be told the nature and purpose of the research and any risks or benefits. In medical research, the participants generally are suffering from a disease or illness, and an additional requirement is that they be informed about alternative treatments. Of course, it is unethical to withhold a treatment known to work.

In section 8.02 of its code of ethics, "Informed Consent to Research," the American Psychological Association provides these guidelines for informed consent in experiments in psychology (http://www.apa.org/ethics/code, Section 8.02):

(a) When obtaining informed consent as required in Standard 3.10, Informed Consent, psychologists inform participants about (1) the purpose of the research, expected duration, and procedures; (2) their right to decline to participate and to withdraw from the research once participation has begun; (3) the foreseeable consequences of declining or withdrawing; (4) reasonably foreseeable factors that may be expected to influence their willingness to participate such as potential risks, discomfort, or adverse effects; (5) any prospective research benefits; (6) limits of confidentiality; (7) incentives for participation; and (8) whom to contact for questions about the research and research participants' rights. They provide opportunity for the prospective participants to ask questions and receive answers.

(b) Psychologists conducting intervention research involving the use of experimental treatments clarify to participants at the outset of the research (1) the experimental nature of the treatment; (2) the services that will or will not be available to the control group(s) if appropriate; (3) the means by which assignment to treatment and control groups will be made; (4) available treatment alternatives if an individual does not wish to participate in the research or wishes to withdraw once a study has begun; and (5) compensation for or monetary costs of participating including, if appropriate, whether reimbursement from the participant or a third-party payor will be sought.

Informed Consent in Medical Research

The United States Department of Health and Human Services has a detailed policy on informed consent practices in medical research. The Department of Health and Human Services Office for Human Research Protections, Office for Protection from Research Risks has provided the following tips and checklist. (Source: http://www.hhs .gov/ohrp/policy/ictips.html.) The acronym IRB stands for Institutional Review Board, which is a board that all research institutions are required to maintain for oversight of research involving human and animal participants.

Tips on Informed Consent

The process of obtaining informed consent must comply with the requirements of 45 CFR 46.116. The documentation of informed consent must comply with 45 CFR 46.117. The following comments may help in the development of an approach and proposed language by investigators for obtaining consent and its approval by IRBs:

- **Informed consent is a process, not just a form.** Information must be presented to enable persons to voluntarily decide whether or not to participate as a research subject. It is a fundamental mechanism to ensure respect for persons through provision of thoughtful consent for a voluntary act. The procedures used in obtaining informed consent should be designed to educate the subject population in terms that they can understand. Therefore, informed consent language and its documentation (especially explanation of the study's purpose, duration, experimental procedures, alternatives, risks, and benefits) must be written in "lay language" (i.e., understandable to the people being asked to participate). The written presentation of information is used to document the basis for consent and for the subjects' future reference. The consent document should be revised when deficiencies are noted or when additional information will improve the consent process.

- **Use of the first person (e.g., "I understand that . . .") can be interpreted as suggestive, may be relied upon as a substitute for sufficient factual information, and can constitute coercive influence over a subject.** Use of scientific jargon and legalese is not appropriate. Think of the document primarily as a teaching tool not as a legal instrument.

- **Describe the overall experience that will be encountered.** Explain the research activity, how it is experimental (e.g., a new drug, extra tests, separate research records, or nonstandard means of management, such as flipping a coin for random assignment or other design issues). Inform the human subjects of the reasonably foreseeable harms, discomforts, inconvenience, and risks that are associated with the research activity. If additional risks are identified during the course of the research, the consent process and documentation will require revisions to inform subjects as they are recontacted or newly contacted.

- **Describe the benefits that subjects may reasonably expect to encounter.** There may be none other than a sense of helping the public at large. If payment is given to defray the incurred expense for participation, it must not be coercive in amount or method of distribution.

- **Describe any alternatives to participating in the research project.** For example, in drug studies, the medication(s) may be available through their family doctor or clinic without the need to volunteer for the research activity.

- **The regulations insist that the subjects be told the extent to which their personally identifiable private information will be held in confidence.** For example, some studies require disclosure of information to other parties. Some studies inherently are in need of a Certificate of Confidentiality which protects the investigator from involuntary release (e.g., subpoena) of the names or other identifying characteristics of research subjects. The IRB will determine the level of adequate requirements for confidentiality in light of its mandate to ensure minimization of risk and determination that the residual risks warrant involvement of subjects.

- **If research-related injury (i.e., physical, psychological, social, financial, or otherwise) is possible in research that is more than minimal risk (see 45 CFR 46.102[g]), an explanation must be given of whatever voluntary compensation and treatment will be provided.** Note that the regulations do not limit injury to "physical injury." This is a common misinterpretation.

- **The regulations prohibit waiving or appearing to waive any legal rights of subjects.** Therefore, for example, consent language must be carefully selected that deals with what the institution is *voluntarily* willing to do under circumstances such as providing for compensation beyond the provision of immediate or therapeutic intervention in response to a research-related injury. In short, subjects should not be given the impression that they have agreed to and are without recourse to seek satisfaction beyond the institution's *voluntarily* chosen limits.

- **The regulations provide for the identification of contact persons who would be knowledgeable to answer questions of subjects about the *research, rights as a research subject,* and *research-related injuries.* These three areas must be explicitly stated and addressed in the consent process and documentation.** Furthermore, a single person is not likely to be appropriate to answer questions in all areas. This is because of potential conflicts of interest or the appearance of such. Questions about the research are frequently best answered by the investigator(s). However, questions about the rights of research subjects or research-related injuries (where applicable) may best be referred to those not on the research team.

These questions could be addressed to the IRB, an ombudsman, an ethics committee, or other informed administrative body. Therefore, each consent document can be expected to have at least two names with local telephone numbers for contacts to answer questions in these specified areas.

- **The statement regarding voluntary participation and the right to withdraw at any time can be taken almost verbatim from the regulations** (45 CFR 46.116[a][8]). It is important not to overlook the need to point out that no penalty or loss of benefits will occur as a result of *both* not participating or withdrawing at any time. It is equally important to alert potential subjects to any foreseeable consequences to them should they unilaterally withdraw while dependent on some intervention to maintain normal function.

- **Don't forget to ensure provision for appropriate additional requirements which concern consent.** Some of these requirements can be found in sections 46.116(b), 46.205(a)(2), 46.207(b), 46.208(b), 46.209(d), 46.305(a)(5–6), 46.408(c), and 46.409(b). The IRB may impose additional requirements that are not specifically listed in the regulations to ensure that adequate information is presented in accordance with institutional policy and local law.

SOURCE: http://ohrp.osophs.dhhs.gov/humansubjects/guidance/ictips.htm

---

Informed Consent Checklist

Basic and Additional Elements

- ☐ A statement that the study involves research
- ☐ An explanation of the purposes of the research
- ☐ The expected duration of the subject's participation
- ☐ A description of the procedures to be followed
- ☐ Identification of any procedures which are experimental
- ☐ A description of any reasonably foreseeable risks or discomforts to the subject
- ☐ A description of any benefits to the subject or to others which may reasonably be expected from the research
- ☐ A disclosure of appropriate alternative procedures or courses of treatment, if any, that might be advantageous to the subject
- ☐ A statement describing the extent, if any, to which confidentiality of records identifying the subject will be maintained
- ☐ For research involving more than minimal risk, an explanation as to whether any compensation, and an explanation as to whether any medical

treatments are available, if injury occurs and, if so, what they consist of, or where further information may be obtained

☐ Research Qs

☐ Rights Qs

☐ Injury Qs

☐ An explanation of whom to contact for answers to pertinent questions about the research and research subjects' rights, and whom to contact in the event of a research-related injury to the subject

☐ A statement that participation is voluntary, refusal to participate will involve no penalty or loss of benefits to which the subject is otherwise entitled, and the subject may discontinue participation at any time without penalty or loss of benefits, to which the subject is otherwise entitled

Additional Elements, as Appropriate

☐ A statement that the particular treatment or procedure may involve risks to the subject (or to the embryo or fetus, if the subject is or may become pregnant), which are currently unforeseeable

☐ Anticipated circumstances under which the subject's participation may be terminated by the investigator without regard to the subject's consent

☐ Any additional costs to the subject that may result from participation in the research

☐ The consequences of a subject's decision to withdraw from the research and procedures for orderly termination of participation by the subject

☐ A statement that significant new findings developed during the course of the research, which may relate to the subject's willingness to continue participation, will be provided to the subject

☐ The approximate number of subjects involved in the study

SOURCE: http://www.hhs.gov/ohrp/policy/ictips.html, accessed July 17, 2013

As you can see, the process of obtaining informed consent can be daunting and some potential participants may opt out because of an excess of information. There are additional issues that arise when participants are in certain categories. For example, young children cannot be expected to fully understand the informed consent process. The regulations state that children should be involved in the decision-making process to the extent possible and that a parent must also be involved. There are special rules applying to research on prisoners and to research on "Pregnant Women, Human Fetuses, and Neonates." These can be found on the website http://www.hhs.gov/ohrp/humansubjects/guidance/45cfr46.html#subpartb.

Research on Animals

Perhaps nothing is as controversial in research as the use of animals. There are some who believe that any experimentation on animals is unethical, whereas others argue that humans have benefited immensely from such research and that justifies its use. There is no doubt that in the past some research with animals, as with humans, has been clearly unethical. But again the professions of medicine and behavioral sciences, which are the two arenas in which animal research is most prevalent, have developed ethical guidelines that their members are supposed to follow.

Nonetheless, animals do not have the rights of humans as participants in experiments. Here are two of the elements of the American Psychological Association's Code of Ethics section on "Humane Care and Use of Animals in Research." Clearly, these statements would not be made regarding human subjects and even these are stated ideals that many researchers may not follow. (*Source:* http://www.apa.org/ethics/code, Section 8.09.)

> *(e) Psychologists use a procedure subjecting animals to pain, stress, or privation only when an alternative procedure is unavailable and the goal is justified by its prospective scientific, educational, or applied value.*

> *(g) When it is appropriate that an animal's life be terminated, psychologists proceed rapidly, with an effort to minimize pain and in accordance with accepted procedures.*

The United States National Institutes of Health also has established the *Public Health Service Policy on Humane Care and Use of Laboratory Animals,* which can be found at http://www.grants.nih.gov/grants/olaw/references/phspol.htm. However, the guidelines for research on animals are not nearly as detailed as they are for human subjects. A research study of the approval of animal-use protocols found strong inconsistences when protocols approved or disapproved by one institution were submitted for approval to another institution's review board. The decisions of the two boards to approve or disapprove the protocols agreed at no better than chance levels, suggesting that the guidelines for approval of animal research are not spelled out in sufficient detail (Plous and Herzog, 2001). For more information and links to a number of websites on the topic of research with animals, visit http://www.socialpsychology.org/methods.htm#animals.

26.2 Assurance of Data Quality

When reading the results of a study, you should be able to have reasonable assurance that the researchers collected data of high quality. This is not as easy as it sounds, and we have explored many of the difficulties and disasters of collecting and interpreting data for research studies in Part 1 of the book.

The quality of data becomes an ethical issue when there are personal, political, or financial reasons motivating one or more of those involved in the research and steps are not taken to assure the integrity of the data. Data quality is also an ethical issue when researchers knowingly fail to report problems with data collection that may distort the interpretation of the results. As a simple example, survey data should always be reported with an explanation of how the sample was selected, what questions were asked, who responded, and how they may differ from those who didn't respond.

United States Federal Statistical Agencies

The United States government spends large amounts of money to collect and disseminate statistical information on a wide variety of topics. In recent years, there has been an increased focus on making sure the data are of high quality. The Federal Register Notices on June 4, 2002 (vol. 67, no. 107) provided a report called *Federal Statistical Organizations' Guidelines for Ensuring and Maximizing the Quality, Objectivity, Utility, and Integrity of Disseminated Information.* (The report can be found on the Web at http://www.fedstats.gov/policy/Stat-Agency-FR-June4-2002.pdf.) The purpose of the report was to provide notice that each of over a dozen participating federal statistical agencies were making such guidelines available for public comment. (A list of federal statistical agencies with links to their websites and the data they provide can be found at www.fedstats.gov.)

Most of the guidelines proposed by the statistical agencies were common-sense, self-evident good practice. As an example, the following box includes guidelines for data collection from the Bureau of Transportation Statistics website.

3.1 Data Collection Operations

Guidelines

- Make the data collection as easy as possible for the collector.
- If interviewers or observers are used, a formal training process should be established to ensure proper procedures are followed.
- Data calculations and conversions at the collection level should be minimized.

For example, if a bus driver is counting passengers, they should not be doing calculations such as summations. The driver should record the raw counts and calculations should be performed where they are less likely to result in mistakes.

- The collection operation procedures should be documented and clearly posted with the data or with disseminated output from the data. If third party data collection is used, procedures used by the third party should be provided as well.

SOURCE: http://www.rita.dot.gov/bts/sites/rita.dot.gov.bts/files/publications/guide_to_good_statistical_practice_in_the_transportation_field/html/chapter_03.html

---

So where does this fit in a discussion of ethics? Although it is laudable that the federal statistical agencies are spelling out these principles, the ethical issues involved with assuring the quality of government data are more subtle than can be evoked from general principles. There are always judgments to be made, and politics can sometimes enter into those judgments.

For instance, a report by the National Research Council (Citro and Norwood, 1997) examined many aspects of the functioning of the Bureau of Transportation Statistics (BTS) and formulated suggestions for improvement. The BTS was created in 1992 and thus had only been in existence for 5 years at the time of the report. One of the data-quality issues discussed in the report was "comparability [of statistics about transportation issues] across data systems and time" (p. 32). One example given was that the definition of a transportation fatality was not consistent across modes of transportation until 1994, when consistency was mandated by the Secretary of Transportation. Before that, a highway traffic fatality was counted if death resulted from the accident within 30 days. But for a railroad accident, the time limit was 365 days. Continuing to use different standards for different modes of transportation would make for unfair safety comparisons. In an era in which various transportation modes compete for federal dollars, politics could easily enter a decision to report fatality statistics one way or another.

Probably the most controversial federal data collection issue surrounds the decennial United States census, which has been conducted every 10 years since 1790. In 1988, a lawsuit was filed, led by New York City, alleging that urban citizens are undercounted in the census process. The political ramifications are enormous because redistricting of congressional seats results from shifting population counts. The lawsuit began a long saga of political and statistical twists and turns that was still unresolved at the taking of the 2000 census. For an interesting and readable account, see *Who Counts? The Politics of Census-Taking in Contemporary America*, by Anderson and Fienberg (2001).

The controversy resulted in a decision by the U.S. Census Bureau to conduct a "post enumeration survey" to estimate the extent of the undercount or overcount of various groups. As a result of the survey in 2001, a decision was ultimately made not to adjust the counts in the 2000 census. The results of the post enumeration survey for the 2010 census were released on May 22, 2012 (http://www.census.gov/newsroom/releases/archives/2010_census/cb12-95.html). The summary of the results was as follows:

> *The results found that the 2010 Census had a net overcount of 0.01 percent, meaning about 36,000 people were overcounted in the census. This sample-based result, however, was not statistically different from zero.*

The report contained additional information about the possible undercount or overcount for various regions and groups and is an interesting illustration of many of the principles outlined in this book.

The Importance of Independence for Statistical Agencies

A report by the National Research Council's Committee on National Statistics (Martin, Straf, and Citro, 2001), called *Principles and Practices for a Federal*

Statistical Agency, enumerated three principles and 11 practices that federal agencies should follow. One of the recommended practices was "a strong position of independence." The recommendation included a clear statement about separating politics from the role of statistical agencies:

> In essence, a statistical agency must be distinct from those parts of the department that carry out enforcement and policy-making activities. It must be impartial and avoid even the appearance that its collection, analysis, and reporting processes might be manipulated for political purposes or that individually identifiable data might be turned over for administrative, regulatory, or enforcement purposes.
>
> The circumstances of different agencies may govern the form that independence takes. In some cases, the legislation that establishes the agency may specify that the agency head be professionally qualified, be appointed by the President and confirmed by the Senate, serve for a specific term not coincident with that of the administration, and have direct access to the secretary of the department in which the agency is located (p. 6).

It should be apparent to you that it is not easy to maintain complete independence from politics; for instance, many of the heads of these agencies are appointed by the president and confirmed by the Senate. Other steps taken to help maintain independence include prescheduled release of important statistical information, such as unemployment rates, and authority to release information without approval from the policy-making branches of the organization.

Experimenter Effects and Personal Bias

We learned in Part 1 that there are numerous ways in which *experimenter effects* can lead to bias in statistical studies. If a researcher has a desired outcome for a study and if conditions are not very carefully controlled, it is quite likely that the researcher will influence the outcome. Here are some of the precautions that may be implemented to help prevent this from happening:

- Randomization done by a third party with no vested interest in the experiment, or at least done by a well-tested computer randomization device.
- Automated data recording without intervention from the researcher.
- Double-blind procedures to ensure that no one who has contact with the participants knows which treatment or condition they are receiving.
- An honest evaluation that what is being measured is appropriate and unbiased for the research question of interest.
- A standard protocol for the treatment of all participants that must be strictly followed.

EXAMPLE 26.2 Janet's (Hypothetical) Dissertation Research

This is a hypothetical example to illustrate some of the subtle (and not so subtle) ways experimenter bias can alter the data collected for a study. Janet is a Ph.D. student and is under tremendous pressure to complete her research successfully. For her study, she hypothesized that role-playing assertiveness training for women would help them learn to say "no" to telephone solicitors. She recruited 50 undergraduate women as volunteers for the study. The plan was that each volunteer would come to her office for half an hour. For 25 of the volunteers (the control group), she would simply talk with them for 30 minutes about a variety of topics, including their feelings about saying "no" to unwanted requests. For the other 25 volunteers, Janet would spend 15 minutes on similar discussion and the remaining 15 minutes on a prespecified role-playing scenario in which the volunteer got to practice saying "no" in various situations. Two weeks after each volunteer's visit, Brad, a colleague of Janet's, would phone them anonymously, pretending to be a telephone solicitor selling a magazine for a good price, and would record the conversation so that Janet could determine whether or not they were able to say "no."

It's the first day of the experiment and the first volunteer is in Janet's office. Janet has the randomization list that was prepared by someone else, which randomly assigns each of the 50 volunteers to either Group 1 or Group 2. This first volunteer appears to be particularly timid, and Janet is sure she won't be able to learn to say "no" to anyone. The randomization list says that volunteer 1 is to be in Group 2. But what was Group 2? Did she say in advance? She can't remember. Oh well, Group 2 will be the control group.

The next volunteer comes in and, according to the randomization list, volunteer 2 is to be assigned to Group 1, which now is defined to be the role-playing group. Janet follows her predefined protocol for the half hour. But when the half hour is over, the student doesn't immediately leave. Just then Brad comes by to say "hello," and the three of them spend another half hour having an amiable conversation.

The second phase of the experiment begins and Brad begins phoning the volunteers. The conversations are recorded so that Janet can assess the results. When listening to volunteer 2's conversation, Janet notices that almost immediately she says to Brad, "Your voice sounds awfully familiar, do I know you?" When he assures her that she does not and asks her to buy the magazine, she says, "I can't place my finger on it, but this is a trick, right? I'm sure I know your voice. No thanks, no magazine!" Janet records the data: a successful "no" to the solicitation.

Janet listens to another call, and although she is supposed to be blind to which group the person was in, she recognizes the voice as being one of the role-playing volunteers. Brad pitches the magazine to her and her response is "Oh, I already get that magazine. But if you are selling any others, I might be able to buy one." Janet records the data: a successful "no" to the question of whether she wants to buy that magazine.

The second phase is finally over and Janet has a list of the results. But now she notices a problem. There are 26 people listed in the control group and 24 listed in the role-playing group. She tries to resolve the discrepancy but can't figure it out. She notices that the last two people on the list are both in the control group and that they both said "no" to the solicitation. She figures she will just randomly choose one of them to move over to the role-playing group, so she flips a coin to decide which one to move. ■

This example illustrates just a few of the many ways in which experimenter bias can enter into a research study. Even when it appears that protocols are carefully in place, in the real world it is nearly impossible to place controls on all aspects of the research. In this case, notice that every decision Janet made benefited her desired conclusion that the role-playing group would learn to say "no." It is up to the researchers to take utmost care not to allow these kinds of unethical influences on the results.

26.3   Appropriate Statistical Analyses

There are a number of decisions that need to be made when analyzing the results of a study, and care must be taken not to allow biases to affect those decisions. Here are some examples of decisions that can influence the results:

- Should the alternative hypothesis be one-sided or two-sided? This decision must be made before examining the data.
- What level of confidence or level of significance should be used?
- How should outliers be handled?
- If more than one statistical method is available, which one is most appropriate?
- Have all appropriate conditions and assumptions been investigated and verified?

One of the easiest ethical blunders to make is to try different methods of analysis until one produces the desired results. Ideally, the planned analysis should be spelled out in advance. If various analysis methods are attempted, all analyses should be reported along with the results.

EXAMPLE 26.3   Jake's (Hypothetical) Fishing Expedition

Jake was doing a project for his statistics class and he knew it was important to find a statistically significant result because all of the interesting examples in class had statistically significant results. He decided to compare the memory skills of males and females, but he did not have a preconceived idea of who should do better, so he planned to do a two-sided test. He constructed a memory test in which he presented people with a list of 100 words and allowed them to study it for 10 minutes. Then the next day, he presented another list of 100 words to the participants and asked them to indicate for each word whether it had been on the list the day before. About half of the words on the new list had been on the previous list, and a participant's score was the number of words that were correctly identified as either having been or not having been on the list the day before. Thus, the scores could range from 0 to 100. The answers were entered into a bubble sheet and scored by computer so that Jake didn't inadvertently misrecord any answers. The scores were as follows:

Males: 69, 70, 71, 61, 73, 68, 70, 69, 67, 72, 64, 72, 65, 70, 100

Females: 64, 74, 72, 76, 64, 72, 76, 80, 72, 73, 71, 70, 64, 76, 70

Jake remembered that they had been taught two different tests for comparing independent samples, called a two-sample *t*-test to compare means and a Mann–Whitney rank-sum test to compare medians. He decided to try them both to see which one gave better results. He found the following computer results:

```
Two-Sample T-Test and CI: Males, Females

Two-sample T for Males vs Females

                 N            Mean            StDev       SE Mean
Males           15           70.73            8.73          2.3
Females         15           71.60            4.75          1.2

Difference = mu Males - mu Females
Estimate for difference: -0.87
95% CI for difference: (-6.20, 4.47)
T-Test of difference = 0 (vs not =): T-Value = -0.34 P-Value =
0.739 DF = 21

Mann-Whitney Test and CI: Males, Females

Males            N = 15           Median = 70.00
Females          N = 15           Median = 72.00
Point estimate for ETA1-ETA2 is           -3.00
95.4 Percent CI for ETA1-ETA2 is (-6.00,1.00)
W = 193.5
Test of ETA1 = ETA2 vs ETA1 not = ETA2 is significant at 0.1103
The test is significant at 0.1080 (adjusted for ties)

Cannot reject at alpha = 0.05
```

Jake was disappointed that neither test gave a small enough p-value to reject the null hypothesis. He was also surprised by how different the results were. The t-test produced a p-value of 0.739, whereas the Mann–Whitney test produced a p-value of 0.108, adjusted for ties. It dawned on him that maybe he should conduct a one-tailed test instead. After all, it was clear that the mean and median for the females were both higher, so he decided that the alternative hypothesis should be that females would do better. He reran both tests. This time, the p-value for the t-test was 0.369, and for the Mann–Whitney test, it was 0.054. Maybe he should simply round that one off to 0.05 and be done.

But Jake began to wonder why the tests produced such different results. He looked at the data and realized that there was a large outlier in the data for the males. Someone scored 100%! Jake thought that must be impossible. He knew that he shouldn't just remove an outlier, so he decided to replace it with the median for the males, 70. Just to be sure he was being fair, he reran the original two-sided hypothesis tests. This time, the p-value for the t-test was 0.066, and the p-value for the Mann–Whitney test (adjusted for ties) was 0.0385. Finally! Jake wrote his analysis explaining that he did a two-sided test because he didn't have a preconceived idea of whether males or females should do better. He said that he decided to do the Mann–Whitney test because he had small samples and he knew that the t-test wasn't always appropriate with sample sizes less than 30. He didn't mention replacing the outlier because he didn't think it was a legitimate value anyway.

Although this hypothetical situation is an exaggeration to make a point, hopefully it illustrates the dangers of "data-snooping." If you manipulate the data and try enough different procedures, something will eventually produce desired results. It is not ethical to keep trying different methods of analysis until one produces a desired result. ∎

As a final example, read Example 20.3 (page 432), "The Debate over Passive Smoking." The issue in that example was if it was ethical to report a 90% confidence interval instead of using the standard 95% confidence level.

What do you think? Should the EPA have used the common standard, and reported a 95% confidence interval? If they had done so, the interval would have included values indicating that the risk of lung cancer for those who are exposed to passive smoke may actually be lower than the risk for those who are not. The 90% confidence interval reported was narrower and did not include those values.

26.4 Fair Reporting of Results

Research results are usually reported in articles published in professional journals. Most researchers are careful to report the details of their research, but there are more subtle issues that researchers and journalists sometimes ignore. There are also more blatant reporting biases that can mislead readers, usually in the direction of making stronger conclusions than are appropriate. We have discussed some of the problems with interpreting statistical inference results in earlier chapters, and some of those problems are related to how results are reported. Let's revisit some of those, and discuss some other possible ethical issues that can arise when reporting the results of research.

Sample Size and Statistical Significance

Remember that whether a study achieves statistical significance depends not only on the magnitude of whatever effect or relationship may actually exist but also on the size of the study. In particular, if a study fails to find a statistically significant result, it is important to include a discussion of the sample size used and the power to detect an effect that would result from that sample size. Often, research will be reported as having found no effect or no difference, when in fact the study had such low power that even if an effect exists the study would have been unlikely to detect it.

One of the ethical responsibilities of a researcher is to collect enough data to have relatively high probability of finding an effect if indeed it really does exist. In other words, it is the responsibility of the researcher to determine an appropriate sample size in advance, as well as to discuss the issue of power when the results are presented. A study that is too small to have adequate power is a waste of everyone's time and money.

As we have discussed earlier in the book, the other side of the problem is the recognition that statistical significance does not necessarily imply practical importance. If possible, researchers should report the magnitude of an effect or difference through the use of a confidence interval, rather than just reporting a p-value.

Multiple Hypothesis Tests and Selective Reporting

In most research studies, there are multiple outcomes measured. Many hypothesis tests may be done, looking for whatever statistically significant relationships may exist. For example, as illustrated by Case Study 24.1 about cereal and birth ratios, a study may ask people to record their dietary intake for many different foods or over a long time period so that the investigators can then look for foods that are correlated with certain health outcomes. The problem is that even if there are no legitimate relationships in the population, something in the sample may be statistically significant just by chance. It is unethical to report conclusions only about the results that were statistically significant, without informing the reader about all of the tests that were done.

The *Ethical Guidelines of the American Statistical Association* (1999) lists the problem under "Professionalism" as follows:

> *Recognize that any frequentist statistical test has a random chance of indicating significance when it is not really present. Running multiple tests on the same data set at the same stage of an analysis increases the chance of obtaining at least one invalid result. Selecting the one "significant" result from a multiplicity of parallel tests poses a grave risk of an incorrect conclusion. Failure to disclose the full extent of tests and their results in such a case would be highly misleading.*

In many cases, it is not the researcher who makes this mistake; it is the media. The media is naturally interested in surprising or important conclusions, not in results showing that there is nothing going on. For instance, the story of interest will be that a particular food is related to higher cancer incidence, not that 30 other foods did not show that relationship. It is unethical for the media to publicize such results without explaining the possibility that multiple testing may be responsible for uncovering a spurious chance relationship.

Even if there is fairly strong evidence that observed statistically significant relationships represent real relationships in the population, the media should mention other, less interesting results because they may be important for people making lifestyle decisions. For example, if the relationship between certain foods and disease are explored, it is interesting to know which foods do not appear to be related to the disease as well as those that appear to be related.

EXAMPLE 26.4 Helpful and Harmful Outcomes from Hormone Replacement Therapy

In July 2002, the results were released from a large clinical trial studying the effects of estrogen plus progestin hormone replacement for postmenopausal women. The trial was stopped early because of increased risk of breast cancer and coronary heart disease among the women taking the hormones. However, many news stories failed to report some of the other results of the study, which showed that the hormones actually decreased the risk of other adverse outcomes and were unresolved about others. The original article (Writing Group for the Women's Health Initiative Investigators, 2002) reported the results as follows:

> *Absolute excess risks per 10,000 person-years attributable to estrogen plus progestin were 7 more CHD [coronary heart disease] events, 8 more strokes, 8 more PEs [pulmonary embolism], 8 more invasive breast cancers, while absolute risk reductions per 10,000 person-years were 6 fewer colorectal cancers and 5 fewer hip fractures.*

These results show that in fact some outcomes were more favorable for those taking the hormones, specifically colorectal cancer and hip fractures. Because different people are at varying risk for certain diseases, it is important to report all of these outcomes so that an individual can make an informed choice about whether to take the hormones. In fact, overall, 231 out of 8506 women taking the hormones died of any cause during the study, which is 2.72%. Of the 8102 women taking the placebo, 218 or 2.69% died, a virtually identical result with the hormone group. In fact, when the results are adjusted for the time spent in the study, the death rate was slightly lower in the hormone group, with an annualized rate of 0.52% compared with 0.53% in the placebo group.

The purpose of this example is not to negate the serious and unexpected outcomes related to heart disease, which hormones were thought to protect against, or the serious breast cancer outcome. Instead, the purpose is to show that the results of a large and complex study such as this one are bound to be mixed and should be presented as such so that readers can make informed decisions. ■

Making Stronger or Weaker Conclusions than Are Justified

As we have learned throughout this book, for many reasons research studies cannot be expected to measure all possible influences on a particular outcome. They are also likely to have problems with ecological validity, in which the fact that someone is participating in a study is enough to change behavior or produce a result that would not naturally occur. It is important that research results be presented with these issues in mind and that the case is not overstated even when an effect or relationship is found. An obvious example, often discussed in this book, is that a cause-and-effect conclusion cannot generally be made on the basis of an observational study.

However, there are more subtle ways in which conclusions may be made that are stronger or weaker than is justified. For example, often little attention is paid to how representative the sample is of a larger population, and results are presented as if they would apply to all men, or all women, or all adults. It is important to consider and report an accurate assessment of who the participants in the study are really likely to represent.

Sometimes there are financial or political pressures that can lead to stronger (or weaker) conclusions than are justified. Results of certain studies may be suppressed while others are published, if some studies support the desired outcome and others don't. As cautioned throughout this book, when you read the results of a study that has personal importance to you, try to gain access to as much information as possible about what was done, who was included, and all of the analyses and results.

CASE STUDY 26.1   **Science Fair Project or Fair Science Project?**

In 1998, a fourth-grade girl's science project received extensive media coverage after it led to publication in the *Journal of the American Medical Association* (Rosa et al., 1998). Later that year, in its December 9, 1998 issue, the journal published a series of letters criticizing the study and its conclusions on a wide variety of issues. There are a number of ethical issues related to this study, some of which were raised by the letters and others that have not been raised before.

The study was supposed to be examining "therapeutic touch" (TT), a procedure practiced by many nurses that involves working with patients through a five-step process, including a sensing and balancing of their "energy." The experiment proceeded as follows. Twenty-one self-described therapeutic touch practitioners participated. They were asked to sit behind a cardboard screen and place their hands through cutout holes, resting them on a table on the other side of the screen. The 9-year-old "experimenter" then flipped a coin and used the outcome to decide which of the practitioner's hands to hold her hand over. The practitioner was to guess which hand the girl was hovering over. Fourteen of the practitioners contributed 10 tries each, and the remaining 7 contributed 20 tries each, for a total of 280 tries.

The paper had four authors, including the child and her mother. It is clear from the affiliations of the authors of the paper as well as from language throughout the paper that the authors were biased against therapeutic touch before the experiment began. For example, the first line read "therapeutic touch (TT) is a widely used nursing practice rooted in mysticism but alleged to have a scientific basis" (Rosa et al., 1998, p. 1005). The first author of the paper was the child's mother and her affiliation is listed as "the Questionable Nurse Practices Task Force, National Council Against Health Fraud, Inc."

The paper concludes that "twenty-one experienced TT practitioners were unable to detect the investigator's 'energy field.' Their failure to substantiate TT's most fundamental claim is unrefuted evidence that the claims of TT are groundless and that further professional use is unjustified." The conclusion was widely reported in the media, presumably at least in part because of the novelty of a child having done the research. That would have been cute if it hadn't been taken so seriously by people on both sides of the debate on the validity of therapeutic touch.

The letters responding to the study point out many problematic issues with how the study was done and with its conclusions. Here are several quotes:

The experiments described are an artificial demonstration that some number of self-described mystics were unable to "sense the field" of the primary investigator's 9-year-old daughter. This hardly demonstrates or debunks the efficacy of TT. The vaguely described recruitment method does not ensure or even suggest that the subjects being tested were actually skilled practitioners. More important, the experiments described are not relevant to the clinical issue supposedly being researched. Therapeutic touch is not a parlor trick and should not be investigated as such (Freinkel, 1998).

To describe this child's homework as "research" is without foundation since it clearly fails to meet the criteria of randomization, control, and valid intervention. . . . Flagrant violations against TT include the fact that "sensing" an energy field is not TT but rather a nonessential element in the 5-step process; inclusion of many misrepresentations of cited sources; use of inflammatory language that indicates significant author bias; and bias introduced by the child conducting the project being involved in the actual trials (Carpenter et al., 1998).

I critiqued the study on TT and was amazed that a research study with so many flaws could be published. . . . The procedure was conducted in different settings with no control of environmental conditions. Even though the trials were repeated, the subjects did not change, thus claims of power based on possible repetitions of error are inappropriate. The true numbers in groups are 15 and 13, thus making a type II error highly probable with a study power of less than 30%.

Another concern is whether participants signed informed consent documents or at least were truly informed as to the nature of this study and that publication of its results would be sought beyond a report to the fourth-grade teacher (Schmidt, 1998).

As can be seen by these reader comments, many of the ethical issues covered in this chapter may cloud the results of this study. However, there are two additional points that were not raised in any of the published letters. First, it is very likely that the child knew that her mother wanted the results to show that the participants would not be able to detect which hand was being hovered over. And what child does not want to please her mother? That would not matter so much if there hadn't been so much room for "experimenter effects" to influence the results. One example is the randomization procedure. The young girl flipped a coin each time to determine which hand to hover over. Coin tosses are very easy to influence, and presumably even a 9-year-old child could pick up the response biases of the subjects. The fact that a proper randomization method wasn't used should have ended any chance that this experiment would be taken seriously.

Is there evidence that experimenter bias may have entered the experiment? Absolutely. Of the 280 tries, the correct hand was identified in 123 (44%) of them. The authors of the article conclude that this number is "close to what would be expected for random chance." In fact, that is not the case. The chance of getting 123 *or fewer* guesses by chance is only 0.0242. If a two-tailed test had been used instead of a one-tailed test, the *p*-value would have been 0.048, a statistically significant outcome. The 9-year-old did an excellent job of fulfilling her mother's expectations.

Exercises

Exercises with numbers divisible by 3 (3, 6, 9, etc.) are included in the Solutions at the back of the book. They are marked with an asterisk (*).

1. A classic psychology experiment was conducted by psychology professor Philip Zimbardo in the summer of 1971 at Stanford University. The experiment is described at the website http://www.prisonexp.org/. Visit the website; if it is no longer operative, try an Internet search on "Zimbardo" and "prison" to find information about this study.

   a. Briefly describe the study and its findings.

   b. The website has a number of discussion questions related to the study. Here is one of them: "Was it ethical to do this study? Was it right to trade the suffering experienced by participants for the knowledge gained by the research?" Discuss these questions.

c. Another question asked at the website is, "How do the ethical dilemmas in this research compare with the ethical issues raised by Stanley Milgram's obedience experiments [described in Example 25.1]? Would it be better if these studies had never been done?" Discuss these questions.

2. In the report *Principles and Practices for a Federal Statistical Agency* (Martin, Straf, and Citro, 2001), one of the recommended practices is "a strong position of independence." Each of the following parts gives one of the characteristics that is recommended to help accomplish this. In each case, explain how the recommendation would help ensure a position of independence for the agency.

a. "Authority for selection and promotion of professional, technical, and operational staff" (p. 6).

b. "Authority for statistical agency heads and qualified staff to speak about the agency's statistics before Congress, with congressional staff, and before public bodies" (p. 6).

*3. Refer to Example 26.2 about Janet's dissertation research. Explain whether each of the following changes would have been a good idea or a bad idea:

*a. Have someone who is blind to conditions and has not heard the volunteers' voices before listen to the phone calls to assess whether a firm "no" was given.

*b. Have Janet flip a coin each time a volunteer comes to her office to decide if they should go into the role-playing group or the control group.

*c. Use a variety of different solicitors rather than Brad alone.

4. Refer to Example 26.3 about Jake's memory experiment. What do you think Jake should have done regarding the analysis and reporting of it?

5. Find an example of a statistical study reported in the news. Explain whether you think multiple tests were done, and, if so, if they were reported.

*6. Marilyn is a statistician who works for a company that manufactures components for sound systems. Two development teams have each come up with a new method for producing one of the components, and management is to decide which one to adopt based on which produces components that last longer. Marilyn is given the data for both and is asked to make a recommendation about which one should be used. She computes 95% confidence intervals for the mean lifetime using each method and finds an interval from 96 hours to 104 hours, centered on 100 hours for one of them, and an interval from 92 hours to 112 hours, centered on 102 for the second one. What should she do? Should she make a clear recommendation? Explain.

Use the following scenario for Exercises 7 to 11: Based on the information in Case Study 26.1, describe the extent to which each of the following conditions for reducing the experimenter effect (listed in Section 26.2) was met. If you cannot tell from the description of the experiment, then explain what additional information you would need.

7. Randomization done by a third party with no vested interest in the experiment or at least done by a well-tested computer randomization device.

8. Automated data recording without intervention from the researcher.

*9. Double-blind procedures to ensure that no one who has contact with the participants knows which treatment or condition they are receiving.

10. An honest evaluation that what is being measured is appropriate and unbiased for the research question of interest.

11. A standard protocol for the treatment of all participants that must be strictly followed.

Exercises 12 to 21: Explain the main ethical issue of concern in each of the following. Discuss what, if anything, should have been done differently to address that concern.

*12. Example 26.1, Stanley Milgram's experiment.

13. In Example 26.2, Janet's decision to make Group 2 the control group after the first volunteer came to her office.

14. In Example 26.2, Janet's handling of the fact that her data showed 26 volunteers in the control group when there should have been only 25.

*15. In Example 20.3 (page 432), the Environmental Protection Agency's decision to report a 90% confidence interval.

16. In Example 26.3, Jake's decision to replace the outlier with the median of 70.

17. In Example 26.4, the fact that many media stories mentioned increased risk of breast cancer and coronary heart disease but not any of the other results.

*18. In Case Study 26.1, the concerns raised in the letter from Freinkel.

19. In Case Study 26.1, the concerns raised in the letter from Carpenter and colleagues.

20. In Case Study 26.1, the concerns raised in the letter from Schmidt.

*21. In Case Study 26.1, the concerns raised about the experimenter effect.

22. Visit the websites of one or more professional organizations related to your major. (You may need to ask one of your professors to help with this.) Find one that has a code of ethics. Describe whether the code includes anything about research methods. If not, explain why you think nothing is included. If so, briefly summarize what is included. Include the website address with your answer.

References

American Statistical Association. (1999). Ethical guidelines for statistical practice. http://www.amstat.org/about/ethicalguidelines.cfm.

Anderson, M. J., and S. E. Fienberg. (2001). *Who counts? The politics of census-taking in contemporary America.* New York: Russell Sage Foundation.

Carpenter, J., J. Hagemaster, and B. Joiner. (1998). Letter to the editor. *Journal of the American Medical Association* 280, no. 22, p. 1905.

Citro, C. F., and J. L. Norwood (eds.). (1997). *The Bureau of Transportation Statistics: Priorities for the future.* Washington, D.C.: National Academy Press.

Martin, M. E., M. L. Straf, and C. F. Citro (eds.). (2001). *Principles and practices for a federal statistical agency,* 2d ed. Washington, D.C.: National Academy Press.

Freinkel, A. (1998). Letter to the editor. *Journal of the American Medical Association* 280, no. 22, p. 1905.

Milgram, Stanley. (1983). *Obedience to authority: An experimental view.* New York: Harper/Collins.

Plous, S., and H. Herzog. (27 July 2001). Reliability of protocol reviews for animal research. *Science* 293, pp. 608–609 http://www.socialpsychology.org/articles/scipress.htm.

Rosa, L., E. Rosa, L. Sarner, and S. Barrett. (1998). A close look at therapeutic touch. *Journal of the American Medical Association* 279, no. 13, pp. 1005–1010.

Schmidt, S. M. (1998). Letter to the editor. *Journal of the American Medical Association* 280, no. 22, p. 1906.

Writing Group for the Women's Health Initiative Investigators. (2002). Risks and benefits of estrogen plus progestin in healthy postmenopausal women. *Journal of the American Medical Association* 288, no. 3, pp. 321–333.

CHAPTER 27

Putting What You Have Learned to the Test

This chapter consists solely of case studies. Each case study begins with a full or partial article from a newspaper or journal. Read each article and think about what might be misleading or missing, or simply misinformation. A discussion following each article summarizes some of the points I thought should be raised. You may be able to think of others. I hope that as you read these case studies you will realize that you have indeed become an educated consumer of statistical information.

CASE STUDY 27.1	**Acting Out Your Dreams**

Sources: Mestel (2009).
Nielsen et al. (2009).

A story in the *Los Angeles Times*, headlined "Acting out dreams is pretty common," reported the following results:

> *"Acting out one's dreams is not uncommon, a study finds. Conducted on that lab rat of human psychology—the undergraduate student—it found that as many as 98% of the more than 1000 subjects reported some kind of dream enactment... The precise percentages depended on how the questions were framed, with a simple "Ever acted out your dreams?" kind of question eliciting the lowest percentage (35.9%) and very specific questions eliciting the highest (98.2%)"* (Mestel, 2009). ■

Discussion

The news story did not present many details about the actual study, but instead focused on amusing anecdotes about the experiences of the author and her friends with acting out a dream. If you don't know what the term "acting out a dream" means, it turns out that you are not alone. The news story provided enough detail that it was possible to find the original journal article, which had been published in the December, 2009, issue of the journal *Sleep*. In consulting the original journal article, we learn what the news story meant by mentioning that the precise percentages depended on how the questions were framed. Participants in the study (psychology students) were divided into three groups. Here are the three versions of what they were asked (Nielsen et al., 2009, p. 1631). Notice the subtle difference between what was asked of Groups 1 and 2.

Group 1 ($n = 443$): "On how many nights did the following occur in the last year?... acting out a dream (while still dreaming)."

Group 2 ($n = 201$): "On how many nights did the following occur in the last year?... acting out a dream while still dreaming (e.g., crying, laughing or arm/leg movements expressing a dream)."

Group 3 ($n = 496$): "The following questions [1–7] concern behaviors that are acted out while you are dreaming about them. The behaviors are different from sleepwalking or sleep-talking behaviors [questions 8–9] which are not accompanied by clear dreams."

For Group 3, a list of behaviors was then presented, such as "How often have you awakened from a dream about talking to find that you are speaking out loud some of the words in the dream?" and "How often have you awakened from a happy dream to find that you are actually smiling or laughing?" Before reading the rest of the discussion, see if you can guess which group had the lowest and highest rate of reported occurrences, based on material you have learned earlier in this book.

If you guessed that Group 1 had the lowest reported rate of acting out a dream and Group 3 had the highest, you are correct. A mere 35.9% of Group 1 responded that they had had such an occurrence, but 76.7% of Group 2 responded that way. The only difference in the two versions is that Group 2 was presented with detailed examples of what the question meant. Simply adding those examples more than doubled the percentage that recognized that they had had this type of experience. As you might guess, the overall response was even higher for Group 3. Almost everyone (98.2%) responded that they had had at least one of those occurrences in the past year. In case you are curious, the most common experience was waking up in fear, which was reported by 92.7% of the respondents in Group 3. Least common was crying, but even that was reported by 54.3% of respondents.

This example illustrates two important points made earlier in this book. First, a small change in wording of questions can have a major impact on how people answer. Second, people rate detailed scenarios more highly than generalities.

CASE STUDY 27.2   **Cranberry Juice and Bladder Infections**

SOURCE: "Juice does prevent infection" (9 March 1994), *Davis* (CA) *Enterprise*, p. A9. Reprinted with permission.

CHICAGO (AP)—A scientific study has proven what many women have long suspected: Cranberry juice helps protect against bladder infections.

Researchers found that elderly women who drank 10 ounces of a juice drink containing cranberry juice each day had less than half as many urinary tract infections as those who consumed a look-alike drink without cranberry juice.

The study, which appeared today in the *Journal of the American Medical Association,* was funded by Ocean Spray Cranberries, Inc., but the company had no role in the study's design, analysis, or interpretation, *JAMA* said.

"This is the first demonstration that cranberry juice can reduce the presence of bacteria in the urine in humans," said lead researcher Dr. Jerry Avorn, a specialist in medication for the elderly at Harvard Medical School.   ■

Discussion

This study was well conducted, and the newspaper report is very good, including information such as the source of the funding and the fact that there was a placebo ("a look-alike drink without cranberry juice"). A few details are missing from the news account, however, that you may consider to be important. They are contained in the original report (Avorn et al., 1994):

1. There were 153 subjects.

2. The participants were randomly assigned to the cranberry or placebo group.

3. The placebo was fully described as "a specially prepared synthetic placebo drink that was indistinguishable in taste, appearance, and vitamin C content, but lacked cranberry content."

4. The study was conducted over a 6-month period, with urine samples taken monthly.

5. The measurements taken were actually bacteria counts from urine, rather than a more subjective assessment of whether an infection was present. The news story claims "less than half as many urinary tract infections," but the original report gave the odds of bacteria levels exceeding a certain threshold. The odds in the cranberry juice group were only 42% of what they were in the control group. Unreported in the news story is that the odds of remaining over the threshold from one month to the next for the cranberry juice group were only 27% of what they were for the control group.

6. The participants were elderly women volunteers with a mean age of 78.5 years and high levels of bacteria in their urine at the start of the study.

7. The original theory was that if cranberry juice was effective, it would work by increasing urine acidity. That was not the case, however. The juice inhibited bacterial growth in some other way.

CASE STUDY 27.3   Children on the Go

SOURCE: Coleman, Brenda C. (15 September 1993), Children who move often are problem prone, *West Hawaii Today*. Reprinted with permission.

CHICAGO—Children who move often are 35 percent more likely to fail a grade and 77 percent more likely to have behavioral problems than children whose families move rarely, researchers say.

A nationwide study of 9915 youngsters ages 6 to 17 measured the harmful effects of moving. The findings were published in today's issue of the *Journal of the American Medical Association.*

About 19 percent of Americans move every year, said the authors, led by Dr. David Wood of Cedars-Sinai Medical Center in Los Angeles. The authors cited a 1986–87 Census Bureau study.

The authors said that our culture glorifies the idea of moving through maxims such as "Go West, young man."

Yet moving has its "shadow side" in the United States, where poor and minority families have been driven from place to place by economic deprivation, eviction, and racism, the researchers wrote.

Poor families move 50 percent to 100 percent more often than wealthier families, they said, citing the Census Bureau data.

The authors used the 1988 National Health Interview Survey and found that about one-quarter of children had never moved, about half had moved fewer than three times and about three-quarters fewer than four times.

Ten percent had moved at least six times, and the researchers designated them "high movers."

Compared with the others, the high movers were 1.35 times more likely to have failed a grade and 1.77 times more likely to have developed at least four frequent behavioral problems, the researchers said. Behavioral problems ranged from depression to impulsiveness to destructiveness.

Frequent moving had no apparent effect on development and didn't appear to cause learning disabilities, they found.

The researchers said they believe their study is the first to measure the effects of frequent relocation on children independent of other factors that can affect school failure and behavioral problems.

Those factors include poverty, single parenting, belonging to a racial minority, and having parents with less than a high school education.

Children in families with some or all of those traits who moved often were much more likely to have failed a grade—1.8 to 6 times more likely—than children of families with none of those traits who seldom or never moved.

The frequently relocated children in the rougher family situations also were 1.8 to 3.6 times more likely to have behavioral problems than youngsters who stayed put and lived in more favorable family situations.

"A family move disrupts the routines, relationships, and attachments that define the child's world," researchers said. "Almost everything outside the family that is familiar is lost and changes."

Dr. Michael Jellinek, chief of child psychiatry at Massachusetts General Hospital, said he couldn't evaluate whether the study accurately singled out the effect of moving. ■

Discussion

There are some problems with this study and with the reporting of it in the newspaper. First, it is not until well into the news article that we finally learn that the reported figures have already been adjusted for confounding factors, and even then, the news article is not clear about this. In fact, the 1.35 relative risk of failing a grade and the 1.77 relative risk of at least four frequent behavioral problems apply *after* adjusting for poverty, single parenting, belonging to a racial minority, and having parents with less than a high school education.

The news report is also missing information about baseline rates. From the original report (Wood et al., 1993), we learn that 23% of children who move frequently have repeated a grade, whereas only 12% of those who never or infrequently move have repeated a grade.

The news report also fails to mention that the data were based on surveys with parents. The results could, therefore, be biased by the fact that some parents may not have been willing to admit that their children were having problems.

Although the news report implies that moving is the cause of the increase in problems, this is an observational study and causation cannot be established. Numerous other confounding factors that were not controlled for could account for the results. Examples are number of days missed at school, age of the parents, and quality of the schools attended.

| CASE STUDY 27.4 | **It Really Is True about Aspirin** |

SOURCE: Reprinted with permission from Aldhous, Peter (7 January 1994), A hearty endorsement for aspirin, *Science* 263, p. 24. Copyright 1994 AAAS.

Aspirin is one of the world's most widely used drugs, but a major study being published this week suggests it is not used widely enough. If everybody known to be at high risk of vascular disease were to take half an aspirin a day, about 100,000 deaths and 200,000 nonfatal heart attacks and strokes could be avoided worldwide each year, the study indicates. "This is one of the most cost-effective drug interventions one could have in developed countries," says Oxford University epidemiologist Richard Peto, one of the coordinators of the study, a statistical overview, or meta-analysis, of clinical trials in which aspirin was used to prevent blood clots.

The new meta-analysis, which covers both aspirin and more expensive antiplatelet drugs, combined the results of 300 trials involving 140,000 patients. Its recommendation: A regime of half a tablet of aspirin a day is valuable for all victims of heart attack and stroke, and other at-risk patients such as angina sufferers and recipients of coronary bypass grafts. . . . The full analysis will be published in three consecutive issues of the *British Medical Journal* beginning on 8 January.

While the results are likely to bring consensus to the field, there are still a few areas of uncertainty. One major disagreement surrounds the study's finding that aspirin can reduce thrombosis in patients immobilized by surgery, where some of the studies were not double-blinded, and therefore could be biased. Worried about this possibility, [researchers] have reanalyzed a smaller set of data. They included only results of trials that had faultless methodology and looked separately at patients in general surgery and those who had had procedures such as hip replacements, where the surgery itself can damage veins in the leg and further increase the risk of thrombosis. Their conclusion: Aspirin is beneficial only following orthopedic surgery.

Discussion

A table accompanying the article summarized the studies [see Table 27.1 on next page].

This article, which is only partially quoted here, is a good synopsis of a major meta-analysis. It does not deal with the influence of aspirin on healthy people, but it does seem to answer the question of whether aspirin is beneficial for those who have had cardiovascular problems. Table 27.1 provides information on the magnitude of the effect. Although no confidence limits are provided, you can get a rough estimate of them yourself because you are told the sample sizes.

It does seem odd that the numbers in the table only sum to 70,000, although the article reports that there were 140,000 patients. No explanation is given for the discrepancy.

TABLE 27.1   Meta-Analysis of Aspirin Studies

	Summary of Results of Aspirin Trials		Proportion Who Suffered Nonfatal Stroke, Nonfatal Heart Attack, or Death During Trial	
Type of Patient	Length of Treatment	Number of Patients	Treatment Group	Control Group
Suspected acute heart attack	1 month	20,000	10%	14%
Previous history of heart attack	2 years	20,000	13%	17%
Previous history of stroke	3 years	10,000	18%	22%
Other vascular diseases	1 year	20,000	7%	9%

Source: Reprinted with permission from Aldhous, Peter (7 January 1994), A hearty endorsement for aspirin, *Science* 263, p. 24. Copyright 1994 AAAS.

The main problem you should have recognized occurs at the end of the reported text. It is indicated that aspirin is not beneficial following surgery except for orthopedic surgery. This should have alerted you to the problem of reduced sample size when only a subset of the data is considered. In fact, the article continued as follows:

> Oxford cardiologist Rory Collins, another coordinator of the study, counters that splitting the data into such small chunks defeats the object of conducting a meta-analysis of many trials. And he notes that if the data for general and orthopedic surgery are considered together, there is still a significant benefit from aspirin, even if methodologically suspect trials are excluded.

In other words, analyzing the studies with faultless methodology only, the researchers concluded that there is still benefit to using aspirin after surgery. If the studies examined are further reduced to consider only those following general surgery, the added benefit of taking aspirin is not statistically significant. Remember, that *does not* mean that there is no effect.

CASE STUDY 27.5   **You Can Work and Get Your Exercise at the Same Time**

SOURCE: White collar commute (11 March 1994), *Davis* (CA) *Enterprise*, p. C5.

One in five clerical workers walks about a quarter mile a day just to complete routine functions like faxing, copying, and filing, a national survey on office

(Continued)

efficiency reports. The survey also shows that the average office worker spends close to 15 percent of the day just walking around the office. Not surprisingly, the survey was commissioned by Canon U.S.A., maker of—you guessed it—office copiers and printers that claim to cut the time and money "spent running from one machine to the next." ■

Discussion

We are not given any information that would allow us to evaluate the results of this survey. Further, what does it mean to say that one in five workers walks that far? Do the others walk more or less? Is the figure based on an average of the top 20% (one in five) of a set of numbers, in which case outliers for office delivery personnel and others are likely to distort the mean?

CASE STUDY 27.6	Sex, Alcohol, and the First Date

SOURCE: "Teen alcohol-sex survey finds the unexpected" (11 May 1994), *Sacramento Bee*, p. A11. Reprinted with permission.

WASHINGTON—Young couples are much more likely to have sex on their first date if the male partner drinks alcohol and the woman doesn't, new research shows.

The research, to be presented today at an American Psychological Association conference on women's health, contradicts the popular male notion that plying a woman with alcohol is the quickest path to sexual intercourse.

In fact, interviews with 2052 teenagers found that they reported having sex on a first date only 6 percent of the time if the female drank alcohol while the male did not. That was lower than the 8 percent who reported having sex when neither partner drank.

Nineteen percent of the teens reported having sex when both partners drank, but the highest frequency of sex on the first date—24 percent—was reported when only the male drank.

The lead researcher in the study, Dr. M. Lynne Cooper of the State University of New York at Buffalo, said that drinking may increase a man's willingness "to self-disclose things about himself, be more likely to communicate feelings, be more romantic—and the female responds to that."

Rather than impairing a woman's judgment, alcohol apparently makes many women more cautious, Cooper said. "Women may sort of smell danger in alcohol, and it may trigger some warning signs," she said. "It makes a lot of women more anxious." ■

Discussion

This is an excellent example of a misinterpreted observational study. The authors of both the original study and the news report are making the assumption that drinking behavior influences sexual behavior. Because the drinking behavior was clearly not randomly assigned, there is simply no justification for such an assumption.

Perhaps the causal connection is in the reverse direction. The drinking scenario that was most frequently associated with sex on the first date was when the male drank but the female did not. If a couple suspected that the date would lead to sex, perhaps they would plan an activity in which they both had access to alcohol, but the female decided to keep her wits.

The simplest explanation is that the teenagers did not tell the truth about their own behavior. In 1994, it would have been less socially acceptable for a female to admit that she drank alcohol and then had sex on a first date than it would be for a male. Therefore, it could be that males and females both exaggerated their behavior, but in different directions.

CASE STUDY 27.7	**Unpalatable Pâté**

SOURCE: Plous, S., (1993), Psychological mechanisms in the human use of animals, *Journal of Social Issues* 49, no. 1, pp. 11–52. Copyright © 1993 Blackwell Publishing. Reprinted with permission.

This article is about human psychological perceptions of the use of animals. Just one paragraph will be used for this example:

> *With the exception of several recent public opinion polls on animal rights and a few studies on vegetarianism and hunting, a computer-assisted literature review yielded only six published research programs specifically investigating attitudes toward the use of animals. The first was an exploratory study in which roughly 300 Australian students were asked whether they approved or disapproved of certain uses of animals (Braithwaite and Braithwaite, 1982). As it turned out, students frequently condemned consumptive practices while endorsing consumption itself. For example, nearly three-fourths of the students disapproved of "force-feeding geese to make their livers swell up to produce pâté for restaurants," but the majority did not disapprove of "eating pâté produced by the force-feeding of geese." The authors interpreted these findings as evidence of an inconsistency between attitudes and behaviors* (p. 13).

Discussion

This is a good example of how a slight variation of the wording of a question can produce very different results. The first version of the question is almost certain to produce more negative results because of the mention of making the geese' livers swell. It would have been more appropriate to ask the students how they felt about "force-feeding geese to produce pâté for restaurants," without mention of the livers swelling. The authors may have overinterpreted the results by claiming that this illustrates a difference between attitudes and behaviors rather than acknowledging the difference in wording.

| CASE STUDY 27.8 | **Nursing Moms Can Exercise, Too** |

SOURCE: "Aerobics OK for breast-feeding moms" (17 February 1994), *Davis* (CA) *Enterprise,* p. A7. Reprinted with permission.

Moderate aerobic exercise has no adverse effects on the quantity or quality of breast milk produced by nursing mothers and can significantly improve the mothers' cardio-vascular fitness, according to UC Davis researchers.

For 12 weeks, the study monitored 33 women, beginning six to eight weeks after the births of their children. All were exclusively breast-feeding their infants, with no formula supplements, and had not previously been exercising. Eighteen women were randomly assigned to an exercise group and 15 to a nonexercising group. The exercise group participated in individual exercise programs, including rapid walking, jogging, or bicycling, for 45 minutes each day, five days per week. . . . At the end of the 12-week study, [the researchers] found:

Women in both groups experienced weight loss. The rate of weight loss and the decline in the percentage of body fat after childbirth did not differ between the exercise and control groups, because women in the exercise group compensated for their increased energy expenditure by eating more.

There was an important improvement in the aerobic fitness of the exercising women, as measured by the maximal oxygen consumption.

There was no significant difference between the two groups in terms of infant breast-milk intake, energy output in the milk, or infant weight gain.

Prolactin levels in the breast milk did not differ between the two groups, suggesting that previously observed short-term increases in the level of that hormone among non-lactating women following exercise do not influence the basal level of prolactin.   ■

Discussion

This is an excellent study and excellent reporting. The study was a randomized experiment and not an observational study, so we can rule out confounding factors resulting from the fact that mothers who choose to exercise differ from those who do not. The mothers were randomly assigned to either the exercise group or a nonexercising control group. To increase ecological validity and generalizability, they were allowed to choose their own forms of exercise. All mothers were exclusively breast feeding, ruling out possible interactions with other food intake on the part of the infants. The study could obviously not be performed blind because the women knew whether they were exercising. It presumably could have been single blind, but was not.

The article reports that there was an important improvement in fitness; it does not give the actual magnitude. The reporter has done the work of determining that the improvement is not only statistically significant but also of practical importance. From the original research report, we learn that "maximal oxygen uptake increased by 25 percent in the exercising women but by only 5 percent in the control women ($P < .001$)" (Dewey et al., 1994, p. 449).

What about the differences that were reported as nonsignificant or nonexistent, such as infant weight gain? There were no obvious cases of misreporting important but not significant differences. Most of the variables for which no significant differences were found were very close for the two groups. In the original report, confidence intervals are given for each group, along with a p-value for testing whether there is a statistically significant difference. For instance, 95% confidence intervals for infant weight gain are 1871 to 2279 grams for the exercise group and a very similar 1733 to 2355 grams for the control group ($p = 0.86$). The p-value indicates that if aerobic exercise has no impact on infant weight gain in the population, we would expect to see sample results differ this much or more very often. Therefore, we cannot rule out chance as an explanation for the very small differences observed. The same is true of the other variables for which no differences were reportedly found.

| CASE STUDY 27.9 | So You Thought Spinach Was Good for You? |

SOURCE: Reprinted with permission from Nowak, R. (22 April 1994), Beta-carotene: Helpful or harmful? *Science* 264, pp. 500–501. Copyright 1994 AAAS.

Over the past decade, it's become a tenet of cancer prevention theory that taking high doses of antioxidant vitamins—like vitamin E or A—will likely protect against cancer. So in light of that popular hypothesis, cancer prevention experts are having to struggle to make sense of the startling finding, published in the 14 April *New England Journal of Medicine,* that supplements of the antioxidant beta-carotene markedly increased the incidence of lung cancer among heavy smokers in Finland.

The result is particularly worrying because it comes from a large, randomized clinical trial—the gold standard test of a medical intervention. And as well as dumbfounding the experts, the Finnish study has triggered calls for a moratorium on health claims about antioxidant vitamins (beta-carotene is converted into vitamin A in the body) and prompted close scrutiny of several other large beta-carotene trials that are currently under way.

What mystifies the experts is that the Finnish trial goes against all the previously available evidence. Beta-carotene's biological activity suggests that it should protect against cancer. It's an antioxidant that can sop up chemicals called free radicals that may trigger cancer. And over a hundred epidemiologic surveys indicate that people who have high levels of beta-carotene in their diet and in their blood have lower risks of cancer, particularly lung cancer. Finally, the idea that beta-carotene would have only beneficial effects on cancer is buttressed by the results of the only other large-scale clinical trial completed thus far. It found that a combination of beta-carotene, vitamin E, and selenium reduced the number of deaths from stomach cancer by 21% among 15,000 people living in Linxian County in China, compared with trial participants who did not take the supplements. . . .

. . . When all the data were in and analyzed at the end of the trial, it became apparent that the incidence of lung cancer was 18% higher among the 14,500 smokers who took beta-carotene than among the 14,500 who didn't. The probability that the increase was due to chance is less than one in one hundred. In clinical trials, a difference is taken seriously when there is less than a one-in-twenty probability that it happened by chance.

(Continued)

The trial organizers were so baffled by the results that they even wondered whether the beta-carotene pills used in the study had become contaminated with some known carcinogen during the manufacturing process. Tests have ruled out that possibility.   ■

Discussion

The researchers seem to be getting very upset about what could simply be the luck of the draw. Notice that the article mentions that "over a hundred epidemiologic surveys indicate that people who have high levels of beta-carotene in their diet and in their blood have lower risks of cancer, particularly lung cancer." Further, the p-value associated with this test is "less than one in one hundred." It is true by the nature of chance that occasionally a sample will result in a statistically significant finding that is the reverse of what is true in the population.

It may sound like a sample of size 29,000 is large enough to rule out that explanation. But from the original report, we find that there were only 876 new cases of lung cancer among all the groups. Further, although "the incidence of lung cancer was 18% higher among the 14,500 smokers who took beta-carotene than among the 14,500 who didn't," a 95% confidence interval for the true percentage in the population ranged from 3% to 36% higher. Although the entire confidence interval still lies in the direction of higher rates for those who took beta-carotene, the lower endpoint of 3% is close to zero.

It is also the case that the researchers performed a multitude of tests, any one of which may have been surprising if it had come out in the direction opposite to what was predicted. For example, when they looked at total mortality, they found that it was only 8% higher for the beta-carotene group (mainly due to the increased lung-cancer deaths), and a 95% confidence interval ranged from 1% to 16%.

It may turn out that beta-carotene is indeed harmful for some people. But the furor caused by this study appears to be overblown, given the number of studies that have been done on this topic and the probability that eventually a study will show a relationship opposite to that in the population just by chance.

CASE STUDY 27.10   **Chill Out—Move to Honolulu**

SOURCE: Friend, Tim (13 May 1994), Cities of discord: Home may be where the heart disease is, *USA Today*, p. 5D. *USA Today*. Copyright May 13, 1994. Reprinted with permission.

Do you count the number of items that the selfish person in front of you is holding in the express lane at the grocery store?

Been fantasizing about slamming your brakes the next time some jerk tailgates you in rush hour?

Yeah? Then you may have a "hostile" personality, on the road to an early death.

"Anger kills," says Dr. Redford Williams, Duke University Medical Center, whose book by the same name is out today in paperback.

Williams also releases today his study revealing the five most hostile and five mellowest cities in the United States. It's based on a poll that measured citizens' hostility levels and compared them with cities' death rates.

"The study was stimulated by research of people with hostile personalities and high levels of cynical mistrust of others—people who have frequent anger and who express that anger overtly," Williams says.

More than two decades of research involving anger and heart disease show that people with higher hostility levels have higher rates of heart disease deaths and overall deaths.

"The hypothesis for the new study was since we know that hostility is a risk factor for high death rates, are cities characterized by high hostility levels also characterized by high death rates? Is what's bad for an individual, bad for a population?" Williams asks.

He and the Gallup Organization found that cities with higher hostility scores consistently had higher death rates. Those with lower hostility scores had lower death rates. The new finding suggests hostile cities may want to chill out.

SOURCE: Friend, Tim (13 May 1994), Philly hostile? "Who asked ya?", *USA Today*, p. 1D. *USA Today*. Copyright May 13, 1994. Reprinted with permission.

Philadelphia—the City of Brotherly Love—appears to be the USA's most hostile town, and may be paying for it with higher death rates.

Dr. Redford Williams, Duke University Medical Center, asked the Gallup Organization to measure the levels of hostility and mistrust in 10 cities from states with the highest and lowest heart disease death rates.

Studies have shown that people with hostile personalities have an increased risk of dying from heart disease, he says.

He wanted to find out if that applied to cities, too.

Hostility levels were measured, ranked, and then paired with cities' death rates:

- Philadelphia had the highest hostility score and highest death rate. New York was second, Cleveland third.
- Honolulu had the lowest hostility score and the lowest death rate. Seattle was second, Minneapolis was third lowest.
- Death rates in the five cities with the highest hostility were 40% higher than in the cities with the lowest score.

Statistically, the probability of the correlation occurring by chance is less than 1 in 10,000. ■

Discussion

These two stories appeared on the same day in *USA Today*. They discuss a study that Dr. Redford Williams commissioned the Gallup Organization to conduct and which he released on the same day the paperback edition of his book was to be released. That sounds like a good advertising gimmick.

The study sounds interesting, but we are not provided with sufficient details to ascertain whether there are major confounding effects. For example, was the interviewer *also* from the city that was being interviewed? Remember that the interviewer can bias responses to questions. Did the interviewers know the purpose of the study? Also, one piece of information is available only by consulting a graph accompanying the articles. The graph, titled "Hostility leads to death," shows a scatterplot of death rate versus hostility for the 10 cities. At the bottom is a footnote: "[Hostility index] adjusted for race, education, age, income, gender." What does that mean?

The main problem with these articles is the implication that if the residents of cities like Philadelphia would lower their hostility index, they would also lower their death rates. But this is an observational study, and there are numerous potential confounding factors. For instance, do you think the weather and pollution levels are better for you in Honolulu or in Philadelphia? Do you think you are more likely to be murdered in New York City or in Seattle? Are the populations equally likely to be elderly in all of the cities?

The implication that these results would occur by chance less than once in 10,000 is also misleading. As admitted in the article, the study purposely used cities that were outliers in both directions on death rates. The graph shows that the death rate (per 1000) in Honolulu is less than half of what it is in Philadelphia. Remember from Chapter 11 that outliers can have a large impact on correlations.

CASE STUDY 27.11   **So You Thought Hot Dogs Were Bad for You?**

SOURCE: Reprinted with permission from Holden, Constance (27 May 1994), Hot dog hazards, *Science* 264, p. 1255. Copyright 1994 by AAAS.

Parents worried about whether electromagnetic fields (EMFs) cause cancer now have a more all-American concern: Hot dogs, warns a study from the University of Southern California (USC), are more than 4 times as likely as EMFs to be linked with childhood leukemia. But the researchers caution that the data may not cut the mustard.

In the May issue of *Cancer Causes and Control,* three research groups report a link between cured-meat consumption and cancer. The most striking evidence comes from a group led by USC epidemiologist John Peters, which earlier found that EMF exposure is associated with a doubling of the risk for childhood leukemia. Among the 232 cases in the study, children who ate 12 or more hot dogs a month were nine times as likely as hot dog-free controls to develop leukemia. Peters also found an increased risk for kids whose fathers ate a lot of hot dogs.

Not that Mom can eat with impunity either: The other two studies linked maternal intake of hot dogs and cured meats during pregnancy with childhood brain tumors. Looking at 234 cases of various childhood cancers, University of North Carolina epidemiologist David Savitz found that children whose mothers downed hot dogs at least once a week were more than twice as likely as controls to develop brain tumors. And in a study at Children's Hospital of Philadelphia, epidemiologist Greta Bunin found a weak link between maternal hot dog intake during pregnancy and increased risk of a brain tumor, astrocytic glioma, in their children.

The epidemiologists say that the effects they observe could be from the N-nitroso compounds in cured meats, such as nitrites, which cause cancer in lab animals. And that makes sense, says Savitz, because vitamins—with their carcinogen-fighting antioxidant properties—appear to have a protective effect in juvenile hot dog eaters.

Savitz warns that the studies are far from conclusive. They all suffer from a lack of data on subjects' exposures to other N-nitroso compounds. And, says Bunin, "the cured-meat association could be an indicator of a diet poor in other ways." What's needed now, Savitz says, is research looking more closely at diet. "Who knows, maybe it's the condiments," he says.   ■

Discussion

From the information available in this story, it appears that the three studies were all of the case-control type. In other words, researchers located children who already had leukemia or other cancers and similar "control" children, and then interviewed the families about their habits.

One of the problems with trying to link diet to cancer in a case-control study is that people have poor memories about what they ate years ago. Parents who are searching for an answer to why their child has cancer are probably more likely to "remember" feeding the child an unhealthful diet than parents whose children do not have cancer.

Another obvious problem, partially addressed in the story, is that a diet that includes 12 or more hot dogs a month is likely to be low in consumption of more healthful foods such as fruits and vegetables. So, the fact that the children who consumed more hot dogs were more likely to have developed cancer could be confounded with the fact that they consumed fewer vitamin-rich foods, had higher fat in their diets, or a variety of other factors. The article does mention that "vitamins . . . appear to have a protective effect in juvenile hot dog eaters," indicating that those who consumed vitamin-rich foods, or vitamin supplements, did not have the same increased risk as those who did not.

One additional problem with the report of these studies is that no baseline information about leukemia rates in children is provided. Because hot dogs are a convenient food that children like to eat, parents should be provided with the information needed to weigh the risks against the benefits. According to an article in the Atlantic City *Press* (1 July 1994, p. C10), leukemia rates for children in New Jersey counties range from about 3.1 to 7.9 per 100,000. Remember that those figures include the whole population, so they do not mean that the rate for hot dog consumers would be nine times those rates. Therefore, although a relative risk of 9.0 for developing leukemia is frighteningly high, the overall rate of leukemia in children is still extremely low.

References

Avorn, J., M. Monane, J. H. Gurwitz, R. J. Glynn, L. Choodnovskiy, and L. A. Lipsitz. (1994). Reduction of bacteriuria and pyuria after ingestion of cranberry juice. *Journal of the American Medical Association* 271, no. 10, pp. 751–754.

Braithwaite, J., and V. Braithwaite. (1982). Attitudes toward animal suffering: An exploratory study. *International Journal for the Study of Animal Problems* 3, pp. 42–49.

Dewey, K. G., C. A. Lovelady, L. A. Nommsen-Rivers, M. A. McCrory, and B. Lönnerdal. (1994). A randomized study of the effects of aerobic exercise by lactating women on breast-milk volume and composition. *New England Journal of Medicine* 330, no. 7, pp. 449–453.

Mestel, Rosie. (2009). Acting out dreams is pretty common. *Los Angeles Times*, http://latimesblogs.latimes.com/booster_shots/2009/11/acting-out-dreams-is-pretty-common-study-finds.html, 30 November 2009.

Nielsen, Tore, Connie Svob and Don Kuiken. (2009). Dream-enacting behaviors in a normal population. *Sleep*, 32(12), pp. 1629–1636.

Wood, D., N. Halfon, D. Scarlata, R. Newacheck, and S. Nessim. (1993). Impact of family relocation on children's growth, development, school function, and behavior. *Journal of the American Medical Association* 270, no. 11, pp. 1334–1338.

Contents of the Appendix and Companion Website

I. News Stories and Sources in the Appendix and on the Companion Website

News Stories: All of the news stories are either printed or summarized in the Appendix, and most of them are on the Companion Website as well.

Additional News Stories: An additional news story accompanies a few of the studies, either in the Appendix, on the Companion Website, or both.

Original Sources: Most of the journal articles and reports listed as Original Sources are on the Companion Website. The ones that are not included may be found through online access at most university and college libraries. Links are provided for some of them in this listing, but cannot be guaranteed to be up to date.

News Story 1

"Mending Mindfully"
Heidi Kotansky, *Yoga Journal*, November 2003, p. 35.

Online at http://www.yogajournal.com/practice/1050

Original Source 1 *(Not on Companion Website)*: Davidson, Richard J., PhD; Kabat-Zinn, Jon, PhD; Schumacher, Jessica, MS; Rosenkranz, Melissa, BA; Muller, Daniel, MD, PhD; Santorelli, Saki F.,EdD; Urbanowski, Ferris, MA; Harrington, Anne, PhD; Bonus, Katherine, MA; and Sheridan, John F., PhD. (2003). "Alterations in Brain and Immune Function Produced by Mindfulness Meditation," *Psychosomatic Medicine* 65, pp. 564–570.

News Story 2

"Research Shows Women Harder Hit by Hangovers"
Lee Bowman, *Sacramento Bee* (from Scripps Howard News Service) 15 September 2003, p. A7.

Original Source 2: Slutske, Wendy S.; Piasecki, Thomas M.; and Hunt-Carter, Erin E. (2003). "Development and Initial Validation of the Hangover Symptoms Scale: Prevalence and Correlates of Hangover Symptoms in College Students." *Alcoholism: Clinical and Experimental Research 27*, pp. 1442–1450.

News Story 3

"Rigorous Veggie Diet Found to Slash Cholesterol"
Sacramento Bee (original by Daniel Q. Haney, Associated Press) 7 March 2003, p. A9.

Additional News Story 3

http://news.utoronto.ca/_bulletin/2003-07-28.pdf, page 5

University of Toronto news release, July 22, 2003: "Diet as good as drug for lowering cholesterol, study says," by Lanna Crucefix.

Original Source 3: Jenkins, David J. A.; Kendall, Cyril W. C.; Marchie, Augustine; Faulkner, Dorothea A.; Wong, Julia M. W.; de Souza, Russell; Emam, Azadeh; Parker, Tina L.; Vidgen, Edward; Lapsley, Karen G.; Trautwein, Elke A.; Josse, Robert G.; Leiter, Lawrence A.; and Connelly, Philip W. (23 July 2003). "Effects of a Dietary Portfolio of Cholesterol-Lowering Foods vs Lovastatin on Serum Lipids and C-Reactive Protein." *Journal of the American Medical Association 290*, no. 4, pp. 502–510.

News Story 4

"Happy People Can Actually Live Longer"
Henry Reed, Venture Inward Magazine, October/November 2003, p. 9.

Original Source 4:

http://today.duke.edu/2003/04/cardiacstudyemotions04.03.html

"*Duke Health Briefs*: Positive Outlook Linked to Longer Life in Heart Patients." 21 April 2003.

News Story 5

"Driving While Distracted Is Common, Researchers Say; Reaching for Something in the Car and Fiddling with the Stereo Far Outweigh Inattentiveness Caused by Cellphone Use, Traffic Study Finds"
Susannah Rosenblatt, *Los Angeles Times* (Los Angeles, Calif.) 7 August 2003, p. A.20.

Online at http://articles.latimes.com/2003/aug/07/nation/na-drivers7

Original Source 5: Stutts et al., "Distractions in Everyday Driving." Technical report prepared for the AAA Foundation for Traffic Safety, June 2003 (129 pages). Available online at:

https://www.aaafoundation.org/distractions-everyday-driving-0

News Story 6

"Music as Brain Builder"
Constance Holden (26 March 1999), *Science* 283, p. 2007.

Original Source 6: Graziano, Amy B.; Peterson, Matthew; and Shaw, Gordon L. (March 1999). "Enhanced Learning of Proportional Math through Music Training and Spatial-temporal Training." *Neurological Research* 21, pp. 39–152.

News Story 7

"State Reports Find Fraud Rate of 42% in Auto Body Repairs"
Edgar Sanchez, *Sacramento Bee*, 16 September 2003, p. B2.

Original Source 7: California Department of Consumer Repairs. (September 2003). "Auto Body Repair Inspection Pilot Program: Report to the Legislature." Department of Consumer Repairs, technical report,

http://www.collisionweek.com/cw/graphics/2003-0911-BARReport.pdf

News Story 8

"Education, Kids Strengthen Marriage"
Marilyn Elias, *USA Today*, 7 August 2003, p. 8D.

Original Source 8: None—story based on poster presentation at the American Psychological Association's annual meeting, Toronto, Canada, August 2003.

News Story 9

"Against Depression, a Sugar Pill Is Hard to Beat"
Shankar Vedantam, *Washingtonpost.com*, 7 May 2002, p. A01.

Original Source 9: Khan, Arif, MD; Khan, Shirin; Kolts, Russell, PhD; and Brown, Walter A., MD. (April 2003). "Suicide Rates in Clinical Trials of SSRIs, Other Antide-pressants, and Placebo: Analysis of FDA Reports." *American Journal of Psychiatry* 160, pp. 790–792.

News Story 10 *(Not on Companion Website)*

"Churchgoers Live Longer, Study Finds"
Nancy Weaver Teichert, *Sacramento Bee*, 24 December 2001, p. 3.

Additional News Story 10

"Keeping the Faith: UC Berkeley Researcher Links Weekly Church Attendance to Longer, Healthier Life"

http://www.berkeley.edu/news/media/releases/2002/03/26_faith.html

UC Berkeley press release.

Original Source 10 *(Not on Companion Website)***:** Oman, Doug; Kurata, John H.; Straw-bridge, William J.; and Cohen, Richard D. (2002). "Religious attendance and cause of death over 31 years." *International Journal of Psychiatry in Medicine* 32, pp. 69–89.

News Story 11

"Double Trouble Behind the Wheel"
John O'Neil, *Sacramento Bee (from New York Times)* 14 September 2003, p. L8.

Online at http://www.nytimes.com/2003/09/02/health/vital-signs-at-risk-double-trouble-behind-the-wheel.html

Original Source 11 *(Not on Companion Website)***:** Horne, J. A.; Reyner, L. A.; and Barrett, P. R. (2003). "Driving impairment due to sleepiness is exacerbated by low alcohol intake." *Journal of Occupational and Environmental Medicine* 60, pp. 689–692.

Online at http://www.ncbi.nlm.nih.gov/pmc/articles/PMC1740622/pdf/v060p00689.pdf

News Story 12

"Working Nights May Increase Breast Cancer Risk"

Sacramento Bee (Associated Press) 17 October 2001, p. A7.

Original Source 12a: Hansen, Johnni. (17 October 2001). "Light at Night, Shiftwork, and Breast Cancer Risk." *Journal of the National Cancer Institute* 93, no. 20, pp. 1513–1515.

Original Source 12b: Davis, Scott; Mirick, Dana K.; and Stevens, Richard G. (17 October 2001). "Night Shift Work, Light at Night, and Risk of Breast Cancer." *Journal of the National Cancer Institute* 93, no. 20, pp. 1557–1562.

Original Source 12c: Schernhammer, Eva S.; Laden, Francine; Speizer, Frank E.; Willett, Walter C.; Hunter, David J.; and Kawachi, Ichiro. (17 October 2001). "Rotating Night Shifts and Risk of Breast Cancer in Women Participating in the Nurses' Health Study." *Journal of the National Cancer Institute* 93, no. 20, pp. 1563–1568.

News Story 13

"3 Factors Key for Drug Use in Kids"

Sacramento Bee (Jennifer C. Kerr, Associated Press) 20 August 2003, p. A7.

Original Source 13: "2003 CASA National Survey of American Attitudes on Substance Abuse VIII: Teens and Parents." A report from the National Center on Addiction and Substance Abuse at Columbia University, August 2003. Available at:

http://www.casacolumbia.org/templates/Publications_Reports.aspx#r27

News Story 14

"Study: Emotion Hones Women's Memory"

Sacramento Bee (from Associated Press by Paul Recer) 23 July 2002, p. A1, A11.

Original Source 14 (*Not on Companion Website*): Canli, Turhan; Desmond, John E.; Zhao, Zuo; and Gabrieli, John D. E. (2002). "Sex differences in the neural basis of emotional memories." *Proceedings of the National Academy of Sciences* 99, pp. 10789–10794.

News Story 15

"Kids' Stress, Snacking Linked"

Jane E. Allen, *Sacramento Bee* (orig. *Los Angeles Times*) 7 September 2003, p. L8.

Online at http://articles.latimes.com/2003/aug/18/health/he-stress18

Additional News Story 15

"Stress Linked to Obesity in School-age Children"

André Picard, *The Globe and Mail [UK]* 2 August 2003

http://www.theglobeandmail.com/life/stress-linked-to-obesity-in-school-age-children/article1020072/

Original Source 15: (*Abstract only on the Companion Website*) Cartwright, Martin; Wardle, Jane; Steggles, Naomi; Simon, Alice E.; Croker, Helen; and Jarvis, Martin J. (August 2003). "Stress and Dietary Practices in Adolescents." Health Psychology 22, no. 4, pp. 362–369.

News Story 16

"More on TV Violence"

Constance Holden (21 March 2003), *Science* 299, p. 1839.

Original Source 16: (*Abstract only on the Companion Website*) Huesmann, L. Rowell; Moise-Titus, Jessica; Podolski, Cheryl-Lynn P.; and Eron, Leonard D. (2003). "Longitudinal Relations Between Children's Exposure to TV Violence and Their Aggressive and Violent Behavior in Young Adulthood: 1977–1992." *Developmental Psychology* 39, no. 2, pp. 201–221.

News Story 17

"Even when Monkeying Around, Life Isn't Fair"

Jamie Talan, *San Antonio Express-News* (original from *Newsday.com*, 18 September 2003) 21 September 2003, p. 18A.

Original Source 17 (*Not on Companion Website*): Brosnan, Sarah F.; and de Waal, Frans B. M. (18 September 2003). "Monkeys reject unequal pay." *Nature* 425, pp. 297–299.

Online at http://www.emory.edu/LIVING_LINKS/publications/articles/Brosnan_deWaal_2003.pdf

News Story 18

"Heavier Babies Become Smarter Adults, Study Shows"

Sacramento Bee (by Emma Ross, Associated Press) 26 January 2001, p. A11.

Original Source 18: (*Not on Companion Website*) Richards, Marcus; Hardy, Rebecca; Kuh, Diana; and Wadsworth,

Michael E. J. (January 2001). "Birth weight and cognitive function in the British 1946 birth cohort: longitudinal population based study." *British Medical Journal* 322, pp. 199–203.

Online at http://www.ncbi.nlm.nih.gov/pmc/articles/PMC26584/

News Story 19

"Young Romance May Lead to Depression, Study Says"

Sacramento Bee (by Malcolm Ritter, Associated Press) 14 February 2001, p. A6.

Additional News Story 19

"Puppy love's dark side: First study of love-sick teens reveals higher risk of depression, alcohol use and delinquency"

http://www.news.cornell.edu/stories/2001/05/love-sick-teens-risk-depression-alcohol-use-delinquency

Cornell University press release.

Original Source 19 *(Not on Companion Website):* Joyner, Kara; and Udry, J. Richard. (2000). "You Don't Bring Me Anything But Down: Adolescent Romance and Depression." *Journal of Health and Social Behavior* 41, no. 4, pp. 369–391.

News Story 20

"Eating Organic Foods Reduces Pesticide Concentrations in Children"

The New Farm, News and Research, online; Copyright The Rodale Institute.

Original Source 20: Curl, Cynthia L.; Fenske, Richard A.; and Elgethun, Kai. (March 2003). "Organophosphorus Pesticide Exposure of Urban and Suburban Preschool Children with Organic and Conventional Diets." *Environmental Health Perspectives* 111, no. 3, pp. 377–382.

II. Applets

1. Sampling Applet: This applet helps you to explore simple random sampling, described in Section 4.4. It shows a simple random sample of size 10 being selected from a population of 100 individuals. You can see how the proportion of females and the mean height change with each sample you take.

2. Empirical Rule Applet: This applet allows you to explore how well the Empirical Rule from Chapter 8 works on data with various shapes. The data were collected from students at the University of California at Davis and Penn State University.

3. Correlation Applet: This applet helps you visualize correlation, described in Chapter 10, and the way it can be influenced by a few points, as described in Chapter 11. It presents you with the challenge of placing points on a scatterplot to try to obtain specified correlation values. You can observe how the correlation changes as a result of outliers and other interesting points.

4. Sample Means Applet: This applet helps you understand the Rule for Sample Means from Section 19.3. It generates random samples from a population in which the mean is 8 and the standard deviation is 5, like the weight loss example in Section 19.3. A histogram of the means from the samples is constructed, and you can see how its shape approaches a bell shape when hundreds of sample means are included.

5. TV Means Applet: Like the Sample Means Applet, this applet helps you understand the Rule for Sample Means from Section 19.3. For this applet, unlike the previous one, the shape of the population is highly skewed; the data represent the hours of TV that college students reported watching in a typical week. The histogram of sample means still becomes bell-shaped when enough (large) samples are chosen, but it takes longer than it did in the Sample Means applet because of the initial skewed distribution.

6. Confidence Level Applet: This applet illustrates the idea of a confidence interval from Chapter 20 and confidence level from Chapter 21. You choose a confidence level. The applet will generate repeated samples from a population representing weights of college-age men and calculate a confidence interval for the mean based on each sample, using the confidence level you specified. The goal is to see if the proportion of intervals that capture the population mean is close to the specified confidence level.

Appendix of News Stories

Mending mindfully: A recent study suggests that practicing meditation can boost the immune system

By Heidi Kotansky

Link to News Story 1:
http://www.yogajournal.com/practice/1050

Summary of News Story 1

This story reported the results of a randomized experiment to study the effects of mindfulness meditation, conducted by researchers at the University of Wisconsin – Madison. The participants in the study were volunteers recruited from a local biotechnology company. They were randomly assigned to the meditation group (25 participants) or the control group (16 participants).

The meditation group received training in mindful-based stress reduction (MBSR), a form of meditation developed by Jon Kabat-Zinn at the University of Massachusetts Medical Center. (MBSR has gained wide popularity but was relatively unknown at the time of this study.) The training, led by Kabat-Zinn, consisted of weekly classes for eight weeks, a one-day retreat, and hourly practice at home at least six days a week. The control group did not meditate during the course of the study, but were offered the training after the study ended.

In the MSBR training program, participants are taught to focus on the moment through techniques such as paying attention to their breathing, bodily sensations, thoughts, and feelings. The method was originally developed to help patients suffering from chronic pain and disease, but was expanded to serve as a general stress-reduction program.

After the meditation training period the researchers, led by Professor Richard Davidson of the University of Wisconsin, measured activity in the region of the brain known to correspond to positive emotion and optimism. The meditation group show significantly more activity in this area of the brain than did the control group.

The researchers also measured the participants' immune system response to a flu vaccine, given to all participants in both groups. Four and eight weeks after the vaccine was given, antibodies against the vaccine in the blood of the participants were measured and compared to their antibody levels at the time of the vaccine. Both groups had a positive immune response, but the increase in antibodies for the meditation group was statistically significantly higher than for the control group.

The researchers concluded that an eight-week training program in mindfulness meditation can have positive effects on the brain and on the immune system. They planned to do more research, including a study of these effects in people who had been practicing meditation for long periods of time, and a study of the effects of the MBSR program on specific health conditions.

SOURCE: From Kotansky, Heidi, "Mending Mindfully." *Yoga Journal*, Nov. 2003, p. 35.

Research shows women harder hit by hangovers

By Lee Bowman, Scripps Howard News Service

In the life-is-not-fair category, new research finds that women not only get drunk on fewer drinks than men but women also suffer from worse hangovers.

A team at the University of Missouri-Columbia developed a new scientific scale for measuring hangover symptoms and severity. Even accounting for differences in the amount of alcohol consumed by men and women, hangovers hit women harder. "This finding makes biological sense, because women tend to weigh less and have lower percentages of total body water than men do, so they should achieve higher degrees of intoxication and, presumably, more hangover per unit of alcohol," said Wendy Slutske, an associate professor of psychology who led the team.

The study, supported by the National Institutes of Health, is being published Monday in the journal *Alcoholism: Clinical and Experimental Research*.

The researchers asked 1,230 drinking college students, only 5 percent of them of legal drinking age, to describe how often they experienced any of 13 symptoms after drinking. The symptoms in the study ranged from headaches and vomiting to feeling weak and unable to concentrate.

Besides women, the study found that the symptoms were more common in students who reported having alcohol-related problems or who had one or both biological parents with a history of alcohol-related problems.

"We were surprised to discover how little research had been conducted on hangover, because the research that does exist suggests that hangover could be an important factor in problem drinking," said Thomas Piasecki, an assistant professor of clinical psychology who took part in the study.

Other research has pinpointed hangover impairment as an important factor in drinkers suffering injury or death, and in economic losses arising from people taking time off to recover from a drinking bout.

The most common symptom reported was dehydration or feeling thirsty. The least common symptom was trembling or shaking. Based on having at least one of the symptoms, most students had been hung over between three and 11 times in the previous year. On average, students had experienced five of the 13 symptoms at least once during that period.

"While hangover is a serious phenomenon among college drinkers, for most of them it occurs rarely enough that it is unlikely to have a major deleterious impact on academic performance," Slutske said.

However, 26 percent of the students reported having experienced hangovers at least once a month in the past year, and the researchers speculate they could be at a higher risk of failure and wondered if they might represent some identifiable segment of the student population, such as fraternity or sorority members.

SOURCE: From Bowman, Lee, "Research Shows Women Harder Hit by Hangovers." Scripps Howard News Service, September 15, 2003. Copyright © 2003 Scripps Howard News Service. Reprinted with permission.

News Story 3

Rigorous veggie diet found to slash cholesterol

By Daniel Q. Haney, Associated Press

People with high cholesterol may lower their levels by a surprising one-third with a vegetarian diet that combines a variety of trendy heart-healthy foods, including plenty of soy and soluble fiber, a study found.

Although a healthy diet is a mainstay of cholesterol control, people typically can reduce their cholesterol only about 10 percent by changing what they eat. As a result, doctors routinely prescribe cholesterol-lowering drugs called statins.

However, a variety of studies suggest certain plant foods are especially good at lowering cholesterol. So a Canadian team put together a diet combining several of these to see what would happen.

"The reductions are surprising," said Cyril Kendall of the University of Toronto. "Most dietitians would not expect that sort of reduction through dietary means."

Whether most people would stick with such a diet is another matter, since it involves daily okra, eggplant and Metamucil, among other things.

Still, Kendall said his preliminary results suggest the diet works about as well as the older statin drugs that are still first-line therapy for people with high cholesterol.

Kendall presented the results of his approach, called the Portfolio diet, at a meeting of the American Heart Association on Thursday in Miami Beach. The research was sponsored by the Canadian government, the Almond Board of California and the food companies Unilever Canada and Loblaw Brands.

"This was a pretty impressive result," said Dr. Stephen Daniels of Children's Hospital Medical Center in Cincinnati. "However, the results need to be replicated. Can this be done in the real world or only in an experiment?"

The diet is based on a low-fat vegetarian regimen that emphasizes foods shown individually to be beneficial—soy, soluble fiber, plant sterols and almonds. Sources of soluble fiber include oats, barley, legumes, eggplant, okra and Metamucil. Some brands of margarine are high in plant sterols.

In the experiment, 25 volunteers ate either a standard low-fat diet or the Portfolio approach, while researchers watched the effects on their LDL cholesterol, which increases the risk of heart disease, and HDL, which lowers it. After a month, LDL levels fell 12 percent in those on the standard diet and 35 percent in those on the Portfolio diet. However, HDL levels were unchanged in people on the Portfolio diet.

Kendall said volunteers found the diet extremely filling, and several have stayed on it after the experiment finished.

"It appears that a Portfolio diet is effective at reducing cholesterol and coronary heart disease risk," he said.

Whether it truly is as good as a statin, though, remains to be seen. Those drugs have been proven to reduce the risk of heart attacks and death, whereas the diet has not been put to that test. And statins may also protect the heart in ways that go beyond their effect on cholesterol.

In the experiment, dieters got foods supplied by the researchers that are all available from supermarkets or health food stores. Every meal contained soy in some form, such as soy yogurt or soy milk.

A typical breakfast included oat bran, fruit and soy milk. Lunch might involve vegetarian chili, oat bran bread and tomato, and a typical dinner was vegetable curry, a soy burger, northern beans, barley, okra, eggplant, cauliflower, onions and red peppers. Volunteers also got Metamucil three times a day to provide soluble fiber from psyllium.

On a 2,000-calorie daily diet, volunteers got two grams of plant sterols from enriched margarine, 16 grams of soluble fiber from oats, barley and psyllium, and 45 grams of soy protein. They also got 200 grams of eggplant and 100 grams of okra daily and 30 grams of raw almonds. Additional vegetable protein was provided by beans, chick peas and lentils.

SOURCE: From Crucefix, Lanna, "Diet as good as drug for lowering cholesterol, study says." University of Toronto news release, July 22, 2003.

News Story 4

Happy people can actually live longer

By Henry Reed

Heart patients who are happy are much more likely to be alive 10 years down the road than unhappy heart patients, according to a study conducted at the Duke University Medical Center. According to the lead investigator, Beverly Brummett, the experience of joy seems a key factor. It has physical consequences and also attracts other people, making it easier for the patient to receive emotional support. Unhappy people, besides suffering from the biochemical effects of their sour moods, also are less likely to take their medicines, eat healthy, or to exercise. The depression has worse consequences than the heart disease itself.

SOURCE: From Reed, Henry, "Happy people can actually live longer." *Venture Inward Magazine*, September/October 2003, p. 9. Reprinted with permission of *Venture Inward Magazine*.

News Story 5

Driving while distracted is common, researchers say; reaching for something in the car and fiddling with the stereo far outweigh inattentiveness caused by cellphone use, traffic study finds

By Susannah Rosenblatt

Link to News Story 5:
http://articles.latimes.com/2003/aug/07/nation/na-drivers7

Summary of News Story 5

This story reported on a study of the kinds of distracting activities drivers engage in while driving. The participants were volunteers in Chapel Hill, North Carolina and Philadelphia, Pennsylvania who agreed to have a small video camera attached to their windshield to record their behavior while driving. The study took place in 2001 and 2002. Seven to eight hours of video were recorded for each participant, from which three hours were randomly selected for analysis.

All participants were observed manipulating the air conditioning or heating controls at some point while driving. Some other behaviors and the percentage of drivers who were observed doing them were: Reaching or leaning (97.1%), changing audio controls (91.4%), external distractions (85.7%), conversing with other occupants (77.1%), eating or drinking (71.4%), preparing to eat or drink (58.6%), grooming (45.7%), reading or writing (40%).

The researchers noted that drivers were engaged in a distracting behavior (excluding talking to passengers) 16.1% of the time their cars were in motion. Cell phone usage was lower than many other distractions [but cell phones were less common in 2001-02, and texting was not an option at all].

The study originally intended to observe the behavior of 144 drivers, but ended up studying only 70 because of time and funding constraints. The lead researcher noted that this reduction in sample size did not change the findings, but did make it more difficult to analyze subgroups (such as different age groups).

The story also presented some background information related to highway fatalities resulting from cell phone use and other distractions, and discussed plans for disseminating information to consumers about the dangers of distracted driving. The report on which the story was based was issued by the non-profit American Automobile Association Foundation for Traffic Safety. (The report is available on the companion website.)

SOURCE: From Rosenblatt, Susannah, "Driving While Distracted Is Common, Researchers Say." Los Angeles Times, LosAngeles, Calif.: Aug 7, 2003. p. A.20.

News Story 6

Music as brain builder

By Constance Holden

Many parents and day-care facilities nowadays expose tots to classical music in hopes of triggering the so-called "Mozart effect"—the sharpening of the brain that some classical music is said to bring about. Now the researchers who started the trend, Gordon Shaw and colleagues at the University of California, Irvine, have come up with evidence that piano lessons hike up children's performance on a test of proportional math.

Six years ago, Shaw's group found that listening to a Mozart two-piano sonata briefly raised college students' spatial skills. They subsequently reported that in preschoolers, piano lessons gave a sustained boost to spatial skills.

In the latest study, Shaw compared three groups of second-graders: 26 got piano instruction plus a math video game that trains players to mentally rotate shapes and to use them to learn ratios and fractions. Another 29 got computer-based English training plus the video game. A control group of 28 got no special training. After 4 months, the results were "dramatic," the authors report in the current issue of Neurological Research.

The piano group scored 15% higher than the English group in a test of what they had learned in the computer game—and 27% higher on the questions devoted to proportional math. These gains were on top of the finding that the computer game alone boosted scores by 36% over the control group.

Shaw says the improvements suggest that spatial awareness and the need to think several steps ahead—both required in piano playing—reinforce latent neuronal patterns. "Music is just tapping into this internal neural structure that we're born with," he says. Piano lessons may well condition the brain just as muscle-building conditions an athlete, says Michael Merzenich, a neuroanatomist at the University of California, San Francisco. Music may be a "skill . . . more fundamental than language" for refining the ability of the brain to make spatial and temporal distinctions, he says.

SOURCE: From Holden, Constance, "Music as Brain Builder." *Science*, Vol. 283, March 26, 1999, p. 2007. Copyright © 1999 American Association for the Advancement of Science. Reprinted with permission.

News Story 7

State report finds fraud rate of 42% in auto-body repairs

By Edgar Sanchez

A car repair that should have been completed in seven weeks took 20 weeks at Folsom Lake Ford, Darin Peets says.

"I was never given a reasonable explanation of why it was taking so long," said Peets, a quadriplegic whose 1987 Ford van needed body repairs after a 2001 traffic accident.

After the $8,800 job was completed, his van was inspected by master mechanics from the Bureau of Automotive Repair, a branch of the state Department of Consumer Affairs.

They found that Folsom Lake Ford had overcharged Peets $395, BAR officials said.

Darryl Fontana, service director for Folsom Lake Ford, said Peets' van had been "totaled," and his insurer initially didn't want to fix it, contributing to the delay.

Explaining the alleged overcharges, Fontana said Monday that "the parts that weren't put on were two Econoline emblems that were not available because the van was old. Mr. Peets opted to take the van without those emblems."

The other service that wasn't performed was a minor paint job to blend paint on a section that included one of the van's doors, Fontana said.

"The van had been modified to accommodate a wheelchair," he said. "You couldn't paint the door because a piece of raw aluminum had been attached to it externally. It was all corroded."

Through BAR's mediation, Peets obtained a $1,395 refund from Folsom Lake Ford. Of that, $1,000 represented the Sacramento man's insurance deductible, which the repair shop paid as a goodwill gesture because of Peets' long wait.

For the past two years, Peets and other consumers have been steered to BAR to determine if their cars had been properly fixed by collision-repair shops across the state.

Of 1,315 vehicles inspected in the two-year BAR study that ended in June, 42 percent were overbilled for labor not performed or parts not supplied, Consumer Affairs Director Kathleen Hamilton said at a news conference last week.

A total of 551 consumers were left "surprised, frightened and alarmed" because they didn't get what they paid for, she said, adding that the average loss was $812.

The 42 percent fraud rate holds steady with BAR's previous assertions that about 40 percent of auto-body repairs are fraudulent.

Officials in the auto-body repair industry blasted the report.

"This was not a true random inspection but a complaint-driven inspection," said David McClune, chief of the California Autobody Association.

The inspected cars belonged to disgruntled drivers, he claimed.

"The results of this study can't be projected upon the industry as a whole," said McClune, whose group represents about 1,000 of the state's estimated 7,000 collision-repair shops. Most repair shops are honest, he said.

John Walcher, vice president of Caliber Collision Centers, a chain of auto-repair shops, called the BAR report "statistically insignificant." During the study period, more than 3 million cars probably were repaired in California's body shops—but BAR "looked at only 1,300 cars that met their criteria," he said.

Mike Luery, a Consumer Affairs Department spokesman, said people who requested inspections were "curious about what happened to their cars" during repairs. "So it's not like these consumers automatically assumed there was a problem," Luery said.

The $2.6 million inspection program—funded entirely by smog-check certification fees and license fees for collision-repair shops—was prompted by legislation written by state Sen. Jackie Speier, D-Hillsborough.

The goal was to determine if the state should more aggressively police auto-body repairs.

"The results confirm the need for BAR to continue its work," Speier said.

BAR is recommending that the Legislature make changes to fight "unfair and illegal practices" in the collision-repair industry.

During the program, BAR obtained more than $500,000 in either direct refunds or redone repairs for consumers.

In addition, Attorney General Bill Lockyer filed 47 administrative actions against auto-body repair shops that face revocation or suspension of their licenses to operate, with more actions expected.

And 46 cases have been sent to local district attorneys for possible criminal or civil prosecution.

Speier expressed concern about an alleged comment by Brad Wilson, a regional manager for Caliber Collision Centers.

According to documents compiled by the attorney general's office, Wilson allegedly told a complaining customer earlier this year, "What fraud? It's only fraud if you get caught."

"If that remark is in fact true, it speaks volumes" as to why so much fraud occurs, Speier said.

Walcher, however, said that Wilson never said such a thing.

"Mr. Wilson has a sterling reputation and a long history in this industry," Walcher said.

The remark is included in fraud allegations filed by Lockyer, who is seeking to suspend or revoke the operating licenses of 10 of Caliber's 38 shops in the state.

An administrative law judge will hear the cases against the 10 shops, which have denied any wrongdoing.

Source: From Sanchez, Edgar, "State reports find fraud rate of 42% in auto body repairs." *Sacramento Bee*, September 16, 2003, p. B2.

News Story 8

Education, kids strengthen marriage

By Marilyn Elias

Long marriages are happiest if spouses are well educated and have raised kids together, but health problems can sour even the longest unions, says a study to be reported this weekend.

Overall, couples approaching their 50th wedding anniversaries are a contented lot.

"They're not ecstatic, but there's not a lot of conflict. They've come to accept and appreciate each other, without unreasonable expectations," says West Virginia University psychologist Jennifer Margrett.

She and co-author Kristopher Kimble will speak at the American Psychological Association meeting in Toronto, which starts today.

Their study of 98 couples married an average of 46 years shows that money, and where adult children live, matters much less than health as couples reach retirement age. Poor health trumps a lot, spurring conflict and weakening feelings of love.

The more education that spouses have, the less ambivalent they feel about partners, possibly because they've waited longer to marry "and seen what's out there," Margrett says.

Most marriage research focuses on young and middle-aged couples, says Stanford University psychologist Laura Carstensen. The most miserable pairs tend to split up early, especially if they're well educated and can afford two households.

Adult kids moving back home, the bane of many middle-aged empty-nesters, may be welcomed by some older couples "because they rely on their kids for help," Carstensen says.

Her studies with John Gottman show that older married couples blend criticism with positive strokes, tempering their anger more than younger couples.

"The woman complaining about her husband messing up her home office will suddenly say, 'But you're a much better lover as you've gotten older.' They criticize, but with affection and perspective."

SOURCE: From Elias, Marilyn, "Education, kids strengthen marriage." USA Today, August 7, 2003, p. 8D. Copyright © August 7, 2003 USA Today. Reprinted with permission.

News Story 9

Against depression, a sugar pill is hard to beat: Placebos improve mood, change brain chemistry in majority of trials of antidepressants

By Shankar Vedantam

Summary of News Story 9

This story reported on research showing that placebos (sugar pills) sometimes work as well as or better than antidepressants for treating depression. In one study designed to compare the herbal supplement St. John's wort to the antidepressant Zoloft, the percentages of depressed patients who were fully cured included 24 percent of those taking St. John's wort, 25 percent of those taking Zoloft, and 32 percent of those taking a placebo.

Researchers also found that taking placebos resulted in profound changes in the brain, in the same areas as the antidepressants. An additional finding was that the results from taking placebos seem to have grown stronger over the past 20 years.

One part of the story described a meta-analysis of 96 studies conducted from 1979 to 1996. The analysis included studies that were published in the medical literature (and noted that they were more likely to be positive) as well as studies that were not made public. One of the results reported was that in 52 percent of the studies the effects of the placebo and the antidepressant were not statistically significantly different. The studies were combined into one analysis and not compared for different settings or subgroups.

Interviews with experts on the research produced some suggestions for why the placebos might work so well in clinical trials (randomized experiments). Most studies are double-blind, and the participants and physicians do not know who is taking the placebos. But all patients in these clinical trials meet with a physician regularly to discuss how they are feeling. Therefore, the researchers speculated, it might be the care and concern of the researchers conducting the study that makes the depressed patients feel better, rather than the pills they are taking.

In contrast, when patients are given antidepressants as part of their routine medical care, rather than as part of a research study, they may not be given that same level of attention as they receive in a clinical trial. Therefore, placebo may not be so effective unless they are combined with the kind of therapeutic care provided during a randomized experiment. In one study, researchers found that when the study ended and participants were told that they had been taking a placebo, they quickly deteriorated.

The news story concluded by encouraging physicians and mental health practitioners to do a better job of integrating biological treatments with the power of belief and the benefits of psychotherapy.

SOURCE: From Vedantam, Shankar, "Against Depression, A Sugar Pill Is Hard to Beat." Washingtonpost.com, May 7, 2002, p. A01.

News Story 10

Churchgoers live longer, study finds

By Nancy Weaver Teichert

Summary of News Story 10

News Story 10 and Additional News Story 10 both report on a study linking church attendance with reduced mortality during the years of the study. Additional News Story 10 (printed below) provides most of the details about the study that were reported in News Story 10. However, there are a few details in News Story 10 that are not mentioned in the additional news story. For instance, the study found that the relationship between church attendance and mortality was stronger for women than for men and that, for women, the reduction in mortality was similar to the reduction associated with regular exercise. The author also provided a quote from the study's lead author noting that a cause and effect relationship cannot be proven with this study.

SOURCE: From Tiechert, Nancy Weaver, "Churchgoers live longer, study finds." *Sacramento Bee*, December 24, 2001, p. A3.

Additional News Story 10

Keeping the faith: UC Berkeley researcher links weekly church attendance to longer, healthier life

http://www.berkeley.edu/news/media/releases/2002/03/26_faith.html

By Sarah Yang

Attending religious services may aid the body in addition to helping the spirit, according to a new study. Researchers in California have found new evidence that weekly attendance at religious services is linked to a longer, healthier life.

In the study, researchers from the Human Population Laboratories of the Public Health Institute and the California Department of Health Services, and from the University of California, Berkeley, found that people who attended religious services once a week had significantly lower risks of death compared with those who attended less frequently or never, even after adjusting for age, health behaviors and other risk factors. The study will be published April 4 in the International Journal of Psychiatry in Medicine.

"We found this difference even after adjusting for factors such as social connections and health behaviors, including smoking and exercising," said Doug Oman, lead author of the study and a lecturer at UC Berkeley's School of Public Health. "The fact that the risk of death by several different causes is lower for those who attend religious services every week suggests that we should look to some psychological factor for answers. Maybe frequent attendees experience a greater sense of inner peace, perhaps because they can draw upon religious coping practices to help them deal with stressful events."

Oman conducted the study while he was a research scientist at the Public Health Institute and a post-doctoral researcher at UC Berkeley's School of Public Health.

Researchers specifically looked at the risk of death from certain diseases, including cancer and heart disease. While they found no significant difference for risk of death by cancer, they did find that people who attended religious services less than once a week or never had a 21 percent greater overall risk of dying, as well as a 21 percent greater risk of dying from circulatory diseases.

There was also a strong trend towards lower mortality from respiratory and digestive diseases, although there is more of a possibility that chance might have played a role in producing those results. When compared with weekly attendees, those who attended less than weekly or never had a 66 percent greater risk of dying from respiratory diseases and a 99 percent greater risk of dying from digestive diseases.

Adherents to Christian religions made up the bulk of the study participants, with 51.9 percent reporting themselves to be liberal Protestants. Twenty-seven percent of the participants were Roman Catholics and 2.5 percent were Jewish. Members of other Western religions made up 7.2 percent of the group, while those practicing non-Western religions made up 0.8 percent of the group. Those with no religious affiliations comprised 10.6 percent of the study participants.

Oman said there are still unanswered questions he hopes will be addressed in future studies, including the significance of spirituality or devoutness. "Several non-Western religions, including Buddhism, place less emphasis on going to a temple or church," he said. "So people of those faiths may be just as devout in their tradition, and that may revolve around a household shrine. They may go to a temple only a few times a year, but they could still be getting the psychological benefits of inner peace."

Still, Oman said this study adds to a growing body of evidence that religious practices are generally linked to better health. He also pointed out that this study is one of the few that has investigated the relationships between religious involvement and several specific causes of death.

Prior studies of Mormons and Seventh Day Adventists noted that beneficial health behaviors, such as abstention from smoking and high fat diets, were strongly integrated in their beliefs. Past studies have also associated positive health effects with the high level of social support found in religious communities.

"The picture that is developing is that religious activity is affecting health through several pathways," said Oman. "Whether it is encouraging better health habits such as exercising, providing a strong social support network, providing a sense of psychological well-being, or all those factors combined, it seems clear that the effects of faith deserves more study."

The researchers used data taken over 31 years from a longitudinal survey of 6,545 adult residents from California's Alameda County. Researchers from the state health department began the survey, which is ongoing, to study the interrelationships among health status and social, familial, environmental and other factors.

Although the study only looked at residents of Alameda County, Oman said the area's ethnic diversity and mix of rural, suburban and urban neighborhoods make the findings relevant to communities throughout the United States.

Data was obtained through mailed questionnaires or through interviews in the participants' homes. Participants aged 21 and over at the start of the study in 1965 were questioned again in 1974, 1983 and 1994. Death records were obtained from state and federal mortality files through 1996.

Co-authors of the study are William Strawbridge and Richard Cohen from the Public Health Institute, and John Kurata from the California Department of Health Services.

The study was supported by grants from the National Institute of Aging, the National Heart, Lung, and Blood Institute, and the California Department of Health Services.

SOURCE: Sarah Yang, "Keeping the faith: UC Berkeley researcher links weekly church attendance to longer, healthier life," http://www.berkeley.edu/news/media/releases/2002/03/26_faith.html. *Reprinted with permission from the UC Berkeley Office of Public Affairs*

News Story 11

Double trouble behind the wheel

By John O'Neil

Link to News Story 11:
www.nytimes.com/2003/09/02/health/vital-signs-at-risk-double-trouble-behind-the-wheel.html

Summary of News Story 11

This story reported on a British study to measure the effect on driving behavior of alcohol, lack of sleep, a combination of the two, and neither of the two. Twelve male volunteers with an average age of 22 participated in the study. There were four phases of the study: Alcohol after normal sleep, alcohol after reduced sleep, no alcohol after normal sleep, and no alcohol after reduced sleep. Each volunteer participated in all four phases in random order, spaced a week apart. For each phase they spent two hours after lunch in a simulated driving situation, and the number of times they drifted out of the lane was recorded.

The alcohol part of the study was double-blind, with the participants having a lunchtime beverage of either beer (for two of the phases), or a non-alcoholic placebo beer (for the other two phases). The reduced sleep condition

could not be double-blind because participants knew how much they had slept. For the reduced sleep phases, they were instructed to sleep only five hours the night before the test.

The results focused on how many times they drifted out of their lane during the two hours of simulated driving. The worst condition was the alcohol and reduced sleep, with an average of 32 lane drifts. Next was the reduced sleep, with 18 lane drifts, followed by the alcohol but full sleep with 15 lane drifts. In the full sleep and no alcohol condition they averaged only 7 lane drifts. The researchers concluded that the combination of fatigue and alcohol was particularly dangerous for drivers.

SOURCE: From O'Neil, John, "Double trouble behind the wheel." New York Times, Sept. 2, 2003, p. F.6.

News Story 12

Working nights may increase breast cancer risk

By Paul Recer

Women who work nights may increase their breast cancer risk by up to 60 percent, according to two studies that suggest bright light in the dark hours decreases melatonin secretion and increases estrogen levels.

Two independent studies, using different methods, found increased risk of breast cancer among women who worked night shifts for many years. The studies, both appearing in the *Journal of the National Cancer Institute*, suggested a "dose effect," meaning that the more time spent working nights, the greater the risk of breast cancer.

"We are just beginning to see evidence emerge on the health effects of shift work," said Scott Davis, an epidemiologist at the Fred Hutchinson Cancer Research Center in Seattle and first author of one of the studies. He said more research was needed, however, before a compelling case could be made to change night work schedules.

"The numbers in our study are small, but they are statistically significant," said Francine Laden, a researcher at Brigham and Women's Hospital in Boston and co-author of the second study.

"These studies are fascinating and provocative," said Larry Norton of the Memorial Sloan-Kettering Cancer Center in New York. "Both studies have to be respected."

But Norton said the findings only hint at an effect and raise "questions that must be addressed with more research."

In Davis' study, researchers explored the work history of 763 women with breast cancer and 741 women without the disease.

They found that women who regularly worked night shifts for three years or less were about 40 percent more likely to have breast cancer than women who did not work such shifts. Women who worked at night for more than three years were 60 percent more likely.

The Brigham and Women's study, by Laden and her colleagues, found only a "moderately increased risk of breast cancer after extended periods of working rotating night shifts."

The study was based on the "medical and work histories of" more than 78,000 nurses from 1988 through May 1998. It found that nurses who worked rotating night shifts at least three times a month for one to 29 years were about 8 percent more likely to develop breast cancer. For those who worked the shifts for more than 30 years, the relative risk of breast cancer went up by 36 percent.

American women have a 12.5 percent lifetime risk of developing breast cancer, according to the American Cancer Society. Laden said her study means that the lifetime risk of breast cancer for longtime shift workers could rise above 16 percent. There are about 175,000 new cases of breast cancer diagnosed annually in the United States and about 43,700 deaths. Breast cancer is second only to lung cancer in causing cancer deaths among women.

Both of the *Journal* studies suggested that the increased breast cancer risk among shift workers is caused by changes in the body's natural melatonin cycle because of exposure to bright lights during the dark hours.

Melatonin is produced by the pineal gland during the night. Studies have shown that bright light reduces the secretion of melatonin. In women, this may lead to an increase in estrogen production; increased estrogen levels have been linked to breast cancer.

"If you exposed someone to bright light at night, the normal rise in melatonin will diminish or disappear altogether," said Davis. "There is evidence that this can increase the production of reproductive hormones, including estrogen." Davis said changes in melatonin levels in men doing nighttime shift work may increase the risk of some types of male cancer, such as prostate cancer, but he knows of no study that has addressed this specifically.

SOURCE: From Recer, Paul, "Night Shift Linked to Breast Cancer Risk." Associated Press, October 17, 2001. Reprinted with permission of the Associated Press.

News Story 13

3 Factors key for drug use in kids: A study says stress, boredom and extra cash boost the risk for substance abuse

By Jennifer C. Kerr

A survey of American children and parents released Tuesday found a mix of three ingredients in abundance for many kids can lead to substance abuse: boredom, stress and extra money.

The annual study by Columbia University's National Center on Addiction and Substance Abuse also found students attending smaller schools or religious schools are less likely to abuse drugs and alcohol.

Joseph Califano Jr., the center's chairman and president, said 13.8 million teens are at moderate or high risk of substance abuse.

"Parental engagement in their child's life is the best protection Mom and Dad can provide," he said.

The study found that children ages 12 to 17 who are frequently bored are 50 percent more likely to smoke, drink, get drunk or use illegal drugs. And kids with $25 or more a week in spending money are nearly twice as likely to smoke, drink or use drugs as children with less money.

Anxiety is another risk factor. The study found that youngsters who said they're highly stressed are twice as likely as low-stress kids to smoke, drink or use drugs.

High stress was experienced among girls more than boys, with nearly one in three girls saying they were highly stressed compared with fewer than one in four boys. One possible factor is social pressure for girls to have sex, researchers said.

Charles Curie, administrator of the Substance Abuse and Mental Health Services Administration, said his agency has found similar risk factors among U.S. youth.

He said the best thing parents can do to steer their kids away from drugs and alcohol is to talk to them and stay involved in their lives. It's also important, he said, to know their children's friends.

For the first time in the survey's eight-year history, young people said they are as concerned about social and academic pressures as they are about drugs. In the past, Califano said, drugs were by far the No. 1 pressure on kids.

There was some encouraging news. The study found that 56 percent of those surveyed have no friends who regularly drink, up from 52 percent in 2002. Nearly 70 percent have no friends who use marijuana.

Among the study's other findings:

- The average age of first use is about 12 for alcohol, 12 ½ for cigarettes and nearly 14 for pot.

- More than five million children ages 12 to 17, or 20 percent, said they could buy marijuana in an hour or less. Another five million said they could buy it within a day.

- Kids at schools with more than 1,200 students are twice as likely as those attending schools with fewer than 800 students to be at high risk for substance abuse.

QEV Analytics surveyed 1,987 children ages 12 to 17 and 504 parents, 403 of whom were parents of interviewed kids. They were interviewed from March 30 to June 14. The margin of error was plus or minus two percentage points for children and plus or minus four percentage points for parents.

[The news article was accompanied by an illustration, titled "Teen drug risks: American teens reporting high stress, boredom or disposable income were more likely to use drugs, a recent survey found." There were three bar graphs illustrating the percent of students who had tried alcohol and marijuana. The first graph showed four levels of weekly disposable income, the next showed stress levels categorized as low, moderate, and high, and the final graph compared students who were not bored and often bored. The illustration was accompanied by this explanation. "About this poll: National telephone survey of 1,987 12- to 17-year-olds was conducted March 30 to June 14. The margin of error was plus or minus 2 percentage points. Source: National Center on Addiction and Substance Abuse at Columbia University."]

SOURCE: From Kerr, Jennifer C., "3 Factors key for drug use in kids." Associated Press, August 20, 2003. Reprinted with permission of the Associated Press.

News Story 14

Study: Emotion hones women's memory

By Paul Recer

Matrimonial lore says husbands never remember marital spats and wives never forget. A new study suggests a reason: Women's brains are wired both to feel and to recall emotions more keenly than the brains of men.

A team of psychologists tested groups of women and men for their ability to recall or recognize highly evocative photographs three weeks after first seeing them and found the women's recollections were 10 percentage points to 15 percentage points more accurate.

The study, appearing in the *Proceedings of the National Academy of Sciences*, also used MRIs to image the subjects' brains as they were exposed to the pictures. It found the women's neural responses to emotional scenes were much more active than the men's.

Turhan Canli, an assistant professor of psychology at SUNY, Stony Brook, said the study shows a woman's brain is better organized to perceive and remember emotions.

The findings by Canli and researchers from Stanford University are consistent with earlier research that found differences in the workings of the minds of women and men, said Diane Halpern, director of the Berger Institute for Work, Family, and Children and a professor of psychology at Claremont McKenna College in California.

Halpern said the study "makes a strong link between cognitive behavior and a brain structure that gets activated" when exposed to emotional stimuli.

"It advances our understanding of the link between cognition and the underlying brain structures," she said. "But it doesn't mean that those are immutable, that they can't change with experience."

Canli said the study may help move science closer to finding a biological basis to explain why clinical depression is much more common in women than in men.

Canli said a risk factor for depression is rumination, or dwelling on a memory and reviewing it time after time. The study illuminates a possible biological basis for rumination, he said.

Halpern said the study also supports findings that women, in general, have a better autobiographical memory for anything, not just emotional events.

She said the study supports the folkloric idea that a wife has a truer memory for marital spats than does her husband.

"One reason for that is that it has more meaning for women and they process it a little more," Halpern said. "But you can't say that we've found the brain basis for this, because our brains are constantly changing."

In the study, Canli and his colleagues individually tested the emotional memory of 12 women and 12 men using a set of pictures. Some of the pictures were ordinary, and others were designed to evoke strong emotions.

Each of the subjects viewed the pictures and graded them on a three-point scale ranging from "not emotionally intense" to "extremely emotionally intense."

As the subjects viewed the pictures, images were taken of their brains using magnetic resonance imaging. This measures neural blood flow and identifies portions of the brain that are active.

Canli said women and men had different emotional responses to the same photos. For instance, the men would see a gun and call it neutral, but for women it would be "highly, highly negative" and evoke strong emotions.

Neutral pictures showed such things as a fireplug, a book case or an ordinary landscape.

The pictures most often rated emotionally intense showed dead bodies, gravestones and crying people. A picture of a dirty toilet prompted a strong emotional response, especially from the women, Canli said.

All the test subjects returned to the lab three weeks later and were surprised to learn they would be asked to remember the pictures they had seen. Canli said they were not told earlier they would be asked to recall pictures from the earlier session.

In a memory test tailored for each person, they were asked to pick out pictures they earlier rated as "extremely emotionally intense." The pictures were mixed among 48 new pictures. Each image was displayed for less than three seconds.

"For pictures that were highly emotional, men recalled around 60 percent and women were at about 75 percent," Canli said.

SOURCE: From Recer, Paul, "Study: Emotion hones women's memory." Associated Press, July 23, 2002. Reprinted with permission of the Associated Press.

News Story 15

Kids' stress, snacking linked

By Jane E. Allen, *Times* Staff Writer

Link to News Story 15:
http://articles.latimes.com/2003/aug/18/health/he-stress18

Summary of News Story 15

This story described a study of 4,320 children in Britain that found a strong relationship between high stress and consumption of fatty foods. The children were students at schools in the London area that were randomly selected to have a range of socioeconomic and other demographic factors. Their students' average age was between 11 and 12 years.

The children were given a standard test of daily stress, and a food questionnaire asking about fatty foods, fruits and vegetables, snacks, and breakfast. The children with the highest stress levels ate almost twice as much fatty food as the lower stress children. They ate larger snacks as well, and were less likely to eat the recommended servings of fruits and vegetables.

SOURCE: From Allen, Jane E., "High stress equals high fat for kids." Los Angeles Times, August 18, 2003, p. F3.

News Story 16

More on TV violence

By Constance Holden

An unusual longitudinal study has strengthened the case that children who watch violent TV become more aggressive adults, but agreement is still elusive on this longsmoldering issue.

L. Rowell Huesmann and colleagues at the University of Michigan, Ann Arbor, studied 557 young Chicago children in the 1970s and found that over a 3-year period their TV habits predicted childhood aggression. Now they've done a 15-year follow-up on 329 of their subjects. In this month's issue of *Developmental Psychology*, they report that people who watched violent shows at age 8 were more aggressive in their 20s. Men who had been in the top 20% of violent TV watchers as children were twice as likely to push their wives around; women viewers were more likely to have thrown something at their husbands. The differences persisted, the researchers say, even when they controlled for children's initial aggression levels, IQs, social status, and bad parenting. Study co-author Leonard Eron thinks the association is now airtight—"just as much as smoking causes lung cancer."

Some researchers agree. The study "shows more clearly than any other that TV is more than just an amplifying factor: It alone can cause increases in aggression," says Duke University biologist Peter Klopfer. But skeptics remain unconvinced. "We already know that exposure to media violence is associated with aggressive behavior," says biostatistician Richard Hockey of Mater Hospital in Brisbane, Australia. And the most plausible explanation is still that "aggressive people like violent TV." Hockey adds that if causation exists, the effect is modest: Correlations of childhood violent TV viewing with adult aggression hover around 0.2, which means that TV contributes just about 5% of the increase in aggressive behavior.

SOURCE: From Holden, Constance, "More on TV Violence." *Science*, Vol. 299, March 21, 2003, p. 1839. Reprinted with permission of the American Association for the Advancement of Science.

News Story 17

Even when monkeying around life isn't fair

By Jamie Talan

It turns out monkeys know when they get a raw deal—which, in this case, is a cucumber.

In a new study, scientists report that for the first time, a species of nonhuman primate knows when it has been treated unfairly.

Scientists at Yerkes National Primate Research Center of Emory University in Atlanta studied capuchin monkeys, known for their social system built on cooperation.

The social groups consist of one dominant male and three to six adult females and their offspring.

The study involved a simple exchange: Two female residents of a colony were each given a small token—a granite pebble.

When they returned it to the researcher, they received a slice of cucumber. Those exchanges were completed 95 percent of the time.

But when one monkey got a grape instead, the rate of completed exchanges fell to 60 percent as the other monkey often refused to accept the cucumber slice, took it and threw it away or handed it to the other capuchin.

Finally, one monkey would get a grape without having to do any work—and the completion rate fell to 20 percent.

When the other monkey recognized the disparity in effort and reward, she sometimes would even refuse to hand over the pebble.

The same findings were observed in each of five pairs. The study appears in Thursday's issue of *Nature*.

The researchers, Frans de Waal, an endowed professor of primate behavior and director of the Living Links Center at Yerkes, and Sarah Brosnan, a graduate student, didn't find the same behavior in male monkeys.

Brosnan said she suspects that's because adult males don't live together cooperatively.

"These female monkeys don't like it when someone gets a better deal," Brosnan said.

The grape-deprived monkeys didn't show any emotional reaction toward their partners—but they did ignore the scientist.

"They knew I was the source of the inequality," she said.

The researchers said their work is aimed at understanding the evolutionary development of social fairness.

They're repeating the study in chimps, a species more closely related to humans.

Robert Frank, an endowed professor of economics at Johnson School of Management at Cornell University, said he had never seen an animal study "where someone turns down an offer because it wasn't good enough"—a trait he said is common in human exchanges.

"People often pass on an available reward because it is not what they expect or think is fair," Brosnan said. "Such irrational behavior has baffled scientists and economists, who traditionally have argued all economic decisions are rational.

Our findings in nonhuman primates indicate the emotional sense of fairness plays a key role in such decisionmaking."

SOURCE: From Talan, Jamie, "Even when monkeying around, life isn't fair." *San Antonio Express-News*, Sept. 21, 2003, p. 18A. Reproduced with permission of *San Antonio Express News*.

News Story 18

Heavier babies become smarter adults, study shows

By Emma Ross

In the biggest study to date examining the influence of birth weight on intelligence, scientists have found that babies born on the heavy side of normal tend to be brighter as adults.

Experts have long known that premature or underweight babies tend to be less intelligent as children.

But the study, published this week in the *British Medical Journal*, found that among children whose birth weight was higher than 5.5 pounds—considered to be normal—the bigger the baby, the smarter it was likely to be.

Scientists think it has something to do with bigger babies having bigger brains, or perhaps with having more connections within their brains.

But the lead researcher on the project said there was no need for parents of smaller infants to despair—the results were averages and size at birth does not necessarily determine intellectual destiny.

"Birth weight is only one of numerous factors that influence cognitive function. It may not actually be a very powerful one," said Marcus Richards, a psychologist at Britain's Medical Research Council who conducted the study. "Parental interest in education—being in the PTA and getting involved in your child's homework—has an enormous impact, one that may even offset the effect of birth weight."

Similarly, Richards said, the head start enjoyed by hefty babies can be squandered. Living in an overcrowded home, breathing polluted air or being caught in the middle of a divorce tend to diminish children's intelligence scores, he said.

The scientists found that birth size influenced intelligence until about the age of 26. After that, it tended to even out, as other factors began to play a more important role.

The study did not offer concrete examples, such as how many IQ points' advantage a 10-pound baby might have over a 7-pound baby.

And of course, there are always exceptions.

The research involved 3,900 British men and women who were born in 1946 and followed since birth. Their intelligence was measured by a battery of tests at the ages of 8, 11, 15, 26 and 43.

Increasing intelligence corresponded with increasing birth weight until the age of 26. By the age of 43, the effect was weaker.

How brainy the children were at 8 seemed to be the most important influence on later intelligence, the study found.

Heavier babies went on to achieve higher academic qualifications. That outcome was mostly linked to how brainy they were at age 8.

"It seems birth weight does what it does by age 8 and that that puts you on a path," Richards said.

But the effect seemed to have waned by the age of 43, by which time the smaller babies apparently caught up.

The results were not affected by birth order, gender, father's social class or mother's education and age. Even after the babies who were underweight were excluded, the link remained strong.

"This is an important finding that shows how strong the link is. We've seen it in low birth-weight babies, but this shows that even if you are a normal weight baby, bigger is better, at least when it comes to intelligence," said Dr. Catherine Gale, who has conducted similar research at Southampton University in England.

Experts don't know exactly what makes a heavy baby, but Gale said well-built, well-nourished mothers tend to produce heavier babies.

Mothers who eat badly, smoke and are heavy drinkers tend to produce smaller babies. However, experts don't know whether those factors influence the relationship between birth weight and intelligence. There are probably several other variables that affect birth weight, but which of those are connected to intelligence is not known, Richards said.

SOURCE: From Ross, Emma, "Heavier Babies become smarter adults, study shows." Associated Press, January 26, 2001. Reprinted with permission of the Associated Press.

News Story 19

Young romance may lead to depression

By Malcolm Ritter

The most famous youthful romance in the English-speaking world, that star-crossed love of Romeo and Juliet, was a tragedy. Now researchers have published a huge study of real-life adolescents in love.

It's also no comedy.

The results suggest that on balance, falling in love makes adolescents more depressed, and more prone to delinquency and alcohol abuse than they would have been if they'd avoided romance.

The reported effect on depression is small, but it's bigger for girls than boys. The researchers suggest it could be one reason teen girls show higher rates of depression than teen boys do, a difference that persists into adulthood.

This is not exactly the view of romance that prevails around Valentine's Day. Researchers who've studied teenage love say that smaller studies had shown teen romance can cause emotional trouble, but that the new work overlooked some good things.

The study was done by sociologists Kara Joyner of Cornell University and J. Richard Udry of the University of North Carolina at Chapel Hill. They presented the results in the December issue of the *Journal of Health & Social Behavior*.

Their results are based on responses from about 8,200 adolescents across the country who were interviewed twice, about a year apart, about a wide variety of things. The kids were ages 12 to 17 at the first interview.

To measure levels of depression, the researchers examined adolescents' answers to 11 questions about the previous week, such as how often they felt they couldn't shake off the blues, felt lonely or sad or got bothered by things that normally wouldn't faze them.

To see what love's got to do with it, the researchers compared responses from adolescents who didn't report any romantic involvement at either interview with those who reported it at both interviews. They looked at how much depression levels changed between interviews for each group.

The finding: The romantically involved adolescents showed a bigger increase in depression levels, or a smaller decrease, than uninvolved teens.

The difference wasn't much. For boys of all ages, it was about one-half point on a 33-point scale. Girls were hit harder, with a 2-point difference for girls who'd been 12 at the first interview, and diminishing with age to about a half-point difference for girls who'd been 17.

The results were a surprise, because studies of adults have shown married people tend to be less depressed than single ones, Joyner said. So why would love lower adolescent mood? By analyzing the adolescents' answers to other questions, Joyner and Udry found evidence for three possible factors: deteriorating relationships with parents, poorer performance in school and breakups of relationships.

In fact, it appeared that for boys, romance made a difference in depression only if they'd had a breakup between interviews. For girls, in contrast, the biggest impact from romance seemed to come from a rockier relationship with Mom and Dad. That was especially so among younger girls, where the bump in depression was biggest.

To Joyner, it makes sense that "if a young daughter is dating, her parents may be concerned about her choice of partner or what she is doing with him. Presumably, their concern leads to arguments. That would be my guess."

But it's only a guess. The study can't prove what caused what. Maybe girls feeling less loved at home were more likely to seek romance with a guy, rather than the other way around.

SOURCE: From Ritter, Malcolm, "Young romance may lead to depression, study says." Associated Press, February 14, 2001. Reprinted with permission of the Associated Press.

News Story 20

Eating organic foods reduces pesticide concentrations in children

Environmental Health Perspectives

A study published today in the March issue of the peer-reviewed journal *Environmental Health Perspectives (EHP)* found that consuming organic produce and juice may lower children's exposure to potentially damaging pesticides.

The researchers compared levels of organophosphorus (OP) pesticide metabolites in 39 children aged 2–5 who consumed nearly all conventional produce and juice versus those who consumed nearly all organic produce and juice. The children eating primarily organic diets had significantly lower OP pesticide metabolite concentrations than did the children eating conventional diets. Concentrations of one OP metabolite group, dimethyl metabolites, were approximately six times higher for the children eating conventional diets. Studies suggest that chronic low-level exposure to OP pesticides may affect neurologic functioning, neurodevelopment, and growth in children.

"The dose estimates suggest that consumption of organic fruits, vegetables, and juice can reduce children's exposure levels from above to below the U.S. Environmental Protection Agency's current guidelines, thereby shifting exposures from a range of uncertain risk to a range of negligible risk," the study authors wrote. "Consumption of organic produce appears to provide a relatively simple way for parents to reduce their children's exposure to OP pesticides."

In this study, families were recruited from both a retail chain grocery store selling primarily conventional foods and from a local consumer cooperative selling a large variety of organic foods. The parents kept a food diary for their children for three days. Urine collected from the children on day three was analyzed for pesticide metabolites. All of the families lived in the Seattle, Washington, metropolitan area.

"Organic foods have been growing in popularity over the last several years," says Dr. Jim Burkhart, science editor for *EHP*. "These scientists studied one potential area of difference from the use of organic foods, and the findings are compelling."

The study team included Cynthia L. Curl, Richard A. Fenske, and Kai Elgethun of the School of Public Health and Community Medicine at the University of Washington.

EHP is the journal of the National Institute of Environmental Health Sciences, part of the U.S. Department of Health and Human Services. More information is available online at http://www.ehponline.org/.

SOURCE: From "Eating Organic Foods Reduces Pesticide Concentrations in Children," http://www.newfarm.org/news/030503/ pesticide_kids.shtml (The New Farm, News and Research, online; Copyright The Rodale Institute). Reprinted with permission of The Rodale Institute.

Solutions to Selected Exercises

CHAPTER 1

3. If the measurements of interest are extremely variable, then a large sample will be required to detect real differences between groups or treatments.

6. a. Observational study. Coffee drinking over many years cannot be randomly assigned.

   b. No.

   c. (i) is justified because it does not imply a causal relationship. (ii) is not justified because it does imply a causal relationship.

9. Visit both numerous times because waiting times vary, then compare the averages.

12. a. No. It would be unethical to randomly assign people to smoke cigars or not.

   b. No. People who smoke cigars may also drink more alcohol, eat differently, or have some other characteristic that is the actual cause of the higher rate of cancer.

15. a. Randomized experiment

   b. Yes, because of the randomization all other features, such as amount of water and sun, should be similar for the two groups of plants.

18. The sample is the few thousand asked; the population is all adults in the nation.

21. a. Observational study

   b. No. It could be that people who choose to meditate tend to have lower blood pressure anyway for other reasons. You cannot establish a causal connection with an observational study.

24. a. Employers who are not satisfied.

   b. The respondents are not representative of all employers. People who voluntarily respond to surveys are those who feel strongly, and in this case, they are more likely to be dissatisfied

CHAPTER 2

3. a. Biased, the guards; unbiased, trained independent interviewers

   b. Biased: You don't have any complaints about how the guards treat prisoners do you? Unbiased: Do you think prisoners are treated fairly or treated unfairly by the guards?

6. Major and grade-point average

9. a. Version 2

   b. Version 1

12. A statistical difference may not have any practical importance.

15. Use Component 3; volunteer respondents versus nationwide random sample

18. No. For example, college major

CHAPTER 3

3. a. Deliberate bias and unnecessary complexity

   b. Do you support or not support banning prayers in schools?

6. Only one-fifth favored forbidding public speeches (version A), but almost one-half did not want to allow them (version B). The word "forbidding" was too strong compared to the wording of "not allowing."

9. Anonymous. No one knows their identity.

12. a. No. It does not allow a response about teachers separate from a response about firemen and policemen.

   b. Biased in favor of not allowing strikes, because it would be dangerous if firemen and policemen went on strike.

   c. Wording may vary, but the question should be about teachers only, and should not suggest a desired response.

15. Here is how the story explains it: "To measure levels of depression, the researchers examined adolescents' answers to 11 questions about the previous week, such as how often they felt they couldn't shake off the blues, felt lonely or sad or got bothered by things that normally wouldn't faze them" (from News Story 19 in the Appendix).

18. a. Example: Gender (male/female)

   b. Example: Time measured on a clock that is always 5 minutes fast

   c. Example: Weight of packages on a postal scale that weighs items 1 oz high half the time and 1 oz low half the time

21. a. Measurement

    b. Categorical

24. a. Discrete

    b. Continuous

27. Yes. Nominal variables are all categorical variables.

30. a. No, there is so much overlap in the times that it would be difficult to make a definite conclusion about which one is really faster based on only five times for each route.

    b. Route 1 is always 14 minutes, and Route 2 is always 16 minutes.

33. a. If taken only once on each student they would vary due to natural variability across individuals. There could also be measurement error.

    b. Natural variability across individuals. Because they are all measured at the same time, and the measurement should be accurate, the other two sources are not involved.

36. a. Natural variability across time and measurement error are both likely to cause variability in blood pressure measurements for the same person across days.

    b. Blood type doesn't change over time for a given individual and is measured accurately, so there would be no variability.

39. Example: Beer consumption measured in bottles (discrete) or ounces (continuous)

42. Likely to be valid because it measures the severity of 13 of the most common hangover symptoms with answers reported by the students who suffer from them.

CHAPTER 4

3. Case study

6. a. Survey

    b. Randomized experiment

9. a. Whether the infant was exposed to the heartbeat sound or to silence

    b. Weight gain over four days

12. a. 7%

    b. 73% to 87%

    c. You would get a different result tomorrow if you sampled a different collection of 200 people. But presumably he is not referring to sampling variability, and what he means is that in two years time the true proportion of parents

in the population who felt uninformed would have changed as well.

15. Yes. The margin of error tells us that the interval from 51% to 55% most likely covers the truth.

18. a. Cluster sampling

    b. It would be difficult to obtain a list of all first-year students from which to select a simple random sample. It would be easier to obtain a list of schools, and then either contact each school and ask them to provide a random sample (as in Exercise 17b) or first select a random sample of schools and then obtain a list of all first-year students at those schools (as in part a).

21. a. Cluster sampling; not biased if everyone responded

    b. Systematic sampling; likely to be biased for a number of reasons.

    c. Stratified sampling; not likely to be biased

    d. Convenience or haphazard sampling; likely to be biased for a number of reasons

24. No; no probability is involved in the selection.

27. Field poll is more likely to represent the population. The SFGate poll was based on a volunteer sample, not likely to represent any larger group.

30. Not reaching the individuals selected (especially for the landlines)

33. A low response rate refers to a survey given to a legitimate random sample, but where many don't respond. A volunteer sample consists of those who respond to a publicly announced survey. A volunteer sample is worse because there is no way to ascertain whether or not it represents any larger group.

36. Used a convenience sample of students in introductory psychology. They probably represent students at colleges similar to that one who take introductory psychology.

CHAPTER 5

3. a. Explanatory is whether or not the person has insomnia; response is whether or not the person has heart problems.

    b. Drinking caffeinated beverages is likely to be related to insomnia (because they keep people awake, and/or because people who are tired need caffeine to help wake up), and caffeine may contribute to heart problems.

6. a. Brand of tire

    b. Remaining tread after 30,000 miles

c. Double-blind

d. Blocks (cars were blocks)

9. The color of the cloth is a confounding variable. Maybe birds prefer red.

12. Explanatory variable is regular church attendance or not, and response variable is age at death.

15. a. Randomized experiment.

b. Observational study.

18. a. No.

b. Yes

c. Yes

21. Yes. Randomly assign treatments to the volunteers. Generalizability may be a problem if volunteers don't represent the population.

24. Ecological validity; experimenter effect

27. a. The Davis study was case control but the Laden study was not.

b. The Davis study was retrospective and the Laden study was prospective.

30. Hawthorne effect

33. a. No. This was an observational study.

b. The teens were randomly selected from a larger population, so results can be extended.

36. It would not be ethical to randomly assign people to either drink alcohol or not.

39. The placebo effect; give placebos instead of nothing to the control group.

42. a. Explanatory is form of exercise; response is weight loss.

CHAPTER 7

3. a. 65

b. 54, 62.5, 65, 70, 78

6. a. 78.5

b. 76.04; median is larger because outlier of 32 pulls down the mean.

c. 14.19

d. Smaller. Scores would be less spread out.

9. Example: Sales prices of cars

12. Example: Incomes for musicians

15. Temperatures for the entire year, with one "cold" mode and one "warm" mode

18. Examples include measurements where the numbers have no meaning, or measurements where there are only a few possible values. Example: area code of telephone numbers for students in your class.

21. a. $48,354

b. $16,768

24. a. Skewed to the right, causing the mean to be much higher than the median

b. Median because half of the users lost that much or more, half that much or less

27. Range because it's determined solely by the two extreme values, the highest and lowest

30. The mean would be higher because of the large outliers.

33. Mean is 10.91, median is 4.5. Median is a better representation.

36. a. Mean would be higher. There would probably be a few expensive ones, pulling the mean up.

b. The mean and median should be about the same because the heights are likely to be approximately bell-shaped.

c. The mean and median should be about the same because shoe sizes are not likely to have more extreme outliers in one direction than the other.

39. The standard deviation is 15 so the variance is 15^2 = 225.

42. a. Both medians are 81.

b. Mother: 81, Father: 75.25

c. Mother: 1.773, Father: 23.55

d. Mother's family is probably more useful because they are much more consistent. Very low ages in father's family could be due to accidents, death in childbirth, etc.

45. a. Organic: 0, 0.2, 0.2, 0.4, 1.3; conventional: 0, 0.5, 1.4, 3.4, 8.3

CHAPTER 8

3. a. 10%

b. 60%

c. 1%

6. a. 2.05

c. −0.67

9. a. z-score is about −0.61, so it's the 27th percentile

b. The standardized score is $(97 − 100)/15 = −0.2$, which is the 42nd percentile.

12. z-score is 1.88, so cholesterol level is 208.8

15. a. Lower quartile is $z = −0.67$ and upper quartile is $z = 0.67$

b. The range from −.67 to +.67 covers 1.34 standard deviations.

c. The interquartile range covers 1.34 standard deviations, so 2×IQR is 2.68 standard deviations. The standardized score of -2.68 corresponds to about the 0.4 percentile and $+2.68$ corresponds to about the 99.6 percentile.

18. a. 68% in 85 to 115, 95% in 70 to 130, 99.7% in 55 to 145

21. 17.56 minutes because 90th percentile has a z-score of 1.28

24. a. 0.75 or 75%

b. No, that's the average. Normal temperatures cover a wide range.

27. The median is much closer to the minimum than to the maximum, so they must be skewed or have large outliers.

CHAPTER 9

3. a. Pie chart (or bar graph)

b. Histogram (or several other choices)

c. Bar graph

d. Scatterplot

6. The horizontal axis does not maintain a constant scale.

9. b. Population has steadily increased, so *numbers* of violent crimes over the years would increase even if the rate stayed constant.

12. In general, not starting at zero allows changes to be seen more readily, while starting at zero emphasizes the actual magnitude of the measurements.

15. a. Positive

b. Nonexistent (ignoring possible global warming)

18. It would be better to adjust for inflation to have a fair comparison over the years.

21. a. Seasonal and cycles

b. Probably trend and seasonal

24. Possible seasonal and cycles. For the seasonal, they may offer attractive rates at certain times of the year, like when college students have just moved to the area and may be looking for a bank.

27. a. Bar graph

CHAPTER 10

3. Yes, about 5 of them

6. a. $y = 0.96\,(x - 0.5) = -0.48 + 0.96x$ (in pints)

b. Intercept $= -0.48$, slope $= 0.96$

c. 1.0

9. A correlation of -0.6 implies a stronger relationship.

12. b. 0.847. Yes it makes sense; the scatterplot shows a positive slope.

15. Predicted ideal weight for a 150-lb woman is 133.9 lb and for a man it is 157.5.

18. a. Negative, because the slope is negative

b. Predicted time is 34.5737, actual time was 0.3363 seconds higher.

c. They decrease by 0.11865 second per year (or about 0.47 second per 4-year Olympics).

21. Yes. The strength of the relationship is the same as for the golfers and the sample is much larger.

24. You would expect a positive correlation.

27. No. It would only make sense if Mother's height of 0 was in the range of possible x values, which it clearly is not.

CHAPTER 11

3. No. People who prefer to exercise (walk) may also be people who keep their weight under control, eat less fat, and so on.

6. Winning time in an Olympic running event and the cost of running shoes are likely to be negatively correlated because prices go up and winning times go down over time. An example of a positive correlation is average salary of teachers and price of an apple, measured yearly over many years, because both have gone up over time.

9. Combining groups inappropriately

12. Salary and years of education in a company where the president makes $1 million a year but has an eighth-grade education

15. a. Because the problem mentions that barbecued foods are known to contain cancer-causing substances, there is likely to be a direct causal link. Either direct cause alone, or contributing cause, would be acceptable answers.

b. There are a few possibilities here. There may just be a third variable causing both, namely personality type. More high-strung people would both see themselves as being under stress and have high blood pressure. Stress could also be a contributing, but not the sole, cause of high blood pressure. A weaker explanation is that high blood pressure causes stress, because people worry about having high blood pressure.

18. No. There are confounding factors, such as more sports cars are red.

21. No, there are too many other possible factors. For instance, if the first winter was very harsh and the second one mild, there would be more fatalities in the first due just to the weather.

24. a. There is an association between the quantity of the explanatory variable, and the level of the response variable.

b. Perhaps the higher the percent fat in someone's diet the higher their blood pressure is likely to be.

c. No.

27. All of reasons 1 to 6 are likely to apply. For instance, for Reason 2, the response variable (unhealthful snacking) may be causing changes in the explanatory variable (stress) because of chemical effects of the poor diet.

30. a. This was a randomized experiment, so the title is justified.

b. This was an observational study, so a cause-and-effect conclusion cannot be made. The title is not justified.

CHAPTER 12

3. a. Explanatory: Male/female; Response: Been in an accident or not.

b.

	Accident yes	Accident no	Total
Male	16	7	23
Female	18	16	34
Total	34	23	57

c. Males: 16/23, Females: 18/34

d. Relative risk = 1.314

e. Odds for males = 16 to 7, or about 2.29 to 1.

6. Relative risk = 10

9. a. Retrospective observational study, probably case-control

b. Reasons 3, 4 and 5 are all possibilities.

12. a. 13.9% **b.** 0.106 **c.** 212/1525 or 0.139 **d.** 465 to 3912

15. Proportion or risk

18. No. It could be that drug use would decline in that time period anyway, or drugs of choice would change.

21. a. 40/285 = 0.14 **b.** 74/1213 = 0.061 **c.** 2.3 **d.** 130%

24. a. Decreased risk **b.** 0.7 **c.** No.

27. b. 76.1% of blacks were approved and 84.7% of whites. Proportions are 0.761 and 0.847.

30. Relative risk

33. The increased risk of dying from circulatory disease is reported to be 21%.

36. Women under the age of 70 who attended religious services less than once a week or never were 1.22 times more likely to die during the course of the study than those who attended religious services at least weekly, after adjusting for factors such as education and income.

CHAPTER 13

3. They are equivalent; the probability is usually set at .05.

6. They are expected if the null hypothesis is true.

9. a. No, $1.42 < 3.84$

b. Yes, $14.2 > 3.84$

c. Yes, $0.02 < 0.05$

d. No, $0.15 > 0.05$

12. a. df = 2, p-value = 0.025, reject null hypothesis.

b. df = 4, p-value = 0.025, reject null hypothesis.

c. df = 6, p-value = 0.25, do not reject null hypothesis.

d. df = 3, p-value = 0.01, reject null hypothesis.

15. a. Null hypothesis: There is no relationship between sex and preferred candidate for the population of voters.

b. The expected counts are

$(450)(500)/1000 = 225$ $(450)(500)/1000 = 225$
$(550)(500)/1000 = 275$ $(550)(500)/1000 = 275$

c. Half of the people asked said they would vote for Candidate A and half for Candidate B. Therefore, if there is no relationship between sex and preferred candidate, we would expect half of the males to vote for each candidate and half of the females to vote for each candidate.

d. 10.1

e. Reject the null hypothesis because $10.1 > 3.84$. There appears to be a relationship between sex and preferred candidate in the population.

18. Null: No population relationship between opinion about banning texting while driving and whether or not individual has texted while driving. Alternative: There is a relationship between opinion about banning texting while driving and whether or not individual has texted while driving.

21. a. 0.28 or 28% for males; 0.23 or 23% for females

    b. 8.64; statistically significant

    c. No. The chi-square statistic would be about 0.86.

24. Null: For the population of Australian couples similar to the ones in the study there is no relationship between wife's parents' divorce status and the couple's separation status.

27. a. Yes. **b.** Yes. **c.** No.

30. First null hypothesis: For the population of college students who drink there is no relationship between gender and experiencing vomiting as a hangover symptom.

33. a. Expected count = (466)(895)/1215

    b. Chi-square statistic = 5.35, p-value = 0.021, reject null hypothesis.

36. Expected count for Male, Yes is 27.

CHAPTER 14

3. Example: The probability that the earth will be uninhabitable by the year 2050

6. Observing the relative frequency

9. a. Relative frequency probability; observe relative frequency. **b.** Personal probability.

12. a. 0.178 **b.** 0.169 **c.** 0.347 **d.** 1 − 0.347 = 0.653

15. a. 0.10, or 10%

    b. (.10)(.10), or .01; assume payments (or not) are independent for the two customers.

18. a. No. **b.** No

21. a. Probably not independent; they are likely to share political preferences.

    b. Probably independent

24. a. Ordering coffee and asking for water are not mutually exclusive.

    b. Ordering coffee and ordering diet soda are not independent.

27. a. 0.2 **b.** (0.8)(0.2) = 0.16 **c.** (0.8)(0.8)(0.2) = 0.128 **d.** 0.2 + 0.16 + 0.128 = 0.488

30. 3.3; you would not expect this each time, but you would expect it in the long run. (Four years of college may not be enough of a long run to reach this exactly.)

33. Discrete data, where the expected value is also the mode; for instance, suppose a small box of raisins contains 19 pieces with probability 0.1, 20 with probability 0.8, and 21 with probability 0.1. Then the expected value is 20 raisins, which is also the most likely value.

CHAPTER 15

3. Number the days from 1 (January 1st) to 365 (December 31st). Choose three numbers from 1 to 365 with replacement to represent the three birthdays.

6. a. Simulate an integer from 1 to 100. A number from 1 to 55 represents "supports candidate," a number from 56 to 100 represents "does not support candidate."

    b. Simulate 100 numbers each from the range 1 to 100 and use the rule in part (a) for each one.

    c. Simulate many polls of 100 and see what proportion of those polls result in 49 or fewer supporting the candidate.

9. a. 1000 **b.** 100 **c.** 1 to 100 **d.** No **e.** Yes **f.** Count how many of the 1000 samples of 100 numbers had 49 or fewer numbers in the range from 1 to 55.

12. a. 100 **b.** 3 **c.** 1 to 8 **d.** Yes **e.** Yes or no is okay **f.** The three numbers represent the lanes you and your friends are in. Find the proportion of the samples that consist of three consecutive numbers.

15. 1 − (918/1000) = 0.082

18. a. Null: Correlation between age and vocabulary score in the population is 0. Alternative: Correlation is greater than 0 in the population.

    b. 0.29

    c. No. The null hypothesis would not be rejected, but that does not mean it would be accepted as true.

    d. About equally likely; true p-value depends on sample size, not on number of simulations.

    e. Include more children in the study.

21. Relative frequency.

CHAPTER 16

3. Anchoring

6. A detailed scenario may increase your personal probability of it happening because of availability. An example is someone trying to sell you a car alarm describing in detail the methods thieves use.

9. Statement A has a higher probability, but people would think statement B did.

12. The certainty effect or the pseudocertainty effect.

15. The rate of the disease for people similar to you

18. Two possible answers: Optimism and the possibility effect.

21. The earthquake scenario would be considered more likely because of the representativeness heuristic.

But that scenario is a subset of the first one, so the first one must be more likely.

24. a. People think nothing will go wrong in the process of remodeling.

   b. Best-case scenarios, in which everything will go according to plan, are readily brought to mind. It's harder to imagine all of the ways in which things might go wrong.

   c. Contractors may anchor consumers on the price of the best-case scenario and fail to mention that something almost always deviates from that scenario.

CHAPTER 17

3. Not surprising; $P(\text{match}) = 1 - P(\text{no matches})$ = .1829.

6. There are numerous meanings to the word *pattern*, and it is easy to find something that fits one.

9. Yes, there are limited choices and people know their friends' calling behaviors.

12. See Exercise 15 for an example.

15. No. Parts don't wear out independently of each other.

18. a. 9,900/10,700 = 0.925

   b. 9,900/99,000 = 0.10

   c. 10,700/100,000 = 0.107

21. Not taking the base rate into account.

24. a. Choice A, because for choice B the expected value is $9.00, less than the $10 gift.

   b. Choice B, because of the possibility effect as shown in Table 16.1.

27. $3.00, so the expected value is $(-\$1.00) \times (3/4) + (\$3.00) \times (1/4)$.

CHAPTER 18

3. 200.29 for 1995 and 401.15 in 2013; the cost of a paperback novel in 1995 was 200.29% of what it was in 1981.

6. a. 1994 salary would be $119,646.

   b. 1994 salary would be $487,356.

9. a. (Partial answer) There was a 72% increase from 1940 to 1950, calculated as $(24.1 - 14.0)/14.0 = 0.72$, or 72%.

   b. Inflation was by far highest during the 1970s.

   c. Inflation was lowest during the 1950s.

12. $56.07

15. $179.9/24.1 = 7.46473$. So one would need $7.464730 \times \$1,000,000 = \$7,464,730$ in 2002 to be the equivalent of a millionaire in 1950. That does not seem like a goal achievable by very many people.

18. Criticism 2: If the price of one item rises, consumers are likely to substitute another.

21. What matters to consumers is how prices are changing, not how current prices compare with the base year prices.

24. "Average duration of unemployment, in weeks, is lagging."

CHAPTER 19

3. b. Virtually impossible; standardized score over 10

   c. Optimism or overconfidence. People think they are likely to be better than average on most things.

6. Example: take a random sample of 1000 students at a large university and find the proportion who have a job during the school year.

9. a. Bell-shaped, mean = 0.20, standard deviation = 0.008

   b. No, the standardized score for 0.17 is −3.75.

12. b. Yes, the standardized score for a mean of 7.1 is about 1.03.

   c. No. There are many ways in which college students differ from all adults in factors affecting sleep, such as age range, flexibility of schedule, perceived need and desire to stay up all night, and so on.

15. a. No, the expected number not meeting standards is only 3.

   b. Yes, conditions are met.

18. Bell-shaped, with a mean of 3.1 and standard deviation of 0.05

CHAPTER 20

3. Population proportion.

6. a. 68%

   b. 90%

   c. 95%

   d. 99%

9. No, the volunteer (self-selected) sample is not representative.

12. $n = 400$, confidence level = 95%. This combination has smallest sample size and greatest confidence level.

15. a. 0.20 to 0.24, or 20% to 24%

b. 0.02 ± 0.002, which is .018 to .022 or 1.8% to 2.2%

18. 25.6% to 48.4%; it almost covers chance (25%).

21. a. Should not. This was a self-selected sample. People dissatisfied with the president may be more likely to respond. **b.** Should. It's doubtful that the handedness characteristics of this sample would be biased toward having a different proportion that is left-handed than in the population of all college students.

24. About .95 × 200 = 190 intervals will cover the population proportion, so about 10 intervals will not.

CHAPTER 21

3. a. SEM is 1/12 hour, or 5 minutes.

b. A 95% confidence interval is 12 hrs 50 min to 13 hrs 10 min.

c. The SEM for the Dutch babies is 0.062, so the standard error of the difference (SED) is the square root of $(0.083)^2 + (0.062)^2 = 0.104$. A 95% confidence interval for the difference is 2 ± 2(0.104) or 1.79 to 2.21 hours.

6. a. The population means are probably different.

b. The population means could be the same.

9. a. 1.97 **b.** 2.01 **c.** 1.65 **d.** 1.68

12. a. Confidence interval is 64.64 ± 0.81, or 63.83 to 65.45.

b. Confidence interval is 63.97 ± 2.0, or 61.97 to 65.97.

c. The interval in part b is wider because the sample size is smaller. We could not conclude that there is a difference in the mean ages because there is overlap in the two intervals.

d. The interval is (64.64 − 63.97) ± 2(1.08) or −1.49 to 2.83.

15. This study exhibited a large placebo effect, meaning that scores changed even for those women taking a placebo instead of calcium. A comparison of the calcium group third cycle scores to their own baseline scores would include both the placebo effect and the actual impact of taking calcium. Thus, it would overrate the influence of taking calcium.

18. Method 1; the difference within couples is desired, not the difference across sexes.

21. a. Neither the women taking the pills or the medical personnel with whom they interacted knew who had which type of pill (calcium or placebo).

b. Because this was a randomized experiment and not an observational study, it is actually possible to make a causal conclusion. Confounding variables, if any, should have been equally present in both groups, so the differences seen between the calcium and placebo-treated groups can be attributed to the calcium.

24. Multiply by 60 minutes to get 2.5 to 20.6 minutes.

27. We can be 95% confident that the risk of dying from all causes (during the time this study) for those not attending church at least once a week was from 1.06 to 1.37 times the risk of dying for those attending church at least once a week.

30. This was an observational study. Thus, it cannot be concluded that attending regular church services causes a change in risk of death.

CHAPTER 22

3. a. <u>Null</u>: Calcium does not have an impact on the severity of premenstrual symptoms. <u>Alternative</u>: Calcium reduces the severity of premenstrual symptoms.

b. Type 1 error

c. Type 1: some women may take calcium because they have been told it would reduce premenstrual symptoms, but in fact it doesn't. Type 2: women are told calcium is not effective in reducing premenstrual symptoms, and thus do not take it, when in fact it would help.

6. a. Null: Psychotherapy and desipramine are equally effective in treating cocaine use. Alternative: One method is more effective than the other.

b. Type 1: There is no difference in the treatments, but the conclusion is that they do differ. Type 2: There is a more effective treatment but the study fails to make that conclusion.

c. Type 1

9. a. 0

b. Population mean GRE score if everyone were to practice meditation is higher than the population mean GRE score if everyone were to take a nutrition class.

c. Meditation does not increase mean GRE score, but we conclude that it does.

d. Meditation does increase mean GRE score, but we fail to make that conclusion.

e. It seems like Type 2 would be more serious because a useful method for increasing test scores would go undiscovered.

12. A type 1 error would be that no ESP is present but the study concludes it is being used.

15. a. Null: There is no link between vertex baldness and heart attacks in men. Alternative: There is a link between vertex baldness and heart attacks in men.

b. There is no link between the two but the study concludes that there is.

c. There is a link between vertex baldness and heart attacks in the population, but the evidence in the sample wasn't strong enough to conclusively detect it.

18. a. Minor disease with serious treatment—for example, tonsillitis and the treatment is surgery

b. Being infected with HIV

21. They should use a higher cutoff for the p-value to reduce the probability of a type 2 error.

24. a. It is less than 0.05.

b. 0.05

c. Reject the null hypothesis. We can conclude that for the population of monkeys the proportion willing to cooperate would be lower after observing another monkey receiving a free grape than after watching another monkey do what they were asked to do, namely, give up a token for a piece of cucumber.

CHAPTER 23

3. One-sided

6. a. No

b. Yes, the p-value would be .04.

9. a. Null: The training program has no effect on test scores.

b. 2.5

c. Alternative; scores would be higher after the program.

12. a. <u>Null:</u> In the population of elderly people, there is no relationship between pet ownership and number of doctor contacts. <u>Alternative:</u> In the population of elderly people, those with pets have fewer doctor contacts than those without pets

b. Reject the null hypothesis. Conclude that elderly people with pets have significantly fewer doctor contacts than those without pets.

15. a. Null: 50% (or more) of adults in the United States think the economy is the most pressing issue. Alternative: Fewer than 50% of adults in the United States think the economy is the most pressing issue.

b. $(.46 - .50)/sqrt[(.5)(.5)/814] = -2.28$

c. The politician is using the word "significantly" to mean statistical significance.

d. The quote is not justified because most people hearing it would think the politician meant practical importance (significance) rather than statistical significance.

18. a. 0.008 **b.** 0.008 **c.** 0.016 **d.** 0.04 **e.** 0.04 **f.** 0.08

21. No. The data would not have supported that alternative because in the sample the dieters lost more fat.

24. For the first test the null hypothesis is: For volunteers who are trained to meditate, there will be no change on average in their negative affect (mood) from before the training to eight weeks after beginning meditation. Similar answer for second test, but at four months.

27. There would be no difference in antibody response to a flu vaccine for meditators and nonmeditators in the population.

30. a. When shown a set of pictures, men and women would rate an equal number to be highly arousing.

b. They stated their conclusion to be consistent with a one-sided alternative hypothesis.

c. A t-test was used. The test statistic was t = 2.41 and it had 22 "degrees of freedom." The p-value for the test was less than 0.025.

d. Statistical version.

CHAPTER 24

3. a. Yes

b. No, the magnitude wasn't much different, but the large sample sizes led to a statistically significant difference.

6. a. $1.637 \times 3 = 4.911$

b. <u>Null:</u> There is no relationship between gender and driving after drinking in the population. <u>Alternative:</u> There is a relationship between them. The alternative would now be chosen because 0.03 is less than 0.05.

c. The power would be higher with the larger sample size.

9. A confidence interval for the relative risk of heart attack during heavy versus minimal physical exertion is from 2.0 to 6.0.

12. The small sample sizes would result in low power, or high probability of a type 2 error even if ESP were present.

15. 1 out of 20, so 1 overall.

18. Warnings 1, 2 and 4.

21. A little variability

24. a. The alternative hypothesis was one-sided.

    b. Yes, a one-sided alternative hypothesis is justified. Based on folklore and previous studies, the researchers were speculating that meditation would increase immune function response.

27. No. The difference was not statistically significant.

30. a. 8200 participants.

    b. Here is the relevant quote: "For boys of all ages, it was about one-half point on a 33-point scale. Girls were hit harder, with a 2-point difference for girls who'd been 12 at the first interview, and diminishing with age to about a half-point difference for girls who'd been 17."

    c. The first two warnings apply. The sample size was very large, so a statistically significant difference was found even though the actual difference in depression levels was very small.

    d. The word "significantly" is being used in the statistical sense, and not in the English sense. This can be seen by the magnitude of the differences quoted in the answer to part (b) of this exercise.

CHAPTER 25

3. a. They would like to make causal conclusions.

    b. Yes. They provide just as much information about the true size of the effect as the studies that did find significance. Excluding them would bias the results in favor of a strong effect.

6. a. The first benefit listed, "detecting small or moderate relationships" appears to be most applicable in this meta-analysis, given the news article quote that meta-analysis "can enable researchers to draw statistically significant conclusions from studies that individually are too small." For the criticism, a number of them could be applied, all of which address the same point - that the patients in these studies were not a representative sample. They tended to be sicker and have more drug use than normal.

b. Statistical significance versus practical importance is definitely not a problem, since if the adverse drug reactions are really that common, knowledge of this problem is of great practical importance.

9. One reason is that some studies used outdated technology.

12. 1000 researchers; 50 could easily be contacted.

15. Vote counting could not detect a relationship, but meta-analysis could, by increasing the overall power.

CHAPTER 26

3. a. Good idea

    b. This may be a good idea, as long as Janet was honest about the coin flip and didn't try to influence it to come out the way she wanted it. She would also have to be sure she knew in advance which coin outcome corresponded to which group.

    c. It is probably better for one person to do them all, for consistency. But, better care should have been taken to make sure Brad did not have any contact with the participants before the phone call.

6. No, the intervals overlap and there is no clear choice. One method produces a slightly higher mean but also has more variability.

9. This condition was not met. The 9-year-old girl was in immediate proximity to the practitioners, and she knew which hand she was hovering over with her hand.

12. The main problem is that participants were deceived into thinking they were hurting others. But there would have been no way to conduct the experiment if they had been told the truth.

15. They chose the confidence level to present the outcome they desired. They should have reported the standard 95% confidence interval.

18. The complaint raised by the letter is that the study was not measuring what it purported to measure.

21. The experimenter effect could have been mitigated by using more careful controls, such as better randomization, better data recording, and not allowing the person who did the randomization to also apply the treatment.

Index